Optical Monitoring of Fresh and Processed Agricultural Crops

Contemporary Food Engineering

Series Editor
Professor Da-Wen Sun, Director
Food Refrigeration & Computerized Food Technology
National University of Ireland, Dublin
(University College Dublin)
Dublin, Ireland
http://www.ucd.ie/sun/

Optical Monitoring of Fresh and Processed Agricultural Crops, *edited by Manuela Zude* (2009)

Food Engineering Aspects of Baking Sweet Goods, *edited by Servet Gülüm Şumnu and Serpil Sahin* (2008)

Computational Fluid Dynamics in Food Processing, *edited by Da-Wen Sun* (2007)

Contemporary Food Engineering Series
Da-Wen Sun, Series Editor

Optical Monitoring of Fresh and Processed Agricultural Crops

Edited by
Manuela Zude

CRC Press
Taylor & Francis Group
Boca Raton London New York

CRC Press is an imprint of the
Taylor & Francis Group, an **informa** business

TP
372.5
.O68
2009

CRC Press
Taylor & Francis Group
6000 Broken Sound Parkway NW, Suite 300
Boca Raton, FL 33487-2742

© 2009 by Taylor & Francis Group, LLC
CRC Press is an imprint of Taylor & Francis Group, an Informa business

No claim to original U.S. Government works
Printed in the United States of America on acid-free paper
10 9 8 7 6 5 4 3 2 1

International Standard Book Number-13: 978-1-4200-5402-6 (Hardcover)

This book contains information obtained from authentic and highly regarded sources. Reasonable efforts have been made to publish reliable data and information, but the author and publisher cannot assume responsibility for the validity of all materials or the consequences of their use. The authors and publishers have attempted to trace the copyright holders of all material reproduced in this publication and apologize to copyright holders if permission to publish in this form has not been obtained. If any copyright material has not been acknowledged please write and let us know so we may rectify in any future reprint.

Except as permitted under U.S. Copyright Law, no part of this book may be reprinted, reproduced, transmitted, or utilized in any form by any electronic, mechanical, or other means, now known or hereafter invented, including photocopying, microfilming, and recording, or in any information storage or retrieval system, without written permission from the publishers.

For permission to photocopy or use material electronically from this work, please access www.copyright.com (http://www.copyright.com/) or contact the Copyright Clearance Center, Inc. (CCC), 222 Rosewood Drive, Danvers, MA 01923, 978-750-8400. CCC is a not-for-profit organization that provides licenses and registration for a variety of users. For organizations that have been granted a photocopy license by the CCC, a separate system of payment has been arranged.

Trademark Notice: Product or corporate names may be trademarks or registered trademarks, and are used only for identification and explanation without intent to infringe.

Visit the Taylor & Francis Web site at
http://www.taylorandfrancis.com

and the CRC Press Web site at
http://www.crcpress.com

Contents

Book Proceedings vii
Series Editor's Preface ix
Preface xi
Series Editor xv
Editor xvii
Contributors xix
Nomenclature xxiii
Introduction xxvii

Chapter 1 What to Measure and How to Measure 1

1.1 Quality Parameters of Fresh Fruit and Vegetable at Harvest
 and Shelf Life 2
 Susan Lurie

1.2 Quality of Processed Plant Food 17
 Sascha Rohn and Lothar W. Kroh

1.3 Optical Sensing 44
 Martina Meinke, Moritz Friebel, and Alessandro Torricelli

Chapter 2 Vision Systems 83

2.1 Machine Vision Systems for Raw Material Inspection 84
 Enrique Moltó Garcia and José Blasco

2.2 Computer Vision for Quality Control 126
 Chaoxin Zheng and Da-Wen Sun

Chapter 3 VIS/NIR Spectroscopy 141

3.1 Spectrophotometer Technology 143
 Bernd Sumpf

3.2 Monitoring and Mapping of Fresh Fruits and Vegetables
 Using VIS Spectroscopy 157
 Bernd Herold

3.3 Near-Infrared Spectroscopy 192
 Kerry B. Walsh and Sumio Kawano

3.4 Network of NIRS Instruments 239
 Peter Tillmann

v

Chapter 4 Fluorescence 251

4.1 Introduction 253
 Michael U. Kumke and Hans-Gerd Löhmannsröben

4.2 Blue, Green, Red, and Far-Red Fluorescence Signatures
 of Plant Tissues, Their Multicolor Fluorescence Imaging,
 and Application for Agrofood Assessment... 272
 Claus Buschmann, Gabriele Langsdorf, and Hartmut K. Lichtenthaler

4.3 Monitoring Raw Material by Laser-Induced Fluorescence
 Spectroscopy in the Production 319
 Yasunori Saito

4.4 Front-Face Fluorescence Analysis to Monitor Food Processing
 and Neoformed Contamination.. 337
 Jad Rizkallaf, Lyes Lakhal, and Inès Birlouez-Aragon

4.5 Integrated System Design 359
 Takaharu Kameoka and Atsushi Hashimoto

Chapter 5 Spectroscopic Methods for Texture and Structure Analyses 377

5.1 Brief Overview on Approaches for Nondestructive Sensing
 of Food Texture 378
 David G. Stevenson

5.2 Spectroscopic Technique for Measuring the Texture
 of Horticultural Products: Spatially Resolved Approach 391
 Renfu Lu

5.3 NMR for Internal Quality Evaluation
 in Horticultural Products.. 423
 *Natalia Hernández Sánchez, Pilar Barreiro Elorza,
 and Jesús Ruiz-Cabello Osuna*

Chapter 6 Process Monitoring 469
 Ali Cinar and Sinem Perk

Index 525

Book Proceedings

An international team of authors was invited to submit contributions from their specific fields of work. Their excellent manuscripts were discussed and the editor reviewed them keeping in mind the objective of the book, which is getting more people involved in the highly innovative field of optical sensing of agro-foods and improving the process management by means of *in situ* analyses. The manuscripts were partially peer-reviewed by Professor Dr. Hartmut K. Lichtenthaler, and within the Leibniz Institute for Agricultural Engineering Potsdam-Bornim, Germany, by Dr. Martin Geyer, Professor Dr. Hans-Jürgen Hellebrand, Dr. Laszlo Baranyai, and Dr. Werner B. Herppich. The initial formatting of all manuscripts was graciously carried out by Andrea Gabbert, and we wish to thank the production team of the publisher for their efforts.

Series Editor's Preface

CONTEMPORARY FOOD ENGINEERING

Food engineering is a multidisciplinary field of applied physical sciences combined with a knowledge of product properties. Food engineers provide the technological knowledge transfer essential to the cost-effective production and commercialization of food products and services. In particular, food engineers develop and design processes and equipment in order to convert raw agricultural materials and ingredients into safe, convenient, and nutritious consumer food products. However, food engineering topics are continuously undergoing changes to meet diverse consumer demands, and the subject is being rapidly developed to reflect market needs.

In the development of food engineering, one of the many challenges is to employ modern tools and knowledge, such as computational materials science and nanotechnology, to develop new products and processes. Simultaneously, improving food quality, safety, and security remains a critical issue in food engineering study. New packaging materials and techniques are being developed to provide more protection to foods, and novel preservation technologies are emerging to enhance food security and defense. Additionally, process control and automation regularly appear among the top priorities identified in food engineering. Advanced monitoring and control systems are being developed to facilitate automation and flexible food manufacturing. Furthermore, saving energy and minimizing environmental problems continue to be important food engineering issues and significant progress is being made in waste management, efficient utilization of energy, and reduction of effluents and emissions in food production.

Consisting of edited books, the Contemporary Food Engineering book series attempts to address some of the recent developments in food engineering. Advances in classical unit operations in engineering applied to food manufacturing are covered as well as topics such as progress in the transport and storage of liquid and solid foods; heating, chilling, and freezing of foods; mass transfer in foods; chemical and biochemical aspects of food engineering and the use of kinetic analysis; dehydration, thermal processing, nonthermal processing, extrusion, liquid food concentration, membrane processes, and applications of membranes in food processing; shelf-life, electronic indicators in inventory management, and sustainable technologies in food processing; and packaging, cleaning, and sanitation. The books are aimed at professional food scientists, academics researching food engineering problems, and graduate-level students.

The editors of the books are leading engineers and scientists from many parts of the world. All the editors were asked to present their books in a manner that would address market need and pinpoint cutting-edge technologies in food engineering. Furthermore, all chapters are written by internationally renowned experts who have both academic and professional credentials. In each chapter authors have attempted to provide critical, comprehensive, and readily accessible information on the art and science of a relevant topic, with reference lists to be used by readers for further information. Therefore each book can serve as an essential reference source to students and researchers in universities and other research institutions.

Da-Wen Sun

Preface

Assuring healthy human nutrition and improving the economic success of farmers and processors are priorities in the context of current global changes. In the age of information technology, process-oriented data analysis is predicted to form the basis for economic growth. The general consensus is that, especially in food economies, new innovative technologies are needed for appropriate process management. This should help to maintain nutritional product quality and decrease losses due to produce decay along the supply chain of perishable agro-foods and, therefore, make the processes more economical.

Product quality during production is determined by the plant genome, environmental conditions, and microclimate as well as the production system (Figure 1). At the point of harvest, fruit quality appears heterogeneous. In postharvest, the quality level generally decreases as a function of time, dependent on conditions that affect the rate of quality decrease. The quality of processed food depends essentially on the

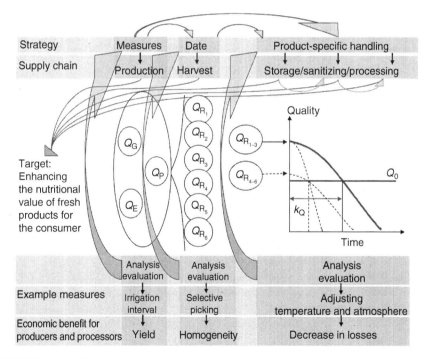

FIGURE 1 Quality of perishable products along the supply chain with the major influencing factors in the production, and as a function of time in the postharvest sector. Q_E, Q_G, and Q_P are the quality-influencing factors (environment, genome, and production system); Q_R is the real product quality; Q_0 is the acceptance limit; and k_Q is the coefficient of shelf life. (From Zude, M. and McRoberts, N., *Agricult. Eng.*, 61, 2, 2006.)

quality of raw materials. Thus, the initial quality level at harvest and subsequent conditions determine postharvest storage quality by determining the length of time that the quality remains above the acceptance level. From an economic point of view, produce is finally lost when quality drops below this level. In principle, precise knowledge of the specific physiological, biochemical, and physical properties of individual products and often complex interactions are essential for process management during harvesting, preparation, storage, and marketing.

In the fresh product, complex attributes such as maturity, taste, and texture are recognized by the consumer. Here, monitoring specific markers such as the decrease of chlorophyll content and carotenoid increase may help to control production and postharvest actions. In postharvest, monitoring of the further development of the product down to senescence can help control the necessary conditions for perishable produce. Processing conditions should be adjusted to avoid the synthesis of neo-formed compounds. In processed food, single compounds are already largely included in quality descriptions due to safety issues and the demand for similarity equivalence with respect to novel food regulations. In conclusion, *in situ* analyses of physiological and physical product properties by means of noninvasive methods are needed in future process management.

Furthermore, due to increasing demands for well-being, better insights into human nutrition, and product assessment by consumers and its resulting market value, individual compounds of the product are getting increased attention, since they may promote health or have negative effects in human nutrition. As a result, the visual appearances as well as levels of single, native and neo-formed compounds need to be quantitatively analyzed by the new methods.

Developing tools for product monitoring aimed at adapted process management for perishable food needs R&D activities in the fields of product monitoring and data analysis:

- For product monitoring, the development of noninvasive analysis methods for physical and biochemical product properties is needed, targeting online information gathering.
- Process control will use the data obtained for the product along the supply chain, with the aim of adapting production measures, postharvest processing, and storage conditions to the physical and nutritional properties as well as the physiological stage of the product in real time (lower part of Figure 1) and developing product-based strategies for maintaining product quality (upper part of Figure 1).

At the leading edge of monitoring, we realize that due to the rapid and precise recording of the product parameter in question, optical methods are well suited for use along the entire supply chain. Machine vision systems have been established for sorting as well as monitoring during processing. Research groups in cooperation with industry have recently developed new sorting lines using spectroscopy in the visible and near-infrared wavelength range, e.g., for grading according to fruit soluble solids and pigment contents or product homogeneity. Also, desktop modules and portable instruments for individual product testing have been developed in the last four years based on the same technology. For instance, fruit and vegetable pigment contents can be assessed directly in the production process and subsequently checked in the entire supply chain. It is precisely this repeated analysis along the supply chain that is essential for developing methods to assess the impact of processes on the product quality in production and postharvest.

Nowadays, the discipline of product monitoring is developing rapidly. Innovative optical compounds, adapted data processing methods, namely chemometrics, and recent modeling approaches of the interactance of light and matter provide the means to reach the next level of knowledge. For understanding and using innovative technologies in the supply chain of raw material and processed food, basic knowledge of the methods is in high demand.

This book is intended to give insight into new technologies in an easy-to-read manner to encourage interdisciplinary understanding. The concept is to keep reading motivation high by presenting the contents the way they naturally occur and in an interesting, relevant context. For instance, in Section 1.1 the basic criteria for assessing quality parameters in the field of agricultural crops are presented, accurately introducing the terms used in this research field in a relevant context along with the flow of the text. Questions arising when processing raw material in the food industry are discussed in Section 1.2. Section 1.3 provides the basis for understanding optical measurements. Here, we start with a brief introduction on optical properties immediately providing an experiment to give the reader real optical data and causal interactions. Subsequently, in Section 1.3.2, the full theory is provided when the reader has realized the problems. The following chapters provide fundamentals on optical methods, which are presently used or introduced in practice.

Each method is addressed by at least two contributions from international experts. These contributions offer deep insight, always directly relating to the questions and applications appearing in the practice of agricultural crop management (ACM). Results and evaluations are given by means of specific, detailed examples from the entire supply chain of agricultural crops. The final chapter, Chapter 6, matches the questions presented in the introductory chapter. In summary, the book provides necessary knowledge on up-to-date ACM.

The expected readership will be professionals in the food industry: agricultural engineers, food scientists and technologists, life sciences researchers, human nutritionists, professionals in extension service, retailers, quality managers, and others. The book will also appeal to the student and academic faculty.

Manuela Zude

Series Editor

Born in Southern China, Professor Da-Wen Sun is a world authority in food engineering research and education. His main research activities include cooling, drying, and refrigeration processes and systems; quality and safety of food products; bioprocess simulation and optimization; and computer vision technology. His innovative studies on vacuum cooling of cooked meats, pizza quality inspection by computer vision, and edible films for shelf-life extension of fruits and vegetables have been widely reported in national and international media. Results of his work have been published in over 180 peer-reviewed journal papers and more than 200 conference papers.

He received a first class BSc honors and MSc in mechanical engineering, and a PhD in chemical engineering in China before working in various universities in Europe. He became the first Chinese national to be permanently employed in an Irish university when he was appointed college lecturer at the National University of Ireland, Dublin (University College Dublin) in 1995, and was then continuously promoted in the shortest possible time to senior lecturer, associate professor, and full professor. Sun is now professor of Food and Biosystems Engineering and director of the Food Refrigeration and Computerized Food Technology Research Group at University College Dublin.

As a leading educator in food engineering, Sun has contributed significantly to the field of food engineering. He has trained many PhD students, who have made their own contributions to the industry and academia. He has also, on a regular basis, given lectures on advances in food engineering in academic institutions internationally and delivered keynote speeches at international conferences. As a recognized authority in food engineering, he has been conferred adjunct/visiting/consulting professorships from ten top universities in China including Zhejiang University, Shanghai Jiaotong University, Harbin Institute of Technology, China Agricultural University, South China University of Technology, and Jiangnan University. In recognition of his significant contribution to food engineering worldwide and for his outstanding leadership in the field, the International Commission of Agricultural Engineering (CIGR) awarded him the CIGR Merit Award in 2000 and again in 2006, and the Institution of Mechanical Engineers based in the United Kingdom named him Food Engineer of the Year 2004.

Dr. Sun is a fellow of the Institution of Agricultural Engineers. He has also received numerous awards for teaching and research excellence, including the President's Research Fellowship, and twice received the President's Research Award of the University College Dublin. He is a member of the CIGR executive board and honorary vice president of CIGR, editor-in-chief of *Food and Bioprocess Technology—An International Journal* (Springer), series editor of the Contemporary Food Engineering book series (CRC Press/Taylor & Francis), former editor of the *Journal of Food Engineering* (Elsevier), and editorial board member for the *Journal of Food Engineering* (Elsevier), the *Journal of Food Process Engineering* (Blackwell), *Sensing and Instrumentation for Food Quality and Safety* (Springer), and the *Czech Journal of Food Sciences*. He is also a chartered engineer registered in the U.K. Engineering Council.

Editor

Manuela Zude has a background in chemistry (Technical University Berlin), but switched to the field of applied plant physiology, which she teaches at the Humboldt-Universität zu Berlin and at the Applied University, Berlin. Her former studies were aimed at defining the effects and interactions of reduced oxygen partial pressure to quantify plant metabolic responses (exchange of volatiles, water relations, and pyridine nucleotides in fruit trees) to abiotic stress. Since 2000, her research focus has been on assessing fruit compounds for determining fruit maturity stage and nutritional value by means of spectroscopic approaches. Her studies target basic research as well as knowledge transfer for product monitoring in the supply chain processes. Dr. habil.  Manuela Zude is a board member of two international journals, the standardization committee for sensor solutions, and in the company CP (Falkensee, Germany).

Contributors

Inès Birlouez-Aragon
Agroparistech
Paris, France
and
Institut Polytechnique
Beauvais, France

José Blasco
Centro de Agroingenieria
Instituto Valenciano de Investigaciones
 Agrarias
Moncada, Spain

Claus Buschmann
Botanical Institute
University of Karlsruhe
Karlsruhe, Germany

Ali Cinar
Department of Chemical and Biological
 Engineering
Illinois Institute of Technology
Chicago, Illinois

Pilar Barreiro Elorza
Physical Properties Laboratory—
 Advanced Technologies in Agrofood
Rural Engineering Department
Universidad Politécnica de Madrid
Madrid, Spain

Moritz Friebel
Laser und Medizin Technologie
Berlin, Germany

Enrique Moltó Garcia
Instituto Valenciano de Investigaciones
 Agrarias
Moncada, Spain

Atsushi Hashimoto
Graduate School of Bioresources
Mie University
Tsu, Japan

Bernd Herold
Horticultural Engineering
Leibniz Institut für Agrartechnik
 Potsdam-Bornim
Potsdam, Germany

Takaharu Kameoka
Graduate School of Bioresources
Mie University
Tsu, Japan

Sumio Kawano
Nondestructive Evaluation Laboratory
National Food Research Institute
Tsukuba, Japan

Lothar W. Kroh
Department of Food Technology and
 Food Chemistry
Chair of Food Chemistry and Food
 Analysis
Technische Universitat Berlin
Berlin, Germany

Michael U. Kumke
Institut für Chemie
Universität Potsdam
Potsdam, Germany

Lyes Lakhal
Institut Polytechnique
Beauvais, France

Gabriele Langsdorf
Botanical Institute
University of Karlsruhe
Karlsruhe, Germany

Hartmut K. Lichtenthaler
Botanical Institute
University of Karlsruhe
Karlsruhe, Germany

Hans-Gerd Löhmannsröben
Institut für Chemie
Universität Potsdam
Potsdam, Germany

Renfu Lu
Sugarbeet and Bean Research Unit
Agricultural Research Service
U.S. Department of Agriculture
East Lansing, Michigan

Susan Lurie
Department of Postharvest Science
Volcani Center
Bet Dagan, Israel

Neil McRoberts
Scottish Agricultural College
Edinburgh, Scotland

Martina Meinke
Medizinische Physik und Lasermedizin
Charité Universitätsmedizin Berlin
Berlin, Germany

Jesús Ruiz-Cabello Osuna
Instituto de Estudios Biofuncionales
Universidad Complutense de Madrid
Madrid, Spain

Sinem Perk
Department of Chemical and Biological
  Engineering
Illinois Institute of Technology
Chicago, Illinois

Jad Rizkallaf
Agroparistech
Paris, France

and

Institut Polytechnique
Beauvais, France

Sascha Rohn
Department of Food Technology and
  Food Chemistry
Chair of Food Chemistry and Food
  Analysis
Technische Universitat Berlin
Berlin, Germany

Natalia Hernández Sánchez
Properties Laboratory—Advanced
  Technologies in Agrofood
Rural Engineering Department
Universidad Politécnica de Madrid
Madrid, Spain

Yasunori Saito
Faculty of Engineering
Shinshu University
Nagano, Japan

David G. Stevenson
National Center for Agricultural
  Utilization Research
U.S. Department of Agriculture
Peoria, Illinois

Bernd Sumpf
Ferdinand Braun Institut für
  Höchstfrequenztechnik
Berlin, Germany

Da-Wen Sun
Department of Biosystems Engineering
University College Dublin
Dublin, Ireland

Peter Tillmann
Centro de Agroingenieria
Verband Deutscher
  Landwirtschaftlicher Untersuchungs
  und Forschungsanstalten
Kassel, Germany

Alessandro Torricelli
Dipartimento di Fisica
Politecnico di Milano
Milano, Italy

Kerry B. Walsh
Centre for Plant and Water Sciences
Central Queensland University
Rockhampton, Queensland, Australia

Chaoxin Zheng
Department of Computer Science
O'Reilly Institute
Trinity College
Dublin, Ireland

Nomenclature

This nomenclature is used throughout the book, if not marked differently. Symbols representing scalars are given in italic, vectors in italic with arrows, and matrices in bold and italic.

^1H	proton
a_w	water activity
A	absorption
Acr	acrylamide
B_0	static magnetic field
c	velocity of light
c_F	concentration of the fluorophore
c_i	concentration of molecule i
cw	continuous wave
CCD	charge-coupled device
CML	carboxymethyllysine
D	diffusion coefficient
E	energy
EDA	emitting diode array
EEM	excitation–emission matrix
F	fluorophore
FFT	fast Fourier transformation
FTIR	Fourier-transform infrared spectrophotometer
g	anisotropic factor
h	Planck's quantum
$H_{in}(z)$	excitation fluence rate at z owing to radiant power incident on the tissue surface at the origin 0 per unit incident power [J m^{-2}]
$H_{out}(z)$	radiant excitance at the origin 0 owing to an isotropic fluorescence source at z per unit power of the source [J m^{-2}]
HMF	hydroxymethylfurfural
I_0	intensity of incident light
I_F	apparent fluorescence excitation or reemission intensity
I_F^0	fluorescence intensity in the absence of a quencher
I_{NIR}	intensity at bandpass in the near infrared
I_R	intensity measured in reflectance or remittance mode
I_{RED}	intensity at bandpass in the red chlorophyll absorption range
I_T	intensity of transmitted light
InGaAs	indium–gallium–arsenic
$J(\vec{r}, t)$	flux vector
k_x	rate constant
K_{SV}	Stern–Volmer constant
L	length/path length

$L(\vec{r}, \vec{s})$	radiation density, radiance, at the position r in the direction s
M	net magnetization
M_{xy}	transverse magnetization
M_z	longitudinal magnetization (z-direction parallel to the magnetic field)
MIR	mid-infrared
MRI	magnetic resonance imaging
MRR	nuclear magnetic resonance relaxometry
MRS	nuclear magnetic resonance spectroscopy
NFC	neoformed compounds
NIR	near infrared
NIRS	near-infrared spectroscopy
NIT	near-infrared transmittance
NMR	nuclear magnetic resonance
NN	neural networks
$p(\vec{s}, \vec{s}\,')$	scattering phase function
P	fluorescence power escaping the matrix per unit area
P_0	irradiance of the excitation light on the tissue surface
PCA	principal components analysis
PCR	principal components regression
PD maps	proton density maps
PDA	photo diode array
PLS	partial least squares regression
Q	quencher
r	radial distance from incidence point
$\vec{r}$	position vector
R_∞	diffuse reflectance from sample with infinite thickness
R_d	diffuse reflectance
R_s	spectral resolution
RF	radio frequency
$\vec{s}$	direction vector
$S(\vec{r}, \vec{s})$	source term
SAXS	small-angle x-ray scattering
Si	silicon
SSC	soluble solids content
SWNIR	short-wave near-infrared radiation
t	time
T	transmittance
T_1	longitudinal relaxation time in MRR
T_2	transverse relaxation time in MRR
T_2^*	effective transverse relaxation time in MRR
T_d	diffuse transmission
T_t	total transmission
TRS	time-resolved laser remittance or diffuse reflectance spectroscopy

UV	ultraviolet
VIS	visible
WAXS	wide-angle x-ray scattering
z	depth within the matrix

GREEK

γ_i	gyromagnetic ratio of a nucleus
ε_i	molar extinction coefficient
ε_f	extinction coefficient of fluorophore
ν	frequency
λ	wavelength
λ_{ex}	excitation wavelength
λ_{em}	emission wavelength
μ_a	absorption coefficient
μ_{eff}	effective attenuation coefficient
μ_s	scattering coefficient
μ_s'	reduced scattering coefficient
μ_t	attenuation coefficient
Φ	azimuthal angle
Φ_F	fluorescence quantum yield
Θ	scattering angle
δ_{eff}	penetration depth
σ	shielding constant
σ_a	absorption cross section
σ_s	scattering cross section
τ	time interval
$\Delta\tau$	gate width
τ_F^0	fluorescence decay time in the absence of a quencher
τ_F	apparent fluorescence decay time
τ_F^N	natural fluorescence lifetime
V	wave number
Ω	solid angle
χ_0	static nuclear magnetic susceptibility
ω_0	Larmor frequency

Introduction

Neil McRoberts

INTRODUCTION TO SUPPLY CHAINS

This introduction gives a conceptual framework for the more detailed and technically demanding information in the rest of the book. There are very good reasons for measuring quality attributes of fresh and processed agricultural crops, not the least of which is to improve the supply of fresh fruit and vegetables in the human diet—an important political issue in both developed and developing countries. In addition to such public policy justifications there are, of course, sound commercial motivations for measuring quality in fresh produce supply chains, since chains that reliably deliver high-quality produce are likely to be more profitable than those that do not.

It is interesting that two different types of problems arise in trying to apply methods for measuring produce quality in supply chains. The first set of problems is formed by the technical issues that must be overcome to make any measurement method useful in practice. Later chapters cover many specific cases of this type with different classes of optical measurement. The second set of problems is generic and concerns the method by which the data gathered from measurement are used to improve management and predict supply chain performance. Here, the difficult issues arise mainly from uncertainty about how different ways of defining quality relate to each other and from the problem of communication among groups of people with very different disciplinary backgrounds.

WHAT A SUPPLY CHAIN IS

The very simplest supply chains are those in which people grow, harvest, and consume their own food. For much of the world's population, this is still the main (or indeed only) supply chain by which they obtain their food, and optical analyses of quality are carried out by the human eye. However, most of the time when we think about supply chains, and the issue of how quality is measured and managed, we consider situations in which the produce being supplied changes hands at least once before it reaches the consumer. With this added complication, the simplest supply chains are those in which a producer grows the produce and sells it directly to the consumer. It is instructive, in preparation for considering how to use modern optical methods in more complex supply chains, to think about when quality measurements might occur in such a simple producer→consumer supply chain.

Let us assume that the grower in our hypothetical simple example is a vegetable producer and that cucumbers are in season. The grower selects a batch of fruits that conform to the quality criteria that are considered to indicate high quality (e.g., depth and uniformity of greenness, firmness, and lack of obvious blemishes). The grower takes the batch of fruits to the local market where they are presented for sale, perhaps after discarding the fruits damaged during transport. A consumer considering buying a cucumber will assess (i.e., measure) quality before purchase (in other words when the commodity is making a transition from one participant in the supply chain to another). The consumer may use exactly the same set, a subset, or a partly or completely different set, of criteria as those used by the grower to judge which of the available cucumbers has the best quality for his/her purposes.

In this simple example, we see the important characteristics about supply chains that we want to emphasize in this chapter and also both of the important issues concerning the measurement of quality along supply chains. First, a supply chain is simply a means of getting something from a point of production to a point of use (or consumption). No matter what level of complication or sophistication is apparent in a particular chain, how much processing is applied, or how far the goods travel and what external conditions they experience along the way, a supply chain remains simply a means of delivering something.

Notice that the concept of a supply chain just given contains no direct statements about quality. Specifying that a supply chain should deliver high quality introduces the two remaining issues that we will discuss in detail. Returning to our simple producer→consumer example, in making a choice to purchase a particular cucumber, the consumer merged a number of quantitative or ordinal quality variables (color, firmness, size, evenness, etc.) into quality categories. Ultimately we must be able to relate our analytical framework to this type of discrete categorization, whatever method we use to measure quality. Given this, when discussing quality generically, it makes sense to use an analytical framework that includes the transition from fine-grain quantitative measurement to course-grain categorization as a matter of course.

The second issue that we must deal with is that if the purpose of a supply chain is to deliver something of an agreed high quality, then quality measurement has an ethical dimension. Measurements should be made in such a way and at such times that the passage of quality along the chain can be recorded and the benefits of good performance shared equitably by all participants. Ideally, this will mean that a minimal set of quality measurements will include observations at each transition point in the chain. If we visualize the chain as a series of compartments with quality flowing from one compartment to the next, it is apparent that this minimal set of measurements will allow quality to be assessed at the beginning and at the end of the period of responsibility of each participant in the chain. In simple terms we will know what went into and came out of each component and hence what each participant did to the quality of the goods in the chain. Individual participants may want to add further measurements of quality within their components for their own purposes. Indeed, if the chain, overall, does not deliver the quality standard expected of it, this will be an absolute necessity to understand and resolve the problem.

MEASUREMENTS AND MODELS: ANALYZING THE DELIVERY OF QUALITY

Having established some basic issues, we now give the concepts under discussion some tangible existence. From what has already been said about how measurement should be embedded within food production/supply chains, two sets of transitions appear to be important: (1) transitions in quality attributes of the individual items, for example, fruits, flowers, etc., resulting from the interaction between internal biochemical and physiological processes and the environment and (2) transitions that the batches of commodities undergo in association with transfers from one chain participant to the next. These sets of processes are organized into a set of hierarchical systems in which the processes occurring within individual items are at the lowest level and the supply chain forms the upper level (Figure 1).

Many of the measurement techniques described in this book are used to assess continuous variables that are directly related to the processes responsible for determining quality, for example, those causing color or texture changes. People, however, do not generally assign quality in a continuous way, but rather integrate a variety of stimuli to make categorical classifications such as "good," "acceptable, but not the best," and "not good enough" (Sloof et al. 1996; Schepers and van Kooten 2006). This consumer's perspective may not seem like a promising starting point for developing a formal approach to measuring and modeling quality. However, it provides not only an intuitive entry point to quality analysis for those who do not have a strong numerical background, but also unites the consumer's categorical perspective with the quantitative perspective provided by many optical measurement methods and more standard dynamic modeling approaches. These issues are discussed again in the last section. The aim here is to introduce the concept.

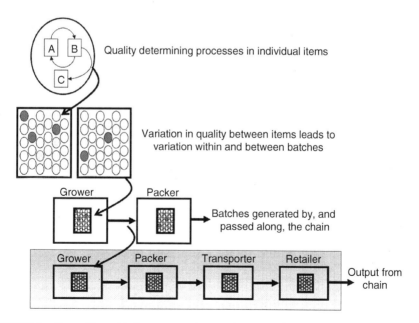

FIGURE 1 Some hierarchical structures that are important in measurements of quality in agricultural supply chains. The hierarchy spans scales from the subcellular to the global and has numerous organizational levels.

McRoberts (2006) suggested an approach that starts with quality categories. The approach was also suggested independently by Leduaphin and coworkers (2006). We assume a simple situation in which quality can be assigned to one of three* mutually exclusive ordinal categories q_1, q_2, and q_3 such that quality decreases from q_1 to q_2 to q_3. Thus, the desired quality is monotonically decreasing as the quality category index increases. We further assume that the quality categories reflect increasing levels of ripeness such that q_1 might be thought of as ideal ripeness, q_2 as full ripeness to slight overripeness, and q_3 as unacceptably overripe. Since the differentiation of produce items into different quality categories (or classes) is dependent on ripening processes, we know that in any interval of time there is a finite probability that an individual item in a quality category q_i will mature to quality class q_j where $j > i$, but that transitions in the opposite direction are not possible. Given this, the system of possible quality transitions can be shown in diagrammatic form, as in Figure 2. The figure, which borrows some inspiration from population demography, can be called a quality chain graph (QCG). If we write p_{ij} to represent the probability that an individual item will progress from q_i to q_j in the observation period, the meaning of the links in the QCG is obvious—they are the transition probabilities between classes. One can think of the transition probabilities as having a conditional statement attached to them, which reads "given the time taken for this chain to operate the probability of quality transition $q_i \rightarrow q_j$, with $i = 1,2\ldots j$ is p_{ij}." Note, in the example of Figure 2, that the observed value of p_{13} may be greater than 0 even though, logically, we expect every item to pass through state q_2 on its way from q_1 to q_3 and therefore that the value of p_{13} is 0. However, if the period between observations is sufficiently long it will be possible for items to apparently mature straight from q_1 to q_3 without passing through q_2.

The QCG represents a set of possible events which arise from underlying processes. By focusing on the events (i.e., the changes in quality class of items) and ignoring the rates of the processes, the QCG provides a way to capture expert knowledge about the performance of supply chains as a first step toward building formal models. The rates of the processes that determine quality change are integrated into the transition probabilities over whatever time step the chain takes to operate.

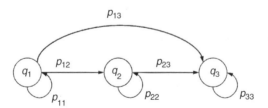

FIGURE 2 QCG depicting the hypothetical possibilities for items to degrade in quality across three quality categories. The probabilities of transition between quality categories over a relevant timescale are indicated by the p_{ij} values.

---

* Note that the analysis is not dependent on the number of quality categories; three is chosen as offering an intuitive separation between good and not acceptable by one intermediate category.

Under this assumption, the QCG contains all the information that is needed to predict what will happen to the composition of a batch of items that is handled by the chain. The transition probabilities between qualities tell us the expected proportions of items in each quality class entering the chain that will either remain in the same class or degrade to a lower class. If the values in the QCG give a true picture of the world, for large batches, the observed proportions should be close to these probabilities, with less agreement in smaller batches where random noise is not masked.

It is now apparent how the interaction between modeling and measurement can be used to improve quality levels in produce supply chains. For example, assume that we had obtained the p_{ij} values by asking the supply chain manager to make estimates based on judgment or historical performance data. By measuring the quality profile of a number of batches at the start and end of the chain, we would be able to test the manager's assumptions about chain performance and possibly identify problems. To analyze performance within the chain, the approach that has just been outlined for the entire chain can be applied to each of its components. If the set of QCGs that represent the chain is written out in order, an informal visual model of the sequence of probable quality changes is obtained. Note that each QCG in the set will represent a different amount of time. As with the analysis of the whole chain, the points at which batches enter and leave each component of the chain represent a minimal set of measurement points that will allow reflective learning by the chain participants to improve chain performance.

Visual QCG models of the chain can be readily turned into formal quantitative models by making use of the fact that each QCG can be translated into a matrix, and the rules of matrix multiplication exploited to use the resulting quality projection matrices (QPMs) to model the flow of quality classes along the chain. Leduaphin and coworkers (2008) describe a method for estimating the p_{ij} values from time-series panel data. The terminology here is, again, inspired by demography where population projection matrices are a central tool in the analysis of future population size and structure. An example based on the results obtained by Nunes and coworkers (2003) is given in Figure 3. There are many introductory textbooks on matrix calculations for readers who are not familiar with them. Caswell (2001) introduces the subject in a demographic context which translates easily to change in quality states.

McRoberts (2006) discussed several ways in which QCGs and QPMs might be used to analyze how quality is handled by supply chains providing known conditions. The important issue, in the context of this book, is that all of these uses depend very much on the interaction between measurements made on the produce and a model (or description) of the expected behavior of the chain, in order to analyze its performance. Having laid out the general framework into which measurement must be placed, the next two sections develop some key points about the use of information arising from measurements in supply chain management.

MEASUREMENT FOR ETHICAL MANAGEMENT

Given the availability of a variety of methods for measuring quality attributes of agricultural commodities, we must ask ourselves what the possible intentions in making such measurements might be. The obvious answer is to gain knowledge of

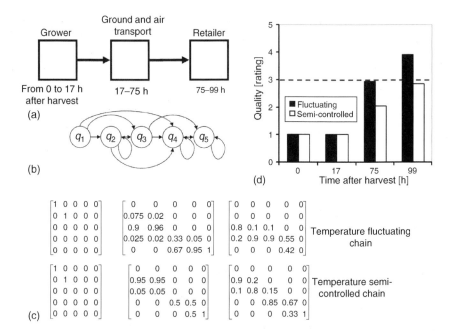

FIGURE 3 Hypothetical example of how QCGs and QPMs can be used to analyze quality change in a supply chain. The example is based on results from Nunes et al. (2003). (a) Two chains for supplying fresh strawberries to the retail market were examined, one in which temperature was semi-controlled along the chain and the other in which it was allowed to fluctuate between ambient conditions. (b) The QCG at the centre of the figure includes the links inferred for the transport component of the fluctuating chain. (c) QPMs: The quality values were fixed to match the values shown in Nunes et al. (2003) (Figure 2B) and values for the p_{ij} within the matrices were found iteratively by trial and error to match the time course of quality change shown by Nunes et al. The QPMs do not represent a unique solution and the example is for illustrative purposes only. (d) The values in the bar chart can be compared with the points shown by Nunes and coworkers at corresponding times. The dashed horizontal line indicates the acceptable quality limit.

the quality state of the commodities being measured. It is almost trivial to point out that information gained from measurements of quality can be used in quality management in supply chains, since there is clearly a connection between the analysis of quality and the management of quality. It was already pointed out as an aspiration, the minimal set of measurements on any chain ought to include observations at the start and end of each link in the chain. However, the issue of the purpose of measurement becomes more closely connected to the sustainability of supply chain performance when the motivation of the supply chain participants who make the measurements is considered. The information gained from measuring quality can be used more or less ethically depending on the motivation for collecting it. The motivation might be to increase cohesion along the chain, including the final consumer, or to allow the chain to be managed to the advantage of one or more participants. The difference between these approaches is exposed by asking whether the purpose for measuring quality is to identify problems, plan mitigation, and avoid the necessity of informing the final consumer (or another chain participant) of problems; or, alternatively, to share information on the performance of the chain with all participants so that collective action can be taken to improve it. The difference between these approaches to the use of information gained from measurement is clearly an ethical one, and mounting evidence on the performance of supply chains suggests that it is an important determinant of long-term business success; successful and sustainable supply chains are those in which inter-participant trust is maximized through the open sharing of information (Bowersox et al. 2000; Collins et al. 2002; Collins 2003).

As measurement techniques become more sophisticated, the opportunities to measure quality nondestructively in real time are increasing. However, not all participants in a chain may be equally placed to exploit new measurement methods, and such inequalities in what might be called the initial conditions of the chain only serve to increase the responsibility on those participants who do have the opportunity to measure quality to share the information openly. It is also important that participants in chains are aware of what information a given set of measurements provides. For example, suppose that the chain depicted in Figure 1 is one for supplying the retailer with bulk freshly washed carrots. The major contractual arrangement in such chains is often between the retailer and the packer/grader, who will offer subcontracts to growers to produce the carrots required to meet the contract with the retailer. Assume that some real-time optical measurement system is used to assess the quality profile of batches of carrots leaving the packer/grader, but not before this. Strictly, these measurements reflect the effects of both the producer and the packer/grader on the quality of carrots in the chain at that point and any negotiation between the two parties on the price to be paid to the grower ought to take this into account.

MATCHING MEASUREMENTS TO THE REQUIRED INFORMATION HIERARCHY

The main aim of this section is to show how the analytical approach described relates to more familiar methods for analyzing quality change over time. There are two main motivations for this. Firstly, data collected from optical measurements of produce are more likely to be used in conjunction with these standard approaches than with those described above. Secondly, in showing how the two sets of approaches are related, we will emphasize the point that the distinction between qualitative and quantitative views of quality is imposed by the observer and reflects the hierarchical structure of the systems that comprise agricultural production/supply chains.

To illustrate our line of argument here we will use a first-order degeneration model (De Ketelaere et al. 2004). Writing $Q(t)$ to stand for quality at time t, this model is given by the equation $Q(t) = Q(0)e^{-kt}$ where $Q(0)$ is the initial quality and k is the rate parameter of quality decline. It is probably the simplest standard dynamic model of quality change that we could use. It is important to stress that it will be correct in its dynamics only for some applications; that is, those in which the decline in quality follows a negative exponential curve over time. This is not an important limitation to the model's illustrative strength, but we should bear its limitations in mind. The model has an underlying mechanistic basis in enzyme kinetics (Tijskens and Polderdijk 1996) and has been widely used in quality modeling (Thai and Shewfelt 1990; De Ketelaere et al. 2004). Although such models are commonly utilized in a deterministic dynamic framework as differential equations describing the rates of biological processes, we will take an alternative view in which the degeneration model is derived as the solution of survival process model, also referred to as hazard or failure time models (Collett 1996). The use of survival models in quality analysis has been explored in detail by Guillermo Hough and coworkers (see, e.g., Hough et al. 2003, 2006; Luz Calle et al. 2006).

Consider individual produce items. Let us assume there is a probability, p, that during a given time interval (e.g., a day), an item will degrade in quality such that it changes from acceptable to unacceptable. Thus, there is a probability $(1-p)$ that quality remains acceptable on any given day. If we assume that p is constant over time, the probability that an item remains acceptable for t days is given by $P(t) = (1-p)^t$. If we further assume that all items in a batch behave similarly and independently of each other and all were of acceptable quality initially, $P(t)$ is an estimate of the expected proportion of the batch that will be of acceptable quality up to any time, t. The quantity $1 - P(t)$ is known as the cumulative risk or cumulative hazard and is the cumulative probability of a failure event in the interval up to time t. If the batch contains N items, the value $n(t) = N[1 - P(t)]$ gives the expected number of acceptable items in the batch over time. For a large collection of batches, each of size N, and all sharing the same dynamics for quality, the time-dependent binomial distribution $B[N,(1-p)^t]$ provides an estimate of the proportion of batches with $0, 1, \ldots, n$ acceptable items per batch over time. Before further elaboration, we can see that this very simple model allows us to link the dynamics of quality change for individual items to the quality properties of individual batches and further to the properties of collections of batches.

Examining the behavior of this simple model, we can make a few useful comments in relation to the issue of quality measurement. First, the average (or expected) quality of the batch changes in a predictable way with time, showing exponential decline. In fact, the first-order degeneration model gives the expectation of the time-dependent stochastic model captured in $B[N,(1-p)^t]$ since, in fact, $k = -\ln(1-p)$; that is, the rate parameter from the degeneration model is related to the instantaneous probability of failure in the survival model. In systems with this type of dynamics, quality changes more rapidly early on, so efforts to sample batches should, perhaps, focus on these early periods. A second point to note is that expected variability among batches will be greater in the middle of the time course of quality change than at the beginning or the end. This is a reflection of the fact that the variance in the number of acceptable items per batch is given by the variance of the time-dependent binomial distribution $B[N,P(t)]$ and is defined as $\{P(t)[1-P(t)]\}/N$. This is a quadratic expression in $P(t)$ (i.e., if $\gamma = 1/N$, we have $\gamma\{P(t)[1-P(t)]\} = \gamma P(t) - \gamma P(t)^2$) which is maximal when $P(t) = 0.5$ and small when $P(t)$ tends to 0 or 1.

How does this dynamic model relate to the QCG and QPM approach we examined earlier? The link comes through the probabilities of state change represented in the models. Consider the transition probability p_{23} in the QPM model in Figure 2. This is the estimated probability of quality change from acceptable to unacceptable over the number of days in which the chain operates. Letting the number of days the chain takes to operate be t, we can see that p_{23} is the cumulative risk of quality failure; that is, $p_{23} = 1 - P(t)$. The transition probabilities in QPM models are cumulative risks in survival models. The survival model assumes a fixed (short) time interval and generates the cumulative risk from the dynamics of the instantaneous risk. The QPM approach describes the consequences of known or assumed cumulative risks for the variable time intervals selected to represent an entire process, and integrates the underlying dynamics of the instantaneous risk.

In swapping from the survival model to the QPM model, we go up one scale in the spatial hierarchy of the system from internal processes to properties of whole items, and switch from continuous variation in properties to differences between discrete quality categories.

A little further elaboration on the survival model is worthwhile because it reveals something about the source and nature of variance, which we can expect to see in batches of produce. This is important for two reasons. First, variance in quality is a major point of commercial and academic interest (Tijskens et al. 2003; Tijskens and van Kooten 2006). Second, following on from the points made above, if continuous measurements of quality are used to categorize items as either acceptable or not acceptable, the variance in acceptability will have statistical properties that relate to the categorical nature of the quality assignment, not to the continuous nature of the measured properties. The relevant statistical properties are entailed in the QPM models.

Recall that the time-dependent behavior of $P(t)$ is defined by the relationship $P(t) = (1-p)^t$. This form of relationship was chosen because it generates simple exponential decay in expected quality over time. There is no a priori reason why the relationship between $P(t)$ and p should take this form, which only arises because we have assumed that p does not vary (1) with time or (2) between individual items. In general, we can say that $P(t)$ is a function of $p[P(t) = f(p)]$ and is found by integrating $f(p)$ over time. What would happen if we relaxed the two assumptions about p?

Allowing p to vary with time will change the dynamics of expected quality over time. Different assumptions will correspond to different formulations of the differential equations in the corresponding dynamic degeneration models. In the field of survival analysis, where this type of modeling is a basic tool, the Weibull distribution is often used as flexible, empirical model, which allows different types of dynamics (e.g., logistic decay), to be captured (Collett 1996; Hough et al. 2006). Note that even when p is time dependent (and hence $P(t) = f[p(t)]$) as long as the behavior of individual items with respect to quality change is independent, the variability in the proportion of acceptable items per batch will still be described by a set of time-dependent binomial distributions $B[N,P(t)]$ and their associated binomial variances. However, if p varies among batches, the binomial distribution will be unlikely to provide a good statistical description of the variation in quality acceptability over batches. Such a situation might arise for one or more plausible reasons. For example, the batches of produce might come from different zones in a field, glasshouse, or orchard where quality varies among zones because of variations in environmental and management factors. Alternatively, some batches may contain inoculum of microbial pathogens (such as the gray mould fungus *Botrytis cinerea*), which cause contagious loss of quality through contact between items in a batch. Finally, for certain types of produce, variations in ripening among items entering the post-harvest chain may result in variations in release of volatile ripening agents, such as ethylene, among individual items, leading to localized variations in ripening among batches. In all of these cases, the variance in quality among batches may be described by a set of time-dependent β-binomial distributions, $\beta B[N,\theta,P(t)]$. The additional parameter, θ, in the β-binomial distribution captures the variation in p among batches. For any value of $\theta > 0$, the variance of β-binomial distribution is greater than that of the binomial distribution with the same mean. The β-binomial distribution has proved useful as a statistical model for contagious processes in plant disease epidemiology (see Madden et al. 2007) and has been widely used to account for extra-binomial variance in taste panel data (Liggett and Delwiche 2005) but has not yet been applied in the context of monitoring produce quality.

CONCLUSION

Understanding the way in which quality changes in food production and supply chains depends on the iterative comparison of observation and prediction. Nondestructive real-time optical measurements of quality attributes offer further advances in our capacity to understand and manage quality. Some fairly simple approaches to analyze quality change along supply chains suggest that measurement of quality attributes at transition points in the chain offers a minimal basis for gathering information and comparing observation with expectation so that chain performance can be improved in an equitable way. It is useful to bear in mind the hierarchical nature of the systems from which supply chains are constructed, and the way the hierarchy influences switches between quantitative and qualitative perspectives when making measurements on these system and in constructing models to predict their behavior.

ACKNOWLEDGMENTS

The Scottish Agricultural College (SAC) is a main research provider for the Scottish Government Rural and Environment Research and Analysis Directorate (RERAD). This chapter is an output from RERAD Workpackage 3.1 (Sustainable Farming Systems). NM thanks Pol Tijskens for inspiration; Manuela Zude and Gareth Hughes for their helpful comments; and Philip Leat and Cesar Revoredo of the SAC, food marketing research team, and the horticulture production chains group at Wageningen University for pitching in with their hard work and providing a rich literature resource.

REFERENCES

Bowersox, D.J., D.J. Closs, and T.P. Stank. 2000. Ten mega-trends that will revolutionize supply chain logistics. *Journal of Business Logistics* 21:1–16.

Caswell, H. 2001. *Matrix Population Models: Construction, Analysis and Interpretation*. Sinauer Associates, Inc. Publishers, Sunderland, MA.

Collett, D. 1996. *Modelling Survival Data in Medical Research* (2nd Edn.). Chapman & Hall, London, UK.

Collins, R. 2003. Supply chains in new and emerging fruit industries: The management of quality as a strategic tool. *Acta Horticulturae* 604:75–84.

Collins, R., A. Dunne, and M. O'Keeffe. 2002. The locus of value: A hallmark of chains that learn. *Supply Chain Management: An International Journal* 7:318–321.

Hough, G., K. Langohr, G. Gómez, and A. Curia. 2003. Survival analysis applied to sensory shelf life of foods. *Journal of Food Science* 68:359–362.

Hough, G., L. Garitta, and G. Gomez. 2006. Sensory shelf life predictions by survival analysis: Accelerated storage models. *Food Quality and Preference* 17:468–473.

de Ketelaere, B., J. Lammertyn, G. Molenberghs, M. Desmet, B. Nicolai, and J. de Baedemaeker. 2004. Tomato cultivar grouping based on firmness change, shelf life and variance during post harvest storage. *Postharvest Biology and Technology* 34:187–210.

Leduaphin, S., D. Pommeret, and E.M. Quannari. 2006. A Markovian model to study products shelf-lives. *Food Quality and Preference* 17:598–603.

Leduaphin, S., D Pommeret, and E.M. Quannari. 2008. Application of hidden Markov model to products shelf lives. *Food Quality and Preference* 19:156–161.

Liggett, R.E. and J.F. Delwiche. 2005. The beta-binomial model: Variability in overdispersion across methods and over time. *Journal of Sensory Perception* 20:48–61.

Luz Calle, M., G. Hough, A. Curia, and G. Gomez. 2006. Bayesian survival analysis modelling applied to sensory shelf life of foods. *Food Quality and Preference* 17:307–312.

Madden, Laurence V., G. Hughes, and F. van den Bosch. 2007. *The Study of Plant Disease Epidemics*. APS Press, St Paul, MN.

McRoberts, N. 2006. Karpography: A generic concept of quality for chain management and knowledge transfer in supply chains. *Acta Horticulturae* 712:153–158.

Nunes, M.C.N., J.P. Emond, and J.K. Brecht. 2003. Quality of strawberries as affected by temperature abuse during ground, in-flight and retail handling operations. *Acta Horticulturae* 712:239–246.

Schepers, H. and O. van Kooten. 2006. Profitability of ready-to-eat strategies. In *Quantifying the Agri-Food Supply Chain*, eds. C.J.M. Ondersteijn, J.H.M. Wijnands, R.B.M. Huirne, and O. van Kooten. Springer, The Netherlands, pp. 117–132.

Sloof, M., L.M.M. Tijskens, and E.C. Wilkinson. 1996. Concepts for modelling the quality of perishable products. *Trends in Food Science and Technology* 7:165–171.

Thai, C.N. and R.L. Shewfelt. 1990. Peach quality changes at different constant storage temperatures: Empirical models. *Transactions of the American Association of Agricultural Engineers* 33:227–233.

Tijskens, L.M.M. and J.J. Polderdijk. 1996. A generic model of keeping quality of vegetable produce during storage and distribution. *Agricultural Systems* 51:431–452.

Tijskens, L.M.M., P. Konopacki, and M. Simcic. 2003. Biological variance, burden or benefit? *Postharvest Biology and Technology* 27:15–25.

Tijskens, L.M.M. and O. van Kooten. 2006. Theoretical considerations on generic modelling of harvest maturity, enzyme status and quality behaviour. *International Journal of Postharvest Technology and Innovation* 1:106–120.

Zude, M. and N. McRoberts. 2006. Product monitoring and process control in the crop supply chain. *Agricultural Engineering* 61:2–3.

1 What to Measure and How to Measure

*Moritz Friebel, Lothar W. Kroh, Susan Lurie,
Martina Meinke, Sascha Rohn,
and Alessandro Torricelli*

CONTENTS

1.1 Quality Parameters of Fresh Fruit and Vegetable at Harvest
and Shelf Life 2
   1.1.1 Definition of Quality 2
       1.1.1.1 Visual 3
       1.1.1.2 Organoleptic 4
       1.1.1.3 Nutrition 5
       1.1.1.4 Hygienic 6
   1.1.2 Harvest Maturity 7
       1.1.2.1 Maturity Measurements 9
       1.1.2.2 Variation within a Tree 11
   1.1.3 Maintenance of Quality during Postharvest Processing 11
       1.1.3.1 Quality Determination from Product Monitoring during
               Sorting, Packaging, and Marketing 11
       1.1.3.2 Quality Determination at the Entrance to Storage,
               Prediction of Quality at the End of Storage,
               and Prediction as to How Long a Product
               Should Be Stored 12
   1.1.4 Advantages of Nondestructive Sensing and Measuring Points
       in the Processing Chain 13
   1.1.5 Conclusions 14
References 15
1.2 Quality of Processed Plant Food 17
   1.2.1 Quality: Definitions and Criteria 17
   1.2.2 Quality of Fresh and Minimally Processed Produce 19
       1.2.2.1 Raw Material 19
       1.2.2.2 First Processing Steps 20
       1.2.2.3 Processing and Preservation 21

1.2.3 Determinants for Quality Aspects: Stage of Maturity,
Technological Functionality, and Health Beneficial Effects............. 24
  1.2.3.1 Glucosinolates.. 26
  1.2.3.2 Terpenes/Carotenoids 27
  1.2.3.3 Phenolic Compounds/Polyphenols.................................... 27
  1.2.3.4 Sulfur-Containing Compounds... 28
  1.2.3.5 Cyanogenic Glycosides 28
  1.2.3.6 Alkaloids.. 29
1.2.4 Quality Changes during Plant Food Processing 29
  1.2.4.1 Enzymatic-Induced Changes 29
  1.2.4.2 Thermal Treatment 30
  1.2.4.3 Nonthermal Food Processing... 33
1.2.5 Concluding for Nondestructive Analysis and Product Monitoring 35
References... 37
1.3 Optical Sensing 44
  1.3.1 Determination of Optical Properties of Turbid Media:
  Continuous Wave Approach 44
    1.3.1.1 Introduction... 44
    1.3.1.2 Radiation Transport Theory.. 45
    1.3.1.3 Kubelka–Munk Theory... 47
    1.3.1.4 Interesting Values Out of the Diffusion Theory.................. 47
    1.3.1.5 Monte Carlo Simulation 48
    1.3.1.6 Principle of Inverse Monte Carlo Simulation (iMCS) 48
    1.3.1.7 Experimental Setup for Measuring
    a Horticultural Product 50
    1.3.1.8 Spectra of Carrots 51
    1.3.1.9 Optical Properties of Carrots 52
References... 54
  1.3.2 Determination of Optical Properties in Turbid Media:
  Time-Resolved Approach... 55
    1.3.2.1 Light Propagation in Diffusive Media: Photon Migration..... 55
    1.3.2.2 Mathematics for Photon Migration..................................... 60
    1.3.2.3 Time-Resolved Instrumentation for Photon Migration 67
    1.3.2.4 Applications of Time-Resolved Reflectance
    Spectroscopy to Fresh Fruit and Vegetable........................ 73
References... 78

1.1 QUALITY PARAMETERS OF FRESH FRUIT AND VEGETABLE AT HARVEST AND SHELF LIFE

SUSAN LURIE

1.1.1 DEFINITION OF QUALITY

Consumer demand for high-quality fresh fruit and vegetable has increased in the past decades, and is still increasing. The word *quality* has many ramifications and can mean

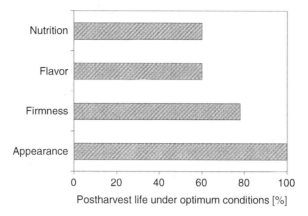

FIGURE 1.1.1 The relative length of shelf life of different quality attributes of a commodity.

many things to different groups in the supply chain from the farmer to the consumer. For the farmer it generally means high yield of a product with few blemishes or decay problems. For the supplier and marketer it can mean a firm product that has good shelf life and which does not decay before being sold. For the consumer it can mean good external appearance and good taste. It has been shown repeatedly (Kader 2002) that a commodity can still appear acceptable when its taste has already been lost (Figure 1.1.1). Therefore, the appearance of a commodity is not sufficient when we speak about quality. Many consumers complain that the taste of fruits and vegetables that they remember from the past has been lost in the commodities being sold in the global marketplace. There is some truth in this, since breeding for longer shelf life generally (not always) is at the expense of some flavor. In addition, fruits and vegetables used to be seasonal and so were consumed only for that part of the year. These products were produced locally, had a short marketing chain, and so could be harvested at an advanced stage of development or ripening. With the increase in global marketing many local varieties have been marginalized and denied shelf space in the supermarket chain. The fruit and vegetable marketing chain in a developed country is shown in Figure 1.1.2. Produce has a number of way stations to go past before reaching the consumer. At each of these points the produce may be loaded, unloaded, cooled/warmed, or sorted/unsorted, on its way to the final destination. The shortest route is the local farmers' market, which accounts for only a small portion of the total market. A positive aspect of global marketing is the marketing of products that have been bred for enhanced nutritional and health-promoting aspects. These include fruits and vegetables with high pigmentation, and increased antioxidant and vitamin content. Hence, it is important to define what is meant by quality. I have divided this parameter into four aspects: visual, organoleptic, nutritional, and hygienic.

1.1.1.1 Visual

Visual quality encompasses the appearance of the product. The product should look fresh, have the normal size and color associated with the particular fruit or vegetable, and be without blemishes or signs of decay. There are often parts of a fruit or

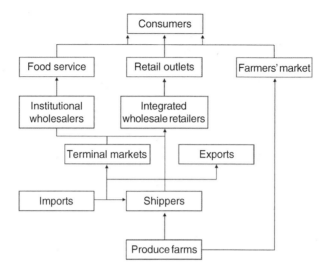

FIGURE 1.1.2 A schematic of the marketing chain of fruit and vegetable.

vegetable that are not edible but are important for appearance. Examples are the stems of cherries and of grape clusters, and the calyx of fruits such as tomatoes and persimmons. The stems should remain green and turgid, not brown and withered. A green calyx makes a fruit look just off the tree or plant.

The absence of blemishes or signs of decay is also of utmost importance. With the long marketing chain of many fruits and vegetables that is currently in place, bruises are a common problem that can develop. This can take the appearance of a discolored, soft area as in apples (Van Zeebroeck et al. 2007), or in pitting as in cherries (Toivonen et al. 2004). Bruising can also lead to lower shelf life and a higher risk of bacterial and fungal contamination. Cuts and scratches can also be a problem, because of poor harvest procedures or poor handling.

Fruits and vegetables marketed after storage may develop storage disorders such as surface scalding or browning, yellowing due to disappearance of chlorophyll (particularly in green fruits and vegetables), and pitting or waterlogging of tissue due to chilling injury. All of these will reduce the visual quality of the fruit.

1.1.1.2 Organoleptic

Eating quality of fruits and vegetables includes the sweetness due to soluble solids content (SSC) (mainly sugars); sourness due to titratable acidity (TA) (mainly organic acids); texture due to cell wall properties; turgor, astringency, or tartness due to polyphenols; and aroma due to volatiles. All these components add up to a characteristic flavor of a commodity.

In the case of non-climacteric fruits and vegetables these components do not change greatly following harvest, until senescence processes begin. However, with climacteric fruits, ripening continues after harvest and it is always a question of when to harvest to achieve the optimum organoleptic quality at the time of consumption. If a fruit is harvested too early it will not develop enough flavor by the time it softens enough for consumption. If it is harvested fully ripe it will soften too quickly or become bruised during marketing, resulting in poor quality. In addition, if a fruit or vegetable is to be stored before marketing it may need to be harvested at a different stage of ripening than if it is to be marketed directly. Therefore, a great deal of work has been done to determine the proper marketing time for various fruits depending on their fate after harvest.

More and more research is being done to determine what are the best organoleptic characteristics of a commodity. In some cases one component is the major one. A study of table grapes found that SSC was the major component determining acceptability (Sonego et al. 2002). In others it may be a combination of SSC and TA. A study in a particular variety of plum found that if the SSC was between 10% and 11.9% and if the TA was 0.6% and not higher then the plum was rated good. However, if the SSC of the fruit was above 12% then the TA played a minor role in taste acceptance, and the fruit was still rated highly even if the TA was 1.0% (Crisosto et al. 2004, 2007). In apples, texture is a very important organoleptic component (Harker et al. 2006). It was found that the apple texture and the way that it releases juice during chewing affected the perception of sweetness of the fruit.

1.1.1.3 Nutrition

The macronutrients are carbohydrates, organic acids, proteins, and fats. Fruits and vegetables are high in carbohydrates; polysaccharides and soluble sugars make up the cell walls. In some cases there are also storage carbohydrates such as starch, which forms part of the commodity to be consumed. The cell wall and storage polysaccharide molecules are important in diet as a source of fiber. In addition, fruits and vegetables are important sources of other dietary nutritional components including antioxidants, pigments and vitamins, and minerals. In some instances where the presence of these compounds is elevated, the fruit or vegetable is termed as functional food.

Functional food is any food or food ingredient that provides a health benefit beyond the traditional nutrients it contains. These are foods that contain bioactive, health-promoting phytonutrients. As any toxicologist may say, the difference between a beneficial effect and a toxic effect is the dose. Too much of a bioactive compound can be toxic. It is possible for a person to consume vitamin X from a functional food, take a vitamin pill containing vitamin X, and also eat additional vitamin X in fortified food. The total exposure could be toxic.

A fruit or vegetable that is considered as a functional food generally contains a number of functional compounds. These compounds are usually secondary metabolites. As an example, soybeans contain phytates, phenolic acids, flavonoids/ isoflavonoids, carotenoids, coumarins, lignans, terpenes, enzyme inhibitors, and saponins (Shahidi 2004). In each family of compounds there may be a number of members present. Epidemiological and clinical studies have found that populations consuming a large proportion of plant-based foods have a lower incidence of cardiovascular diseases, certain types of cancer, immune system decline, and certain neurological disorders (Shahidi and Naczk 2003). Initial research examined the benefit of plant-based foods by considering individual constituents, such as vitamin C, vitamin E, or carotenoids. However, it was found that benefits correlated with more than one component. The compounds in a functional food have a cooperative and sometimes synergistic action in disease prevention. Some of the compounds in fruits and vegetables that are active in cancer prevention inhibit the initiation of cancer while others inhibit tumor growth and tissue damage.

Endogenous phenolics in plant foods are the largest group of secondary metabolites. Phenolic compounds are among the phytochemicals in fruits and vegetables that may render their effects via antioxidation and relief from oxidative stress. The antioxidative effect of phenols is due to direct free radical scavenging activity, and also indirectly from chelation of pro-oxidant metal ions. It is the antioxidative capacity in fruits and vegetables that has been identified as one of the major mechanisms contributing to the maintenance of human health (Vinson et al. 2001; Hodges and Kalt 2003).

Another aspect of nutrition in fruits and vegetables is whether organic or conventionally grown produce is more nutritious (Brandt et al. 2004; Lester 2006). There are increasing numbers of reports detailing studies of organically and conventionally grown fruits and vegetables. The conclusions of the studies are sometimes contradictory. Magkos and coworkers (2003) found that there were no differences in vitamins, minerals, or heavy metals between conventionally grown crops and organic crops. However, Worthington (2001) found that organic apples contained significantly more vitamin C, iron, magnesium, and phosphorus and significantly less protein, nitrates, and lower amounts of heavy metals than conventional fruit. Rembialkowska (2003) also found that organic crops contained higher total sugars, more vitamin C, calcium iron, magnesium, phosphorus, potassium, less protein and nitrates, and no clear impact on β-carotene.

1.1.1.4 Hygienic

Fruit and vegetable are treated with various chemicals during their growth in the field to prevent attack due to insects and microorganisms. These chemicals should be applied long enough before harvest so that their residues will not be present at harvest, or will not be above the allowed level. In good agricultural practice, the grower will be aware of how soon before harvest a particular chemical can be applied. However, there should be measures taken to check compliance with the correct treatments, and ensure that the commodity does not have elevated residues of any chemicals (Baker et al. 2002).

After harvest chemical treatments are also often applied, particularly if the product will be stored before being marketed. The treatments may be to prevent decay or to prevent storage disorders from developing. An example is a dip of diphenylamine (DPA) given to apples to prevent superficial scald, a storage disorder developing in some cultivars of apples that leads to skin browning. This is an oxidative process and the DPA is a chemical antioxidant. However, since the apples may be harvested with fungal spores on the fruit, the dip may cause cross-infection of healthy fruit. Therefore, to the dip is added an antifungal chemical as well. These compounds will disappear during the storage time if it is long enough, but if, because of problems with the storeroom or for other reasons, the apples are marketed within the first few months of treatment, residues may be higher than the allowable limit.

The other side of the coin is food safety from harmful microorganisms. These include *Salmonella*, *Shigella*, and *Escherichia coli* O157:H7. Outbreaks of sickness due to eating contaminated fruits and vegetables have been documented in recent years, particularly in the United States, and the disease vector traced back to a specific commodity from a particular location. The contamination can be due to impure water sources, proximity to livestock, or poor hygiene among the farm workers. The disease causing organisms can survive the packaging and shipping conditions. An example is netted muskmelon (Hodges and Lester 2006). Netted muskmelon has been associated with large outbreaks of human illness (more than 25,000 individual cases since 1990) in the United States and Canada (Castillo et al. 2004). Even surface pasteurization does not prevent the survival of human illness pathogens, probably because of the inaccessible sites of the netting (Annous et al. 2004; Beuchat and Scouten 2004).

Organic fruit and vegetable production is increasing. Both the amount of organic produce being grown worldwide and the land area in production have increased. As production grows the market becomes more sophisticated and horticultural products need to be stored longer, transported over greater distances, and/or processed more intensively. The need for disease-free fruits and vegetables is as important for organic as for conventional products. However, the tools to combat insects and microorganisms are more limited, and the storage and shelf life of these products may be shorter than that of conventionally grown food (Bourn and Prescott 2002). The appearance of signs of decay or disorders that may be controlled in the orchard or the field in conventional agriculture by sprays that are not allowed in organic agriculture can be a problem that must be addressed.

1.1.2 Harvest Maturity

Harvest maturity is the stage when a fruit or vegetable should be picked so that following the marketing chain it will remain of high quality. Figure 1.1.3 shows the horticultural maturity of different classes of fruits and vegetables compared to a timeline of the plant development. As mentioned above, non-climacteric vegetables and fruits do not continue to ripen after harvest and should be picked at a proper stage for consumption. Climacteric fruits, such as pome and stone fruits, as well as subtropical and tropical fruits such as avocados, bananas, and mangos do continue their ripening processes and can be harvested at a stage where consumption should be delayed. Table 1.1.1 gives a list of climacteric and non-climacteric fruit. Ripening of climacteric fruit includes changes in peel color, the conversion of storage starch to glucose and sucrose, the development of aromatic compounds, and disassembly of the cell wall polysaccharides leading to softening. These two divisions between climacteric and non-climacteric are not hard and fast, since non-climacteric fruits, such as grapes, citrus, and strawberries, will have alterations in some of their quality parameters after harvest. These include a decrease in titratable acidity, softening, and increased volatile production. However, the processes are not accompanied by a climacteric rise in respiration or ethylene production, which is the characteristic of climacteric fruit. Vegetables that are mainly non-climacteric will also show changes after harvest, such as softening or wilting and color changes, particularly loss of

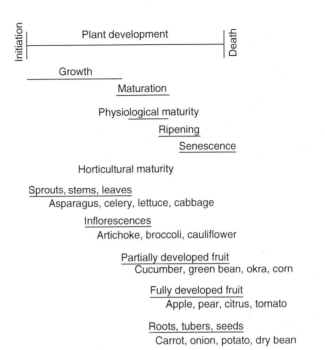

FIGURE 1.1.3 Horticultural maturity of fruits and vegetables in relation to developmental stages of the plant.

TABLE 1.1.1
Fruits Classified According to Climacteric or Non-Climacteric Respiratory Behavior during Ripening

Climacteric		Non-Climacteric	
Apple	Muskmelon	Blackberry	Lychee
Apricot	Nectarine	Cacao	Mandarin
Avocado	Papaya	Caramola	Okra
Banana	Passion fruit	Cashew apple	Olive
Biriba	Peach	Cherry	Orange
Blueberry	Pear	Cranberry	Pea
Breadfruit	Persimmon	Cucumber	Pepper
Cherimoya	Plantain	Date	Pineapple
Durian	Plum	Eggplant	Pomegranate
Feijoa	Quince	Grape	Prickly pear
Fig	Rambutan	Grapefruit	Raspberry
Guava	Sapodilla	Jujube	Strawberry
Jackfruit	Sapote	Lemon	Summer squash
Kiwifruit	Soursop	Lime	Tomarillo
Mango	Sweetsop	Longan	Tangerine
Mangosteen	Tomato	Loquat	Watermelon

chlorophyll. These will be more rapid if the vegetable is harvested at a very mature stage. Therefore, determining harvest maturity is important for all commodities to guarantee the maximum storage or shelf life before loss of quality.

One aspect of tree crops is that there is generally a large variation of ripening among the fruit on each tree. This is due to the position on the tree, the amount of light interception of the canopy, the number of leaves supplying metabolites to the developing fruit, and additional factors (Caruso et al. 2001). In grape bunches there is also a gradient of ripening within a bunch, and the berries closest to the stem are more mature than the berries at the bottom of the bunch. The difference in SSC can be more than 1% (Tarter and Keuter 2005). Generally, selective harvests must be conducted for tree crops with set maturity indices determining which fruits are harvested in which harvest.

1.1.2.1 Maturity Measurements

The concept of horticultural maturity implies the use of a measurable character, changes in which can be used to indicate when a commodity should be harvested for a particular purpose. This characteristic is known as a maturity or harvesting index (Kader 2002). Maturity indices for various horticultural crops have relied on different features of the commodity, such as duration of development, for example, in days from flowering or anthesis, size, starch or sugar (SSC) content, color, and firmness (Shewfelt 1993). Studies have associated high consumer acceptance with high SSC in many commodities (Parker et al. 1991; Kader 2002), but there are more factors involved, such as acidity (TA) (Peterson and Ivans 1988; Kader 2002), SSC/TA ratio (Nelson 1985; Kader 2002), phenolics (Roberson and Meredith 1989), and volatiles (Chapman et al. 1991).

Maturity measurements often begin with fruit or vegetable size, to which are added other parameters, depending on the crop. For apples and pears firmness is an additional measurement, since these commodities are stored. Fruit that is too soft at harvest cannot be held in extended storage. Firmness is measured destructively with a penetrometer on a peeled section of the fruit. Starch is also determined destructively by dipping a slice of the fruit into an iodine solution as seen in Figure 1.1.4. As the apple fruit matures on the tree, starch is converted to sucrose, fructose, and glucose and the amount of starch remaining is an indication of its maturity. SSC may also be determined by squeezing a drop from the fruit onto a handheld refractometer. These three measurements are often combined into the Streif factor, which is used for many apple cultivars as a determination of fruit maturity (Streif 1996). Nowadays, new maturity indices are being proposed, such as including a nondestructive measurement following chlorophyll decrease in the peel of a commodity (Zude-Sasse et al. 2000).

Stone fruits also have a number of measurements that have been used as maturity indices. In addition to fruit size these include firmness, peel background color, SSC, TA, SSC/TA ratio, and flesh color. Figure 1.1.5 shows a chart for apricot harvest stages. The optimum harvest of these fruits will be stages 3 or 4. In many places the two easiest, and nondestructive, measurements are fruit size combined with a color maturity chart (Delwiche 1987). These two parameters do not give any

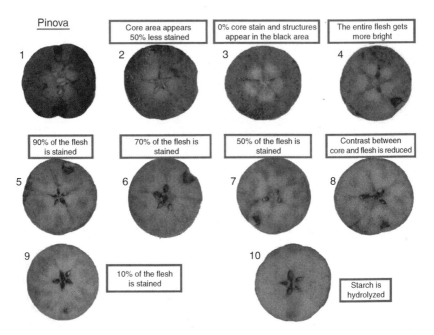

FIGURE 1.1.4 A starch–iodine index chart for apples (*Malus x domestica* 'Pinova') based on the percentage of stained tissue. Firmness was determined before starch and some slices show holes from the penetrometer.

information on the internal characteristics of the fruit. However, a study comparing different maturity indices of different peach cultivars with sensory panel determinations found that peel color and firmness of the fruit cheek and blossom end correlated best with flavor aspects (Brovelli et al. 1998).

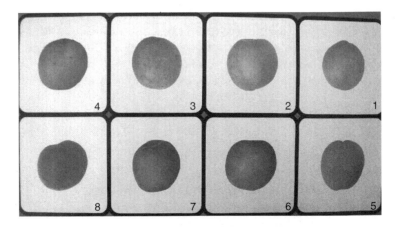

FIGURE 1.1.5 A chart for apricot harvest.

The decision when to harvest a horticultural product is an important one. Overripe fruits are extremely susceptible to injury and decay. Immature fruits may not ripen to standards required by consumers, and, if stored under an inappropriate temperature regime or for too long, may have storage disorders. It is therefore very important to determine when a commodity has reached horticultural maturity and can reach the consumer in satisfactory condition after having undergone a specified storage and handling period.

1.1.2.2   Variation within a Tree

In any harvest there will be variation among the fruit picked, even when they conform to the harvest parameters that have been decided upon. In stone fruits, the initial fruit quality of several peach, nectarine, and plum cultivars were evaluated according to fruit canopy position (Marini 1991; Saenz 1991). Differences in SSC, acidity, and fruit size were detected between fruit obtained from the outside and from the inside canopy positions of the tree. For these fruits, those grown under a high light environment had a longer shelf life than those from a low light environment from the tree interior (Crisosto et al. 1997). It was also found that fruit that developed in the shaded inner canopy developed more storage disorders than fruit from the other canopy (Crisosto et al. 1997). This is also true of other tree fruits. Apples from the inner canopy were more susceptible to superficial scald than apples from the other canopy (Lurie et al. 1989). Similarly, avocados that were shaded were more prone to chilling injury during storage than those exposed to the sun (Woolf et al. 2000).

1.1.3   MAINTENANCE OF QUALITY DURING POSTHARVEST PROCESSING

Once a commodity has been harvested, the best postharvest procedures will only maintain quality. In virtually all cases, fresh produce quality is set at harvest and then inevitably declines during postharvest lifetime. The commodity will be subjected to a number of stresses due to its removal from the plant or tree. These include water stress due to moisture loss, temperature stress from storage conditions, anaerobic stress if the product is stored in modified or controlled atmosphere, biotic stress from pathogens present at harvest or that enter small wounds during the postharvest sorting, and mechanical stress from transport, sorting, and packing. The ability to monitor the response of commodities to these stresses can be important in designing the best postharvest management procedures.

1.1.3.1   Quality Determination from Product Monitoring during Sorting, Packaging, and Marketing

Most studies in product quality during postharvest handling focus on one or two steps in the handling system, generally the quality at harvest and the quality at the end of the process. A few studies have utilized a systems approach, whereby the effect of each handling step is assessed on the final product quality. This allows for establishment of coordination and control of quality within the operation. This approach has been employed to study quality changes of peas (Shewfelt et al. 1985), peaches

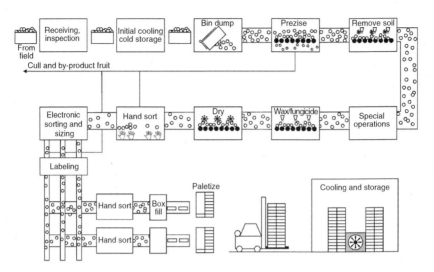

FIGURE 1.1.6 A schematic of a packinghouse line.

(Shewfelt et al. 1987), tomatoes (Shewfelt et al. 1989), and green beans (Shewfelt et al. 1986). The emphasis has been on maintenance of quality during realistic handling conditions rather than ideal laboratory conditions. This type of monitoring for quality changes during the whole postharvest chain is an area that nondestructive methods can contribute to greatly. Figure 1.1.6 shows a schematic of a typical unit operation in a mechanized packinghouse. There are a number of points along the pathway where produce can be damaged because of drops or contact with other fruit or the sides of the packing line. However, there are also points along this pathway where nondestructive methods can complement or replace other sorts of sorting. A recent study of changes in quality during a postharvest chain was conducted with harvested carrots (Zude et al. 2007). Carrot quality was assessed in terms of sensory attributes such as sugar content, an important component of taste, and secondary plant compounds such as α- and β-carotene, which show provitamin A activity in humans. The study monitored changes in sugar content and carotenes from harvest and through different postharvest treatments until consumption using a noninvasive spectrophotometric technique (see Chapter 3). Another study of storage time of lamb's lettuce used chlorophyll fluorescence to monitor quality changes (Ferrante and Maggiore 2007). This type of monitoring can provide a quality control tool in modern supply chain monitoring.

1.1.3.2 Quality Determination at the Entrance to Storage, Prediction of Quality at the End of Storage, and Prediction as to How Long a Product Should Be Stored

In many commodities, measurements of maturity at harvest can give an indication of how long they can be stored. Apples and pears below a certain firmness (see Chapter 5.2) are not recommended for long-term storage. Other measurements used

What to Measure and How to Measure 13

for apples to determine storage potential are internal ethylene concentration and starch content (Kader 2002). In stone fruits, apricots, nectarines, peaches, and plums, a range of firmness for each cultivar determines how long they should be stored (Lurie and Crisosto 2005). Fruit above this firmness range are immature and will not be tasty or ripen fully, while fruit below the range are soft and will soften too rapidly after storage to be marketed. Avocados are judged for maturity by dry weight or oil content. Many other fruits are judged by peel color. All except the last measurement are destructive tests.

During long-term storage of fruits, particularly apples that can be stored in controlled atmosphere for up to a year, samples must be examined periodically to make sure that quality is maintained. The normal practice is to have boxes of fruit in a window that can be opened without breaking the low oxygen in the room. These apples are held in shelf life to determine the quality of the fruit that will reach the consumer. A new application of a well-established measurement is to monitor the fluorescence emitted from chlorophyll in the fruit peel excited by a low powered light source (Stephens and Tanner 2005). The fluorescence signal (see Chapters 4.1 and 4.2) is measured periodically and analyzed to determine the response of the fruit to low oxygen.

Another fruit that is stored for four or five months is persimmons. Following storage the fruit often softens too rapidly to be effectively marketed. Examination of firmness changes during storage did not show softening during storage. However, by examining fruit texture with an acoustic method, it was shown that texture changed during storage and above a certain value the fruit could be marketed while below this value the fruit softened excessively during marketing (Ben-Arie et al. 2001). This technology has also been used to predict the maximum storage period of apples (Shmulevich et al. 2003). These examples show that different methods of nondestructive monitoring can play an important role in monitoring quality of fruit during storage.

1.1.4 Advantages of Nondestructive Sensing and Measuring Points in the Processing Chain

Determination of quality is important throughout the harvest and marketing chain. It begins with the determination of the best time to harvest for optimum quality for the consumer. Nondestructive methods have an advantage over destructive methods in this situation. Destructive methods are based on taking a small sample of the commodity to be harvested and testing it and then discarding the sample. This tendency to use as few samples as possible often results in increased lot to lot variability in the parameter measured. Nondestructive methods would allow for the sample measured to remain part of the harvest, and so larger numbers of a commodity can be measured (see Chapter 3). During growth, all kinds of small differences in growing conditions, such as position of fruit on the tree relative to leaves and other fruit, microclimate, hormonal and nutritional and irrigation effects, can lead to considerable variation in maturity at harvest. This biological variation can be measured nondestructively and be modeled so that the fruit can be graded into different maturity levels (Tijskens et al. 2007; Valero et al. 2007). The more mature fruit can be marketed before the less mature fruit, or could be sent to closer markets.

At the sorting or packinghouse line, whether the fruit are designated for direct marketing or for storage before marketing, in addition to grading for size and color, most sorters try to weed out fruits with imperfections, either bruises, disorders, or decay. Currently, most packinghouse lines use a combination of automatic methods (for weight, size, and sometimes color and shape) and visual inspection (for decay, bruises, and other disorders). Nondestructive methods can be very effective at this stage for both sizing and color grading and to identify imperfections. It is also possible to grade fruit for higher or lower SSC using nondestructive means (see Chapter 3), something that is not possible with visual inspection (see Chapter 2).

After harvest, quality losses occur as a function of time, mechanical impacts, and storage conditions. Low temperature storage will decrease respiratory activity and positively affect the keeping quality of fresh, perishable produce. However, the cool chain from harvest to consumption is never completely maintained and this can have a negative effect on product quality. A study of the postharvest marketing chain of papaya found that fluctuating temperatures, which simulated the conditions of air shipment of papaya fruit caused greater loss of quality compared to fruit held at a constant temperature (Nunes et al. 2006). The problems included increased decay and also loss of nutrition (see Chapter 4) including decrease in vitamin C. The study found that fluctuating temperatures that are often encountered for as little as a few hours during handling operations may result in considerable amount of rejected fruit. This type of monitoring during the whole supply chain (see Chapter 6) can have an important impact on maintaining quality and delivering better fruits and vegetables to the consumer.

1.1.5 Conclusions

Maintaining quality of a commodity is important throughout the chain of harvest, handling, and marketing. Nondestructive methods have a major role to play in monitoring different aspects of quality. Of increasing importance are the nutritional components of fruits and vegetables. This aspect was not well developed prior to the current emphasis on proper nutrition as a means of prevention of disease. However, with many studies showing correlations between diet and heart disease and cancer, and with the description of the "French paradox" as well as the "Mediterranean diet," consumers are becoming more aware of the properties of fresh fruit and vegetables. These healthful components of fruits and vegetables are one of the areas that nondestructive methods can help to quantify (see Chapter 4).

Another aspect that is growing in importance with the increasing globalization of food supply is food safety. The presence of harmful chemicals or microorganisms must be monitored, and nondestructive methods can play a role in their detection. With increasing mechanization and high-throughput of packinghouse and processing lines, these methods will be utilized to a greater extent than currently. In conclusion, quality monitoring is a sector that will be an increasingly important area for application of nondestructive methods.

REFERENCES

Annous, B.A., A. Burke, and J.E. Sites. 2004. Surface pasteurization of whole fresh cantaloupes inoculated with *Salmonella* Poone or *Escherichia coli*. *Journal of Food Protection* 67:1876–1885.

Baker, B.P., C.M. Benbrook, and E. Groth. 2002. Pesticide residues in conventional, integrated pest management (IPM)-grown and organic foods: Insights from three US data sets. *Food Additives and Contaminants* 19:427–446.

Ben-Arie, R., A. Gizis, O. Nerya et al. 2001. Predicting the storage life of 'Triumph' persimmons. *Alon Hanotea* 55:201–204 (in Hebrew).

Beuchat, L.R. and A.J. Scouten. 2004. Factors affecting survival, growth and retrieval of *Salmonella* Poona in intact and wounded cantaloupe rind and stem scar tissue. *Food Microbiology* 21:683–694.

Bourn, D. and J. Prescott. 2002. Comparison of the nutritional value, sensory qualities, and feed safety of organically and conventionally produced foods. *Critical Reviews in Food Science and Nutrition* 42:1–34.

Brandt, K., L.P. Christensen, J. Hansen-Moller et al. 2004. Health promoting compounds in vegetables and fruits: A systematic approach for identifying plant components with impact on human health. *Trends in Food Science and Technology* 15:384–393.

Brovelli, E.A., J.K. Brecht, W.B. Sherman, and C.A. Sims. 1998. Potential maturity indices and developmental aspects of melting-flesh and nonmelting-flesh peach genotypes for the fresh market. *Journal of the American Society of Horticultural Science* 123:438–444.

Caruso, T., P. Inglese, C. DiVaio, and L.S. Pace. 2001. Effect of different fruit-thinning patterns on crop efficiency and fruit quality for greenhouse-forced 'May Glo' nectarines trees. *HortTechnology* 11:412–415.

Castillo, A., I. Mercado, I.M. Lucia et al. 2004. *Salmonella* contamination during production of cantaloupe: A binacional study. *Journal of Food Protection* 67:713–720.

Chapman, G.W., R.J. Horvat, and W.R. Forbus. 1991. Physical and chemical changes during the maturation of peaches. *Journal of Agriculture and Food Chemistry* 39:867–870.

Crisosto, C.H., R.S. Johnson, T. DeJong, and K.R. Day. 1997. Orchard factors affecting postharvest stone fruit quality. *HortScience* 32:820–823.

Crisosto, C.H., D. Garner, G.M. Crisosto, and E. Bowerman. 2004. Increasing 'Blackamber' plum (*Prunus salicina* Lindell) consumer acceptance. *Postharvest Biology and Technology* 34:237–244.

Crisosto, C.H., G.M. Crisosto, G. Echeverria, and J. Puy. 2007. Segregation of plum and pluot cultivars according to their organoleptic characteristics. *Postharvest Biology and Technology* 44:271–276.

Delwiche, M.J. 1987. Grader performance using a peach ground color maturity chart. *HortScience* 22:87–89.

Ferrante, A. and T. Maggiore. 2007. Chlorophyll *a* fluorescence measurements to evaluate storage time and temperature of *Valeriana* leafy vegetables. *Postharvest Biology and Technology* 45:73–80.

Harker, F.R., F.A. Gunson, and C.M. Triggs. 2006. Apple firmness: Creating a tool for product evaluation based on a sensory–instrumental relationship. *Postharvest Biology and Technology* 39:327–330.

Hodges, D.M. and W. Kalt. 2003. Health functionality of small fruit. *Acta Horticulturae* 626:17–23.

Hodges, D.M. and G.E. Lester. 2006. Comparisons between orange- and green-fleshed non-netted and orange fleshed netted muskmelons: Antioxidant changes following different harvest and storage periods. *Journal of the American Society of Horticultural Science* 131:110–117.

Kader, A.A. 2002. *Postharvest Technology of Horticultural Crops*. Third edition. University of California Publication 3311, ISBN 1-879906-51-1.

Lester, G.E. 2006. Organic versus conventionally grown produce: Quality differences and guidelines for comparison studies. *HortScience* 41:296–300.

Lurie, S., S. Meir, and R. Ben-Arie. 1989. Superficial scald on Granny Smith apples: The effect of preharvest ethephon spray. *HortScience* 24:104–106.

Lurie, S. and C.H. Crisosto. 2005. Chilling injury in peach and nectarine: A review. *Postharvest Biology and Technology* 37:195–208.

Magkos, R., F. Arvaniti, and A. Zampelas. 2003. Organic food: Nutritious food of food for thought? *International Journal of Food Science and Nutrition* 54:357–371.

Marini, R.P., D. Sowers, and M.C. Marini. 1991. Peach fruit quality is affected by shade during final swell of fruit growth. *Journal of the American Society of Horticultural Science* 116:383–389.

Nelson, K.E. 1985. *Harvesting and Handling California Table Grapes for Market*. University of California Division of Agriculture and Natural Resources, Bulletin 1913 ISBN 0-931876-33-8.

Nunes, M.C.N., J.P. Edmond, and J.K. Brecht. 2006. Brief deviations from set point temperatures during normal airport handling operations negatively affect the quality of papaya (*Carica papaya*) fruit. *Postharvest Biology and Technology* 41:328–340.

Parker, D., D. Ziberman, and K. Moulton. 1991. How quality relates to price in California fresh peaches. *California Agriculture* 45:14–16.

Peterson, J.E. and E.M. Ivans. 1988. Variability in early season navel orange clone maturity and consumer acceptance. *Citriculture: Proceedings of the Sixth International Citrus Congress* 1631–1635.

Rembialkowska, E. 2003. Organic farming as a system to provide better vegetable quality. *Acta Horticulturae* 607:473–479.

Roberson, J.A. and J. Meredith. 1989. Characteristics of fruit from high and low quality peach cultivars. *HortScience* 23:1032–1034.

Saenz, M.V. 1991. Effect of position in the canopy on the postharvest performance and quality of stone fruit. In: *1990 Report on Research Projects for California Peaches and Nectarines*. California Tree Fruit Agreement, Sacramento, California.

Shahidi, F. and M. Naczk. 2003. *Phenolics in Food and Neutraceuticals*. CRC Press ISBN 0-12-639990-5.

Shahidi, F. 2004. Functional foods: Their role in health promotion and disease prevention. *Journal of Food Science* 69:R146–149.

Shewfelt, R.L. 1993. Measuring quality and maturity. In: *Postharvest Handing: A Systems Approach*, eds. R.L. Shewfelt, and G.L. Staby, pp. 99–123. Academic Press ISBN 0-931876-33-8.

Shewfelt, R.L., S.E. Prussia, W.C. Hurst, and J.L. Jordan. 1985. A systems approach to the evaluation of changes in quality during postharvest handling of southern peas. *Journal of Food Science* 50:769–774.

Shewfelt, R.L., S.E. Prussia, J.L. Jordan, W.C Hurst, and A.V.A. Resureccion. 1986. A systems analysis of postharvest handling of fresh snap beans. *HortScience* 21:470–472.

Shewfelt, R.L., S.C. Meyers, S.E. Prussia, and J.L. Jordan. 1987. Quality of fresh-market peaches within the postharvest handling system. *Journal of Food Science* 52:361–364.

Shewfelt, R.L., J.K. Brecht, R.B. Beverly, and J.C. Garner. 1989. Modification of conditions at the wholesale warehouse to improve quality of fresh-market tomatoes. *Journal of Food Quality* 11:397–409.

Shmulevich, I., R. Ben-Arie, N. Sendler, and Y. Carmi. 2003. Sensing technology for quality assessment in controlled atmospheres. *Postharvest Biology and Technology* 29:145–154.

Sonego, L., Y. Zuthi, T. Kaplunov, I. Kosto, and S. Lurie. 2002. Factors affecting taste scores of early season seedless table grapes Mystery and Prime. *Journal of Agriculture and Food Chemistry* 50:121–125.
Stephens, B.E. and D.J. Tanner. 2005. The harvest watch system—measuring fruit's healthy glow. *Acta Horticulturae* 687:363–364.
Streif, J. 1996. Optimum harvest date for different apple cultivars in the "Bodensee" area. In: *Proceedings COST 94, The Postharvest Treatment of Fruit and Vegetables. Determination and Prediction of Optimum Harvest Date of Apples and Pears*, eds. A. de Jager, D. Johnson, and E. Hohn, pp. 15–20. Lofthus.
Tarter, M.E. and S.E. Keuter. 2005. Effect of rachis position on size and maturity of Cabernet Sauvignon berries. *American Journal Enology Viticulture* 56:86–89.
Tijskens, L.M.M., P.E. Zerbini, and R.E. Schouten. 2007. Assessing harvest maturity in nectarines. *Postharvest Biology and Technology* 45:204–213.
Toivonen, P., F. Kappel, S. Stan, D.L. Mackenzie, and R. Hocking. 2004. Firmness, respiration and weight loss of 'Bing', 'Lapins' and 'Sweetheart' cherries in relation to fruit maturity and susceptibility to surface pitting. *HortScience* 39:1066–1069.
Valero, C., C.H. Crisosto, and D. Slaughter. 2007. Relationship between nondestructive firmness measurements and commercially important ripening fruit stages for peaches, nectarines and plums. *Postharvest Biology and Technology* 44:248–253.
Van Zeebroeck, M., V. Van Linden, H. Ramon, J. De Baerdemaeker, B.M. Nicolai, and E. Tijskens. 2007. Impact damage of apples during transport and handling. *Postharvest Biology and Technology* 45:157–167.
Vinson, J.A., X. Su, L. Zubik, and P. Bose. 2001. Phenol antioxidant quality and quality in foods: Fruits. *Journal of Agriculture and Food Chemistry* 49:5315–5321.
Woolf, A., A. Wexler, D. Prusky, I. Kobiler, and S. Lurie. 2000. Direct sunlight influences postharvest temperature responses and ripening of five avocado cultivars. *Journal of the American Society of Horticultural Science* 125:370–376.
Worthington, V. 2001. Nutritional quality of organic versus conventional fruits, vegetables and grains. *Journal of Alternative and Complementary Medicine* 7:161–173.
Zude-Sasse, M., B. Herold, and M. Geyer. 2000. Comparative study on maturity prediction in 'Elstar' and 'Jonagold' apples. *European Journal of Horticultural Science* 65:260–265.
Zude, M., I. Birlouez-Aragon, P.J. Paschold, and D.N. Rutledge. 2007. Non-invasive spectrophotometric sensing of carrot quality from harvest to consumption. *Postharvest Biology and Technology* 45:30–37.

1.2 QUALITY OF PROCESSED PLANT FOOD

SASCHA ROHN AND LOTHAR W. KROH

1.2.1 QUALITY: DEFINITIONS AND CRITERIA

What is quality? The word quality originates from the Latin term *qualis* (meaning of what kind, of such a kind) (*Oxford Advanced Learner's Dictionary* 2005) and describes the state, the grade, or the value of an object or a product, in this case a food product. According to this definition, quality should be an objective, neutral term that only describes and does not grade.

Food quality is a concept and in most cases it is a compromise between varying expectations of all the players involved in the supply chain. The players are first of all not only the consumers but also the producers and the traders of food, the scientists, as well as the legislatives; all of them having different interests and possibilities of intervention.

In the literature, one can often find subdivisional quality terms and corresponding criteria; all of them are product orientated and therefore depend on different expectations. Quality terms can describe the market value, the utilization value, the sensory value, the nutritional and health value, the ecological value, and the non-material value. The following are some of the criteria (Kader 1999; Shewfelt 1999; Huyskens-Keil and Schreiner 2003):

Market value: Hygiene, labeling, composition, shape, color, size, freshness, consistency, represent parameters that serve often as minimum requirements defined in norms aiming at the harmonization of the market. Also some non-material values are important. Here, cultural and social aspects come into the picture that might severely influence the products' prestige, for instance, by religion, origin/local importance of the product, and traditional processing methods.

Utilization value: Technological parameters such as susceptibility to temperature as well as mechanical properties, for instance, in cutting processes, shelf life, storability, and expenditure of work, are of importance for the food industry, especially for producers.

Sensory value: The shape, color, size, freshness, consistency, aroma, and flavor are the most important criteria for consumers.

Nutritional and health value: Hygiene; absence of residues and contaminants (nitrate, pesticides, heavy metals, toxins including mycotoxins, drugs, pollutions); composition; energy density; presence of desired ingredients (vitamins, minerals and trace elements, carbohydrates, dietary fiber, pre- and probiotics, other functional compounds).

Ecological impact: Here, the cultivation methods (integrated plant production, organic production) and the adaptation of the production system to the resources in the production site are the main criteria. If all these aspects are recognized and the conditions of the workers are suitable, we obtain a sustainable supply chain.

Presently, beside the appearance of the product, the products' nutritional and health properties are getting more and more important for consumers, especially in the case of beneficial compounds related to well-being as well as residues and neo-formed compounds (NFC) related to safety issues. There are defined prescriptive limits and maximum residual levels for residues and contaminants. But, even these aspects are the result of a scientific and political negotiation process in which economical interests with food safety aspects are weighed. This is shown by the variation of maximum residual levels (e.g., for pesticides) in different countries. For beneficial compounds, no regulations exist up to now.

However, changes in best practice rules or regulations, reflecting not only the rising awareness of environmental, economical, and health concerns of the society but also short-term product trends, lead to a dynamic process of food quality.

To overcome subjectivity in the assessment of products, new attempts are necessary to define quality more comprehensively. Quality managements systems using specific quality parameters and more complex quality models would be valuable to combine the aspects of product-orientated, process-orientated, and especially consumer-orientated food quality (Huyskens-Keil and Schreiner 2003). Further, "... it would be highly advantageous if the practical definition of [food] quality was capable of being linked in with standards such as the [DIN EN] ISO series and through them to regulations, codes of practice, objective measurements, and analytical methods" (Bremner 2000). With regard to valuable secondary plant metabolites, analytical tools already become more and more important, reflecting the consumers' requirements concerning not only health-promoting food properties but also environmental benefits (Huyskens-Keil and Schreiner 2003; Zude and Geyer 2005).

Furthermore, nowadays, also the quality of the process in which the raw material is transformed is of major interest. Here, the main question is: How is the food produced and processed? The kind of farming (organic or conventional) plays a role, as well as the degree of processing, the processing conditions, application of additives, or methods of conservation and sanitizing treatments. Indirectly, marketing and trade strategies (multinational trading, regional marketing, and fair trade) contribute to the term of process quality. Ecological aspects are the amount of waste (especially wastewater), energy use, and efficiency during the transport from the farm to the households (von Koerber et al. 1999).

We discuss the different influencing factors on the produce quality associated with the process parameters according to the processing degree in the following sections.

1.2.2 Quality of Fresh and Minimally Processed Produce

1.2.2.1 Raw Material

It is evident that achieving plant food product quality is always a tightrope walk, as the requirements that the product has to fulfill should be considered. Should the product have a full developed aroma profile? Then, the raw material has to be harvested in a fully ripe stage. In the case of a desired good storability, harvest should be as early as possible. This is even more important for climacteric fruits. From technological point of view, maturity is often considered as "that stage at which a commodity has reached a sufficient stage of development that after harvesting and postharvest handling, its quality be at least the minimum acceptable to the ultimate consumer" (Reid 1992).

Production practices have a tremendous effect on the quality of perishable fruit and vegetable at harvest and on postharvest quality and related shelf life (see Section 1.1). In almost all cases, the inherent quality of the plant product cannot be improved after harvest but can only be maintained for a period until the acceptance limit (Zude and McRoberts 2006) is reached (shelf life). It is important to harvest fruit and vegetable at a proper stage and size to reach peak quality when being sold and consumed. Postharvest technologies target the stabilization and storability of unprocessed foods (Bourne 2004). To achieve a maximum shelf life there are some major aspects to keep in mind: Respiration of the product has to be kept at a minimum. Otherwise, metabolic activity of the product would lead to degradation of their constituents such as lipids, proteins, acids, and sugars, resulting in a decreased food value. Wounding, bruising, and breaking during harvest and postharvest handling have to be avoided. Especially, crops destined for storage should be as free as possible even from minor skin breaks, bruises, spots, rots, decay, and other deterioration (Suslow 2000).

Packaging may provide an effective step in postharvest handling. It should be designed to prevent mechanical damage to the product as it protects the product from mechanical impacts such as vibration, abrasion, and pressure impact. Additionally, packaging may help to adjust conditions that reduce exposure to contaminants and pathogens. Modified atmosphere packaging (MAP) of fresh fruit and vegetable refers to the technique of sealing respiring products in polymeric film packages to modify O_2 and CO_2 levels within the package atmosphere. Similarly, edible coatings can be applied for changing the intracellular gas composition.

1.2.2.2 First Processing Steps

1.2.2.2.1 Fresh-Cut Produce

In convenience produce, first degrees of processing such as cutting allow consumers to have fresh-like quality fruit and vegetable in modern urban life. Definitions for both terms imply that fresh-cut products have been freshly cut, washed, packaged, and maintained with refrigeration. They remain in a raw state, even though minimally processed. In such cases, shelf life is a very sensitive issue. As the tissues are still "living," enzymatic reactions including respiration and accumulation of unfavorable secondary plant metabolites make them more susceptible to microbial spoilage (Legnani and Leoni 2004).

MAP is commonly used for minimally processed products and for highly perishable, high-value commodities such as small berries, broccoli florets, or asparagus tips. It is desirable to generate an atmosphere limiting the respiration rate and ethylene evolution of the produce as well as the growth of microorganisms. On the other hand, anaerobic conditions have to be prevented as fermentation would be induced. Modification of the atmosphere can be established inside the package either passively by the respiratory activity of produce or actively by applying a desired gas mixture. Such mixtures can be further improved by the use of adsorbing substances that scavenge O_2 or CO_2 (Kader and Watkins 2000).

1.2.2.2.2 Minimally Processing

Minimally processing as a second processing step further reduces accelerated quality loss by a combination of processing steps including sanitation, mild heat treatments (blanching), followed by freezing or packaging (Rico et al. 2007).

Temperature control is the most important tool for maintaining postharvest quality. Quick cooling down reduces further softening and color changes of the product. Undesirable metabolic changes and respiratory activity, causing further heat, are slowed down. The higher the temperature, the higher the respiration rate will be. High temperature leads to low air water potential resulting in moisture loss due to enhanced water potential gradients from the product to the environment and corresponding wilting. Low temperatures will prevent the growth of microorganisms (bacteria, fungi, and yeasts) that may cause spoilage (Suslow 2000).

What to Measure and How to Measure

Although temperature is the primary concern in the postharvest handling of fruit and vegetable, optimum relative humidity is also important. Most spoilage vectors accelerate as temperature and humidity increase; this makes it more difficult to control spoilage in tropical than in temperate climate (Bourne 2004). The easiest way to cool down the product and ensure good humidity is the combination of cooling with washing (hydrocooling). The fruits are dumped into cold (ice) water, which is an efficient way to remove heat and serves as a means of cleaning at the same time. In addition, water loss and wilting are reduced. The possibility of the parallel use of disinfectants in the water, if in agreement with regulations, reduces the spread of diseases (Suslow 2000). Sanitation protects not only the produce from postharvest diseases but also the consumer from food-borne illnesses. There are several microbial hazards that may be transferred to the human organism via fresh fruit and vegetable (e.g., *Salmonella*, *Shigella*, *Listeria*, hepatitis virus). The most commonly used disinfectant is chlorine/hypochlorite, which can be applied with the washing water (Suslow 2000). However, use of chlorination reagents during food processing has been controversially discussed, as formation of potentially carcinogenic chlorinated by-products may occur. Such undesired compounds can be halomethanes, chlorophenols, or neo-formed reaction products with other food constituents, for example, N-chloroderivatives with proteins (Fukayama et al. 1986). Alternatives or modified methods are necessary. Ozonation is another technology that can be used to sanitize plant food products (Suslow 2000; Hassenberg et al. 2007). Ozone has a high redox potential that oxidizes organic molecules on fruit and vegetable including microorganisms. It can be applied as gas, which is useful during packaging, or as ozonated water. An expert panel recommended that ozone be classified as generally recognized as safe (GRAS), which was officially accepted by the Food and Drug Administration (FDA) in 2001. Although reaching GRAS classification, the formation of all kinds of radicals resulting from the spontaneous decomposition of the ozone molecules might affect other food constituents and lead to reaction products of which a risk classification is not given. For example, ozone treatment of strawberries showed a 40% reduced emission of volatiles, aroma-relevant esters (Perez et al. 1999). Schreiner and Huyskens-Keil (2006) described in a review that gamma and ultraviolet radiation are also good sanitation tools that may extend the shelf life of fresh and processed fruits and vegetables by destroying food-borne pathogens such as fungi and bacteria. Irradiation is also applied to various tropical and subtropical fruits to suppress microbiological decay causing severe losses during long distance overseas transport.

1.2.2.3 Processing and Preservation

The preparation of food may be one of the most important developments in the history of humankind. The majority of food products consumed are provided by processing. Some raw material even has to be prepared to reduce toxicity and to achieve a higher bioavailability of compounds. Furthermore, food processing can improve sensory properties by releasing the aroma/flavor during the preparation process or via the formation of new compounds as a result of different technological treatments. However, these benefits are by far not that important as the need for extending the shelf life of food and securing the consumer from microbiological hazards caused by food spoilage: After harvesting, the number of microorganisms begins to exceed. They can rapidly spoil food by causing the food to lose flavor, by changing the color and texture of the food, and by releasing harmful toxins. Furthermore, enzymes that are also present in all raw foods promote degradation and chemical changes affecting the product texture.

Traditionally, thermal treatment is used to prevent food from spoilage. Cooking, roasting, baking, frying, and deep-frying are capable of eliminating microorganisms and inactivate enzymes by denaturing their protein structure. These processing techniques lead to a transformation of aroma/flavor compounds that gives the food its characteristic sensory properties.

For cold storage, refrigeration or deep-freezing short hot-water treatment (blanching) is used to inactivate enzymes and to rinse off microorganisms. For long-term storage, food products are canned. The process of canning is often called sterilization, because here the heat treatment eliminates all microorganisms that can spoil the food and those that are harmful to humans, including direct pathogenic bacteria and those that produce lethal toxins. Most commercial canning operations are based on the principle that destruction of bacteria increases approximately 10-fold for each 10°C increase in temperature. The D-value (decimal reduction time) gives the time period (in seconds) of the heat treatment that is necessary to destroy one-tenth of the microorganisms (Kessler 2002).

As microorganisms do not grow in a water-free environment and enzymes need water to be active, removal of water offers an excellent protection against the common causes of food spoilage. Most chemical reactions are greatly retarded. Water activity is given as a_w value. It can be used for the prediction of shelf life (e.g., dried fruits 0.72, honey 0.75, milk 0.97, and distilled water 1). The lower this value, the longer is the shelf life (Krämer 2002). Drying is a traditional method to decrease the product water content. An advanced technique is freeze-drying, a combination of freezing and dehydration under vacuum conditions. Compared with canning, dehydration by drying is preferred if the product is to be stored at high temperature. However, reabsorption of water from the air has to be prevented.

A relatively new food processing technique to maintain quality and shelf life is the irradiation of food. Briefly, food particles are exposed to ionization radiation (α-, β-, or γ-rays). Rays penetrate deeply into the food and interact with all constituents, causing chemical reactions that decrease the rate of the food decay.

Further, nonthermal food processing technologies have been developed in recent years. Treatment of food with nonthermal techniques is gentler and should prevent the food from a possible loss of nutrients as a result of thermal degradation. Undesirable browning and formation of new, possibly hazardous compounds do not take place. Two examples of these innovative and emerging technologies are high-pressure processing and treatment of food with electric fields. High-pressure processing (HPP), also known as high hydrostatic pressure processing (HHP) or ultra-high-pressure processing (UHP), is a method of food treatment, where food is subjected to elevated pressures (100–1000 MPa) to achieve degradation of microorganisms, inactivation of enzymes, or even altering structural food properties to gain a consumer-desired quality. Pressure-induced unfolding, aggregation, and gelation of food proteins take place depending on the pressure effects on non-covalent protein interactions. Resulting from these effects, enzymes may be inactivated (Cheftel 1995). The physicochemical properties of products may change dramatically. For example, high-pressure treatment at ambient temperature partially inactivated polygalacturonase from tomato puree. In parallel, the color and viscosity of the product was improved compared to heat pasteurization. HPP resulted in a push back of the natural tomato microbiota to a level below the detection limit (Krebbers et al. 2003). Pressure between 300 and 660 MPa can inactivate yeasts, molds, and most vegetative bacteria including most infectious food-borne pathogens. Bacterial spores can only be killed by high pressures running a certain protocol. First, these spores have to be forced to germinate by lower pressures before they can be destroyed by pressure or in combination with heat treatment (Smelt 1998). In Europe, practical application of high pressure in the food industry is pending. There are several aspects to be considered with regard to novel food legislation.

Pulsed electric field (PEFs) is another emerging technique in food processing. During the treatment, electric fields are generated by abrupt discharges of capacitors and application of the resulting impulses via electrodes to the food. Arrangement and geometry of the electrodes influence the course of the streamlines and therefore the homogeneity of the electric field are important. Other process parameters are the strength of the electric field, duration of the impulse, number of repetitions, warming of the product, and distribution of temperature during the treatment. Especially liquids, for example, fruit juices, are good substrates for the PEF treatment (Heinz et al. 2001). It has been shown that a wide variety of microorganisms can be eliminated using PEF, while spores are not influenced (Liang et al. 2002). The mechanism of the destruction is a continuous induction of pores in the cell membrane, leading to a loss of its semipermeability.

PEF may also inactivate enzymes. Here, the mechanisms are not yet fully understood. It is assumed that the protein structure is affected either by electrochemical reactions (oxidation of disulphide bridges) or local occurring heat development that leads to thermal-induced denaturation (van Loey et al. 2001).

Besides inactivation of potential food-spoiling activities, PEF is capable of improving the extraction/isolation of food constituents. Similar to the inactivation of microorganisms, the plant cell walls can be loosened and as a result of the pore widening, the cell content can be extracted more easily. For example, oil yield from rapeseed was significantly increased after PEF treatment (Guderjan et al. 2007).

When using innovative and emerging technologies during food production, food safety plays a major role. In Europe, such technologies fall within the scope of the European Commission regulation EC 258/97, which considers safety of novel food products. In this regulation, novel food is defined as "foods and food ingredients to which it has been applied a production process not currently used, where that process gives rise to significant changes in the composition or structure of foods or food ingredients which affect their nutritional value, metabolism or level of undesirable substances." In most countries, regulative authorities agreed that high-pressure treated food stuffs have not revealed any evidence of any microbial, toxicological, or allergenic risks of high pressure treatment" (Eisenbrand 2005).

A conjoint study on consumer perceptions of foods processed by innovative and emerging technologies illustrated that high-pressure processing produced the most positive effect (Cardello et al. 2007). In Brazil, fruit juices processed with HPP also showed good acceptance when information on the benefits was presented on the product labels (Deliza et al. 2005).

However, PEF as well as HPP are lacking the confirmation by regulative authorities in Europe, presently.

1.2.3 Determinants for Quality Aspects: Stage of Maturity, Technological Functionality, and Health Beneficial Effects

Food has to be fully evaluated during production and further processing to ensure that the consumers have access to a food product that meets their high expectations with regard to food safety, sensory attributes, visual attractiveness, and nutritional value. In all processing steps, the perishable product would be graded as above or below the acceptable threshold. Hereby, analytical techniques for recording the mechanical, chemical, and microbiological aspects are employed.

Following typical parameters are used for the analysis of fresh and processed food:

Color: It is an important attribute of food, since the consumer realizes the product first by its appearance. Particularly in fruits and vegetables, color changes from green to red or yellow occur during product development. Attractive red color is one of the most important quality characteristics for many fruits, especially berries. Color changes during fruit ripening implicate both synthesis and degradation of plant pigments including chlorophyll, carotenoids, and flavonoids.

To overcome subjectivity, measurement of color by means of either color cards or diffuse reflectance readings were standardized. In the instrumental approach, color is often determined in terms of a three-dimensional space, where one axis represents the lightness (L^*) and the other two coordinates a^* and b^* represent redness-greenness and yellowness-blueness, respectively (Giese 2000). The optimum methods for receiving data on the specific profile of a product are the chromatographic pigment analyses. However, due to their complexity and relatively high costs of equipment, such methods cannot be applied by the producers of fresh and processed food.

Firmness: It can serve as a good indicator of maturity. This parameter is also closely related to the consumer's eating experience. Fruit and vegetable firmness is affected by variety, crop loads, water supply, nutrition, pruning, and seasonal conditions. It can be measured using a penetrometer with a plunger. The force required to drive the plunger through the flesh is measured in kilograms Newton. Major sources for a lack in reproducibility can be the temperature influence, varying velocity of the plunger, and sampling time (since fruits will be firmer in the morning than in the afternoon) (Lorimer and Hill 2006).

Besides physical parameters, the analysis of the chemical composition is of interest. The major constituents such as sugars, lipids, and proteins characterize a food product. Owing to its complexity, the determination of the whole profile is not possible. From the viewpoint of a fast analysis, sensitive markers for the decision making need to be addressed preferably by means of nondestructive, rapid, but reproducible methods.

With regard to sensory quality, sugars are very important, because they are responsible for taste. A sensitive method for assessing the total amounts of sugars, organic acids, and other soluble components is the soluble solids content (SSC). It is easily determinable in the fresh or processed juice using a refractometer, calibrated on equivalent percentage of sucrose. Values are expressed as °Brix or %Brix (Lorimer and Hill 2006). For many fruits and vegetables, the SSC is a sensitive indicator for maturity and minimum °Brix values are defined in quality standards, for example, for kiwi, melon.

Free amino acid composition: It changes diversely. As an example, in grapes arginine, tyrosine and proline content increases gradually with maturity, while threonine and glutamine drop sharply after fruit set (Lamikanra and Kassa 1999).

Lipids: These can be found in all fresh and processed crops. Free fatty acids and triglycerides are capable of undergoing oxidative degradation, and appear to be converted to volatile components responsible for the flavor/aroma after the beginning of ripening and reach a peak in the stage of full ripeness (Belitz et al. 2001). On the other hand, especially unsaturated fatty acids are oxidized due to thermal impact resulting in a quality decrease.

Organic acids: These (e.g., tartaric, malic, citric, and lactic acid) are found in relatively high concentrations in fruits and vegetables. Therefore, they play also a very important role for the taste of plant-based food products. As their profile changes during the development of raw material and total acidity decreases, the analysis of organic acids can be a good indicator for the stage of maturity or even for the use of inferior, unripe material for food processing. A simplified and quick measure for organic acids is the titratable acidity. Here, the total amount of organic acids is determined via titration with a solution of a standard base such as sodium hydroxide. However, in most cases it will not be possible to measure this parameter directly in the field. Measurements of SSC in combination with titratable acidity (SSC:acid ratio) have been found to be more closely related to the quality of fruits than just acid or SSC alone (Lorimer and Hill 2006).

Aroma/flavor: It is the result of a high number of volatile compounds, which arise from different biosynthetic pathways. Aroma compounds can be divided into two main groups: (1) compounds that are direct metabolites of intracellular biosynthesis and (2) secondary products that are not present in intact cells and are formed as a result of stress and tissue damage. Important biosynthetic pathways for both classes are the degradation of lipids, terpenoid synthesis, and the amino acid and phenylpropanoid metabolism. Via the lipoxygenase pathway, volatile components are formed out of linolic acid following plant cell damage. The resulting compounds (small aldehyds and ketones) are responsible for the fresh green aroma impact (Teranishi et al. 1999). Terpenoids, especially monoterpenes are responsible for characteristic aroma profiles of all kinds of fruits and vegetables, for example, citrus fruits, black currants, tomatoes. There are hundreds of volatiles belonging to many different substance classes and fruit varieties differ greatly in the type and amount of volatiles they produce, leading to characteristic flavor/aroma profiles. Examples are limonene in citrus fruits, C6-aldehydes for "green" flavor impressions, isothiocyanates in *Brassicaceae*, sulfur-containing compounds in *Alliaceae*. Owing to the complexity and high specificity, aroma profiles can be taken into account for analyzing authentification of variety or origin using chromatographic analyses (Belitz et al. 2001).

A consumer survey in 1997 revealed that for 32% of the consumers, health aspects are the main decision criterion when purchasing food (Lennernas et al. 1997). At a first glance, dietary health aspects are associated with a reduction of fat and sugars in the diet, and the consumption of high amount of vitamins, both realizable by a diet rich in fruits and vegetables. In this context, a formerly neglected food constituent is nowadays highly associated with health beneficial effects of fruits and vegetables: dietary fiber. Important examples for this class of food compounds are cellulose, hemicelluloses, β-glucanes, pectin, and inulin. According to the physiological definition, dietary fiber includes all molecules that are not or only partly hydrolyzed by degradation enzymes of the upper gastrointestinal tract. As a result, their absorption in the small intestine is low and intestinal transit time is very fast. Because of their ability to bind xenobiotics they may also reduce the transit time of potentially harmful substances, for example, carcinogens. Furthermore, fibers are good substrates for the colonic microbiota. Owing to their enzymatic activity, selected bacteria are capable of hydrolyzing "exotic" polysaccharide bonds resulting in the formation of highly bioactive metabolites such as short-chain fatty acids, which are proposed to be fuel for colonic cells and protective for the intestinal mucosa (Schulze and Bock 1993).

In the last two decades, health beneficial effects of food were more and more attributed to further compounds. Among the macronutrients (sugars, proteins, lipids, vitamins, and water), plant-based food products provide a high number of substances with biological activity. These substances are synthesized in the secondary plant metabolism and are therefore called secondary plant metabolites. They are nonnutritive, as they do not provide energy, but they may exert several biological actions. There are thousands of secondary plant metabolites known belonging to different chemical classes. Depending on the dietary habits, the amount consumed with the diet is relatively low and ranges between some milligrams to one gram per day. It has been shown for a large number of the secondary plant metabolites that there are positive correlations to physiological beneficial effects. However, among these substances, there are other compounds that may exert negative or even toxic effects (Kroll et al. 2003a).

1.2.3.1 Glucosinolates

Sources for glucosinolates are primarily the varieties of cabbage (e.g., broccoli, cauliflower, green, red, and white cabbage), spice plants (e.g., mustard, horseradish), tubers (e.g., kohlrabi, radish), and rapeseed. The average dietary intake is ~40 mg per person per day, depending on dietary habits. In plant cells they are found as thioglucosides. After destroying the cell structure during food processing, a release of enzymes (myrosinases, thioglucosidases) takes place that is capable of degrading the glucosinolates. As a result, degradation products are formed (e.g., nitriles, thiocyanates, isothiocyanates), some of them providing sensory attributes. Glucosinolates and their degradation products are associated with positive as well as negative physiological effects. Although an anticarcinogenic activity was reported for the glucosinolates, the degradation products may exert, beside a positive antimicrobial activity, negative effects such as mutagenic, goitrogenic, cytotoxic, and hepatotoxic effects (Kroll et al. 2003a).

1.2.3.2 Terpenes/Carotenoids

Terpenes are naturally occurring derivatives of isoprene. Depending on the number and distribution of several isoprene molecules, there are mono-, sesqui-, di-, and triterpenes. Some of these terpenes (especially the monoterpenes) are, as described above, part of the aroma/flavor of agricultural crops (e.g., limonene in citrus fruits). Terpenes possess antimicrobial activity and serve as substances for plant communication and defence (deterrents). An antioxidative activity was proven for eugenol, thymol, carvacrol, and allylphenol with values comparable to major antioxidants such as α-tocopherol and butylated hydroxytoluene (BHT) (Lee et al. 2005).

The carotenoids are tetraterpenes. Presently, almost 600 carotenoids are known. Most of them are responsible for the color of yellow, orange, or red commodities, for example, tomatoes, peppers, carrots, maize, apricots, and peaches. Nearly 50 molecules exert pro-vitamin A activity. Carotenoids have a high antioxidative activity, and particularly have the ability of quenching of singlet oxygen ($^1O_2$) (Kroll et al. 2003a).

1.2.3.3 Phenolic Compounds/Polyphenols

With more than 6000 food-relevant substances, the plant phenolic compounds are the largest class of the secondary plant metabolites. All phenolic compounds are structures bearing one or more hydroxyl groups on an aromatic ring system. Most of these hydroxyl groups are derivatized, preferentially glycosidized. With regard to their structure, the class of phenolic compounds can be divided into subclasses, whereas the hydroxybenzoic acids (C_6-C_1-backbone; e.g., gallic acid), the hydroxycinnamic acids (C_6-C_3; e.g., caffeic acid, ferulic acid, chlorogenic acid), and the flavonoids (C_6-C_3-C_6) are the most important ones. The very large class of the flavonoids is further subdivided into flavones, flavonols, flavanols, catechins, anthycyanins, and isoflavones. The most common flavonoids belong to the flavonols (quercetin, myricetin, kaempferol), which are found ubiquitously in fruits and vegetables (Kroll et al. 2003a).

Some of the phenolic compounds are responsible for the color of food, while others may exert, due to their reactivity, negative effects on food, such as browning reactions, inhibition of enzymes, clouding of beverages, and astringent taste (e.g., rutin, procyanidins, brown polymers) (Luck et al. 1994; Kroll et al. 2003a). With regard to plant food, polyphenols have been considered to be effective antioxidants for the human organism. In recent years, antioxidants have been the subject for many epidemiological studies that related consumption of fruit and vegetable with a reduction in the incidence of cardiovascular diseases and several cancers, whereby especially (cereal) fiber and phenolic compounds play an important role. The health beneficial effects of the phenolic compounds are thought to result from their ability to scavenge reactive oxygen and nitrogen species, which are formed in high amounts during intracellular oxidative stress induced by the extraneous attack of pro-oxidants. A decrease of reactive oxygen and nitrogen species is associated with a reduced risk of all kinds of degenerative diseases such as cancer and coronary heart diseases (Halliwell 1996).

Isoflavones possess estrogenic activity. Almost 60 compounds have been isolated preferentially from legumes. Genistein and daidzein are the most common representatives and are found mainly in soy and soy products (Kroll et al. 2004). In many studies, isoflavones showed a diverse bunch of physiological activities such as antioxidative, estrogenic, and antiestrogenic activities, anticancerogenic properties, cytotoxicity, etc. (Rüfer and Kulling 2006).

However, it was shown for highbush blueberries that in parallel to the increase of anthocyanins, further classes of phenolic compounds (flavonols and hydroxycinnamic acids) and correspondingly the antioxidant activity of the fruits are decreasing during maturation and ripening (Rodarte et al. 2006), which may underline the impression that choosing the ideal time point for harvest is difficult when assessing health beneficial food quality.

1.2.3.4 Sulfur-Containing Compounds

In most cases, the beneficial role of *Allium* species is considered to result from sulfur-containing compounds. For human consumption, the organosulfur compounds are of particular interest, because they have several good attributes assigned. They are suggested to be biologically active as antibiotics, as agents in reducing the risk factors of cardiovascular disease, and as blood lipid-reducing agents. They also are gaining growing interest as potential anticarcinogenic agents (Rose et al. 2005). These health- and flavor-related organosulfur phytochemicals are at most S-alk(en)yl-L-cysteine sulfoxides, which are activated by the enzyme alliinase to produce pyruvic acid, ammonia, and sulfenic acids when *Allium* tissue is damaged (Jones et al. 2004).

1.2.3.5 Cyanogenic Glycosides

Cyanogenic glycosides are widespread in the plant kingdom. As food constituents, amygdalin (bitter almond, kernels of apricots, peaches, and cherries), linamarin (lima beans, linseed, and cassava), and dhurrin (millet) are of certain interest. From a nutritional economic point of view, millet and cassava are highly important for world nutrition as they are the basis of diets in many tropical regions, whereas lima beans are a main constituent of the diet particularly in Latin America. Similar to the glucosinolates, cyanogenic glycosides are degraded when cell tissue is damaged and enzymes from different cell compartments are released (linamarases, hydroxyl nitrile lyases). The main degradation product is hydrogen cyanide, which is highly toxic. Thermal food processing techniques (cooking, steaming) may help to remove the hydrogen cyanide due to its high volatility (Rawel and Kroll 2003).

1.2.3.6 Alkaloids

Some of the almost 8000 alkaloids known have importance as constituents of plant food products. The purin alkaloids caffeine, theobromine, and theophylline can be found in coffee, cacao, and tea. Piperine and capsaicine are the main alkaloids in different kinds of peppers (Kroll et al. 2003a). The potato alkaloids solanine and chaconine, as well as tomatine from tomatoes belong to the group of steroid alkaloids. They have been associated with pharmacological or even toxicological properties. The potential human toxicity resulted in the definition of guidelines limiting the glycoalkaloid content of potatoes (Friedman and McDonald 1997).

1.2.4 QUALITY CHANGES DURING PLANT FOOD PROCESSING

Traditionally, the main aim of food processing is the preservation of foods by destroying potentially pathogenic microorganisms and inactivation of enzymes that lead to food spoilage. Especially enzymatic reactions are responsible for a high number of changes of the food constituents. Consequently, such reactions may on the one hand lead to a simple loss of nutrients (e.g., health beneficial compounds such as vitamins or secondary plant metabolites), on the other hand, even more important for food quality, may result in the formation of reaction products that gain a toxicological risk (e.g., hydrogen cyanide from cyanogenic glycosides).

Enzyme-catalyzed reactions occur in perishable products during storage, handling, and especially processing such as peeling, cutting, sanitizing, etc.

1.2.4.1 Enzymatic-Induced Changes

One of the most important enzyme-induced changes is the softening of the fruits. During the fruit development, the cell walls provide physical stability while the cell is growing. After differentiation, they may offer protection against plant pathogens. The most obvious change in cell wall composition that accompanies fruit softening is a decrease in wall bound uronic acids (pectins) that is closely matched by an increase in soluble uronide. The cell walls become less connected due to the formation/activation of hydrolytic enzymes (pectin esterases and polygalacturonases) that cleave especially the pectins of the cell wall (Labavitch 1981). For example, firmness of fresh-cut tomatoes decreased exponentially during storage, highly depending on the stage of maturity and temperature (Lana et al. 2005).

In the senescence period of fruit and vegetable development, breakdown and decay processes get dominant. Sacher (1973) defined, "... being the final phase in ontogeny of the organ in which a series of normally irreversible events is initiated that leads to cellular breakdown and death of the organ." Enzymes involved are respiratory enzymes (e.g., peroxidases), starch degrading enzymes, and hydrolases (e.g., pectin esterases, polygalacturonases), which have the potential for catalyzing the degenerative aspects of the senescence (Sacher 1973).

Besides endogenic hydrolases, hydrolytic activity of spoilage bacteria can also contribute to a faster degradation. It was shown that in such a case, MAP might not be suitable for diminishing pectinase activity (Hao and Brackett 1994).

Enzymatic browning is a major factor contributing to quality loss in foods and beverages that usually impairs a decrease in sensory properties because of the associated changes in color, flavor, and taste. Enzymatic browning is caused by the enzymatic oxidation of phenolic compounds by polyphenol oxidases (PPO). These copper-containing enzymes are also known as catechol oxidases, catecholases, diphenol oxidases, o-diphenolases, phenolases, and tyrosinases. PPO catalyze two reactions involving molecular oxygen. The first reaction is the hydroxylation of monophenols, leading to the formation of o-dihydroxy compounds. The following reaction is the oxidation of the o-dihydroxy structure to the corresponding quinones. Browning resulting from PPO can be prevented by the addition of sulfites, ascorbic acid and its analogues, cysteine, as well as heat treatment and elimination of oxygen (Nicolas 1994).

Lipid oxidation is a further important enzymatic reaction during the processing of food. Lipoxygenases and hydroperoxide lyases cause the degradation of unsaturated fatty acids, which have been shown to decrease blood plasma cholesterol levels and therefore reduce the risk of cardiovascular diseases (Harris 2003). In parallel, they are responsible for the synthesis of volatile compounds contributing to green and fresh notes in ripe fruits. However, concentrations of such reaction products above certain limits may result in the development of off-flavors (Belitz et al. 2001).

1.2.4.2 Thermal Treatment

Since the discovery of fire by humankind, treatment of food with heat is used for preservation of food and for prevention of the above-mentioned enzymatic reactions. Here, the chemical (nonenzymatic) reactions are the main factors, due to the high input of energy. Many of the reactions that occur during the thermal treatment are desired. Depending on the kind of processing (cooking, roasting, and frying), the treatment results in characteristic food properties, which are pre-dominantly sensory impacts (smell, taste, and color).

Most of the new compounds formed are based on nonenzymatic browning reactions. The Maillard reaction is one of the most important browning processes during food processing due to the complex mixture of reaction products resulting from different formation pathways. At the beginning, a condensation between a carbonyl group of a reducing sugar and an amino substituent (from an amino acid, preferentially lysine, etc.) takes place, following a cyclization to the N-substituted glycosylamine and a rearrangement to the so-called Amadori product.

Further reactions, depending on different factors such as temperature, pH, water, and oxygen give rise to different compounds that include dicarbonylic compounds, reductones, furfurals, and a bunch of other cyclic substances. Finally, melanoidins, brown polymers, can be regarded as the end products of this reaction (Hodge 1953). They are associated with the formation of color (Fiedler et al. 2006) and high-antioxidant activity (Caemmerer and Kroh 2006).

Beside the determination of the increase in brown color, the course of the Maillard reaction can be followed by the analysis of hydroxymethylfurfural (HMF), which is the main decomposition product of sugar hydrolysis catalyzed by acid. It is often used as an indicator for heat treatments of fruits, juices, as well as honey. HMF is thought to have genotoxic potential (Zhang et al. 1993). A more sensitive indicator for following the Maillard reaction is given by analyzing dicarbonylic compounds. They are precursors for the formation of melanoidins and can be measured after trapping with *o*-phenylendiamine (Fiedler and Kroh 2007). Carboxymethyllysine and pentosidine are markers for an advanced Maillard reaction (Henle et al. 1997; Charissou et al. 2007). However, fluorescence may provide a nondestructive method for measuring HMF (see Section 4.4).

A few years ago, a further Maillard reaction product gained much attention in food science: acrylamide. Considered to be a potential carcinogen, it is present at elevated concentrations in different types of heat-treated foods such as cereal products and processed potatoes. As the major amino compound for its formation, the amino acid asparagine was identified, being most abundant in cereals and potatoes, as a result leading to high concentrations of acrylamide in strong thermally treated potato products such as crisps or French fries (Mottram et al. 2002). Research is still necessary on how to obtain potato products with low content of acrylamide by diminishing precursor content (asparagines and reducing sugars) (Jung et al. 2003).

Already occurring during storage, but even more important for food processing, a decrease in the concentration of vitamins is observable in many food products. Especially vitamin C is susceptible for degradation as a result of exposure to air and respectively oxidation. The kind of postharvest techniques, storage conditions as well as packaging may influence this decrease. Dramatically, losses of vitamin C take place during thermal food processing. Drying, blanching, cooking, and pasteurizing are critical processing steps. Temperature and duration seem to be the major influence factors (Ramesh et al. 2002). Beside degradation, water-soluble vitamins such as vitamin C may also be leached out into the processing water.

Compared with vitamins, knowledge on the behavior of secondary plant metabolites during food processing is scarce. An exception is the cyanogenic glucosides. Cassava, very rich in cyanogenic glucosides, is traditionally prepared with techniques using high temperatures and large amounts of water/steam to force the highly toxic, but volatile hydrogen cyanide out of the cassava bulbs and products (Rawel and Kroll 2003).

In contrast, glycoalkaloids such as the potato alkaloids solanine and chaconine cannot be removed that easily. Following harvest, the alkaloid content can increase during storage and transportation as well as under the influence of light, temperature, and sprouting (Dimenstein et al. 1997). However, thermal processing (boiling, cooking, frying) have small and variable effects on the glycoalkaloids. Boiling potatoes reduced their levels only up to 4% (Bushway and Ponnampalam 1981). Rytel and coworkers (2005) showed that during French fries production several steps are necessary to remove the alkaloids during peeling, blanching, and frying.

For glucosinolates, it was shown that microwave cooking led to a dramatical loss of 74% of total glucosinolates whereas conventional boiling resulted in a loss of 55%. Almost all glucosinolates and degradation products leached out into the cooking water. Steaming showed only minimal effects (Vallejo et al. 2002). Equally, glucosinolates were only partly degraded to isothiocyanates (Song and Thornalley 2007). However, the breakdown of the glucosinolates and their processing behavior is highly dependent on the kind of vegetable (Cieslik et al. 2007).

As carotenoids are poorly soluble in water, their degradation seems to take a different course. Results presented in the literature are often inconsistent. In minimally processed kale, monitored during storage, the major carotenoids (β-carotene, lutein, violaxanthin) were reduced up to 31% (De Azevedo and Rodriguez-Amada 2005). Khachik and coworkers (1992) reported that total carotenoids remained unchanged during steaming and microwave cooking, whereas Zhang and Hamauzu (2004) found a dramatic decline of total carotenoids in broccoli florets. Particularly β-carotene showed low heat stability. When processing methods were used cumulatively (peeling, chopping, boiling), β-carotene was lost up to 61% (Padmavati et al. 1992). Schweiggert and coworkers (2007) showed an increased processing stability of mono- and di-carotenoid esters compared to their nonesterified counterparts in red and hot chili peppers. Capsanthin, capsorubin, zeaxanthin, and β-cryptoxanthin fatty acid esters displayed similar stability, whereas susceptibility of nonesterified pigments to degradation differed considerably. Literature on degradation products of carotenoids resulting from thermal-induced degradation is rare, but cyclization might play an important role in the reaction mechanism (Perez-Galvez et al. 2005).

Most of the plant phenolic compounds are water soluble rendering them susceptible to leaching (Andlauer et al. 2003). Furthermore, they are often found in significant amounts in the peels, so that peeling and cutting results in a loss. Removal of peach peel showed a loss up to 48% depending on the stage of maturity (Asami et al. 2003). Investigations on changes of phenolic compounds in small berries revealed that thermal treatment diminished the amounts of flavonols, whereby myricetin and kaempferol showed higher stability against thermal treatment than quercetin (Hakkinen et al. 2000). Asami and coworkers (2003) showed that high temperature sterilization of peaches led to a decrease of phenolic compounds (primarily procyanidins), persisting during further storage. Depending on the preparation techniques (cutting, chopping), and processing temperature, concentration of phenolic compounds decreased for several vegetables (Andlauer et al. 2003). Besides leaching out into the cooking water, several reaction products occurred (Price et al. 1997).

Phenolic compounds are highly reactive and susceptible to various chemical reactions during food processing. Apart from the already mentioned enzymatic browning reaction they are substrates for nonenzymatically hydrolytic and oxidative reactions. Such reactions are predominantly induced thermally. The resulting compounds may have an impact on all quality attributes (color changes, nutritional aspects, etc.) of agricultural crops.

There are many model studies on the oxidation of phenolic compounds. In most cases, these processes are radical-induced degradation mechanisms. As a result of free exposure to air, reactive oxygen species (ROS) play an important role. Examples are superoxide anion radicals, hydroperoxide radicals, hydroxyl radicals, and singlet oxygen. In the first step of the oxidation, a hydrogen atom of the phenolic compound is abstracted. In the case of dihydroxy structures, this leads to the formation of semi-quinones (Cilliers and Singleton 1991). The second step of the oxidation is finished with the formation of quinoid structures, which are susceptible to further reactions.

The oxidative degradation of flavonols preferentially leads to changes of the C-ring of the basic skeleton. Cleavage of the C-ring leads to smaller phenolic compounds resulting from the A- and the B-rings. For example, oxidation of

3,7,4′-trihydroxyflavone under alkaline conditions led to protocatechuic acid and 4-hydroxybenzoic acid (Nishinaga et al. 1979).

In the case of onion, it was shown that cooking forced quercetin and its glycosides to degrade to a whole bunch of smaller reaction products depending on pH value, temperature, time, and oxidative conditions. Here, the major degradation product was also protocatechuic acid (Price et al. 1997; Buchner et al. 2006). In contrast, under conditions of drying or roasting, onion flavonol glucosides showed a release of quercetin that remained stable during further thermal treatment (Rohn et al. 2007). As the antioxidant activity of quercetin is higher compared to its glycosides (Rösch et al. 2003), roasting resulted in an increase of overall antioxidant activity.

In this context, an almost unknown research field is the investigation of interactions between phenolic compounds and further food constituents. Proteins, especially, seem to be the preferred interaction partners of phenolic compounds. Diverse interaction mechanisms are possible such as formation of hydrogen and π-bonds, hydrophobic and ionic interactions as well as covalent reactions (Macholz and Lewerenz 1989; Kroll et al. 2003b; Papadopoulou and Frazier 2004).

As already mentioned, oxidation of phenolic compounds leads to very reactive semiquinoid intermediates and quinones, both capable of undergoing further reactions with other food constituents, even till polymerization. The chemical properties of such browning products are comparable to those resulting from enzymatic browning (Rouet-Mayer et al. 1990). The covalent reaction between phenolic compounds and proteins leads to a change in selected physicochemical protein properties. Protein structure was affected. The antioxidant activity of the phenolic compounds attached is diminished (Kroll et al. 2003b). From the nutritional point of view, the reactions between food proteins and polyphenols are assessed as critical. Some of the aspects are (1) the astringency of such complexes leads to decrease in food consumption; (2) protein digestibility and quality is decreased; and (3) indispensable amino acids such as lysine are adversely affected, as the amino group of lysine is a major reactive site (Macholz and Lewerenz 1989; Petzke et al. 2005; Rohn et al. 2006).

The examples presented may give an impression of the complexity of the changes of food constituents during traditionally thermal food processing. Besides the formation of a characteristic sensory profile (color, aroma, flavor), constituents can be either degraded to potentially toxic compounds, or they may improve nutritional quality of foods, as seen for the melanoidins and the phenolic degradation products due to enhanced antioxidant activity. An immense number of new substances are formed, all having the possibility to serve as markers for the quality of plant-based food products.

1.2.4.3 Nonthermal Food Processing

Feeling insecure by the discovery of food-borne toxicants such as acrylamide, consumers nowadays look for foods with not only a long shelf life but also high quality in terms of beneficial nutritional effects. Traditional thermal techniques of food processing do not seem to fulfill this anymore. Owing to high temperatures used, a lot of undesirable changes take place. To meet these demands, manufacturers are developing new and emerging technologies that are less influencing such properties of food associated with freshness and health aspects. However, little is known about how these processes alter the nutrients in processed plant food.

High-pressure is a new processing technique aimed at achieving consumer demands for fresher products with reduced microbiological levels. When taking a look at the continuously increasing number of studies in this research field, it is obvious that the described demands are highly depending on the conditions applied (pressure, duration of treatment, temperature). Following some changes of quality determinants are presented:

Especially vitamins are a class of compounds that are essential for human beings. Therefore, an effective preservation is very important. Selected vitamins of alfalfa sprouts showed only marginal changes after high-pressure treatment (Gabrovska et al. 2007). It is well known that sprouting/germination improves nutritional quality of seeds (Vidal-Valverde et al. 2002). Vitamin C content of germinated cowpeas decreased only slightly after high-pressure treatment. However, the shelf life of sprouts is short and the application of minimal processing techniques such as high pressure may preserve them fresh and stable (Doblado et al. 2007).

This is also true for vegetable juices and their glucosinolates. High pressure preserved the broccoli glucosinolates from degradation (Mandelova and Totusek 2007). Isothiocyanates (ITC), the main degradation products of glucosinolates, showed varying amounts in high-pressure treated juices prepared from a bunch of cruciferous vegetables. ITC's content was lowest in white cabbage. For broccoli high-pressure preserved juice, the ITC's content was comparable with frozen juice (Triska et al. 2007).

For some raw materials, use of high pressure depends on the product demanded. For intact vegetable organs, a compromise is targeted to gain optimum texture of the end product. High pressure sometimes affects vegetable quality, because the tissues of some vegetables can easily be damaged (Prestamo and Arroya 1998). When heading for a juice or a puree, loss of texture resulting from high-pressure treatment is advantageous, as secondary plant metabolites may be released easier into the product. This was shown for selected carotenoids of tomato puree (Sanchez-Moreno et al. 2003) and orange juice (Sanchez-Moreno et al. 2005a). For plant phenolic compounds, an increase during orange juice production was observed after high-pressure treatment (Sanchez-Moreno et al. 2005a).

However, in some cases, high-pressure treatment may also have an eliciting effect on the fruit and vegetable tissues, resulting in a stress-induced increase of phenolic compounds. This increase may be accompanied by higher activities of phenoloxidases and peroxidases (Dörnenburg and Knorr 1997). From the viewpoint of quality this is somehow frightening, as both enzyme classes are responsible for oxidative reactions associated with undesirable browning and formation of off-flavors. Since phenoloxidases and peroxidases are very stable toward pressure, a combination of high-pressure and temperature treatment seems to be inevitable. Stability against high pressure has also been shown for pectinesterases of several fruits and enzymes responsible for lipid oxidation in tomato (lipoxygenase and hydroperoxide lyase). In these cases, combined high-pressure and temperature treatment showed good results for inactivation (Bayindirli et al. 2006; Rodrigo et al. 2007a). For tomato puree no color degradation was observed in combined thermal and high-pressure treatment; a maximum increase of almost 9% in $L^*a^*b^*$ parameters was found for strawberry juices (Rodrigo et al. 2007b).

However, elevated temperature might lead to the Maillard reaction. Especially for foods rich in carbohydrates and proteins, color is predominantly affected by high-pressure treatment (Ahmed and Ramaswamy 2006). The complexity of the Maillard reaction seems to vary even further under high-pressure conditions, as every step of the reaction significantly differs in its pressure dependence (Hill et al. 1996; Moreno et al. 2003). For example, the formation of volatiles in the course of the Maillard reaction is diminished by the application of pressure (Hill et al. 1996; Bristow and Isaacs 1999). Pentosidine is elevated with increasing pressure while formation of pyrraline is reduced (Moreno et al. 2003).

The second emerging technology for minimal food processing discussed is the treatment with PEFs. Owing to the technical reasons, PEF is only applicable for liquids and is therefore preferentially used for the extraction and preservation of fruit and vegetable juices. Compared with the high-pressure treatment, PEF seems to be mild with regard to changes of food compounds.

Quality parameters such as color, pH, °Brix, and titratable acidity of selected juices were not influenced by the PEF treatment. Nonenzymatic browning and HMF were also not affected (Yeom et al. 2000; Cserhalmi et al. 2006). The volatile flavor compounds of treated juices were present in equal amounts compared with the unprocessed ones (Yeom et al. 2000; Cserhalmi et al. 2006).

With regard to vitamins, there were only marginal losses of vitamin C in orange juice (Yeom et al. 2000; Sanchez-Moreno et al. 2005a). Bioavailability of vitamin C from a PEF-treated vegetable soup was affected after PEF treatment (Sanchez-Moreno et al. 2005b). Single or total carotenoids as well as phenolic compounds were not modified in fruit juices (Sanchez-Moreno et al. 2005a).

Zulueta and coworkers (2007) treated an orange juice–milk beverage using PEF and analyzed the fatty acid profile and levels of lipid oxidation. Non-significant changes in the content of saturated fatty acids, monounsaturated fatty acids, and polyunsaturated fatty acids were observed. Neither peroxides nor intolerable levels of furfurals were detected.

With regard to food quality, minimal processing using new and emerging technologies shows good promises for the future. In most cases optimized conditions have to be found to successfully maintain shelf life and quality parameters of the individual products. Considering undesired, maybe toxic, neoformed compounds, research is proceeding.

1.2.5 Concluding for Nondestructive Analysis and Product Monitoring

The quality of agricultural crops is highly dependent on the question of an optimized maturity of the raw products respectively an optimum harvest date and product-specific postharvest as well as processing techniques.

The assessment of maturity providing acceptable quality to the consumer implies measurable parameters in the fruit and vegetable development and the need for analytical techniques. A maturity index for a plant food product is a measurement that can be used to determine whether a particular commodity is mature (Chrisosto 1994).

There are several possibilities to analyze all the characteristics during postharvest processing. Besides the classical wet-chemical methods and chromatographic techniques, molecular biological methods may be involved by determination of differential protein expressions (Kroh and Baltes 2004). For measurements of quality parameters to be carried out in the field or during processing by the producers and quality control personnel, methods need to be readily performable in the field and should require relative inexpensive equipment.

Number of scientific studies dealing with nondestructive noninvasive optical methods in food and life sciences is continuously increasing. The basic assumption behind the application of all spectroscopic techniques to this relies on the specific absorbance and scattering properties of the food (see Section 1.3).

The further practical advantages of a nondestructive analytical technique seem to be obvious (Scotter 1997): direct analysis is the more "true" analysis for food attributes; no toxic reagents are required; rapid results, when once calibrated; suitable for online measurements, required during food processing; analysis of more than one component with one spectral acquisition, etc.

Some of the approaches still have academic value or are still in development, while others have their fixed position in the quality control of plant food production or processing. Following, selected examples should briefly give an impression of the potential to use optical methods for the analysis of the quality parameters introduced in this chapter. For detailed descriptions and applications the reader is referred to Chapters 3 through 6.

As already mentioned during the production of plant food, assessment of maturity and an optimum harvest date is inevitable. Commonly used parameters are color, fruit firmness, soluble solid content, changes in chlorophyll content, etc. Using optical methods, flesh pigments can be measured with devices based on light-emitting diodes (LEDs). (Li et al. 1997; Valero et al. 2003; Zude 2003; Zude et al. 2007). A second method for chlorophyll measurement uses its fluorescence (Georgieva and Lichtenthaler 1999).

Flesh firmness can be followed measuring water content or water distribution with nuclear magnetic resonance (NMR). Magnetic resonance imaging (MRI) was applied successfully for assessing mealiness in apples (Barreiro et al. 2000). For apples, firmness and SSC, which is a measure for sugars, was predicted with laser light backscattering image analysis. Here, laser diodes were used emitting at five wavelength bands (Lu 2007; Qing et al. 2007).

Carotenoids, as important bioactive compounds, can be estimated using spectroscopy in the visible and near-infrared wavelength range (Quilitzsch et al. 2005; Baranska et al. 2006; Zude et al. 2007). Other food constituents such as lipids can also be determined by near infrared spectroscopy (NIRS) (Afseth et al. 2005).

Processing of food is often accompanied by enzymatically or nonenzymatically browning and the development of undesired compounds (e.g., HMF, acrylamide) (Birlouez-Aragon and Zude 2004). Recently, the potential of front-face fluorescence spectroscopy (FFFS) to monitor the development of Maillard browning during thermal processing has been assessed (Schamberger and Labuza 2006). Segtnan and coworkers (2006) used a combination of visible spectroscopy (VIS) and NIRS with partial least squares (PLS) regression to follow acrylamide formation in potato crisps. Changes of extra virgin olive oil induced by thermal treatment were successfully monitored using fluorescence spectroscopy in combination with multivariate regression (Cheikhousman et al. 2005).

Finally, two statements are worth citing in this context. In a review on spectroscopic approaches for measurement of food quality, Belton (1997) stated

> *The problem for the spectroscopist in the elucidation of food quality is that food originates from a very wide variety of plant and animal sources. These may be eaten fresh but much more frequently are processed or combined in some way. The processing may take simple form of cooking, but more usually there are mixing or other mechanical treatment stages designed to change the physical and/or chemical state of the food... The delicate balance of physical, chemical and biological forces is critical to the quality of the final product and may be easily changed by invasive methods. The challenge to spectroscopy and the spectroscopist is therefore a daunting one.*

Also Scotter (1997) claimed "that it is ironic that when a spectroscopic [...] is calibrated against a traditional technique is described as the 'direct' method and the spectroscopic analysis as 'indirect'".

REFERENCES

Afseth, N.K., V.H. Segtnan, B.J. Marquardt, and J.P. Wold. 2005. Raman and near infrared spectroscopy for quantification of fat composition in a complex food model system. *Applied Spectroscopy* 59:1324–1332.

Ahmed, J. and H.S. Ramaswamy. 2006. Changes in colour during high pressure processing of fruits and vegetables. *Stewart Postharvest Review* 2:1–8.

Andlauer, W., C. Stumpf, M. Hubert, A. Rings, and P. Fürst. 2003. Influence of cooking process on phenolic marker compounds of vegetables. *International Journal for Vitamin and Nutrition Research* 73:152–159.

Asami, D.K., Y.J. Hong, D.M. Barrett, and A.E. Mitchell. 2003. Processing-induced changes in total phenolics and procyanidins in clingstone peaches. *Journal of the Science of Food and Agriculture* 83:56–63.

Baranska, M., W. Schutze, and H. Schulz. 2006. Determination of lycopene and beta-carotene content in tomato fruits and related products: Comparison of FT-Raman, ATR-IR, and NIR spectroscopy. *Analytical Chemistry* 78:8456–8461.

Barreiro, P., C. Ortiz, M. Ruiz-Altisent et al. 2000. Mealiness assessment in apples and peaches using MRI. *Magnetic Resonance Imaging* 18:1175–1181.

Bayindirli, A., H. Alpas, F. Bozoglu, and M. Hizal. 2006. Efficiency of high pressure treatment on inactivation of pathogenic microorganisms and enzymes in apple, orange, apricot and sour cherry juices. *Food Control* 17:52–58.

Belitz, H.-D., W. Grosch, and P. Schieberle. 2001. *Lehrbuch der Lebensmittelchemie.* Springer, Heidelberg, ISBN: 978-3-540-41096-1.

Belton, P.S. 1997. Spectroscopic approaches to the measurement of food quality. *Pure and Applied Chemistry* 69:47–50.

Birlouez–Aragon, I. and M. Zude. 2004. Fluorescence fingerprints as a rapid predictor of the nutritional quality of processed and stored food. *Czech Journal of Food Sciences* 22:68–71.

Bourne, M.C. 2004. Selection and use of postharvest technologies as a component of the food chain. *Journal of Food Science* 69:R43–R46.

Bremner, H.A. 2000. Towards practical definitions of quality for food science. *Critical Reviews in Food Science and Nutrition* 40:83–90.

Bristow, M. and N.S. Isaacs. 1999. The effect of high pressure on the formation of volatile products in a model Maillard reaction. *Journal of the Chemical Society Perkin Transactions 2* 10:2213–2218.

Buchner, N., A. Krumbein, S. Rohn, and L.W. Kroh. 2006. Effect of thermal processing on the flavonols Rutin and Quercetin. *Rapid Communications in Mass Spectrometry* 20:3229–3235.

Bushway, R.J. and R. Ponnampalam. 1981. α-Chaconine and α-solanine content of potato products and their stability during several modes of cooking. *Journal of Agricultural and Food Chemistry* 29:814–817.

Caemmerer, B. and L.W. Kroh. 2006. Antioxidant activity of coffee brews. *European Food Research and Technology* 223:469–474.

Cardello, A.V., H.G. Schutz, and L.L. Lesher. 2007. Consumer perceptions of foods processed by innovative and emerging technologies: A conjoint analytic study. *Innovative Food Science and Emerging Technologies* 8:73–83.

Charissou, A., L. Ait-Ameur, and I. Birlouez-Aragon. 2007. Kinetics of formation of three indicators of the Maillard reaction in model cookies: Influence of baking temperature and type of sugar. *Journal of Agricultural and Food Chemistry* 55:4532–4539.

Cheftel, J.C. 1995. Review: High-pressure, microbial inactivation and food preservation. *Food Science and Technology International* 1:75–90.

Cheikhousman, R., M. Zude, D.J.R. Bouveresse, C.L. Leger, D.N. Rutledge, and I. Birlouez-Aragon. 2005. Fluorescence spectroscopy for monitoring deterioration of extra virgin olive oil during heating. *Analytical and Bioanalytical Chemistry* 382:1438–1443.

Chrisosto, C.H. 1994. Stone fruit maturity indices: A descriptive review. *Postharvest News and Information* 5:65N–68N.

Cieslik, E., T. Leszczynska, A. Filipiak-Fiorkiewicz, E. Sikor, and P.M. Pisulewski. 2007. Effects of some technological processes on glucosinolate contents in cruciferous vegetables. *Food Chemistry* 105:976–981.

Cilliers, J.J.L. and V.L. Singleton. 1991. Characterization of the products of nonenzymic autoxidative phenol reactions in a caffeic acid model system. *Journal of Agricultural and Food Chemistry* 39:1298–1303.

Cserhalmi, Z., A. Sass-Kiss, M. Toth-Markus, and N. Lechner. 2006. Study of pulsed electric field treated citrus juices. *Innovative Food Science and Emerging Technologies* 7:49–54.

De Azevedo, C.H. and D.B. Rodriguez-Amaya. 2005. Carotenoid composition of kale as influenced by maturity, season and minimal processing. *Journal of the Science of Food and Agriculture* 85:591–597.

Deliza, R., A. Rosenthal, F.B.D. Abadia, C.H.O. Silva, and C. Castillo. 2005. Application of high pressure technology in the fruit juice processing: Benefits perceived by consumers. *Journal of Food Engineering* 67:241–246.

Dimenstein, L., N. Lisker, N. Kedar, and D. Levy. 1997. Changes in the content of steroidal glycolalkaloids in potato tubers grown in the field and in the greenhouse under different conditions of light, temperature and daylength. *Physiological and Molecular Plant Pathology* 50:391–402.

Doblado, R., J. Frias, and C. Vidal-Valverde. 2007. Changes in vitamin C content and antioxidant capacity of raw and germinated cowpea (*Vigna sinensis* var. *carilla*) seeds induced by high pressure treatment. *Food Chemistry* 101:918–923.

Dörnenburg, H. and D. Knorr. 1997. Evaluation of elicitor- and high-pressure-induced enzymatic browning utilizing potato (*Solanum tuberosum*) suspension cultures as a model system for plant tissue. *Journal of Agricultural and Food Chemistry* 45:4173–4177.

Eisenbrand, D. 2005. Safety assessment of high pressure treated foods—Opinion of the Senate Commission on Food Safety (SKLM) of the German Research Foundation (DFG). *Molecular Nutrition and Food Research* 49:1168–1174.

Fiedler, T., T. Moritz, and L.W. Kroh. 2006. Influence of alpha-dicarbonyl compounds to the molecular weight distribution of melanoidins in sucrose solutions: Part 1. *European Food Research and Technology* 223:837–842.

Fiedler, T. and L.W. Kroh. 2007. Formation of discrete molecular size domains of melanoidins depending on the involvement of several alpha-dicarbonyl compounds: Part 2. *European Food Research and Technology* 227:473–481.

Friedman, M. and G.M. McDonald. 1997. Potato glycoalkaloids: Chemistry, analysis, safety, and plant physiology. *Critical Reviews in Plant Sciences* 16:55–132.

Fukayama, M.Y., H. Tan, W.B. Wheeler, and C. Wei. 1986. Reactions of aqueous chlorine and chlorine dioxide with model food compounds. *Environmental Health Perspectives* 69:267–274.

Gabrovska, D., J. Strohalm, I. Paulickova et al. 2007. Nutritional and sensory quality of selected sprouted seeds. *High Pressure Research* 27:143–146.

Georgieva, K. and H.K. Lichtenthaler. 1999. Photosynthetic activity and acclimation ability of pea plants to low and high temperature treatment as studied by means of chlorophyll fluorescence. *Journal of Plant Physiology* 155:416–423.

Giese, J. 2000. Color measurement in foods as a quality parameter. *Food Technology* 54:62–63.

Guderjan, M., P. Elez-Martinez, and D. Knorr. 2007. Application of pulsed electric fields at oil yield and content of functional food ingredients at the production of rapeseed oil. *Innovative Food Science and Emerging Technologies* 8:55–62.

Halliwell, B. 1996. Oxidative stress, nutrition and health. Experimental strategies for optimization of nutritional antioxidant intake in humans. *Free Radical Research* 25:57–74.

Hakkinen, S.H., S.O. Karenlampi, H.M. Mykkanen, and A.R. Torronen. 2000. Influence of domestic processing and storage on flavonol contents in berries. *Journal of Agricultural and Food Chemistry* 48:2960–2965.

Hao, Y.Y. and R.E. Brackett. 1994. Pectinase activity of vegetable spoilage bacteria in modified atmosphere. *Journal of Food Science* 59:175–178.

Harris, W.S. 2003. N-3 long-chain polyunsaturated fatty acids reduce risk of coronary heart disease death: Extending the evidence to the elderly. *American Journal of Clinical Nutrition* 77:279–280.

Hassenberg, K., C. Idler, E. Molloy, M. Geyer, M. Plöchl, and J. Barnes. 2007. Use of ozone in a lettuce washing process—An industrial trial. *Journal of the Science of Food and Agriculture* 87:914–919.

Heinz, V., I. Alvarez, A. Angersbach, and D. Knorr. 2001. Preservation of liquid foods by high intensity pulsed electric fields—Basic concepts for process design. *Trends in Food Science and Technology* 12:103–111.

Henle, T., U. Schwarzenbolz, and H. Klostermeyer. 1997. Detection and quantification of pentosidine in foods. *Zeitschrift fuer Lebensmittel-Untersuchung und Forschung* A 204:95–98.

Hill, V.M., D.A. Ledward, and J.M. Ames. 1996. Influence of high hydrostatic pressure and pH on the rate of Maillard browning in a glucose-lysine system. *Journal of Agricultural and Food Chemistry* 44:594–598.

Hodge, J.E. 1953. Dehydrated foods. Chemistry of browning reactions in model sytems. *Journal of Agricultural and Food Chemistry* 1:928–943.

Huyskens-Keil, S. and M. Schreiner. 2003. Quality of fruits and vegetables. *Journal of Applied Botany* 77:147–151.

Jones, M.G., J. Hughes, A. Tregova, J. Milne, A.B. Tomsett, and H.A. Collin. 2004. Biosynthesis of the flavour precursors of onion and garlic. *Journal of Experimental Botany* 55:1903–1918.

Jung, M.Y., D.S. Choi, and J.W. Ju. 2003. A novel technique for limitation of acrylamide formation in fried and baked corn chips and in French fries. *Journal of Food Science* 68:1287–1290.

Kader, A.A. 1999. Fruit maturity, ripening and quality relationships. *Acta Horticulturae* 485:203–207.

Kader, A.A. and C.B. Watkins. 2000. Modified atmosphere packaging—Toward 2000 and beyond. *HortTechnology* 10:483–486.

Khachik, F., B. Goli, G.R. Beecher et al. 1992. Effect of food preparation on qualitative and quantitative distribution of major carotenoid constituents of tomatoes and several green vegetables. *Journal of Agricultural and Food Chemistry* 40:390–398.

Kessler, H.-G. 2002. *Food and Bio Process Engineering—Dairy Technology*. Verlag A. Kessler, Munich, ISBN: 3-9802378-5-0.

Krämer, J. 2002. *Lebensmittel-Mikrobiologie*. UTB, Stuttgart, ISBN: 3-8252-1421-4.

Krebbers, B., A.M. Matser, S.W. Hoogerwerf, R. Moezelaar, M.M.M. Tomassen, and R.W. van den Berg. 2003. Combined high-pressure and thermal treatments for processing of tomato puree: Evaluation of microbial inactivation and quality parameters. *Innovative Food Science and Emerging Technologies* 4:377–385.

Kroh, L.W. and W. Baltes. 2004. *Schnellmethoden zur Beurteilung von Lebensmitteln und ihren Rohstoffen*. Behr's Verlag, Hamburg, ISBN: 3-89947-120-2.

Kroll, J., S. Rohn, and H.M. Rawel. 2003a. Secondary plant metabolites as functional constituents of foods. *Deutsche Lebensmittel-Rundschau* 99:259–270.

Kroll, J., H.M. Rawel, and S. Rohn. 2003b. Reactions of plant phenolics with food proteins and enzymes under special consideration of covalent bonds. *Food Science and Technology Research* 9:205–218.

Kroll, J., H. Ranters, H.M. Rawel, and S. Rohn. 2004. Isoflavones as constituents of plant foods. *Deutsche Lebensmittel-Rundschau* 100:211–224.

Labavitch, J.M. 1981. Cell-wall turnover in plant development. *Annual Review of Plant Physiology and Molecular Biology* 32:385–406.

Lamikanra, O. and A.K. Kassa. 1999. Changes in the free amino acid composition with maturity of the noble cultivar of *Vitis rotundifolia* Michx. grape. *Journal of Agricultural and Food Chemistry* 47:4837–4841.

Lana, M.M., L.M.M. Tijskens, and O. van Kooten. 2005. Effects of storage temperature and fruit ripening on firmness of fresh cut tomatoes. *Postharvest Biology and Technology* 35:87–95.

Lee, S.J., K. Umano, T. Shibamoto, and K.G. Lee. 2005. Identification of volatile components in basil (*Ocimum basilicum* L.) and thyme leaves (*Thymus vulgaris* L.) and their antioxidant properties. *Food Chemistry* 91:131–137.

Legnani, P.P. and P. Leoni. 2004. Effect of processing and storage conditions on the microbiological quality of minimally processed vegetables. *International Journal of Food Science and Technology* 39:1061–1068.

Lennernas, M., C. Fjellstrom, W. Becker et al. 1997. Influences on food choice perceived to be important by nationally-representative samples of adults in the European Union. *European Journal of Clinical Nutrition* 51:S8–S15.

Li, M., D.C. Slaughter, and J.F. Thompson. 1997. Optical chlorophyll sensing system for banana ripening. *Postharvest Biology and Technology* 12:273–283.

Liang, Z.W., G.S. Mittal, and M.W. Griffiths. 2002. Inactivation of *Salmonella typhimurium* in orange juice containing antimicrobial agents by pulsed electric field. *Journal of Food Protection* 65:1081–1087.

Lorimer, S. and S. Hill. 2006. Maturity testing of stone fruit. *Agriculture Notes* AG1147, ISSN:1329–8062.

Lu, R. 2007. Nondestructive measurement of firmness and soluble solids content for apple fruit using hyperspectral scattering images. *Sensing and Instrumentation for Food Quality and Safety* 1:19–27.

Luck, G., H. Liao, N.J. Murray et al. 1994. Polyphenols, astringency and proline-rich proteins. *Phytochemistry* 37:357–371.

Macholz, R. and H.J. Lewerenz. 1989. *Lebensmitteltoxikologie*. Akademie-Verlag, Berlin, ISBN: 3-05-500378-0.

Mandelova, L. and J. Totusek. 2007. Broccoli juice treated by high pressure: Chemoprotective effects of sulphoraphane and indole-3-carbinol. *High Pressure Research* 27:151–156.

Moreno, F.J., E. Molina, A. Olano, and R. Lopez-Fandino. 2003. High-pressure effects on Maillard reaction between glucose and lysine. *Journal of Agricultural and Food Chemistry* 51:394–400.

Mottram, D.S., B.L. Wedzicha, and A.T. Dodson. 2002. Acrylamide is formed in the Maillard reaction. *Nature* 419:448–449.

Nicolas, J.J. 1994. Enzymatic browning reactions in apple and apple products. *Critical Reviews in Food Science and Nutrition* 34:109–157.

Nishinaga, A., T. Tojo, H. Tomita, and T. Matsuura. 1979. Base-catalyzed oxygenolysis of 3-hydroxyflavones. *Journal of the Chemical Society Perkin Transactions* 1:2511–2516.

Oxford Advanced Learner's Dictionary, 7th Edition 2005. Oxford University Press, Oxford, ISBN: 978-3-06-800202-5.

Padmavati, K., S.A. Udipi, and M. Rao. 1992. Effect of different cooking methods on β-carotene content of vegetables. *Journal of Food Science and Technology* 29:137–140.

Papadopoulou, A. and R.A. Frazier. 2004. Characterization of protein–polyphenol interactions. *Trends in Food Science and Technology* 15:186–190.

Perez, A., C. Sanz, J.J. Rios, R. Olias, and J.M. Olias. 1999. Effects of ozone treatment on postharvest strawberry quality. *Journal of Agricultural and Food Chemistry* 47:1652–1656.

Perez-Galvez, A., J.J. Rios, and M.I. Minguez-Mosquera. 2005. Thermal degradation products formed from Carotenoids during a heat-induced degradation process of paprika oleoresins (*Capsicum annuum* L.). *Journal of Agricultural and Food Chemistry* 53:4820–4826.

Petzke, K.J., S. Schuppe, S. Rohn, H.M. Rawel, and J. Kroll. 2005. Chlorogenic acid moderately decreases quality of whey proteins in rats. *Journal of Agricultural and Food Chemistry* 53:3714–3720.

Prestamo, G. and G. Arroya. 1998. High hydrostatic pressure effects on vegetable structure. *Journal of Food Science* 63:878–881.

Price, K.R., J.R. Bacon, and M. Rhodes. 1997. Effect of storage and domestic processing on the content and composition of flavonol glucosides in onion (*Allium cepa*). *Journal of Agricultural and Food Chemistry* 45:938–942.

Qing, Z., B. Ji, and M. Zude. 2007. Predicting soluble solid content and firmness in apple fruit by means of laser light backscattering image analysis. *Journal of Food Engineering* 82:58–67.

Quilitzsch, R., M. Baranska, H. Schulz, and E. Hoberg. 2005. Fast determination of carrot quality by spectroscopic methods in the UV-VIS, NIR and IR range. *Journal of Applied Botany and Food Quality* 79:163–167.

Ramesh, M.N., W. Wolf, D. Tevini, and A. Bognar. 2002. Microwave blanching of vegetables. *Journal of Food Science* 67:390–398.

Rawel, H.M. and J. Kroll. 2003. The importance of cassava (*Manihot esculenta* Crantz) as the main staple food in tropical countries. *Deutsche Lebensmittel-Rundschau* 99:102–108.

Reid, M.S. 1992. Maturation and maturity indices. In: *Peaches, Plums and Nectarines: Growing and Handling for Fresh Market*, La Rue, J.H. and R.S. Johnson (eds.), pp. 21–28. University of California, Department of Agriculture and Natural Resources Publication No. 3331.

Rico, D., A.B. Martin-Diana, J.M. Barat, and C. Barry-Ryan. 2007. Extending and measuring the quality of fresh-cut fruit and vegetables: A review. *Trends in Food Science and Technology* 18:373–386.

Rodarte, A.D., I. Eichholz, S. Rohn, L.W. Kroh, and S. Huyskens-Keil. 2008. Phenolic profile and antioxidant activity of highbush blueberry (*Vaccinium corymboium* L.) during fruit maturation and ripening. *Food Chemistry* 109:554–563.

Rodrigo, D., R. Jolie, A. van Loey, and M. Hendrickx. 2007a. Thermal and high pressure stability of tomato lipoxygenase and hydroperoxide lyase. *Journal of Food Engineering* 79:423–429.

Rodrigo, D., A. van Loey, and M. Hendrickx. 2007b. Combined thermal and high pressure colour degradation of tomato puree and strawberry juice. *Journal of Food Engineering* 79:553–560.

Rohn, S., K.J. Petzke, H.M. Rawel, and J. Kroll. 2006. Reactions of chlorogenic acid and quercetin with a soy protein isolate–Influence on the in vivo food protein quality in rats. *Molecular Nutrition and Food Research* 50:696–704.

Rohn, S., N. Buchner, G. Driemel, M. Rauser, and L.W. Kroh. 2007. Thermal degradation of onion quercetin glucosides under roasting conditions. *Journal of Agricultural and Food Chemistry* 55:1568–1573.

Rösch, D., M. Bergmann, D. Knorr, and L.W. Kroh. 2003. Structure-antioxidant efficiency relationships of phenolic compounds and their contribution to the antioxidant activity of sea buckthorn juice. *Journal of Agricultural and Food Chemistry* 51:4233–4239.

Rose, P., M. Whiteman, P.K. Moore, and Y.Z. Zhu. 2005. Bioactive S-alk(en)yl cysteine sulphoxide metabolites in the genus *Allium*: The chemistry of potential therapeutic agents. *Natural Product Reports* 22:351–368.

Rouet-Mayer, M.A., J. Ralambosoa, and J. Philippon. 1990. Roles of *o*-quinones and their polymers in the enzymatic browning of apples. *Phytochemistry* 29:435–440.

Rüfer, C.E. and S.E. Kulling. 2006. Phytoestrogene. In: *Handbuch der Lebensmitteltoxikologie–Belastungen, Wirkungen, Lebensmittel sicherheit, Hygiene*, Dunkelberg, H., T. Gebel, and A. Hartwig (eds.), Volume 5, pp. 2623–2680. Wiley-VCH, Weinheim, ISBN: 978-3-527-31166-8.

Rytel, E., G. Glubowska, G. Lisinska, A. Peksa, and K. Aniolowski. 2005. Changes in glycoalkaloid and nitrate contents in postatoes during French fries processing. *Journal of the Science of Food and Agriculture* 85:879–882.

Sacher, J.A. 1973. Senescence and postharvest physiology. *Annual Review of Plant Physiology and Molecular Biology* 24:197–224.

Sanchez-Moreno, C., L. Plaza, B. de Ancos, and M.P. Cano. 2003. Effect of combined treatments of high-pressure and natural additives on carotenoids extractability and antioxidant activity of tomato puree (*Lycopersicum esculentum* Mill.). *European Food Research and Technology* 219:151–160.

Sanchez-Moreno, C., L. Plaza, P. Elez-Martinez, B. de Ancos, O. Martin-Bellosos, and M.P. Cano. 2005a. Impact of high pressure and pulsed electric fields on bioactive compounds and antioxidant activity of orange juice in comparison with traditional thermal processing. *Journal of Agricultural and Food Chemistry* 53:4403–4409.

Sanchez-Moreno, C., M.P. Cano, B. de Ancos et al. 2005b. Intake of Mediterranean vegetable soup treated by pulsed electric fields affects plasma vitamin C and antioxidant biomarkers in humans. *International Journal of Food Sciences and Nutrition* 56:115–124.

Schamberger, G.P. and T.P. Labuza. 2006. Evaluation of front-face fluorescence for assessing thermal processing of milk. *Journal of Food Science* 71:69–74.

Schreiner, M. and S. Huyskens-Keil. 2006. Phytochemicals in fruit and vegetables: Health promotion and postharvest elicitors. *Critical Reviews in Plant Sciences* 25:267–278.

Schulze, J. and W. Bock. 1993. *Aktuelle Aspekte der Ballastofforschung*. Behrs Verlag, Hamburg, ISBN: 978-3-86022-083-2.

Schweiggert, U., C. Kurz, A. Schieber, and C. Carle. 2007. Effects of processing and storage on the stability of free and esterified carotenoids of red peppers (*Capsicum annuum* L.) and hot chilli peppers (*Capsicum frutescens* L.). *European Food Research and Technology* 225:261–270.

Scotter, C.N.G. 1997. Non-destrcutive spectroscopic techniques for the measurement of food quality. *Trends in Food Science and Technology* 8:285–292.

Segtnan, V.H., A. Kita, M. Mielnik, K. Jorgensen, and S.H. Knutsen. 2006. Screening of acrylamide contents in potato crisps using process variable settings and near-infrared spectroscopy. *Molecular Nutrition and Food Research* 50:811–817.

Shewfelt, R.L. 1999. What is quality? *Postharvest Biology and Technology* 15:197–200.

Smelt, J.P.P.M. 1998. Recent advances in the microbiology of high pressure processing. *Trends in Food Science and Technology* 9:152–158.

Song, L.J. and P.J. Thornalley. 2007. Effect of storage and cooking on glucosinolate content of Brassica vegetables. *Food and Chemical Toxicology* 45:216–224.

Suslow, T. 2000. *Postharvest Handling of Organic Crops*. Oakland: University of California. ANR Publication 7254, ISBN: 978-1-60107-045-6.

Teranishi, R., E.L. Wick, and I. Hornstein. 1999. *Flavor Chemistry. Thirty Years of Progress*. Kluwer Academic/Plenum Publishers, New York, ISBN: 978-0-306-46199-6.

Triska, J., N. Vrchotova, M. Houska, and J. Strohalm. 2007. Comparison of total isothiocyanates content in vegetable juices during high pressure treatment, pasteurization and freezing. *High Pressure Research* 27:147–149.

Vallejo, F., F.A. Tomas-Barberan, and C. Garcia-Viguera. 2002. Glucosinolates and vitamin C content in edible parts of broccoli florets after domestic cooking. *European Food Research and Technology* 215:310–316.

Valero, C., C. Crisosto, D. Garner, E. Bowerman, and D. Slaughter. 2003. Introducing nondestrcutive flesh color and firmness sensors to the tree fruit industry. *Acta Horticulturae* 604:597–600.

Van Loey, A., B. Verachtert, and M. Hendrickx. 2001. Effects of high electric field pulses on enzymes. *Trends in Food Science and Technology* 12:94–102.

Vidal-Valverde, C., J. Frias, I. Sierra, I. Balzquez, F. Lamein, and Y.H. Kuo. 2002. New functional legume foods by germination: Effect of nutritive value of beans, lentils and peas. *European Food Research and Technology* 215:472–477.

von Koerber, K., T. Männle, and C. Leitzmann. 1999. *Vollwert-Ernährung - Konzeption einer zeitgemäßen Ernährungsweise*. Haug Verlag, Heidelberg, ISBN: 3-8304-0573-8.

Yeom, H.W., C.B. Streaker, Q.H. Zhang, and D.B. Min. 2000. Effects of pulsed electric fields on the quality of orange juice and comparison with heat pasteurization. *Journal of Agricultural and Food Chemistry* 48:4597–4605.

Zhang, D. and Y. Hamauzu. 2004. Phenolics, ascorbic acid, carotenoids and antioxidant activity of broccoli and their changes during conventional and microwave cooking. *Food Chemistry* 88:503–509.

Zhang, X.M., C.C. Chan, D. Stamp, S. Minkin, M.C. Archer, and W.R. Bruce. 1993. Initiation and promotion of colonic aberrant crypt foci in rats by 5-Hydroxymethyl-2-furanaldehyde in thermolyzed sucrose. *Carcinogenesis* 14:773–775.

Zude, M. 2003. Comparison of indices and multivariate models to non-destructively predict the fruit chlorophyll by means of visible spectrometry in apples. *Analytica Chimica Acta* 481:119–126.

Zude, M. and M. Geyer. 2005. *Non-Destructive Methods for Detecting Health Promoting Compounds*. Workshop Proceedings. Leibniz-Institut für Agrartechnik Potsdam-Bornim e.V., Potsdam-Bornim, ISSN 0947-7314.

Zude, M. and N. McRoberts. 2006. Product monitoring and process control in the crop supply chain. *Agricultural Engineering* 61:2–3.

Zude, M., I. Birlouez, J. Paschold, and D.N. Rutledge. 2007. Nondestructive spectral-optical sensing of carrot quality during storage. *Postharvest Biology and Technology* 45:30–37.

Zulueta, M., M.J. Esteve, I. Frasquet, and A. Frigola. 2007. Fatty acid profile changes during orange juice-milk beverage processing by high-pulsed electric field. *European Journal of Lipid Science and Technology* 109:25–31.

1.3 OPTICAL SENSING

1.3.1 Determination of Optical Properties of Turbid Media: Continuous Wave Approach

MARTINA MEINKE AND MORITZ FRIEBEL

1.3.1.1 Introduction

In a transparent medium, where no scattering appears, the transmittance follows the Beer–Lambert law (Figure 1.3.1). By contrast, the calculation of the light distribution and propagation in turbid media, such as biological tissue, is much more complicated (Figure 1.3.2).

The light can be reflected on the interface of the sample (Fresnel reflection). Light that enters the sample can be scattered backward (diffuse reflectance), or forward (diffuse transmission), or can be absorbed directly or after some scattering events. If the sample is transparent or thin enough, ballistic photons can pass the sample without being scattered or absorbed (collimated transmission).

The collimated and the diffuse transmissions give the total transmission. In the case of transparent media, all light that is not transmitted is absorbed. Only Fresnel reflection appears at the interface of the sample or cuvette, which can be considered by reference measurements on the solvent without the absorber. The absorption coefficient (μ_a) is easy to determine by the Beer–Lambert law.

$$\ln(I/I_0) = \mu_a L \tag{1.3.1}$$

If the sample thickness L is known, and the transmission (I/I_0), which is collimated in this case, can be measured. In the case of very thin sections of a turbid medium (no multiple scattering), the transmitted light, which is not detected as collimated transmittance, can be absorbed and/or scattered. Therefore, the equation must be changed using the attenuation coefficient (μ_t), which includes the absorption and scattering coefficients

$$\ln(I/I_0) = \mu_t L \tag{1.3.2}$$

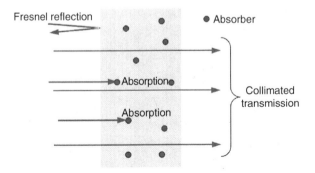

FIGURE 1.3.1 Interaction of light with a transparent media.

What to Measure and How to Measure

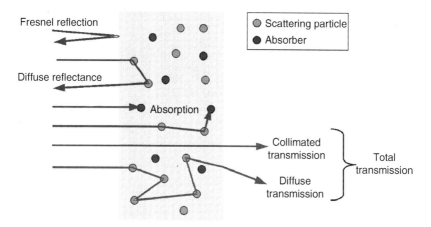

FIGURE 1.3.2 Interaction of light with turbid media.

It is not possible to distinguish between absorption and scattering processes by measurement. Thus, there is no possibility of a direct measurement to calculate the absorption and scattering coefficient separately.

1.3.1.2 Radiation Transport Theory

The distribution of light in turbid media can be described by the stationary radiation transfer (Equation 1.3.3). The equation was developed for the transport of neutrons in solids and liquids (Chandrasekhar 1950) and later adopted for the photon transport in turbid media (atmosphere, biological tissue) (Ishimaru 1978).

$$\frac{dL(\vec{r},\vec{s})}{ds} = -(\mu_a + \mu_s) L(\vec{r},\vec{s}) + \frac{\mu_s}{4\pi} \int_{4\pi} p(\vec{s},\vec{s}') L(\vec{r},\vec{s}) \, d\Omega + S(\vec{r},\vec{s}) \quad (1.3.3)$$

where
 $L(\vec{r},\vec{s})$ is the radiation density, radiance, at the position $\vec{r}$ in direction $\vec{s}$ in W cm^{-2} sr^{-1}
 $p(\vec{s},\vec{s}')$ is the scattering phase function
 μ_a is the absorption coefficient in mm^{-1}
 μ_s is the scattering coefficient in mm^{-1}
 Ω is the solid angle in sr
 $S(\vec{r},\vec{s})$ is the source term (fluorescence, light source) in W cm^{-2} sr^{-1} mm^{-1}

The transport equation only describes the particle behavior of the photons, while effects of the wave character, such as interference or polarization, are not considered.

The absorption coefficient is defined as the product of absorption cross section σ_a and the concentration of absorber

$$\mu_a = c_a \sigma_a \quad (1.3.4)$$

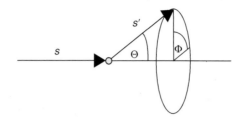

FIGURE 1.3.3 Angular definition of a scattering event.

and the scattering coefficient as product of scattering cross section σ_s and the concentration of the scattering particles

$$\mu_s = c_s \sigma_s \tag{1.3.5}$$

For the complete description of a scattering event, the scattering phase function $p(\vec{s},\vec{s}')$ is needed. It describes the probability that a photon with direction $\vec{s}$ is scattered in direction $\vec{s}'$ (van de Hulst 1957; Kerker 1969).

The scattering phase function can be determined by goniometric measurements on very thin samples or single particles (Figure 1.3.3; Forster et al. 2006). In isotropic media, which have no anisotropic structures such as most biological tissue, the scattering can be assumed to be independent of Φ. To disrobe scattering processes, the anisotropy factor (g) is used. It is the expectation of the cosine of the scattering angle (Θ):

$$\int_0^{2\pi} d\Phi \int_0^{\pi} p(\vec{s},\vec{s}') \cos\Theta \sin\Theta \, d\Theta = \langle \cos\Theta \rangle = g \tag{1.3.6}$$

The anisotropy factor can have values from -1 to $+1$. Backward scattering is described by negative values, forward scattering by positive values, and isotropic scattering by an anisotropy factor of 0 (Figure 1.3.4).

There is no analytical solution for the radiation transport equation, if it is used for complex problems such as the light distribution in biological tissue (Ishimaru 1978). Different approaches were developed such as the diffusion theory (see also Section 1.3.2), and the Kubelka–Munk approximation (Kubelka 1948).

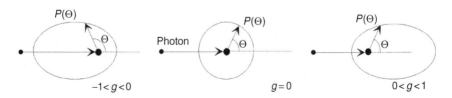

FIGURE 1.3.4 Anisotropy factor g.

What to Measure and How to Measure

However, these theories can only be used under special boundary conditions such as isotropic scattering and weak absorption compared to the scattering (Farell et al. 1992). Different numerical models were developed to solve the radiation transport equation, such as successive-scattering method (Irvine 1964) or the adding-doubling method (van de Hulst 1980), but for complex geometries and boundary conditions they cannot give exact results. For small samples, the time-resolved diffuse optical spectroscopy (Taroni et al. 2007) can be used to calculate optical parameters as μ_a and reduced scattering coefficient $\mu_s' = \mu_s(1-g)$ (see Section 1.3.1.4). Spatially resolved reflectance measurements (see Section 5.2) can also be used to determine these optical parameters noninvasively (Kienle et al. 1996).

1.3.1.3 Kubelka–Munk Theory

The Kubelka–Munk theory can be used to calculate the first set of μ_a, μ_s, and g in the numerical, inverse Monte Carlo simulation (iMCS) and it will be introduced at this point. The theory is a very simplified solution of the radiation transport theory under restricted boundary conditions and represents a special case of the diffusion theory (van Gemert and Star 1987; Graaf et al. 1989). The theoretical approach of this solution is to separate the radiation density $L(\vec{r},\vec{s})$ into diffuse forward and backward flows. The optical properties μ_a, μ_s, and g can be calculated by considering these boundary conditions. The main problem with applying the Kubelka–Munk theory is that the boundary conditions do not normally reflect the real situation. The theory is only applicable to samples when $\mu_a \ll \mu_s$, usually only valid for near-infrared (NIR) spectroscopy and isotropic scattering, which is not the case for biological tissue. Furthermore, the illumination of the sample has to be diffuse and the sample should have plane parallel geometry with infinite lateral dimensions without refractive index changes. Therefore, the application of this theory to real problems is mostly associated with major errors, but is helpful for quick estimations such as the determination of start parameters for iMCSs.

1.3.1.4 Interesting Values Out of the Diffusion Theory

It could be shown that if there is no source term, and interfaces are far away, different data sets of μ_s and g could give similar solutions of the diffusion equation (van de Hulst 1980; Star et al. 1988; Wymann et al. 1989) if $\mu_s' = \mu_s(1-g)$ have always the same value (Groenhuis et al. 1983). The μ_s' is the reduced scattering coefficient. If, for example, g is close to 1 (strong forward scattering) and the number of scattering events is high (high scattering coefficient), the comparable scattering effect (μ_s') can be found as if g is 0 (isotropic scattering) and the number of scattering events (μ_s) is much lower (low μ_s).

The penetration depth of the light into the tissue depends on the optical parameters of the tissue. Using the diffusion theory ($\mu_s > \mu_a$), the effective attenuation coefficient (μ_{eff}) can be calculated to

$$\mu_{\text{eff}} = \sqrt{3\mu_a[\mu_a + \mu_s(1-g)]} \tag{1.3.7}$$

which is inversely correlated with the penetration depth (δ_{eff}):

$$\delta_{eff} = \frac{1}{\mu_{eff}} \quad (1.3.8)$$

$$\delta_{eff} = \frac{1}{\sqrt{3\mu_a[\mu_a + \mu_s(1-g)]}} \quad (1.3.9)$$

1.3.1.5 Monte Carlo Simulation

The most flexible and exact method to calculate the light distribution in scattering and absorbing media with complex geometry and complex boundary conditions is the Monte Carlo simulation (MCS), which can also be used for agriculture products. The MCS is an indirect method, which uses the approaches of the radiation transfer equation and the optical properties μ_a, μ_s, and the phase function for simulating the light distribution through the tissue. It is a numerical method based on the probability of the interaction of the medium and the light and was first used in research on the atmosphere (Metropolis and Ulam 1949). The transfer to biological samples was done by Wilson (Wilson and Adam 1983) and used by several other authors (Groenhuis et al. 1983; Prahl 1988; Keijzer et al. 1989; Gardner and Welch 1992; Kienle and Steiner 1994). The principle of the MCS is a computer simulation of injecting consecutively a number of photons into the medium. The medium can be characterized by probability density functions of the optical properties described by the parameters μ_a, μ_s, $p(\vec{s},\vec{s}')$, or g that are identical to those used in the radiation transport equation (Equation 1.3.3) (Patterson et al. 1991).

The parameters of the simulation such as starting position of the photon, initial direction, free pathlength between the interactions (scattering, absorption), scattering angle, and the probability of reflection or transmission at an interface are calculated using random numbers given by the probability density functions. The randomly generated pathways of the individual photon are calculated until it is absorbed or leaves the sample (Figure 1.3.5).

1.3.1.6 Principle of Inverse Monte Carlo Simulation (iMCS)

The iMCS uses forward Monte Carlo simulations iteratively to calculate optical parameters, for example, μ_a, μ_s, and g, on the basis of measured values such as diffuse reflectance, R_d, total transmission, T_t, and diffuse transmission, T_d, for a selected phase function (Roggan et al. 1999). The reflectance and transmission spectra R_d, T_t, and T_d can be measured using an integrating sphere spectrophotometer as shown in Figure 1.3.6. For the transmission measurements, the sample is positioned in front of the sphere. The total transmitted light of the sample is detected if the backward aperture of the sphere is closed. If the aperture is open, the collimated photons can leave the sphere undetected, and the diffuse transmission is measured. For the measurement of the diffuse reflectance, the sample is positioned behind the sphere and the backscattered light is measured.

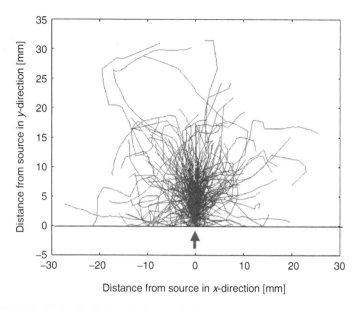

FIGURE 1.3.5 Light distribution in tissue calculated by Monte Carlo simulation.

The iMCS, shown schematically in Figure 1.3.7, uses an estimated set of start parameters μ_a, μ_s, and g from the Kubelka–Munk theory to calculate the resulting values R_d, T_t, and T_d. These are then compared to the R_d, T_t, and T_d values that have been measured experimentally. In the case of a significant deviation, all three parameters are varied slightly and a new forward simulation is carried out. After

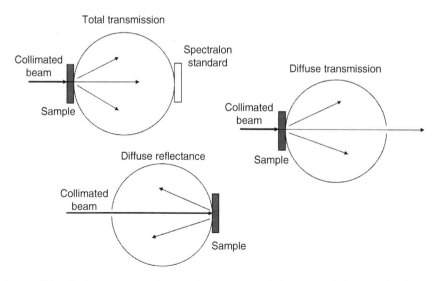

FIGURE 1.3.6 Measurement of the total transmission, diffuse transmission, and the diffuse reflectance using an integrating sphere.

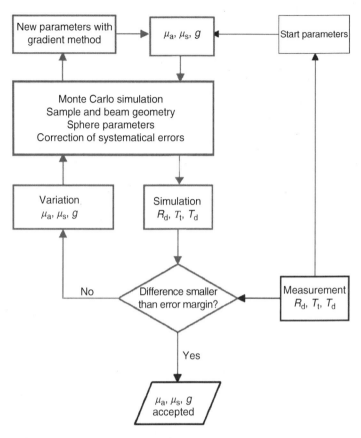

FIGURE 1.3.7 iMCS was carried out using the Reynolds–McCormick phase function Friebel et al. 2006.

this procedure a gradient matrix is built up, allowing the calculation of a new set of optical parameters, which makes a better fit to the measured quantities. This procedure is repeated until the deviation between measured and calculated R_d, T_t, and T_d values is below the given error threshold.

1.3.1.7 Experimental Setup for Measuring a Horticultural Product

The macroscopic optical parameters, diffuse reflectance (R_d), the total transmission (T_t), and the diffuse transmission ($T_d = T_t -$ transmission within an aperture of 5.3°) of different carrot samples were measured every 5 nm in the spectral range from 250 to 1100 nm using an integrating sphere spectrophotometer (Lambda 900, Perkin Elmer, United States), which is a two-beam scanning spectrophotometer with double monochromator system (Figure 1.3.8; Meinke et al. 2005).

The light source consists of a deuterium lamp for the UV-range and a wolfram halogen lamp for the visible/NIR range. The light intensity was adapted to the absorption behavior of the sample to maintain the intensity of the signal in the optimal range of the detectors. The cuvette can be fixed in a defined position at

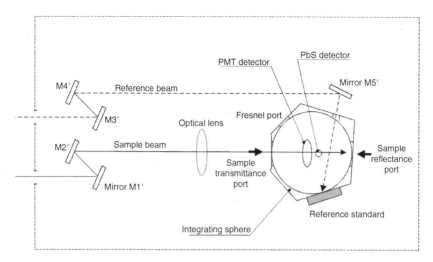

FIGURE 1.3.8 Schematic diagram of the integrating sphere spectrophotometer. The reference beam goes via mirror M3–M5 into the integrating sphere. The sample beam reaches the sample via mirror M1 and M2.

a constant distance to the sphere aperture, in front of or behind the integrating sphere, to measure the transmittance or reflectance spectra. For the measurement of T_t, the reflectance port is closed with a diffuse reflecting Spectralon standard (Labsphere, United States). The T_d is measured after the standard is removed so that non-scattered and forward scattered transmitted light leaves the sphere within an angle of 5.3°. The R_d is measured relative to the reflectance standard by replacing the Spectralon standard with a certified, known reflectance spectrum by the sample that is inclined at an angle of 8° to the incoming light. The Fresnel reflectance of the cuvette glass leaves the sphere through the open Fresnel port to avoid superposition with the diffuse reflectance. This experimental setup allows the measurement of macroscopic radiation distribution with an extremely reduced error potential. The accuracy of repetitive measurements of the reflectance and transmission measurements of a sample phantom with defined optical properties (e.g., polystyrene spheres) was smaller than 0.01%.

For the iMCS an error threshold of 0.15% was used for the simulation of the intrinsic optical parameters of a carrot.

1.3.1.8 Spectra of Carrots

In Figure 1.3.9, upper panel, the spectra of the diffuse reflectance (R_d), the total transmission (T_t), and the diffuse transmission (T_d) of a section of carrot xylem sample are shown. The section has a sample thickness of 1400 μm. To calculate the optical properties from these spectra, all three measurement data must be different. The collimated transmission can be calculated by $T_t - T_d$ that should be higher than zero, and is presented in Figure 1.3.9, lower panel.

The absorption band can be estimated from the transmission spectra in the region between 400 and 600 nm, where the transmission is reduced to 20%. The question is,

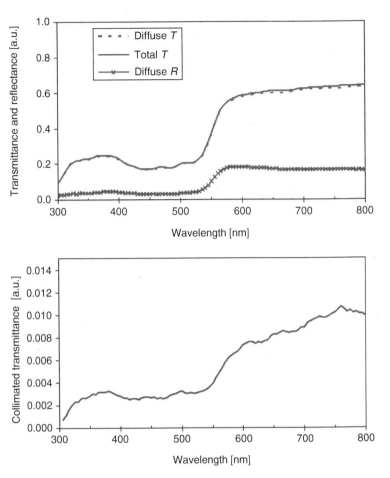

FIGURE 1.3.9 Total transmission, diffuse transmission, collimated transmittance, and diffuse reflectance of a carrot section of 1.4 mm.

whether the loss in transmitted light is evoked by absorption or by scattering? The same question can be asked for the transmittance at 700 nm. From the backscattered signal, we observe that above 550 nm the scattering signal increases. This could be the reason for the loss in transmission of about 40% at 700 nm. An answer to this question can only be given by assessing the optical properties of the carrot sample.

1.3.1.9 Optical Properties of Carrots

Figure 1.3.10 shows the optical properties of a carrot xylem calculated from the spectra in Section 1.3.1.5 using the Monte Carlo simulation. The upper panel shows the absorption coefficient and the reduced scattering coefficient.

The absorption coefficient of $\mu_a = 0.5$ mm^{-1} at 450 nm shows that a photon, which is injected into the carrot sample, undergoes on average one absorption

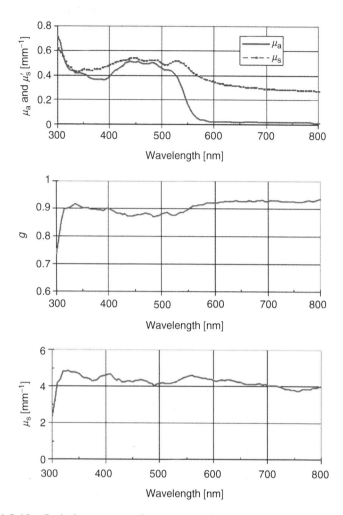

FIGURE 1.3.10 Optical parameters of a carrot sample.

process in 2 mm. The absorption bands of β-carotene and lycopene appear at 420, 480, and 520 nm. The absorption of the carrot matrix is close to zero above 550 nm where scattering is predominant. On average, μ'_s is of the same dimension as μ_a indicating that diffusion and the Kubelka–Munk theory are not applicable in this case. Furthermore, the tissue has a strong forward scattering with a mean anisotropy factor of $g = 0.9$. In principle, the probability of a scattering event (4 mm^{-1}) is higher than the probability of an absorption event, but because of the low scattering angle, the reduced scattering is on average less than a half an event per millimeter. From investigations on blood (Friebel et al. 2006), it is known that the scattering parameters are affected by the absorption, which can be seen in the region of the absorption bands. This should be kept in mind when very simple approaches are used where the scattering is corrected by a linear function in biological tissue.

REFERENCES

Chandrasekhar, S. 1950. *Radiative Transfer*. Oxford University Press, London, United Kingdom.
Farell, T.J., B.C. Wilson, and M.S. Patterson. 1992. A diffusion theory model of spatially resolved, steady-state diffuse reflectance for the non invasive determination of tissue optical properties. *Medical Physics* 19: 879–888.
Forster, F.K., A. Kienle, R. Michels, and R. Hibst. 2006. Phase function measurements on nonspherical scatterers using a two-axis goniometer. *Journal of Biomedical Optics* 11: 024018, 9 pages.
Friebel, M., A. Roggan, G. Müller, and M. Meinke. 2006. Determination of optical properties of human blood in the spectral range 250–1100 nm using Monte Carlo simulations with hematocrit-dependent effective scattering phase functions. *Journal of Biomedical Optics* 11: 031021, 10 pages.
Gardner, G.M. and A.J. Welch. 1992. Improvements in the accuracy and statistical variance of the Monte Carlo simulation of light distribution in tissue. *Proceeding SPIE 1646*, 400–409.
Graaf, R., F. Aamoudse, F.F.M. de Mul, and H.W. Jentink. 1989. Light propagation parameters for anisotropically scattering media based on a rigorous solution of the transport equation. *Applied Optics* 28: 2273–2279.
Groenhuis, R.A.J., H.A. Ferwerda, and J.J. ten Bosch. 1983. Scattering and absorption of turbid material determined from reflection measurements. *Applied Optics* 22: 2456–2462.
Irvine, W.M. 1964. The formation of absorption bands and the distribution of photon optical paths in a scattering atmosphere. *Bulletin of the Astronomical Institute of the Netherlands*, 17: 266.
Ishimaru, A. 1978. *Wave Propagation and Scattering in Random Media*. Academic Press, New York, Vols. 1 and 2.
Keijzer, M., S.L. Jacques, S. Prahl, and A.J. Welch. 1989. Light distribution in artery tissue: Monte Carlo simulation for finite-diameter laser beams. *Lasers in Surgery and Medicine* 9: 148–154.
Kerker, M. 1969. *Scattering of Light and Other Electromagnetic Radiation*. Academic Press, New York.
Kienle, A. and R. Steiner. 1994. Determination of the optical properties of tissue by spatially resolved transmission measurements and Monte Carlo calculation. *Proceeding SPIE 2077*, 142–152.
Kienle, A., L. Lilge, M.S. Patterson, R. Hibst, R. Steine, and B.C. Wilson. 1996. Spatially resolved absolute diffuse reflectance measurements for noninvasive determination of the optical scattering and absorption coefficients of biological tissue. *Applied Optics* 35: 2304–2314.
Kubelka, P. 1948. New contributions to the optics of intensely light scattering materials, Part I. *Journal of Optical Society of America* 38: 448.
Meinke, M., I. Gersonde, M. Friebel, J. Helfmann, and G. Müller. 2005. Chemometric determination of blood parameters using VIS–NIR-spectra. *Applied Spectroscopy* 59: 826–835.
Metropolis, N. and N. Ulam. 1949. The Monte Carlo method. *Journal of the American Statistical Association* 44: 335–341.
Patterson, M.S., B.C. Wilson, and D.R. Wyman. 1991. The propagation of optical radiation in tissue 1. Models of radiation transport and their applications. *Laser in Medical Science* 6: 155.
Prahl, S. 1988. *Light Transport in Tissue*. PhD Dissertation, University of Texas at Austin, TX.
Roggan, A., M. Friebel, K. Dörschel, A. Hahn, and G. Müller. 1999. Optical properties of circulating human blood in the wavelength range 400–2500 nm. *Journal of Biomedical Optics* 4: 36–46.
Star, M.W., J.P.A. Marijnissen, and M.J.C. van Gemert. 1988. Light dosimetry in optical phantoms and in tissues: 1. Multiple flux and transport theory. *Physics in Medicine and Biology* 33: 437–454.

Taroni, P., D. Comelli, A. Farina, A. Pifferi, and A. Kienle. 2007. Time-resolved diffuse optical spectroscopy of small tissue samples. *Optics Express* 15: 3302–3311.
van de Hulst, H.C. 1957. *Light Scattering by Small Particles*. John Wiley & Sons, New York.
van de Hulst, H.C. 1980. *Multiple Light Scattering*. Academic Press, New York, Vols. 1 and 2.
van Gemert, M.J.C. and W.M. Star. 1987. Relations between the Kubelka–Munk and the transport equation models for anisotropic scattering. *Lasers in the Life Sciences* 1: 287–298.
Wilson, B.C. and G. Adam. 1983. A Monte Carlo model for the absorption and flux distribution of light in tissue. *Medical Physics* 10: 824–830.
Wymann, D.R., M.S. Patterson, and B.C. Wilson. 1989. Similarity relation for the interaction parameters in radiation transport. *Applied Optics* 28: 5243–5249.

1.3.2 DETERMINATION OF OPTICAL PROPERTIES IN TURBID MEDIA: TIME-RESOLVED APPROACH

ALESSANDRO TORRICELLI

1.3.2.1 Light Propagation in Diffusive Media: Photon Migration

The term "turbid" or "diffusive" medium refers to many substances, which naturally possess the characteristic of being opaque to visible light (paints, powders, clouds, foams, and biological tissues). Light absorption and light scattering are the natural phenomena responsible for opacity. Absorption depends on the presence of endogenous or exogenous chromophores within the medium, whereas (elastic) scattering depends on microscopic discontinuities in the dielectric properties of the medium.

Suppose to inject a short pulse of monochromatic light within a diffusive medium. By using a simplified description, we can consider the medium as made up of scattering centers and absorbing centers, and the light pulse as a stream of particles, called photons, move ballistically within the medium. Whenever a photon strikes a scattering center, it changes its trajectory and keeps on propagating in the medium, until it is eventually remitted across the boundary, or it is definitely captured by an absorbing center.

The characteristic parameters of light propagation within the diffusive medium are the scattering length, L_s, and the absorption length, L_a, typically expressed in units of millimeter or centimeter, representing the photon mean free pathlength between successive scattering events and absorption events, respectively. Equivalently, and more frequently, the scattering coefficient, $\mu_s = 1/L_s$, and the absorption coefficient, $\mu_a = 1/L_a$, typically expressed in units of per millimeter or per centimeter, can be introduced to indicate the scattering probability per unit pathlength and the absorption probability per unit pathlength, respectively.

In a diffusive medium in the visible and near-infrared spectral regions, light scattering is naturally stronger than light absorption, even if the latter can be non-negligible. This implies that light can be scattered many times before being either absorbed or remitted from the medium. The phenomenon is therefore called multiple scattering of light. Multiple scattering of light in a diffusive medium introduces an uncertainty in the pathlength traveled by photons in the medium. Light propagation in turbid medium is termed photon migration (Ishimaru 1978; Chance 1989).

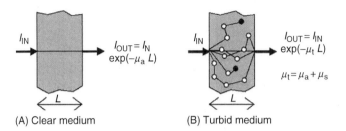

FIGURE 1.3.11 Photon paths (A) in a clear medium and (B) in a turbid medium.

When considering conventional absorption spectroscopy measurements in a collimated geometry (Figure 1.3.11), results may be impeded by the difficulty in discriminating absorption and scattering events. The transmitted intensity through a clear medium can be related by the Beer–Lambert law to the absorption coefficient, μ_a, since the distance traveled by light in the medium equals the source–detector distance, L. Conversely, in a turbid medium, an intensity measurement yields the attenuation coefficient, $\mu_t = \mu_a + \mu_s$, representing the photon loss because of absorption and of photons scattered into different directions than the one of observation.

Moreover, imaging through a scattering medium suffers decreased contrast and spatial resolution because of the blurring introduced by the spread of photon paths in the medium.

The effect of multiple light scattering can be properly taken into account by direct measurement of the photon pathlength. Since photon pathlength is directly related to the time-of-flight in the medium, the natural choice is to perform time-resolved measurements (Chance et al. 1988; Delpy et al. 1988; Jacques 1989a; Wilson et al. 1992).

Because of the many scattering events, photons lose memory of their initial directions; therefore, it is possible to observe remitted light also in the back-scattered geometry. Depending on the geometry of the sample, we can have transmittance owing to reflectance of photons. According to Figure 1.3.12, we refer to time-resolved reflectance if the temporal distribution of remitted photons is collected on the same surface as the injected light, while the collection of the temporal distribution of photons on the opposite side of the medium will be referred to as a time-resolved

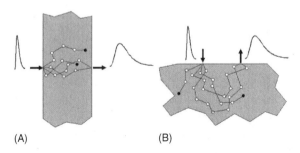

FIGURE 1.3.12 Photon migration: (A) photons paths, scattering and absorbing centers for time-resolved transmittance and (B) diffuse backward reflected photons (remittance).

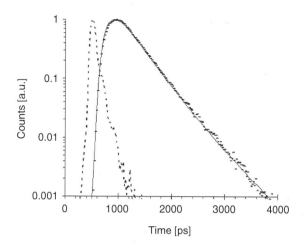

FIGURE 1.3.13 Photon migration: injected and remitted pulse.

transmittance measurement (Jacques 1989b; Patterson et al. 1989; Wilson and Jacques 1990; Cubeddu et al. 1994).

Following the injection of the light pulse within a turbid medium, the temporal distribution of the remitted photons at a distance ρ from the injection point will be delayed, broadened, and attenuated. A typical time-resolved reflectance curve is shown in Figure 1.3.13. As a first approximation, the delay is a consequence of the finite time light takes to travel the distance between source and detector. Broadening is mainly due to the many different paths that photons undergo because of multiple scattering. Finally, attenuation appears because absorption reduces the probability of detecting a photon, and diffusion into other directions within the medium decreases the number of detected photons in the considered direction.

Figure 1.3.14 depicts the effects of changes in source–detector distance (Figure 1.3.14A and B), scattering (Figure 1.3.14C and D), and absorption (Figure 1.3.14E and F) on time-resolved reflectance curves. Increasing the source–detector distance yields an increased delay in the temporal distribution of detected photons and a decrease of their number. Similar behavior is observed when the scattering increases. Finally, absorption affects both the signal intensity and the slope of the tail of the curves, while leaving the temporal position of the curves substantially unchanged.

Whenever the geometrical parameters are known, a rough estimate of the scattering and absorbing characteristics of the medium can be obtained from the evaluation of the peak position and of the slope of the tail, respectively. More quantitative and accurate estimates can be obtained by interpreting the curves with a proper theoretical model (see Section 1.3.2.2).

As shown here above, time-resolved measurements enable the noninvasive optical characterization of diffusive media by assessing the absorbing and the scattering contributions. It is interesting to note that in principle the investigation of diffusive media by means of the analysis of the remitted light can be also accomplished through different modalities such as frequency domain and continuous wave (CW) in the space domain or angle domain.

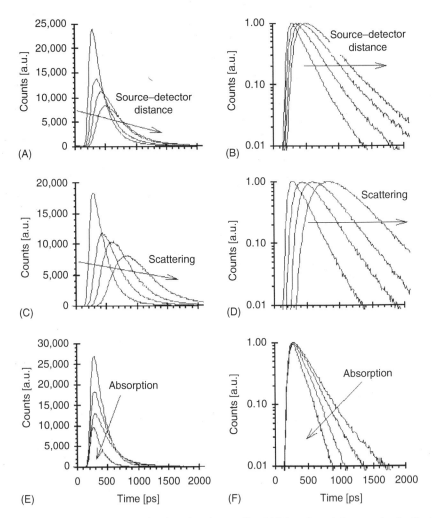

FIGURE 1.3.14 Absolute (left column) and normalized (right column) time-resolved reflectance curves as a function of ρ, μ_s', and μ_a. (A) and (B): $\mu_a = 0.05$ cm^{-1}, $\mu_s' = 20$ cm^{-1}, $\rho = 0.5$, 1.0, 1.5, and 2.0 cm; (C) and (D): $\mu_a = 0.09$ cm^{-1}, $\mu_s' = 4$, 8, 16, and 32 cm^{-1}, $\rho = 1.0$ cm; (E) and (F): $\mu_a = 0.04, 0.08, 0.16$, and 0.32 cm^{-1}, $\mu_s' = 20$ cm^{-1}, $\rho = 1.0$ cm.

Photon migration in the frequency domain (Lakowicz and Berndt 1990; Patterson et al. 1991; Pogue and Patterson 1994) is the natural alternative to the time-domain approach. An intensity modulated light wave is injected into the medium, and the remitted wave is detected (Figure 1.3.15). The optical properties of the traversed medium can be recovered by the analysis of the phase shift and of the demodulation signals.

The frequency-domain approach to photon migration also introduces the idea of diffuse photon density wave (DPDW) (Svaasand et al. 1993; Tromberg et al. 1993). In fact, amplitude-modulated light is dispersed through homogeneous

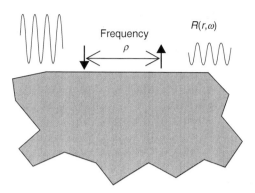

FIGURE 1.3.15 Photon migration in the frequency domain.

multiple-scattering media as diffuse waves with a coherent front. Microscopically individual photons undergo a random walk within the medium, but collectively, a spherical wave of photon density is produced and propagates outward from the source. DPDWs are scalar, damped, traveling wave, and they have been observed to exhibit properties one normally associates with conventional electromagnetic radiation such as refraction, diffraction, interference, and dispersion. Nonetheless, DPDWs bear no relationship to corresponding electromagnetic wave behavior in a turbid media, as for example, phase relationships between optical waves in a diffusive medium vary in a rapid stochastic manner.

Continuous-wave optical methods have been used for a long time to investigate natural and biological media, thanks to the availability of low-cost, easy-to-use light sources with different spectral and power characteristics. However, the main disadvantage of CW methods, when applied to diffusive media, is the impossibility of discriminating between the scattering and the absorbing contributions of the diffusive medium. Measurements of the total intensity attenuation are in fact affected by both the absorbing and the scattering properties of the investigated sample. It is true that, at least in principle, these properties could also be assessed by CW methods, if one is able to reduce the measurement procedures to the basic definition of the parameters to be determined. Nonetheless, this always requires the manipulation of the sample to produce a particular geometry, that is, typically thin sections, and get rid of the multiple scattering (see Section 1.3.1). In many cases, this is not possible and, whenever it is possible, great care has to be taken to avoid artifacts in the measured optical properties that result from the sample preparation procedures. In recent years, CW optical methods have thus evolved to be more complicated configurations such as noninvasively determining the optical parameters of a diffusive medium. The CW space-resolved (Farrell et al. 1997) or CW angle-resolved (Wang and Jacques 1995) configurations are shown in Figure 1.3.16. Remitted photons are collected simultaneously at different source–detector distances or at different angles with respect to the surface of the medium. In this way, it is possible to discriminate photon paths, thus yielding the absorption and the scattering contribution (see also Section 5.2).

Up to now, it is not clear which method is the best. In principle, time-resolved data should have the highest information content, as compared to frequency-resolved

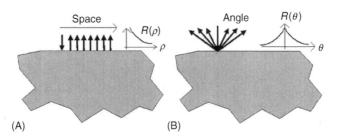

FIGURE 1.3.16 CW photon migration: (A) space-resolved and (B) angle-resolved.

or CW data. In fact in the time-domain approach, the pathlength is determined for each detected photon, as opposed to ensemble averages, which are determined by frequency-domain approaches. In the time-domain approach, a delta-like pulse is injected in the medium, and this corresponds to an infinite series of modulation frequencies, not a single one, as usually performed by the frequency-domain approaches. Moreover, the idea of DPDW has introduced many convenient analytical expressions, which can be used also to interpret time-resolved data, after a proper Fourier transformation.

Depending on the geometry (reflectance/transmittance, semi-infinite/slab) and the technique (space, time, frequency domains), theoretical models and laboratory instrumentation will be notably different. Therefore, accuracy, precision, and robustness of measurements will be characteristic of the particular system used. In the following, we will always refer to photon migration in the time domain, either in reflectance geometry from a semi-infinite medium or in transmittance geometry measuring remitted photons from an infinite slab. We will now discuss the various mathematical theories and time-resolved instrumentation for photon migration studies.

1.3.2.2 Mathematics for Photon Migration

The problem of light propagation in diffusive media has become increasingly important in recent years, particularly in the area of communications, remote sensing and detection, biology, and medicine. Natural media (rain, fog, smog, hail, aerosols, ocean particles, and polymers) and biological media (blood, tissues) can in fact be described as random distributions of many particles or random continua, whose characteristics vary randomly in space and time. Their main feature is the turbid appearance, not transparent at visible wavelengths. This is due to the strength of the phenomenon of light scattering rather than light absorption.

Historically, two distinct theories have been developed in dealing with wave or pulse propagation in random media: electromagnetic theory (Ishimaru 1978; Van de Hulst 1980) and radiative transport theory (Ishimaru 1978; Duderstadt and Martin 1979).

Electromagnetic theory starts with basic differential equations (Maxwell equations, or wave equations), obtains solutions for a single particle, introduces the interaction effects of many particles, and then considers the statistical averages. Electromagnetic theory is mathematically rigorous; however, in practice, it is impossible to obtain a formulation, which completely includes all the physical effects (multiple scattering, diffraction, and interference). Various theories, which yield advantageous solutions, are all approximate, each being useful in a specific range of the parameters. The most important parameter is the density of particles: when the density of particles is tenuous (single scattering regime), electromagnetic theory formulation is quite simple and has been used extensively in many applications including weather radar or ocean acoustics. As the particle density is increased (dense or turbid media, multiple-scattering regime), the electromagnetic theory gives no readily applied solutions because of the complexity of the formulation and of the lack of knowledge on the random properties of media.

The development of radiative transport theory is heuristic and lacks the rigor of electromagnetic theory; however, the formulation is flexible and capable of treating many physical phenomena. It has been successfully employed for the problems of atmospheric and underwater visibility, and very recently applied to biomedical optics. The basic assumption in radiative transport theory is that we can ignore the wave nature of light and simply consider the flow of energy associated with the propagation of photons (photon migration) in the medium.

1.3.2.2.1 Time-Dependent Radiative Transport Equation
The time-dependent radiative transport equation (RTE) can be derived by considering the radiant energy balance in an arbitrary elemental volume:

$$\frac{1}{v}\frac{dL(\vec{r},\vec{s},t)}{dt} = \vec{s}\cdot\nabla L(\vec{r},\vec{s},t) - (\mu_s + \mu_a)L(\vec{r},\vec{s},t) + \mu_s \int_{4\pi} p(\vec{s},\vec{s}')L(\vec{r},\vec{s}',t)d\vec{\Omega}' + S(\vec{r},\vec{s},t)$$

(1.3.10)

We observe the following:

- $L(\vec{r},\vec{s},t)$ is the radiance (the energy per unit time, per unit area, per unit solid angle) with units [W m^{-2} sr^{-1}] at position $\vec{r}$, traveling in direction $\vec{s}$ at time t.
- μ_s [m^{-1}] is the scattering coefficient, that is, the probability of photon scatter per unit pathlength.
- μ_a [m^{-1}] is the absorption coefficient, that is, the probability of photon absorption per unit pathlength.
- $\mu_t = \mu_a + \mu_s$ is the transport coefficient or attenuation coefficient, that is, $1/\mu_t$ is the mean free path in the medium.
- $\vec{s}\cdot\nabla L(\vec{r},\vec{s},t)$ represents the net change owing to energy flow.
- $\mu_a L(\vec{r},\vec{s},t)$ represents radiance loss owing to absorption.
- $\mu_s L(\vec{r},\vec{s},t)$ represents radiance loss owing to scattering.
- Integral term represents the gain in radiance owing to scattering from all other directions, and $p(\vec{s},\vec{s}')$ is the scattering phase function, which defines the probability of a photon moving in the direction $\vec{s}$ to be deflected in the direction $\vec{s}'$.
- $S(\vec{r},\vec{s},t)$ represents the spatial and angular distribution of the light source [W m^{-3} sr^{-1}].

Exact solutions of the RTE have been obtained only for a limited few cases. For a tenuous distribution of scatterers, the radiative transport theory in the first-order multiple-scattering approximation gives solutions equivalent to the electromagnetic theory. For a dense distribution of scatterers, the diffusion approximation (Ishimaru 1978; Arridge 1999) has been commonly used both in the time domain or frequency domain and in the steady-state regime, since analytical results are available for simple geometry (infinite medium, semi-infinite medium, infinite slab, sphere, and cylinder) and the numerical solution (finite difference, finite element method) is straightforward (Arridge et al. 1993).

For improved accuracy, numerical solutions of the RTE have been obtained with many different methods in recent years, for example, the Monte Carlo method (see Section 1.3.1 and references therein) has been extensively used. Nevertheless, since in practice neither of these approaches yields closed-form expressions that can be easily used to manipulate experimental data, the diffusion approximation is the most widely adopted to describe photon migration in turbid media.

1.3.2.2.2 Diffusion Approximation

Mathematically, the diffusion equation can be derived from the RTE by first expanding the radiance $L(\vec{r},\vec{s},t)$ and the source term $S(\vec{r},\vec{s},t)$ into spherical harmonics $Y_{l,m}$, truncating the expansion to the N term (P_N approximation), and retaining only the lowest orders ($l = 0, m = 0$ and $l = 1, m = -1, 0, 1$) (Kaltenbach and Kaschke 1993).

$$L(\vec{r},\vec{s},t) = \frac{1}{4\pi}\Phi(\vec{r},t) + \frac{3}{4\pi}\vec{J}(\vec{r},t) \cdot \vec{s} \tag{1.3.11}$$

$$S(\vec{r},\vec{s},t) \cong \frac{1}{4\pi}S_0(\vec{r},t) + \frac{3}{4\pi}S_1(\vec{r},t) \cdot \vec{s} \tag{1.3.12}$$

From a physical point of view, this is equivalent to the linear anisotropy approximation, which considers the angular distribution of the radiance as almost isotropic because of the many scattering events in the medium. Thus, the radiance $L(\vec{r},\vec{s},t)$ can be expressed as an isotropic fluence rate $\Phi(\vec{r},t)$, that is, the energy per unit time, unit area [W m^{-2}], plus a small-directional flux vector $\vec{J}(\vec{r},t)$ [W m^{-2}]. The definitions of $\Phi(\vec{r},t)$ and $\vec{J}(\vec{r},t)$ are reported below:

$$\Phi(\vec{r},t) = \int_{4\pi} L(\vec{r},\vec{s},t)\, d\vec{s} \tag{1.3.13}$$

$$\vec{J}(\vec{r},t) = \int_{4\pi} L(\vec{r},\vec{s},t)\vec{s}\, d\vec{s} \tag{1.3.14}$$

Substituting Equations 1.3.13 and 1.3.14 in the RTE, with the further assumptions that

- Scattering phase function $p(\vec{s},\vec{s}')$ depends only on the scalar product $\vec{s} \cdot \vec{s}'$, and moreover, $p(\vec{s},\vec{s}') \approx 1 + g$, where $g = \cos(\vec{s} \cdot \vec{s}')$ is the anisotropy factor (Henyey and Greenstein 1941)

What to Measure and How to Measure

- Source is almost isotropic
- Change in the flux vector in the time interval necessary for the light to travel along one mean free path is small compared with the flux vector itself

The time-dependent diffusion equation for a homogeneous medium is obtained

$$\frac{1}{v}\frac{d\Phi(\vec{r},t)}{dt} - D\nabla^2\Phi(\vec{r},t) + \mu_a\Phi(\vec{r},t) = S_0(\vec{r},t) \quad (1.3.15)$$

where
$D = 1/(3\mu_s')$ is the diffusion coefficient
$\mu_s' = \mu_s(1 - g)$ is the reduced scattering coefficient (or transport scattering coefficient)
$S_0(\vec{r},t)$ is the isotropic light source term

With the same assumptions, the quantities $\Phi(\vec{r},t)$ and $\vec{J}(\vec{r},t)$ also satisfy the equation (Fick's law)

$$\vec{J}(\vec{r},t) = -D\nabla\Phi(\vec{r},t) \quad (1.3.16)$$

Note that D is independent of the absorption coefficient as was recently noted (Furutsu and Yamada 1994).

In practical applications, the photon fluence rate $\Phi(\vec{r},t)$ is not measured directly. It is in fact important that these techniques be noninvasive if they are to be clinically useful. Therefore, the optical fibers delivering the laser light must be placed on the surface of the tissue. The measurable quantity is therefore the number of photons that reach the surface per unit area, per unit time, at a given source–detector distance, calculated as the absolute value of the flux vector at the surface boundary, as depicted in Figure 1.3.17:

$$R(r,t) = |\vec{J}(r,z,t)|_{z=0} \quad (1.3.17)$$

$$T(r,d,t) = |\vec{J}(r,d,t)|_{z=d} \quad (1.3.18)$$

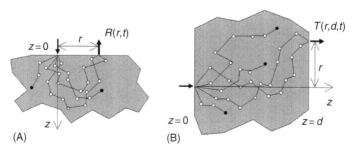

FIGURE 1.3.17 Measurement geometry: (A) for reflectance from a semi-infinite medium and (B) for transmittance from an infinite slab.

Analytical solutions of the time-dependent diffusion equation have been written for the case of an infinitesimally narrow pulse $\delta(\vec{r},t)$ incident on a homogeneous semi-infinite medium or a homogeneous slab, and for cylindrical and spherical geometry (Patterson et al. 1989; Arridge et al. 1992). The finite element method has been used to tackle more complex geometry and inhomogeneous media (Arridge et al. 1993).

In all cases, since the presence of a boundary is inevitable, a great deal of attention has been devoted to the problem of the influence of boundary conditions on the accuracy of diffusion approximation (Haskell et al. 1994; Aronson 1995; Hielscher et al. 1995; Contini et al. 1997).

For a diffusing medium bounded by either a totally absorbing or a transparent surface (Σ), the RTE requires that the light leaving the medium through the surface Σ does not return into the medium, that is, the radiance $L(\vec{r},\vec{s},t)$ is zero for every $\vec{r}$ on Σ and for every direction $\vec{s}$ pointing inward. This condition cannot be satisfied with the simple angular distribution assumed for $L(\vec{r},\vec{s},t)$ by the linear anisotropy approximation. Therefore, an approximate condition is considered, which sets to zero the total inwardly directed diffuse flux. Moreover, since a refractive index mismatch occurs at the boundary in practical applications, a more exact boundary condition should take into account that a significant fraction of the radiant energy incident on the boundary from inside will be reflected back into the medium. The total diffuse inwardly directed flux is not zero but it is the part of the outwardly directed flux reflected at the interface.

This condition is usually referred to as the partial current boundary (PCB) condition. It can be equivalently expressed in terms of the fluence rate and its normal derivative calculated at the surface Σ:

$$0 = (1 - 2QDn \cdot \nabla)\Phi(\vec{r},t)|_{\text{boundary}} \qquad (1.3.19)$$

where Q depends on the Fresnel reflection coefficient for unpolarized light that is dependent on the relative refractive index n (ratio of the refractive index of the diffusive medium to that of the surrounding medium).

The diffusion equation with PCB can be solved analytically, but it has been used very little in practical applications. The two most commonly applied boundary conditions are in fact the extrapolated boundary condition (EBC), and the zero boundary condition (ZBC). The EBC assumes the fluence rate to be equal to zero on an extrapolated flat surface at a distance $z_e = 2QD$ from the actual geometrical boundary of the medium depending on the refractive index mismatch. The ZBC sets the fluence rate as equal to zero at the actual geometrical boundary and the eventual refractive index mismatch is not considered.

The method of images can be used to construct fluence rate solutions that satisfy the boundary conditions PCB, EBC, and ZBC, respectively. Figure 1.3.18 shows the source and image configurations for the semi-infinite geometry.

Although EBC and ZBC are somehow unphysical, they are mathematically simple and some studies have shown that they give sufficiently good approximation for many circumstances of practical interest in biomedical applications. Comparison with Monte Carlo simulations (Hielscher et al. 1995) for a semi-infinite geometry has in fact shown that all three approaches give similar results when the source–detector distance is larger than 3 cm or $\mu'_s/\mu_a > 50$. Comparison with Monte Carlo simulations for an

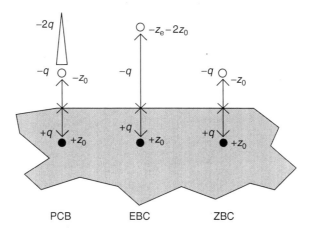

FIGURE 1.3.18 PCB, EBC, and ZBC: source and image configurations for the semi-infinite geometry.

infinite slab (Contini et al. 1997) shows that PCB and EBC give similar results, while ZBC presents discrepancies depending on the refractive index mismatch.

1.3.2.2.3 Workable Solutions for Time-Resolved Reflectance and Transmittance

In the following, we report the analytical expressions for time-resolved reflectance from a semi-infinite homogeneous medium and time-resolved transmittance from a homogeneous infinite slab as derived within the diffusion approximation with EBC and ZBC condition, and the random walk model. As discussed in the previous sections, these are the most widely used models, both for their simplicity and their effective accuracy in practical situations.

For the derivation of these expressions, I refer to the original papers listed in the References at the end of this section. It is not the intention here to discuss the theoretical limitations of these models; rather I focus attention on the experimental test of the efficacy of these theoretical models.

1. Time-resolved reflectance from a semi-infinite homogeneous medium:
   a. Diffusion approximation: zero boundary condition (Patterson et al. 1989)

$$R(\rho,t) = AD^{-3/2}t^{-5/2}\exp(-\mu_a vt)\exp\left(-\frac{\rho^2}{4Dvt}\right)z_0\exp\left(-\frac{z_0^2}{4Dvt}\right) \quad (1.3.20)$$

where
$D = 1/(3\mu'_s)$
$z_0 = 3D$
$v = c/n$
$A = (4\pi v)^{-3/2}$

b. Diffusion approximation: extrapolated boundary condition (Contini et al. 1997)

$$R(r,t) = AD^{-3/2}t^{-5/2}\exp(-\mu_a vt)\exp\left(-\frac{r^2}{4Dvt}\right)$$

$$\times \left[z_0\exp\left(-\frac{z_0^2}{4Dvt}\right) - z_p\exp\left(-\frac{z_p^2}{4Dvt}\right)\right] \qquad (1.3.21)$$

where

$$A = \frac{1}{2}(4\pi v)^{-3/2}$$

$$z_p = z_0 + 2z_e$$

$$z_e = 2QD$$

$$Q = \frac{1+r_d}{1-r_d}$$

$$r_d = \frac{r_\phi + r_\gamma}{2 - r_\phi + r_\gamma}$$

$$r_\varphi = \int_0^{\pi/2} 2\cdot\sin\theta\cdot\cos\theta\cdot R_{\text{Fresnel}}(\theta)\cdot d\theta$$

$$r_\gamma = \int_0^{\pi/2} 3\cdot\sin\theta\cdot(\cos\theta)^2\cdot R_{\text{Fresnel}}(\theta)\cdot d\theta$$

Note that r_d can be approximated by $r_d = -1.440n^{-2} + 0.710n^{-1} + 0.668 + 0.0636n$.

2. Time-resolved transmittance through an infinite homogeneous slab:
  a. Diffusion approximation: zero boundary condition (Patterson et al. 1989)

$$T(\rho,d,t) = AD^{-3/2}t^{-5/2}\exp(-\mu_a vt)\exp\left(-\frac{\rho^2}{4Dvt}\right)\sum_{k=0}^{\infty} z_-\exp\left(-\frac{z_-^2}{4Dvt}\right)$$

$$- z_+\exp\left(-\frac{z_+^2}{4Dvt}\right) \qquad (1.3.22)$$

where

$$z_\pm = (2k+1)d \pm z_0$$

$$A = (4\pi v)^{-3/2}$$

b. Diffusion approximation: extrapolated boundary condition (Contini et al. 1997)

$$T(\rho,d,t) = AD^{-3/2}t^{-5/2}\exp(-\mu_a vt)\exp\left(-\frac{\rho^2}{4Dvt}\right)\sum_{k=-\infty}^{+\infty} z_-\exp\left(-\frac{z_-^2}{4Dvt}\right)$$
$$- z_+\exp\left(-\frac{z_+^2}{4Dvt}\right) \quad (1.3.23)$$

where

$$z_- = d(1-2k) - 4kz_e - z_0$$
$$z_+ = d(1-2k) - (4k-2)z_e + z_0$$
$$z_e = 2QD$$
$$A = \frac{1}{2}(4\pi v)^{-3/2}$$

1.3.2.3 Time-Resolved Instrumentation for Photon Migration

Photon migration measurements in the time domain rely on the ability to extract the information encoded in the temporal distribution of the remitted light, following the injection of a short monochromatic pulse in a diffusive medium. Typical values of the optical parameters in the red and near-infrared part of the spectrum set the time scale of photon migration events in the range 1–10 ns, and fix the ratio of detected to injected power at about −80 dB.

The two key points in the designing of a system for time-resolved measurements are thus temporal resolution and high sensitivity. Temporal resolution is mainly affected by the width of the light pulse and by the response of the detection apparatus. Pulsed laser, which produce short (10–100 ps) and ultrashort (10–100 fs) light pulses with repetition frequency up to 100 MHz, and photo-detection systems with temporal resolution in the range of 10–150 ps are nowadays widely available. With regard to sensitivity, we must observe that the power of the injected light pulse is fixed to proper values to avoid possible damage or injury to the sample. In the case of biological tissues, the safety regulations (Anonymous 1985) are set for laser pulses in the wavelength range 600–1000 nm to the maximum permissible value of 2 mW/mm^2.

The technique of choice must therefore deal with fast and weak light signals. For this purpose, time-correlated single-photon counting (TCSPC) and streak camera (SC) detection techniques have been extensively used for time-resolved measurements of light in the picosecond and nanosecond regime. More recently, TCSPC (Chance et al. 1988; Patterson et al. 1989; Andersson-Engels et al. 1990; Wilson and Jacques 1990; Benaron and Stevenson 1993; Cubeddu et al. 1994) and SC (Delpy et al. 1988; Ho et al. 1989) have been introduced into photon-migration measurements. A detection technique based on time-gated intensified charge-coupled device (ICCD) sensors has been introduced in the nanosecond regime, and also more recently in the picosecond regime for imaging purposes.

1.3.2.3.1 Time-Correlated Single-Photon Counting

Over the last 20 years, TCSPC (O'Connor et al. 1984; Becker 2006) has been widely used, for example, for the measurements of excited states lifetime of free atoms, for the determination of the fluorescence lifetime from biological molecules and cells, and for the assessment of luminescence decay from semiconductor devices and materials. The basic principle of TCSPC is that the probability distribution for remittance of a single photon after an excitation event yields the actual intensity against time distribution of all the photons remitted as a result of the excitation. For this purpose, it is fundamental that the intensity at the detector is low enough so that the probability of detecting more than one photon per excitation event can be neglected.

In a TCSPC experiment, the temporal profile of the remittance curve is not directly measured but is retrieved by repeatedly measuring the delay between an excitation event and a remittance event for a statistically significant number of photons. A key parameter in this is the count rate, that is, the number of photons per second (counts per second, cps), which can be processed without exceeding the single-photon statistics. Standard TCSPC systems have a maximum count rate of about 5×10^6 cps. Figure 1.3.19 shows the typical electronic chain for TCSPC measurements. A fast photodiode (PD) followed by a constant fraction discriminator (CFD) generates an electrical pulse (START) at a time exactly correlated with the time of generation of the excitation pulse, without being affected by the eventual amplitude variation of the optical signal. The START signal is then fed into a time-to-amplitude converter (TAC) through a proper delay line (DL). When a remitted photon is detected by a photomultiplier tube (PMT), an electronic pulse (STOP) is generated. The STOP signal is routed through a constant fraction discriminator (CFD) to the input of the TAC. This device produces an output signal that is proportional to the time delay between START and STOP pulses. Finally, a multi-channel analyzer (MCA) classifies the TAC output signal by storing a count in a proper memory channel. This procedure is repeated until the histogram of the number of counts against channel address represents to the required precision the temporal profile of the remitted photons.

In practice to increase the count rate, TCSPC system commonly operates in the reverse mode with the START and STOP signal provided by the PMT and the PD, respectively. In this way, it is possible to avoid the pileup of START signals because of the higher repetition rate of the excitation signal with respect to the remittance rate of photons from the sample.

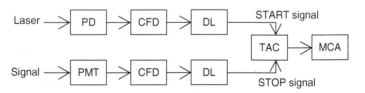

FIGURE 1.3.19 Scheme of a TCSPC system: photodiode, PD; photomultiplier tube, PMT; constant fraction discriminator, CFD; delay line, DL; time-to-amplitude converter, TAC; multichannel analyzer, MCA.

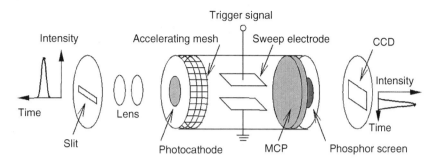

FIGURE 1.3.20 Operating principle of the streak camera system.

Standard TCSPC nuclear instrumentation module (NIM) modules have a typical temporal resolution of about 3 ps/channel and a count rate of up to 10^5 cps. Novel TCSPC personal computer (PC) boards reach up to 5×10^6 cps at a lower temporal resolution of 25 ps/channel.

1.3.2.3.2 Streak Camera

The streak camera is a complex device used to directly detect ultrafast light phenomena with an extreme temporal resolution (Anonymous 1994). The operating principle of the streak camera is shown in Figure 1.3.20.

The light being measured passes through an adjustable slit, which is imaged onto the photocathode of the streak tube. The incident light is converted into a proportional number of photoelectrons that are accelerated toward a phosphor screen by a first high-voltage stage. A second high-voltage stage, positioned transversely to the path of the photoelectrons, applies a high-speed sweep at a time synchronized with the incident pulses. The photoelectrons all arriving at slightly different times are thus bombarded at different points onto the phosphor screen, where they are converted into light again. A microchannel plate (MCP) is usually placed in front of the phosphor screen and used to multiply the photoelectrons before they hit the phosphor screen. A high-sensitive CCD camera detects the images generated onto the phosphor screen. The image intensity correlates with the intensity of the light passing through the slit. The image direction transverse to the entrance slit corresponds to the time axis. The image direction parallel to the entrance slit can be used to measure the time variation of the incident light with respect to either position when used in combination with proper optics, or the wavelength when used in combination with a spectroscope.

In the single-sweep operation, the SC system has a maximum temporal resolution of 0.6 ps, while in the synchroscan operation, it usually offers a maximum temporal resolution of 10 ps.

1.3.2.3.3 Gated Intensified CCD Camera

Novel ICCD cameras are characterized by ultimate sensitivity down to single-photon detection. The basic scheme of this instrument is shown in Figure 1.3.21. It basically consists of a photocathode, a microchannel plate photomultiplier tube (MCP-PMT), and a phosphor screen like a streak camera system.

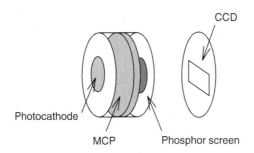

FIGURE 1.3.21 Operating principle of the intensified camera system.

In addition, high-temporal resolution can be achieved with ICCD by fast gating of the intensifier cathode. Conventional gating schemes give rise to irising problems caused by the RC time constant of the cathode to MCP capacitance and the cathode resistivity. This RC time constant gives rise to a diffusion of charge toward the tube axis and the diffusion rate cannot be increased by external circuitry; consequently, the tube gating time achievable is limited. Until recently, the time resolution was limited to about 1 ns restricting the use of ICCD for the investigation of faster processes. This limitation is avoided with a new gating scheme. A highly conducting mesh or ring is placed in front of the input window of the tube and this is pulsed to a high voltage. Most of the voltage appears between the mesh and the cathode, but sufficient voltage also appears between the cathode and the MCP input surface to gate the tube quickly. Ultrashort gates can be achieved by using smaller image tubes. Commercial ICCD is now available with 80 ps at 1 kHz or 300 ps at 100 MHz (Dowling et al. 1998).

The gated ICCD system is thus potentially able to measure the spatial and temporal profiles of the remitted light from a diffusive medium by acquiring different images synchronized for different time delays with respect to the excitation. Every image contains the spatial information at a certain time instant, while the successive values stored in the memory and referring to the same pixel determine the temporal distribution of the detected signal.

1.3.2.3.4 Discussion and Choice

The TCSPC is the system used in our experimental apparatus for photon-migration studies. The choice has been in some way driven by the former availability of TCSPC systems in our laboratories for time-resolved fluorescence spectroscopy measurements. Nonetheless, TCSPC has certain advantages compared to the SC. First, TCSPC systems offer a superior light collection area at a substantially reduced cost. Moreover, the photon count rate is no longer a drawback of TCSPC system, which is in fact currently around 5×10^6 cps using state-of-the-art electronics. A reasonable signal-to-noise ratio can therefore be achieved with a minimum acquisition time of about 100 ms. However, temporal resolution of the TCSPC system is limited by the transit time spread of PMTs or MCP-PMTs (about 50 ps at best), but this can be managed by deconvolution procedures (see section below). In addition, TCSPC offers further advantages over the SC, thanks to a better dynamic range and a lower sensitivity of the measured signals to intensity fluctuations.

Up to now, a gated ICCD system has been much more expensive than a TCSPC system, but it offers the intrinsic advantage of direct two-dimensional imaging compared to broad area, a feature that is not easily available in TCSPC system without resorting to a scanning device. The presence of a gated ICCD system for fluorescence lifetime imaging (Cubeddu et al. 1995) allowed us to perform very preliminary studies in the field of photon migration. However, the limited temporal resolution of the instrument (500 ps gate width) resulted in a lower efficacy compared to the TCSPC system. A new gated ICCD system with a better performance (300 ps gate width, 100 MHz repetition rate) has been also developed.

The possible use of single-photon avalanche diode (SPAD) (Giudice et al. 2007) for time-resolved measurements is also promising. These devices are characterized by extended spectral sensitivity (up to 1 μm without the need for water cooling), low-voltage supply (+5 V), and a relatively low cost. The main drawback until now has been the interplay between dimensions of the detection area and temporal resolution. Because of technological problems, to achieve a temporal resolution of about 30 ps, a small detection area (50 μm in diameter) is required. The collection efficiency will be thus dramatically reduced.

1.3.2.3.5 Instrument Response Function and Data Analysis

The noninvasive assessment of the optical properties of a diffusive medium by the analysis of the remitted light has the disadvantage that none of the mathematical expressions reported in the previous sections for time-resolved remittance (TRR, either reflectance or transmittance) can be inverted to directly extract either the absorption coefficient or the scattering coefficient.

A rough estimate of the optical properties can be obtained by the evaluation of the peak position and the long-time asymptotic slope of the TRR curve, which are in some way dependent on the scattering coefficient and the absorption coefficient, respectively, as is described in former section. Approximate formulas have been derived (Matcher 1997) and used, but they are characterized by significant errors.

It is thus necessary to resort to a nonlinear fitting procedure to get μ_a and μ'_s from the TRR curve. The problem is not trivial since the TRR curve is broadened by the instrument response function (IRF). As shown in Figures 1.3.22 and 1.3.23, the situation is analogous to the characterization of a linear, time-invariant system by the analysis of the impulse response. The IRF can be recorded by facing the injection and the collection fiber so as to send the light pulse directly to the detection apparatus.

When probing the medium, the light pulse $x(t)$ is first altered into $y(t)$ by the function $h(t)$, which identifies the optical properties of the medium, and is then changed into the TRR curve by the function $s(t)$, which identifies the detection apparatus.

FIGURE 1.3.22 Scheme for the acquisition of the IRF. $x(t)$ is the light pulse, $s(t)$ is the unknown response of the detection apparatus.

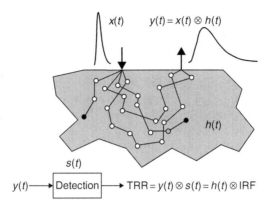

FIGURE 1.3.23 The TRR as the convolution ($\otimes$) of the IRF and the medium function $h(t)$.

In an ideal system, both the laser pulse $x(t)$ and the response of the detection apparatus $s(t)$ are delta-like pulses. As a consequence, the TRR curve is coincident with the function $h(t)$.

In a nonideal system, the broadening of the TRR curve can be neglected if the IRF is notably narrower than the TRR curve. This is, for example, the case for many fluorescence lifetime measurements where the time evolution of the phenomena is on the scale of 10–100 ns, and the IRF is about 80 ps. However, as has already been mentioned, the time scale of photon-migration events is in the range 1–10 ns, while the IRF can be as broad as 160 ps. Moreover, the IRF can introduce distortions into the shape of the TRR curve, thus leading to errors in the estimates of the optical parameters that have an overall influence on the TRR curve.

In this work, the solution that has been chosen to take into account of IRF problem is to first calculate the convolution TRR_c of the expression of the theoretical model TRR_0 with the IRF, and then to use the result to fit the experimental TRR curve (Figure 1.3.24).

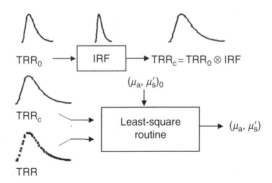

FIGURE 1.3.24 Scheme of the fitting procedure. TRR_0, theoretical expression for time-resolved remittance; TRR_c, convolution between TRR_0 and IRF; TRR, experimental time-resolved remittance curve; $(\mu_a, \mu'_s)_0$, initial guess for the optical parameters; (μ_a, μ'_s), final estimate.

What to Measure and How to Measure 73

The procedure is more robust than the alternative method based on the deconvolution of the TRR curve with the IRF. The two methods are, in principle, equivalent, but the latter has been found to be strongly dependent on the noise level of the experimental TRR curve.

1.3.2.4 Applications of Time-Resolved Reflectance Spectroscopy to Fresh Fruit and Vegetable

We present a short collection of results obtained by applying time-resolved reflectance spectroscopy for the nondestructive evaluation of the internal properties of fruits and vegetables. In particular, we will show the empirical estimate of the penetration depth of time-resolved reflectance spectroscopy (TRS) measurements and the use of absorption and reduced scattering spectra to nondestructively assess information on tissue components and structure. Further examples can be found in the literature (Cubeddu et al. 2001a,b; Cubeddu et al. 2002; Eccher Zerbini et al. 2002; Cubeddu et al. 2003; Valero et al. 2004a,b, 2005; Eccher Zerbini et al. 2006; Tijskens et al. 2007).

1.3.2.4.1 Penetration Depth of Time-Resolved Reflectance Measurements
Penetration depth of a TRS measurement can be empirically determined. It is well known that the volume probed by a TRS measurement is a banana-shaped region connecting the injection and collection points (Feng et al. 1995). It is not easy to define the measurement volume since the photon paths are more densely packed in the banana region but can be distributed in the whole medium. Attempts were made to determine the maximum depth in the tissue that yields a detectable contribution to the TRS curve. A series of measurements were performed on a 'Starking Delicious' apple, where slices of flesh were cut from opposite sides of the measurement site. Spectra were taken on the whole apple, and then slices were removed to yield a total thickness of 4.1, 2.7, 2.1, and 1.5 cm. The fitted absorption and scattering spectra are shown in Figure 1.3.25. The absorption coefficient μ_a is unchanged down to a thickness of 2.7 cm. For 2.1 cm, μ_a starts deviating from the measurement on the whole apple with a discrepancy of 25% at 680 nm, whereas for a thickness of 1.5 cm, the discrepancy increases up to 50%. The highest variations are observed on the tails of the spectrum, where the absorption is lower. The results of the scattering coefficient show similar behavior, with almost no changes down to a thickness of 2.7 cm, and discrepancies of 15% and 25% for a 2.1 and 1.5 cm thickness, respectively. Overall, these data show that the TRS measurement is probing a depth of at least 2 cm in the pulp. This is, of course, only a rough estimate, yet it confirms that the TRS measurement is not confined to the surface of the fruit. Moreover, the penetration depth can be in some way dependent on the optical properties, and we expect deeper penetration in less absorbing and scattering fruits.

1.3.2.4.2 Absorption Spectra and Tissue Components
Typical absorption spectra of different fruits are reported in Figure 1.3.26A. The absorption spectrum of the apple is dominated by the water peak, centered around 970 nm, with an absolute value of about 0.4 cm^{-1}. Minor absorption features of water are usually detected around 740–835 nm, where the absorption coefficient is

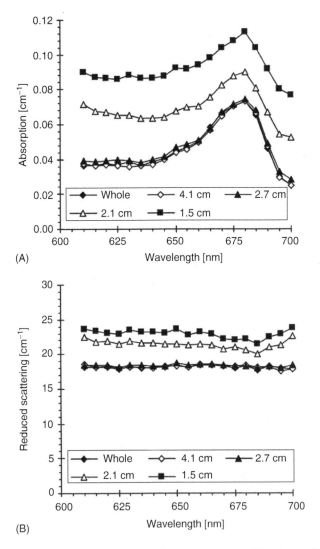

FIGURE 1.3.25 (A) Absorption and (B) scattering spectra of a 'Starking Delicious' apple. Different curves correspond to measurements on the whole apple, and on slices of the same apple obtained by cutting the fruit on the opposite side of the measurement site.

low (0.05 cm^{-1}). A significant absorption peak (0.12–0.18 cm^{-1}) is found at 675 nm, which corresponds to chlorophyll-*a* (see Section 3.2). Both the line-shape and the absolute values of the absorption spectra of peach and tomato are quite similar to those of apples. However, for kiwifruit, chlorophyll-*a* absorption is considerable, as expected from the visual appearance of its flesh, with a maximum value up to two or three times the water maximum in the near infrared.

Information on the water content can be obtained by considering the absolute values of the absorption at 970 nm. In agreement with the different water/fibers ratio

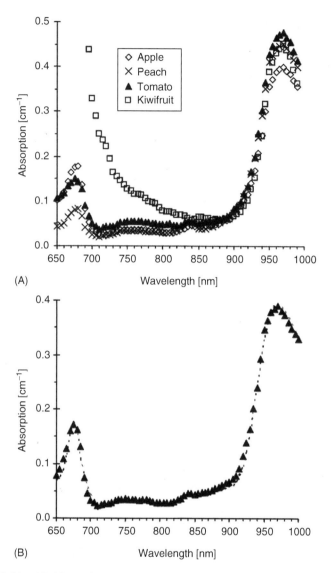

FIGURE 1.3.26 (A) Absorption spectra of apple, peach, tomato, and kiwifruit; (B) best fit of chlorophyll-*a* and water line shape to the absorption spectrum of a 'Starking' apple.

in distinct species, a higher absorption was detected in tomatoes ($\sim$0.5 cm^{-1}), than in peaches and kiwifruits ($\sim$0.45 cm^{-1}), and in apples ($\sim$0.4 cm^{-1}).

The absorption at 675 nm provides information about the chlorophyll-*a* content, and preliminary data obtained from apples suggests that this could be a useful parameter to test the ripening stage. A series of measurements performed on the same fruits showed a progressive decrease in red absorption, in agreement with the gradual reduction in the chlorophyll content with postharvest ripening (Cubeddu et al. 2001b).

TABLE 1.3.1
Chlorophyll-a and Water Content in Different Fruits

Fruit	Chlorophyll-a [μM]	Water [%]
'Starking Delicious' apple	0.96	82.6
Peach	0.49	93.8
Tomato	0.52	95.0
Kiwifruit	6.91	98.8

To quantify the percentage volume of water and the chlorophyll-a content in the bulk of the intact fruits, a best fit of the absorption spectrum was performed using the line shape of water (Hale and Querry 1973) and of chlorophyll-a (Shipman et al. 1979). To account for the presence of other native chromophores of fruit, such as carotenoids and anthocyanines, which exhibit characteristic peaks at shorter wavelengths than 650 nm, a flat background spectrum of arbitrary amplitude was used as a free parameter in the fit.

Figure 1.3.26B shows a typical example of fit for the absorption spectrum of a 'Starking Delicious' apple to the line shape of water and chlorophyll-a. Table 1.3.1 reports the chlorophyll-a and water content in different fruits. In all cases, a 0.02–0.03 cm^{-1} contribution was added by the flat background spectrum.

1.3.2.4.3 Scattering Spectra and Tissue Structure
The scattering properties, of all the species considered, showed no particular spectral features. The value of the transport scattering coefficient decreased progressively with increasing wavelength. Typical examples are shown in Figure 1.3.27A for the 'Starking Delicious' apple, peach, tomato, and kiwifruit. The transport scattering spectrum of the kiwifruit was noisier than the spectrum of other fruits, particularly in the 675 nm region where the high absorption of chlorophyll reduced the accuracy of the evaluation of transport scattering by TRS measurements.

Even though marked variations in the absolute values were noticed depending on variety and ripeness, kiwifruits and tomatoes were usually characterized by a lower scattering than other species, while carrots, for instance, show relatively high absorbance coefficients (Zude et al. 2008).

Further information could be obtained by interpreting the transport scattering spectra with Mie theory. For a homogeneous sphere of radius, r, Mie theory predicts the wavelength dependence of the scattering and the relation between scattering and sphere size. Under the hypothesis that the scattering centers are homogeneous spheres behaving individually, the relationship between μ'_s and wavelength (λ) can be empirically described (Mourant et al. 1997) as follows:

$$\mu'_s = ax^b \qquad (1.3.24)$$

where the size parameter x is defined as $x = 2\pi r\, n_m \lambda^{-1}$, with the refraction index of the medium n_m chosen to be 1.35, and a and b are free parameters. In particular, a is proportional to the density of the scattering centers, and b depends on their size.

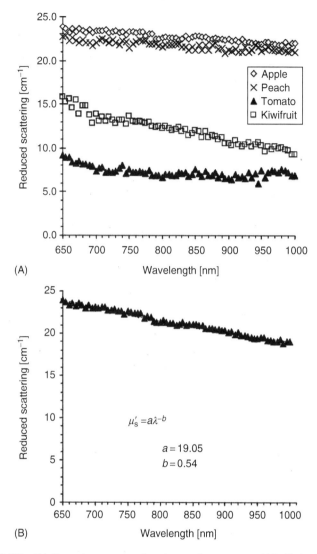

FIGURE 1.3.27 (A) Scattering spectra of apple, peach, tomato, and kiwifruit; (B) best fit of Mie theory to the scattering spectrum of a 'Starking' apple.

Moreover, b can be empirically expressed as a third-order polynomial function of r; therefore the estimate of b can yield the sphere radius r (Nilsson et al. 1998).

Figure 1.3.27B shows a typical transport scattering spectrum of a 'Starking Delicious' apple and the best fit to Mie theory. The estimated average size of scattering centers in different fruits is shown in Table 1.3.2. It was observed that a and b varied in the range 1.54–15.85 cm^{-1} and 0.18–1.83, respectively. This suggests that different fruits have different density and average dimensions of scattering centers (the range for r is 0.15–0.78 µm). It is worth noting that being tissues, they are a complex distribution of cells tissue such as the xylem, phloem,

TABLE 1.3.2
Parameters *a* and *b* for Different Fruits

Fruit	a [cm^{-1}]	b	r [μm]
Starking delicious apple	17.4	0.12	0.759
Peach	14.4	0.20	0.740
Tomato	2.9	0.48	0.591
Kiwifruit	4.5	0.95	0.266

mesokarp, exokarp, etc. and these parameters do not assess the real size of scattering centers in the tissue. They are in fact average equivalent parameters, which could eventually be related to physical fruit characteristics such as firmness or useful for correction approaches when calibrating on contents in sugar or pigments (Zude et al. 2008).

REFERENCES

Andersson-Engels, S., R. Berg, S. Svanberg, and O. Jarlman. 1990. Time-resolved transillumination for medical diagnostics. *Optics Letters* 15: 1179–1181.
Anonymous. 1985. *Compliance Guide for Laser Products*. U.S. Department of Health and Human Services, FDA, MD, HHS Publication FDA86-8260, FDA, Rockville, Maryland, USA.
Anonymous. 1994. *Guide to Streak Cameras*. Hamamatsu Photonics Inc, Hamamatsu City, Japan.
Aronson, R. 1995. Boundary conditions for diffusion of light. *Journal of the Optical Society of America A-Optical Physics* 12: 2532–2539.
Arridge, S.R. 1999. Optical tomography in medical imaging. *Inverse Problems* 15: R41–R93.
Arridge, S.R., M. Cope, and D.T. Delpy. 1992. The theoretical basis for the determination of optical pathlengths in tissue: Temporal and frequency analysis. *Physics in Medicine and Biology* 37: 1531–1560.
Arridge, S.R., M. Schweiger, M. Hiraoka, and D.T. Delpy. 1993. A finite element approach for modelling photon transport in tissue. *Medical Physics* 20: 299–309.
Becker, W. 2006. *Advanced TCSPC*. Springer, Berlin, Germany.
Benaron, D.A. and D.K. Stevenson. 1993. Optical time-of-flight and absorbance imaging of biologic media. *Science* 259: 1463–1466.
Chance, B. 1989. *Photon Migration in Tissues*. Plenum Press, New York.
Chance, B., J.S. Leigh, H. Miyake, et al. 1988. Comparison of time-resolved and -unresolved measurements of deoxyhaemoglobin in brain. *Proceedings—National Academy of Sciences USA* 85: 4971–4975.
Contini, D., F. Martelli, and G. Zaccanti. 1997. Photon migration through a turbid slab described by a model based on diffusion approximation. I. Theory. *Applied Optics* 36: 4587–4599.
Cubeddu, R., M. Musolino, A. Pifferi, P. Taroni, and G. Valentini. 1994. Time-resolved reflectance: A systematic study for application to the optical characterization of tissues. *IEEE Journal of Quantum Electronics* 30: 2421–2430.
Cubeddu, R., A. Pifferi, P. Taroni, G. Valentini, and G. Canti. 1995. Tumor detection in mice by measurements of fluorescence decay time matrices. *Optics Letters* 20: 2553–2555.

Cubeddu, R., C. D'Andrea, A. Pifferi, et al. 2001a. Non-destructive quantification of chemical and physical properties of fruits by time-resolved reflectance spectroscopy in the wavelength range 650–1000 nm. *Applied Optics* 40: 538–543.

Cubeddu, R., C. D'Andrea, A. Pifferi, et al. 2001b. Non-destructive measurements of the optical properties of apples by means of time-resolved reflectance spectroscopy. *Applied Spectroscopy* 55: 1368–1374.

Cubeddu, R., A. Pifferi, P. Taroni, and A. Torricelli. 2002. Measuring fresh fruit and vegetable quality: Advanced optical methods. In: *Fruit and Vegetable Processing*, ed. W. Jongen, 150–169. CRC Press-Woodhead Publishing, Cambridge, United Kingdom.

Cubeddu, R., A. Pifferi, P. Taroni, and A. Torricelli. 2003. Spectroscopic techniques for analysing raw material quality. In: *Rapid and Online Instrumentation for Food Quality Assurance*, ed. E. Tothill, 150–169. CRC Press-Woodhead Publishing, Cambridge, United Kingdom.

Delpy, D.T., M. Cope, P. van der Zee, S.R. Arridge, S. Wray, and J.S. Wyatt. 1988. Estimation of optical pathlength through tissue from direct time of flight measurement. *Physics in Medicine and Biology* 33: 1433–1442.

Dowling, K., M.J. Dayel, M.J. Lever, P.M.W. French, J.D. Hares, and A.K.L. Dymoke-Bradshaw. 1998. Fluorescence lifetime imaging with pico second resolution for biomedical applications. *Optics Letters* 23: 810–812.

Duderstadt, J.J. and W.R. Martin. 1979. *Transport Theory*. J. Wiley, New York.

Eccher Zerbini, P., M. Grassi, R. Cubeddu, A. Pifferi, and A. Torricelli. 2002. Nondestructive detection of brown heart in pears by time-resolved reflectance spectroscopy. *Postharvest Biology and Technology* 25: 87–99.

Eccher Zerbini, P., M. Vanoli, M. Grassi, et al. 2006. A model for the softening of nectarines based on sorting fruit at harvest by time-resolved reflectance spectroscopy. *Postharvest Biology and Technology* 39: 223–232.

Farrell, T.J., M.S. Patterson, and B.C. Wilson. 1997. A diffusion theory model of spatially resolved, steady-state diffuse reflectance for the non-invasive determination of tissue optical properties in vivo. *Medical Physics* 19: 879–888.

Feng, S., F.A. Zeng, and B. Chance. 1995. Photon migration in the presence of a single defect: A perturbation analysis. *Applied Optics* 34: 3826–3837.

Furutsu, K. and Y. Yamada. 1994. Diffusion approximation for a dissipative random medium and the applications. *Physiological Reviews* E50: 3634–3640.

Giudice, A., M. Ghioni, R. Biasi, F. Zappa, and S. Cova. 2007. High-rate photon counting and picosecond timing with silicon-SPAD based compact detector modules. *Journal of Modern Optics* 54: 225–238.

Hale, G.M. and M.R. Querry. 1973. Optical constants of water in the 200 nm to 200 mm wavelength region. *Applied Optics* 12: 555–563.

Haskell, R.C., L.O. Svaasand, T.T. Tsay, T.C. Feng, M.S. McAdams, and B.J. Tromberg. 1994. Boundary conditions for the diffusion equation in radiative transfer. *Journal of the Optical Society of America A-Optical Physics* 11: 2727–2741.

Henyey, L.G. and J.L. Greenstein. 1941. Diffuse radiation in the galaxy. *Astrophysics Journal* 93: 70–83.

Hielscher, A.H., S.L. Jacques, L. Wang, and F.K. Tittel. 1995. The influence of boundary conditions on the accuracy of diffusion theory in time-resolved reflectance spectroscopy of biological tissues. *Physics in Medicine and Biology* 40: 1957–1975.

Ho, P.P., P. Baldeck, K.S. Wong, K.M. Yoo, D. Lee, and R.R. Alfano. 1989. Time dynamics of photon migration in semiopaque random media. *Applied Optics* 28: 2304–2310.

Ishimaru, A. 1978. Wave propagation and scattering in random media. *Single Scattering and Transport Theory*, Vol. 1. Academic Press, New York.

Jacques, S.L. 1989a. Time resolved propagation of ultrashort laser pulses within turbid tissues. *Applied Optics* 28: 2223–2229.

Jacques, S.L. 1989b. Time-resolved reflectance spectroscopy in turbid tissues. *IEEE Transactions on Biomedical Engineering* 36: 1155–1161.

Kaltenbach, J. and M. Kaschke. 1993. Frequency- and time-domain modelling of light transport in random media. In: *Medical Optical Tomography: Functional Imaging and Monitoring*, Vol. IS11 of Institute Series of SPIE, eds. G. Mueller, B. Chance, R.R. Alfano, et al. 65–86. SPIE Bellingham, WA.

Lakowicz, J.R. and K.W. Berndt. 1990. Frequency-domain measurements of photon migration in tissues. *Chemical Physics Letters* 166: 246–252.

Matcher, S.J. 1997. Closed-form expression for obtaining the absorption and scattering coefficients of a turbid medium with time-resolved spectroscopy. *Applied Optics* 36: 8298–8302.

Mourant, J.R., T. Fuselier, J. Boyer, T.M. Johnson, and I.J. Bigio. 1997. Predictions and measurements of scattering and absorption over broad wavelength ranges in tissue phantoms. *Applied Optics* 36: 949–957.

Nilsson, M.K., C. Sturesson, D.L. Liu, and S. Andersson-Engels. 1998. Changes in spectral shape of tissue optical properties in conjunction with laser-induced thermotherapy. *Applied Optics* 37: 1256–1267.

O'Connor, D.V. and D. Philip. 1984. *Time Correlated Single Photon Counting*. Academic Press, London, United Kingdom.

Patterson, M.S., B. Chance, and B.C. Wilson. 1989. Time resolved reflectance and transmittance for the non-invasive measurement of tissue optical properties. *Applied Optics* 28: 2331–2336.

Patterson, M.S., J.D. Moulton, B.C. Wilson, K.W. Berndt, and J.R. Lakowicz. 1991. Frequency-domain reflectance for the determination of the scattering and absorption properties of tissue. *Applied Optics* 30: 4474–4476.

Pogue, B.W. and E. Patterson. 1994. Frequency-domain optical absorption spectroscopy of finite tissue volumes using diffusion theory. *Physics in Medicine and Biology* 39: 1157–1180.

Shipman, L.L., T.M. Cotton, J.R. Norris, and J.J. Katz. 1979. An analysis of the visible absorption spectrum of chlorophyll a monomer, dimer and oligomer in solution. *Journal of the American Chemical Society* 98: 8222–8230.

Svaasand, L.O., B.J. Tromberg, R.C. Haskell, T.-T. Tsay, and M.W. Berns. 1993. Tissue characterization and imaging using photon density waves. *Optical Engineering* 32: 258–266.

Tijskens, L.M.M., P. Eccher Zerbini, R. Schouten, et al. 2007. Assessing harvest maturity in nectarines. *Postharvest Biology and Technology* 45: 204–213.

Tromberg, B.J., L.O. Svaasand, T.-T. Tsay, and R.C. Haskell. 1993. Properties of photon density waves in multiple-scattering media. *Applied Optics* 32: 607–616.

Valero, C., M. Ruiz-Altisent, R. Cubeddu, et al. 2004a. Detection of internal quality in kiwi with time-domain diffuse reflectance spectroscopy. *Applied Engineering in Agriculture* 20: 223–230.

Valero, C., M. Ruiz-Altisent, R. Cubeddu, et al. 2004b. Selection models for the internal quality of fruit, based on time domain laser reflectance spectroscopy. *Biosystems Engineering* 88: 313–323.

Valero, C., P. Barreiro, M. Ruiz-Altisent, et al. 2005. Mealiness detection in apples using time resolved reflectance spectroscopy. *Journal of Texture Studies* 36: 439–458.

Van de Hulst, H.C. 1980. *Multiple Light Scattering: Tables, Formulas and Applications*. Academic Press, New York.

Wang, L. and S.L. Jacques. 1995. Use of a laser beam with an oblique angle of incidence to measure the reduced scattering coefficient of a turbid medium. *Applied Optics* 34: 2362–2366.

Wilson, B.C. and S.L. Jacques. 1990. Optical reflectance and transmittance of tissues: Principles and applications. *IEEE Journal of Quantum Electronics* 26: 2186–2199.

Wilson, B.C., E.M. Sevick, M.S. Patterson, and K.P. Chan. 1992. Time-dependent optical spectroscopy and imaging for biomedical applications. *Proceeding of the IEEE* 80: 918–930.

Zude, M., L. Spinelli, and A. Torricelli. 2008. Approach for nondestructive pigment analysis in model liquids and carrots by means of time-of-flight and multi-wavelength remittance readings. *Analytica Chimica Acta* 623: 204–212.

2 Vision Systems

José Blasco, Enrique Moltó Garcia, Da-Wen Sun, and Chaoxin Zheng

CONTENTS

2.1	Machine Vision Systems for Raw Material Inspection	84
	2.1.1 Basics of Commercial Machine Vision	85
	2.1.1.1 Lighting Sources	85
	2.1.1.2 CCD Cameras and Video Signals	92
	2.1.1.3 Basic Steps of Machine Vision Applications	95
	2.1.2 Current Commercial Sorters Based on Machine Vision	105
	2.1.2.1 Elements of a Packing Line	105
	2.1.2.2 Major Constraints of Machine Vision Systems in Commercial Applications	113
	2.1.3 Future of Machine Vision Applications for Raw Material Inspection	114
	2.1.3.1 Exploitation of Fluorescent Properties	114
	2.1.3.2 Hyperspectral Images	115
	2.1.3.3 Look at the Internal Quality	117
	2.1.3.4 Combining Several Sensors for Complete Quality Assessment	120
References		120
2.2	Computer Vision for Quality Control	126
	2.2.1 Introduction	126
	2.2.2 Image Analysis and Machine Learning	127
	2.2.2.1 Image Analysis	127
	2.2.2.2 Machine Learning	130
	2.2.3 Applications	132
	2.2.3.1 Quality Control of Fruits and Vegetables	132
	2.2.3.2 Quality Control of Grain	134
	2.2.3.3 Quality Control of Other Foods	135
	2.2.4 Image Features for Quality Evaluation: Usefulness and Limitation	136
	2.2.5 Conclusions	138
References		138

2.1 MACHINE VISION SYSTEMS FOR RAW MATERIAL INSPECTION

ENRIQUE MOLTÓ GARCIA AND JOSÉ BLASCO

The application of machine vision in agriculture has increased considerably in recent years. There are many fields in which machine vision is involved: terrestrial and aerial mapping of natural resources, crop monitoring, precision agriculture, robotics, automatic guidance, nondestructive inspection of product properties, quality control and classification in processing lines and, in general, process automation. Many authors have successfully applied some of the techniques developed in one of these fields to the others, since all of them try to mimic the human sense of sight (Chen et al. 2002; Brosnan and Sun 2004; Sun 2007).

The breadth of applications depends, among many other things, on the fact that machine vision systems provide substantial information about the nature and attributes of the objects present in a scene. Another important feature of such systems is that they open the possibility of studying these objects in regions of the electromagnetic spectrum, where human eyes are unable to operate, as in the ultraviolet or infrared regions. In general, the need for the development of nondestructive automatic techniques for quality control of fruits and vegetables has been emphasized due to the following issues (Bellon et al. 1992):

- Greater reliability and objectivity of these techniques than human inspection, as the decisions made by operators, are affected by factors such as fatigue, acquired habits, etc. A study on the reliability of the measurements taken from apples concluded that the measurements of the damaged surface area carried out by image analysis provided results, which were five times less disperse than those performed by ordinary methods (Studman and Ouyang 1997).
- Increase in productivity and the regional specialization of crops lead from diversified small-scale agriculture to large-scale, highly specialized agriculture, which requires automation to manage and provide an outlet for a growing product volume.
- Increasing problems with the availability of seasonal labor, although this is also affected by the economic and social cycles characteristic to each region.
- Inconvenience of destructive quality control methods, which increase inspection times and do not ensure the individual quality of the agricultural products.
- Development of new sensors allows the detection of injuries and anomalies, which are not visible to human eyes.

The quality of a particular fresh or processed fruit and vegetable is defined by a series of characteristics, which make it more or less attractive to the consumer, such as its ripeness, size, weight, shape, color, presence of blemishes and diseases, presence or absence of fruit stems, presence of seeds, etc. In summary, they cover all of the factors that exert an influence on the product's appearance, on its nutritional and organoleptic qualities or on its suitability for preservation. Most of these factors have traditionally been assessed by visual inspection performed by trained operators, but currently many of them are estimated with commercial vision systems, or these will be soon incorporated to the sorting machines.

The high risk of human error in classification processes has been underlined and it is one of the most important drawbacks that machine vision can avoid. In a study carried out with different varieties of apple, where various parameters of shape, size, and color were compared, one of the conclusions reached was the limited human capacity to reproduce the estimation of quality, which they define as "inconsistency" (Paulus et al. 1997). Moreover, as the number of parameters considered in the decision-making process increases, so does the error in classification.

Furthermore, automatic inspection allows for the generation of precise statistics on aspects related to the quality of the inspected perishable product, which leads to greater control over the product and facilitates its traceability.

Nevertheless, the automation of agricultural produce inspection shows certain particularities and problems, as for example:

- Great variability, due to differences between species and varieties, as well as individual differences in shape, color, size, etc.
- Product's fragility, which conditions the mechanisms and technologies that can be used for the automatic separation of the product in categories.
- Physiological evolution that fruit and vegetable continue to undergo after harvest and during handling and storage.
- Seasonal nature of production and the expected long-term return on invest obtained limit investments in automation and, therefore, the technology that can be employed. For instance, sorting machines are used only a few months a year in packinghouses of certain commodities, thus yearly recovery of investment is low.

Nevertheless, nowadays fresh and processed food manufacturers benefit enormously from machine vision techniques, as they allow them to reduce costs, to homogenize the quality of products, and to reduce manual handling.

2.1.1 Basics of Commercial Machine Vision

2.1.1.1 Lighting Sources

The perceived color of an object in a scene basically depends on the lighting source, the reflective characteristics of its surface, and the spectral response of the observer. For this reason, international standards define average values for the human observer (CIE Standard S 014-1/E:2006) and describe standard illuminants (CIE Standard S 014-2/E:2006). However, these standards are difficult to implement in industrial applications, since common commercial illuminants change their emission spectra during their life span or are affected by dust. Moreover, the angle of observation is affected by mechanical and economical constraints related to the design of an inspection chamber in a packing house.

The success of a machine vision inspection system largely depends on the lighting system, more than on a sophisticated image analysis (Brown et al. 1993). The characteristics of the lighting source have a major influence on the system's performance and final cost and exert a decisive influence on the time needed for the system to process the images. Under unstable or variable lighting conditions the preprocessing of the images needed to obtain reasonable results increases rapidly in complexity. The effects of a good lighting system are therefore twofold: on the one hand, it allows the user to maximize the quality of the final results of the image analysis and, on the other, it also enables the user to perform a more efficient data analysis by avoiding the costly preprocessing stages required to eliminate noise or to correct the lighting (Zheng et al. 2006a).

The quality of the image depends, among other factors, on the spectral emission of the light source, the way the light is applied to the scene, the characteristics of the surface that receives the light (reflective capacity, roughness, etc.), and the spatial relationships (distance and angles) between cameras, surfaces, and sources.

Owing to the biological nature of horticultural products, they show considerable variation in sizes, forms, textures, and colors, which not only emphasize the importance of the accurate design of illumination systems but also make it difficult to establish a universal system of illumination that would be effective for all products and in all kinds of situations (Du and Sun 2004).

An inefficient lighting system can prevent defects from being detected because they can be confused with healthy areas or because they coincide with glossy or shadowy areas (Bennedsen et al. 2005). The darkening of the edges can even lead to this area of sound peel frequently being mistaken for damaged skin. Inappropriate illumination can also lead to confusion with regard to perceived colors. When the lighting system cannot be improved due to mechanical, spatial, or cost limitations, it becomes necessary to correct the negative effects produced by the reflection of light on the fruit by adding preprocessing stages or by excluding the poorly lighted areas from the analysis. Different illumination techniques have been envisaged for minimizing the problems related to the reflection of the light on the object. The following describes some of the most common in the automatic inspection of horticultural products.

The objective of diffuse lighting is to create a uniformly illuminated area, in which shadows disappear and where the negative effects for image analysis caused by spectral reflection are minimized. Light diffusion is normally produced by focusing the light source onto a surface with certain properties of reflection or transmission. By illuminating the object in this way, the appearance of unwanted shadows and bright patches is reduced (Paulsen and McClure 1986). Different configurations are commonly employed to produce diffuse lighting. Figure 2.1.1 shows a method for producing it from direct lighting with several fluorescent tubes, put close to each other, and increasing light scattering by placing a translucent screen beneath the tubes plus polarized filters. This is the scheme that is most commonly employed in sorting machines in fruit and vegetable packing houses, because it allows the construction of a wide inspection chamber on top of several lanes that carry the product. The lack of uniformity of this type of lighting is one of its major drawbacks; however, these inspection chambers are compact, easy, and cheap to build.

Vision Systems

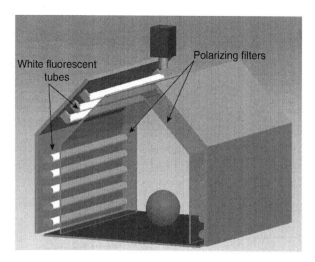

FIGURE 2.1.1 Direct lighting in an inspection chamber. Polarizing filters are frequently added to avoid specular reflection.

An alternative method is to produce diffuse illumination from indirect lighting (Figure 2.1.2). In this case, a circular fluorescent lamp is located under the target object or level with it, but a panel prevents the light from reaching the surface of the object directly. Light is directed toward the upper semi-spherical surface that reflects light uniformly onto the fruit. This system is very well fitted for quasi-spherical objects, since they receive quite uniform lighting on their entire surface. The uniformity of lighting decreases when the object is far from the center of the semi-sphere. Such inspection chambers are employed in small sorting machines for olives, cherries, or prunes, since these fruits are also small and circulate on narrow conveyor

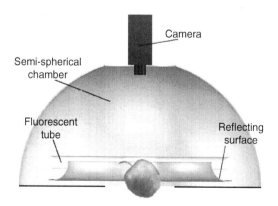

FIGURE 2.1.2 Schematic of an inspection chamber. Diffuse illumination is produced thanks to the reflection of light on the upper semi-spherical chamber. Direct lighting of the object is avoided by the reflecting surface that surrounds the circular fluorescent tube.

FIGURE 2.1.3 (See color insert following page 376.) Back illumination of satsuma segments. This method provides a highly contrasted image that facilitates the detection of seeds and the analysis of size and shape.

belts or rollers. Semi-spherical chambers are normally more expensive and difficult to build.

In some cases, the object to be inspected is translucent or the most important characteristic to be assessed by machine vision is associated with an accurate determination of its contour. In such circumstances, backlighting can be the most appropriate technique to employ: light is applied so that the object is situated between the light source and the camera. Figure 2.1.3 shows a picture obtained in an inspection chamber of a sorter of segments of mandarins. By illuminating the segments from behind, they clearly contrast with the background and their size and shape are easily determined. At the same time, seeds appear dark and highly contrasted, which facilitates their detection.

In addition to all the lighting methods described above, cross-polarization can avoid unwanted specular reflection in specific areas of the scene. For this purpose, the object is illuminated with polarized light and a polarizing filter is fitted on the camera lens. Total elimination of reflections is achieved when the filter polarizes the light perpendicular to the light source. Such filtering has been employed in a machine for inspecting citrus fruits (Moltó et al. 2000a).

The three methods described above (direct, indirect, and backlighting) are used in conventional inspection machines based on image analysis (Table 2.1.1).

Other illumination systems have been developed in laboratory and are under research. For instance, illumination with structured light consists of the projection of a pattern of light on the fruit's surface so that the resulting differences between the original and the projected pattern can be measured and used to obtain volumetric information (Table 2.1.2). This type of illumination enables the detection of

TABLE 2.1.1
Summary of Major Advantages, Disadvantages, and Commercial Use of Different Lighting Systems

	Advantages	Disadvantages	Common Commercial Application
Direct lighting in rectangular inspection chambers	Cheap, easy to build, compact (easy adaptation to a packing line) Suitable for wide inspection zones	Medium lighting uniformity	Large fruit and vegetables (oranges, apples, peaches, etc.)
Indirect lighting, spherical chambers	Adequate for spherical objects Uniform lighting in the center of the semi-sphere	Difficult to build, expensive. Require more space or produce smaller illuminated areas	Small fruit and vegetable (olive, prunes, cherries, etc.)
Back lighting	Adapted to the inspection of internal tissues Improved accuracy when detecting contours	Distortion of colors Not suited for inspection of external damage	Internal inspection of translucent produce (mandarin segments) Fast assessment of size and shape of opaque produce (rice)

TABLE 2.1.2
Summary of Applications of Different Light Sources

Light Source	Application
Visible	Detection of most damage that a human observer can perceive
Near infrared	Detection of invisible rottenness Stem identification Background removal Identification of particular skin damage Maturity estimation (under research)
Ultraviolet	Detection of skin breakage Detection of oleocellosis Detection of initial rottenness
Structured light	3-D estimation Detection of shape irregularities Detection of stem
Laser	Maturity (under research) Skin thickness

irregularities in the fruit's skin, and has been used to detect some quality defects or to locate the stems in apples (Yang 1993).

Non-visible areas of the light spectrum, such as ultraviolet or near infrared, are under investigation for use as an aid in the estimation of maturity and for the detection of certain defects that are invisible or difficult to detect by standard machine vision.

Near-infrared radiation has been widely reported as a tool for the detection of damage in fruit and vegetable. A large part of the bibliography refers to spectrophotometric studies, which reveal the spectral characteristics of different plant tissues from the same product or of the areas affected by different types of blemishes or diseases. From the data obtained, the selection of the most effective wavelengths for each case enables the development of vision systems (Table 2.1.2).

As an example, spectrophotometric studies on citrus peel in the visible and near-infrared wavelength range revealed that the reflectance curves of peel areas having certain injuries or diseases showed significant differences with respect to the healthy areas (Gaffney 1973). Wavelengths between 750 and 2200 nm are interesting for the quality evaluation of various biological materials according to criteria of maturity, sugar content, and signs of damage (Binenko et al. 1989). For example, spectral changes due to mechanical damage produced during the storage of apples give rise to lower reflectance from the affected zones, which increases with the storage time (Upchurch et al. 1994).

As a consequence of these or similar studies, some authors have explored the possibility of inspecting fruits and vegetables in several spectral bands at the same time (Blasco et al. 2007a). Figure 2.1.4 illustrates the basics of multiband inspection systems. It shows images of the same orange with damage caused by *Penicillium digitatum* (green mould) using cameras sensitive in the band passes of ultraviolet (200–400 nm), visible (400–750 nm), and near-infrared (750–1800 nm) illumination.

An image acquisition system in the visible spectrum, with which it is possible to measure size and shape, combined with infrared spectrometry have been used to determine the ratio between leaves and stems in alfalfa plants (Patil and Sokhansanj 1992).

Laser lighting has been also used to emphasize features associated with the quality of fresh products. The image size produced by laser's dispersal near its point of incidence has been related to apple firmness and maturity (Duprat et al. 1995). Coherent polarized laser emissions diffused by the sub-cuticular layers of the peel, filtered and captured in a matrix charge-coupled device (CCD) camera have demonstrated to be a nondestructive technique for measuring the peel thickness of oranges destined for fresh market consumption. Resulting correlations suggest that this method may be a successful tool in real-time food processing operations, giving an objective evaluation of peel thickness, and subsequently, edible volume, juice content, and the ease with which the peel can be removed (Affeldt and Heck 1993). More recent works use image analysis techniques to study laser light backscattering to predict soluble solid contents and firmness of apples (Lu 2004; Qing et al. 2007).

In many cases, it is not easy to obtain an illumination system that is diffuse and uniform. Often correction procedures are needed to reduce the heterogeneity due to the lighting system. In most cases, these methods are based on plane models and

Vision Systems

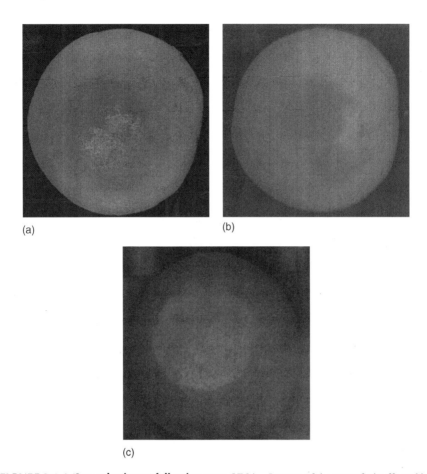

FIGURE 2.1.4 (See color insert following page 376.) Images of the same fruit affected by green mould in (from left to right) visible, near-infrared, and ultraviolet bands.

use specific plates with known reflectance spectra to modify the impression that light reflected from the scene produces on the camera (Kleynen et al. 2005). Then a problem occurs when the object to be inspected can no longer be considered flat. These techniques definitively fail to correct the effect introduced by the reflection of light on quasi-spherical surfaces. This generally results in a darkening of the object's edges, while the central part appears brighter (Gómez et al. 2008). Such difficulties arise in the inspection of many fresh products that are more or less spherical in shape, as in the case of oranges, peaches, tomatoes, mandarins, apples, and so forth.

Various attempts have been made to solve this problem by applying spherical models with a constant curvature and variable radii (Tao and Wen 1999). Their ultimate purpose is to increase the rate of correct classification of damaged and healthy fruit, without an in-depth correction of the image that makes it possible to differentiate between different types of damage.

2.1.1.2 CCD Cameras and Video Signals

Video cameras convert the light that they receive from the scene into electronic signals. Depending on the application, different image acquisition devices are encountered in the literature. Most popular cameras are based on a CCD that consists of an array of sensors (pixels), each of which is a photocell and a condenser (Peterson 2001). The load acquired by the condenser depends on the amount of light received by the photocell. In the CCD, these charges are converted into voltage and subsequently converted into a video signal.

Some cameras are based on a linear CCD, composed of a one-dimensional array of cells, which acquire a narrow strip of the scene; these are called line scan cameras. They are suitable for use in machines where the object is moved below the camera, or the camera moves above the object, so that the complete image of its surface is gradually obtained, line by line. The main advantage of these cameras is the higher resolution of the acquired images compared with other standard systems. Because only one line of pixels is captured each time, the size of the images can be varied. High resolutions can be obtained by joining the lines in further processing. On the other hand, perfect synchronization between the advance of the object and the acquisition of each line is required to avoid problems related to the spacing or overlapping of the objects, and this may require a relatively slow advance speed. Inspection systems based on these cameras have been described, for instance, for the detection of defects in prunes (Delwiche et al. 1990).

Matrix cameras are the most widespread in commercial applications. They acquire images by using a bidimensional CCD (Figure 2.1.5). The first cameras based on this technology to appear on the market were monochromatic. These only perceive the amounts of total light received by each pixel, thus producing gray level images. Monochromatic cameras are suitable for estimating the size and shape of objects, as well as for detecting defects that show a great contrast with the healthy skin.

FIGURE 2.1.5 Frontal view of a matrix camera after removing the lenses. A 1/3″ CCD can be observed.

The use of appropriate filters allowed an approximation to color estimation and, subsequently, the possibility of finding external defects on the surface of different products. A monochromatic camera and a filter centered on 550 nm have been proposed for sorting cucumbers by color (Lin et al. 1993). For example, filters centered on the red, green, and blue bands have been employed to inspect asparagus (Rigney et al. 1992).

Color CCD cameras opened up new possibilities in the estimation of quality parameters related to ripeness, and increased precision in the detection of defects. They are commercially available with a single CCD sensor, sensitive to the red, green, and blue components of light by means of a Bayer filter (Bayer 1976).

The most advanced cameras have three different monochromatic CCD sensors. Light enters through the lens and is divided into three equal parts by a prism, then directed to each CCD. Before reaching the CCD, the light is conveniently filtered, so that each CCD represents the red, green, and blue bands (Figure 2.1.6). Greater color fidelity is obtained with these cameras than with those based on a single CCD, but they are more expensive. Three-CCD cameras are currently the most widely used in research and many examples are found in the scientific literature (Du and Sun 2006a).

Multi-spectral cameras have been developed to analyze the scenes in different particular bands, since, as we described in the previous paragraph, lighting in different zones of the spectrum may reveal or underline the presence of some defects or blemishes on the surface of the product. These usually decompose the incident light into parts that are directed toward specific sensor devices. For instance, a camera composed of a color and a monochrome CCD sensor, this last fitted with a filter centered on 750 nm (NIR), has been proposed for inspecting fresh fruit (Moltó et al. 2000b). By means of an adapted optical system, the same scene was captured with both sensors, thus obtaining four images (infrared, red, green, blue), which is called a multi-spectral image (Figure 2.1.7).

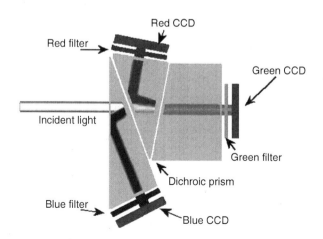

FIGURE 2.1.6 Schematic of a 3 CCD camera.

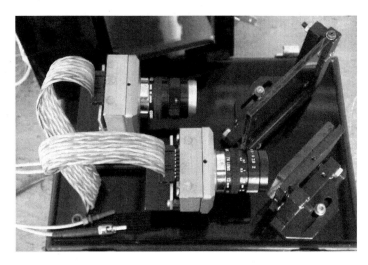

FIGURE 2.1.7 Multi-spectral camera composed of a color and a monochrome CCD sensor with a filter centered on 750 nm (NIR).

Another special camera, aimed at acquiring images at the 740 and 950 nm wavelengths, in combination with visible imaging, has been proposed for classifying apples in real time (Throop et al. 2005). Other authors use four interference bandpass filters centered on 450, 500, 750, and 800 nm for detecting defects in apples (Unay and Gosselin 2006).

The images captured by CCD cameras, which are named frames, can be interlaced (composed of two fields) or non-interlaced (progressive scan cameras). Interlaced cameras first scan odd lines of pixels and then even ones, building what are known as the odd and even fields. The first CCD cameras were interlaced for compatibility with standard television signals. The reason for producing interlaced images was to preserve the sensation of movement in the human eye while watching a video on a screen. However, when inspecting agricultural products, which travel at a high speed under the camera, the objects move between the acquisitions of both fields so they are displaced in relation to each other, thereby deforming the shape of the objects and complicating image analysis. To avoid these problems, progressive scan cameras produce non-interlaced images. This is combined with a high electronic shutter speed, which decreases the effect of the object's movement by reducing the time that the CCD is exposed to the light coming from the scene. As the shutter speed increases, the intensity of lighting must be increased to avoid underexposure.

More modern cameras are based on complementary metal oxide semiconductor (CMOS) circuits (Ghazanfari et al. 1996). Currently, their use is not as widespread as the CCD, but they have a promising future because of their lower energy consumption.

The increase in computational power has boosted the enlargement of images. Current cameras have surpassed the traditional Comite Consultatif International des RadioCommunications (CCIR) video standard (768×576 pixels), to 1024×768 pixels or even greater sizes. This has brought about an increase in the resolution of the images together with an improvement in the performance of machine vision systems, since nowadays it is possible to detect small defects of only a few square millimeters. Another important advance is related to the implementation of high-speed protocols for data transfer between external devices and computers, like universal serial bus (USB), FireWire IEEE (Institute of Electrical and Electronics Engineers) 1394, or Giga-Ethernet. These have modified the traditional configuration of a video camera plus a frame grabber, which initially was basically an analogical to digital converter, to direct communication between the camera and the computer.

"Intelligent cameras" are also reaching the market. They incorporate a microprocessor and an operating system on which the user can develop image-processing algorithms that run inside the camera. Also, special high-speed camera systems capable of capturing even thousands of images are also on the market, but they are too expensive and they are limited to capturing sequences of only a few seconds, which are stored in built-in memory. Although these cameras can be applied to other agricultural processes, they are not intended for inspection purposes.

The spreading of the Internet has increased the availability of the so-called IP-cameras, which transfer the images using a network standard like Transmission Control Protocol (TCP/IP), and allow the control of the camera via Internet.

2.1.1.3 Basic Steps of Machine Vision Applications

After its acquisition, the image is stored in a memory. It can be considered as an n-dimensional array ($n > 2$), in which the first two dimensions are the vertical and horizontal position of the pixels, and the rest ($n - 2$) are gray level values corresponding to the different sorts of sensors employed. For instance, a monochromatic image is a three-dimensional array in which the third coordinate is the gray level value of the pixel, which has a position determined by the first two coordinates. Therefore, a color image is a five-dimensional array, in which the last three coordinates are the gray level values obtained on the red, green, and blue sensors. Machine vision applications deal with these arrays to extract valuable information for the user.

2.1.1.3.1 Image Preprocessing
The objectives of most preprocessing techniques used in fresh and processed food inspection are normally focused toward eliminating the noise introduced by electromagnetic causes or correcting the effects of non-optimized lighting. Image preprocessing transforms one image into another, in which some relevant characteristics are highlighted to facilitate its understanding by the operator or by subsequent automation procedures. Contrastingly, another objective of preprocessing could be to homogenize the observed color or light intensity in the image, to avoid the heterogeneity caused, for instance, by irregular lighting. However, in any case, preprocessing does not imply an increase in the available information present in the acquired image.

The color of a pixel in an image is often expressed in RGB (red, green, blue) values, although colors can be expressed in different three-dimensional data structure, called color spaces. RGB coordinates are the most widely used, since they work in a way quite similar to our eyes, and for this reason were initially developed for television broadcast. Other spaces like HSB (hue, saturation, brightness) and HSI (hue, saturation, intensity) are also commonly used in research. Mathematical expressions for the transformations between color spaces are standardized. However, RGB, HSB, or HSI are nonuniform color spaces, because equal distances from two points in different regions of these spaces do not produce the same sensation of difference on a standard observer. For this reason uniform spaces like CIE (Commission Internationale d'Eclairage) 1976 (L^*, a^*, b^*) color space have been defined and are commonly used for color comparison (Berns 2000). Figure 2.1.8 shows the decomposition of two RGB and HSI images into its individual components.

Depending on the way in which information is managed, preprocessing techniques are classified as follows:

- Individual pixel operators: These are operations that modify the gray level of each of the pixels of a region in an image without taking into account the neighboring pixels. They are functions that assign a new gray level, v, to an existing one, u, using the transformation: $v = J(u)$. In technical jargon this is known as pixel remapping. Thresholding is a particular case of one of these

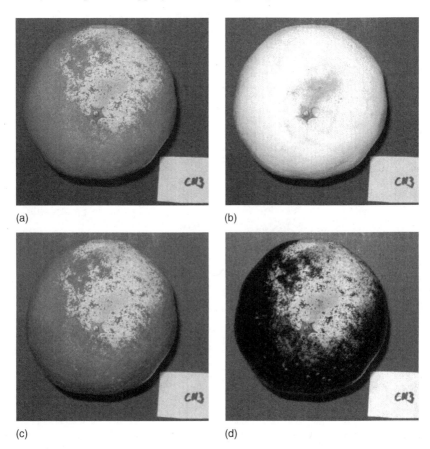

(a) (b)

(c) (d)

FIGURE 2.1.8 (See color insert following page 376.) Original color image decomposed into its individual components RGB (above) and HSI (below).

Vision Systems

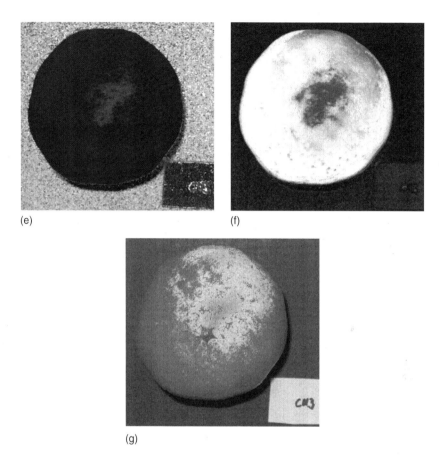

FIGURE 2.1.8 (continued)

techniques. Here, a constant gray level (normally the one that corresponds to white or black) is assigned to all the pixels having a gray value above the defined threshold, and a different gray level (the opposite to the one that has been chosen before) is assigned to the remaining pixels. The resulting image is, or can be considered, binary. Binarization accelerates many subsequent algorithms.

Transformations of the color space can also be considered as a special application of individual pixel operators. In this case, the color coordinates of a pixel in a certain space (R,G,B) are converted into R′,G′,B′ = u(R,G,B). This is used, for instance, for color normalization or for changing from RGB to HSI coordinates.

- Spatial operators: These modify the gray level of a pixel depending on the gray level of its neighboring pixels. They are frequently used not only to eliminate or to reduce noise (often perceived as blur) but also to increase the contrast between the objects of interest and the background (sharpening of objects, edge detection). They are based on mathematical operations called convolutions, and are implemented by using masks or windows (3 × 3 and 5 × 5 pixels are the most common) that define the size of the surrounding area that is used for the transformation and how its pixels are combined. Convolutions produce a new gray level value in each of the pixels of the image. This value is generally a linear combination of the gray levels of the pixels included in the window. These techniques can be used to separate touching objects and to build the skeleton of a shape.

- Techniques based on the modification of the histogram scaling: These modify the gray values of the pixels in the image to bring its histogram as close as possible to a predefined one. These operations are useful, for instance, when the images have very low contrast (and consequently have a very narrow histogram). There are many techniques that expand the histogram, which results in an increase in the perceived contrast. These techniques are not used in commercial sorting machines because they have an important computational cost.

As presented in the previous point, many preprocessing techniques have been used to correct the defective illumination of the object surface. One common solution is to erode the borders of the fruit (Blasco et al. 2003), but this involves a loss of quality in the inspection because sometimes an important part of the object surface is not analyzed. To avoid this problem some authors assume that the part of the object's surface that is the closest to the border and the most centered part belong to different regions of interest. For example, also in the same work, a supervised segmentation algorithm was used in which more classes to the sound skin (e.g., light skin, normal skin, and dark skin) were assigned while considering all three classes as the same one for some image analysis procedures. The major drawback of these techniques in multi-class problems is that most classifiers reduce its success rate as the number of classes increases (Duda et al. 2000).

2.1.1.3.2 Image Segmentation
Image segmentation techniques are a set of procedures for separating the object of interest from the background of the image (Figure 2.1.9) (Fu and Mui 1981). The simplest one is called thresholding, which consists of assigning all the pixels that have a gray level under a certain value (the threshold) to the same class (object or background). It is particularly useful in monochromatic images, but a generalization of this technique has also been proposed for color images.

In many cases, the threshold is chosen from the study of the histogram. This normally occurs when the histogram shows a bimodal distribution because background pixels have similar gray levels. This makes a peak appear on one side of the histogram, while the pixels belonging to the object have different gray levels, which are similar to each other, producing another peak in the histogram (Figure 2.1.10). In these cases, finding an optimal gray value between the two peaks facilitates the distinction between the background and the objects of interest. Examples in which these techniques have been implemented works for detecting cracking in maize (Gunasekaran 1987) or for finding defects in peaches (Miller and Delwiche 1989).

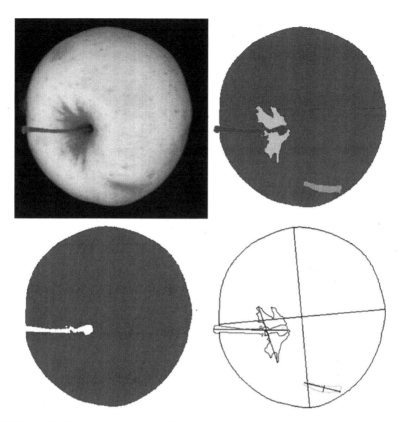

FIGURE 2.1.9 Image segmentation of an apple (top left). The regions of interest, corresponding to stem, russeting, and blemish have been differentiated (top right). Special attention must be paid to distinguish between the stem and the blemishes (bottom left). The final step is to characterize such regions by mean of morphological features (bottom right).

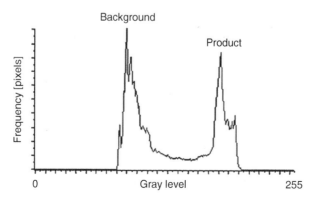

FIGURE 2.1.10 Histogram showing a bimodal distribution of the pixels in the image. Darker pixels belong to a conveyor belt. Brighter pixels belong to pomegranate arils.

In color images thresholding in one or various bands is equally employed. Sometimes, the color image is converted to a monochromatic one. In this case, the original RGB image is converted into the HSI color coordinates and subsequently the histogram of the intensity (I) is used to segment the horticultural product from the image's background (Felföldi et al. 1996). Other authors estimate apple maturity by analyzing the histogram of the saturation component (Varghese et al. 1991). Use of thresholds in the three histograms of the HSI image at the same time has also been reported (Cerruto et al. 1996).

More sophisticated techniques have been developed for color image segmentation. Discriminant analysis makes it possible to distinguish between the colors of different regions of the image. In this technique, the color coordinates are used as independent variables. Although initially applied to segment images of oranges in the tree canopy (Harrel 1991), this technique has been used in grading applications to segment the stem from the skin of oranges (Ruiz et al. 1996) or to detect damage in pistachios (Pearson and Schatzki 1998). A more advanced step is the use of nonlinear discriminant analysis (Fukunaga 1972). Other statistical techniques like cluster analysis are also used, because they can automatically divide the image into a predefined number of regions with similar characteristics, for example, similar color or texture, one or several of these parts belonging to the object of interest.

All the previous techniques are pixel oriented: their aim is to classify each pixel as belonging to a particular region of interest (Blasco et al. 2002). The color of each individual pixel, expressed either as a gray level in a monochromatic image or as two- or three-dimensional coordinates in a particular color space, is used as the only feature to segment the image. In most cases, segmentation procedures are supervised techniques, which mean that the system should be properly trained off-line in a calibration step before the on-line image analysis. In contrast, unsupervised methods, such as region-oriented segmentation, analyze information about connected areas of the image before classifying the pixels into regions of interest. Since the information used to classify a pixel is obtained from its surrounding area, these methods are robust against color variations. One method consists of assigning some pixels to particular classes between the possible objects that can be present in the image. These pixels are known as seed pixels. The pixels that surround a seed are added to the region if they previously fulfill defined similarity criteria. Neighboring pixels are then added to the region, making it grow. Regions with similar characteristics can also merge. The procedure is repeated until all the pixels are assigned to a class. Various authors use region-oriented algorithms to segment images of fresh and processed fruit (Blasco et al. 2007b); however, these techniques normally require intensive computing and are not applied in real-time applications. Figure 2.1.11 shows an example of successive steps followed by a growing region algorithm to segment different regions of interest (background, sound skin, defects) in an image of an orange.

2.1.1.3.3 Feature Extraction
The following step in image analysis is to extract some features from the objects of interest present in the image that allows their description or characterization. Commonly, the features that are calculated during the inspection of fruit and vegetable are mainly related to size, shape, and color. There are many ways of representing these

Vision Systems

in image analysis (Zheng et al. 2006b). Some authors estimate shape and size by calculating the principal inertia axes and the relationship between them (Tillett and Batchelor 1991).

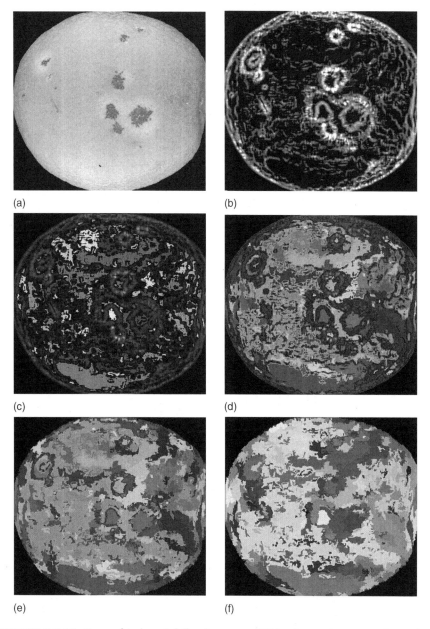

FIGURE 2.1.11 (See color insert following page 376.) Successive steps of a region growing algorithm: from the original image (top left) to the segmented image (bottom right).
(*continued*)

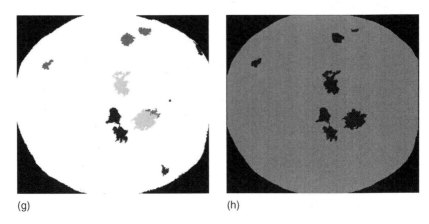

(g) (h)

FIGURE 2.1.11 (continued)

Fourier descriptors are also helpful for depicting the shape of objects. They have been used to characterize the shape of raisins (Okamura et al. 1991), or potatoes (Tao et al. 1995b). The polar signature of the perimeter of the object is decomposed with a fast Fourier transform and combined with more traditional descriptors, like length, circularity, roundness, elongation, or symmetry, in a machine to classify Satsuma segments before they are canned (Blasco et al. 2007c).

The curvature of the perimeter has also been frequently used to describe shape. The maximum curvatures of contour segments combined with coefficients of symmetry and with the relationship between the main axes of inertia have been reported to estimate the shape of maize grains (Liao et al. 1994). Some algorithms have been developed to measure length and upper curvature to classify carrots (Howarth et al. 1992).

Fractal theory is widely used for image compression and it is an important tool for image analysis (Fisher 1998). Fractal dimension can be used for describing the complexity of the shape of an object and has been employed for apple classification (Li et al. 2002).

Hough's linear and circular transforms (Hough 1962) have been used to determine the orientation shape of fruits or, with some adaptations, to detect seeding lines for a seed drill (Leemans and Destain 2006).

The morphological characteristics of peppers have been assessed using thinning techniques (Van Eck et al. 1998). Using parameters of shape as a base, some authors have attempted to estimate volume by making potatoes rotate while acquiring a series of images (Wright et al. 1984).

Often the analysis of color is not enough to detect defects or other characteristics of the object and the analysis of surface roughness is required. In these cases, texture features are employed. Texture tries to discriminate different patterns in the images by extracting the dependency of intensity between pixels and their neighboring pixels or by obtaining the variance of intensity across pixels (Haralick et al. 1973). However, textural patterns in food images are generally irregular and thus difficult to describe by some regular texture elements. Nevertheless, texture analysis has been used to classify currants, to grade apples after dehydration, and to predict the sugar content of oranges (Zheng et al. 2006c). Texture features are also important tools for describing the changes of microstructure of food surface, including potatoes, bananas, pumpkins, and carrots (Fernández et al. 2005).

2.1.1.3.4 Object Classification

When the features of each object have been extracted from the image, these are used to classify them into categories. The simplest classification method consists of determining thresholds of particular features and has been applied for sorting peaches (Singh et al. 1993). Sometimes these thresholds are set based on national or international quality standards (Blasco et al. 2003). More complex methods of classification are also described in the literature; for example, expert systems for classifying mushrooms (Heinemann et al. 1991). However, one of the most widely used techniques for object classification is statistical discriminant analysis. Linear discriminant analysis has been used in a grading system to classify fruits into different classes (Tao et al. 1995a), while nonlinear discriminant analysis has been used for the classification of peaches (Steinmetz et al. 1999).

Other methods are based on artificial neural networks (NNs). These are an interconnected group of artificial neurons that use a mathematical or computational model for information processing. An artificial neuron is an abstraction of biological neurons. It receives one or more inputs and combines them using an activation or transfer function to produce an output. NNs are adaptive systems that change their structure based on external or internal information that flows through the network. They can be considered as nonlinear statistical data modeling tools and are used to model complex relationships between inputs and outputs or to find patterns in data.

There are two most commonly used learning paradigms, which are the supervised and the unsupervised learning. In supervised learning, a given set of examples is shown to the network and the aim is to find a function that matches the examples. The most common algorithm is called back-propagation. This algorithm is used for pattern recognition, classification, or function approximation. Unsupervised learning produces an organization of data and is used as a clustering tool, for filtering and for data compression. Excellent results are reported in the classification of melons (Ozer et al. 1995) and tomatoes (Guedalia and Edan 1994). Sensors that measure color, size, shape, and defects have been integrated by using an NN to estimate apple quality (Heinemann et al. 1995).

Many authors compare the efficiency of discriminant analysis and NNs (Marchant and Onyango 2003; Díaz et al. 2004). Reviews of these and other intelligent learning techniques for feature extracting can be found in the literature (Du and Sun 2006b).

Support vector machines (SVMs) are other supervised methods that simultaneously minimize the empirical classification error and maximize the separation between two classes (Recce et al. 1996). The algorithms that use this technique map input vectors to a higher dimensional space where a maximal separating hyperplane is constructed. Their major drawback is related to the fact that they only deal with two classes. SVMs have been successfully employed to classify fruit by shape (Vapnik 1999).

2.1.1.3.5 Acceleration of Algorithms and Applications in Real Time

One of the main requirements of automatic inspection systems is to work in real time, an objective which in many cases is difficult to fulfill because of the time-consuming image analysis algorithms. Therefore, in general, very simple processing techniques are used, specific hardware is applied, or certain tasks are optimized.

Digital signal processors (DSPs) are fast microprocessors specialized in particular calculations that are sometimes included in image acquisition boards. With such a system for real-time orange classification, a speed of 5–10 fruits/s was obtained (Gui et al. 2007). It classified the color histograms obtained with six perpendicular images of the fruit using NNs. If the classification indicated a possible defect in any of the images, this image was analyzed with five kinds of independent masks that provide the inputs of another NN for identifying the defects.

Real-time classification of maize grains has been achieved with an image-processing board with a specific microprocessor, implementing certain operations in hardware. This work uses a bidimensional look-up table (LUT) in the R-G (red-green) space, to take into account aspects such as color, grain cracking, or germ damage (Liao et al. 1994).

Microprocessors and pipeline techniques for real-time identification of various types of defect in peaches are also reported. In this case, erosion was employed to discard border pixels and to avoid errors produced by defective lighting. Fruit shape was determined from the estimated eccentricity and defective areas were located from the analysis of the near-infrared histogram. The simplification of certain algorithms allowed for real-time application (Singh et al. 1993).

Machines that use linear cameras with a high-speed optic fiber for the transmission of the video signal capable of working at 5–10 fruits/s have also been described. This specific hardware allowed for the use of NNs and principal components analysis to classify the fruits into categories (Lev 1998).

Optimization of software also enables real-time segmentation of the color image. Simple algorithms to implement a Bayesian classifier the R-G color space on an LUT can enable the estimation of maturity at a rate of 5–10 fruits/s (Singh et al. 1992).

Fast algorithms are especially important when inspecting small fruits, for example, when (Moltó et al. 1998) developing a machine vision system capable of inspecting 250,000 olives/h. In this case, each image contained a bidimensional array of olives, which turned and advanced past the vision system. Three images of each olive in different positions were acquired for analysis. The use of an LUT and optimization of the algorithm to search the objects in the image allowed real-time requirements to be achieved (Hwang et al. 1998).

Simple segmentation techniques based solely on gray levels can be used in specific applications. This is the case of a sorting machine for dried wild mushrooms. For this prototype to function in real-time, low resolution images were acquired, and very simple calculations were implemented to determine parameters related to shape. The algorithms were based on NNs whose input was gray levels on the upper and lower views of the mushroom. The average mushroom classification time was 0.7 s.

A combination of hardware simplification and software optimization is parallel processing, which allows the use of complex algorithms. An algorithm divided into two parallel processes that run on two DSPs—one dedicated to the estimation of size

Vision Systems

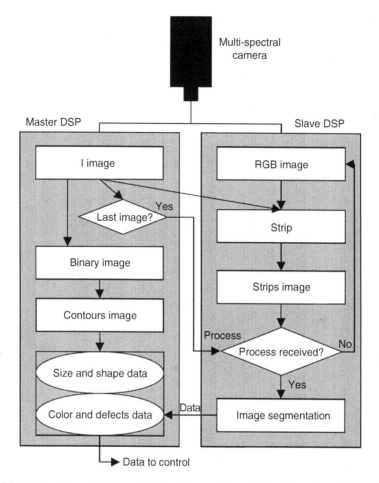

FIGURE 2.1.12 Example of parallel algorithm used for real-time inspection of citrus fruit.

and the other to color estimation and blemish detection—have been described on a sorting machine for citrus sorting (Aleixos et al. 2002). Figure 2.1.12 describes how the inspection algorithm can be parallelized. Each of the branches was processed by one DSP.

2.1.2 CURRENT COMMERCIAL SORTERS BASED ON MACHINE VISION

2.1.2.1 Elements of a Packing Line

Virtually all electronic classifiers currently available have a series of elements in common. Basically, they all consist of a feeding system that individualizes the fruit, a transport system, an inspection system formed by sensors that measure parameters related to product quality, a system that processes these measurements and makes decisions on quality, a system for synchronization, a system for separating the product into categories, and an user interface and a software that manage the whole machine (Figure 2.1.13).

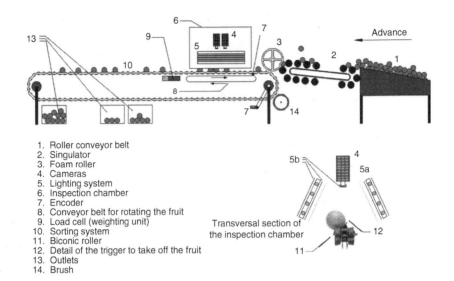

1. Roller conveyor belt
2. Singulator
3. Foam roller
4. Cameras
5. Lighting system
6. Inspection chamber
7. Encoder
8. Conveyor belt for rotating the fruit
9. Load cell (weighing unit)
10. Sorting system
11. Biconic roller
12. Detail of the trigger to take off the fruit
13. Outlets
14. Brush

FIGURE 2.1.13 Scheme of a standard machine vision sorter.

However, there are many differences between electronic classifiers of the same brand, which are essential due to their particular adaptation to the product, and are related to the number of objects processed per unit of time, the mechanism used for separating them or to their capacity for assessing one or another quality parameter. These differences increase when machines made by different manufacturers are compared. In this case, technical elements such as the transport systems, the mechanisms used for fruit singulation, the maximum number of categories that can be used to grade the product, or the technology of the sensors used to inspect them vary.

2.1.2.1.1 Individual Feeding
The performance of a sorting machine is closely related to the way it is fed. If fed too quickly, the product may accumulate in the transport and inspection systems, which normally encumbers product singulation, hindering both inspection and separation of the product by categories. Underfeeding, on the other hand, reduces performance levels.

To determine the individual quality of each inspected object, it is essential for the sensors to be able to analyze the objects separately, without any contact between them that could influence measurements. For example, if an image of two touching objects is taken (Figure 2.1.14), the vision system could confuse them for a single fruit, therefore making a mistake in the estimation of size or average color. Likewise, if two fruits of different quality travel together (Figure 2.1.15), the system cannot separate them into the different outlets; hence at least one of them will be wrongly classified. If mechanical singulation fails, these and other problems (Figure 2.1.16) may be solved with specific machine vision algorithms (Aleixos et al. 2002).

Vibration tables (Figure 2.1.17) are employed to individualize soft products (like canned fruit) before inspection. V belt singulators (Figure 2.1.18) are commonly

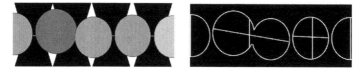

FIGURE 2.1.14 Fruit in contact during transport (left) confuses a computer vision system (right).

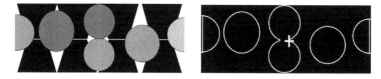

FIGURE 2.1.15 Inadequately singulated fruit transport on the line (left). The machine vision system must detect the contact point to detect this malfunctioning (right).

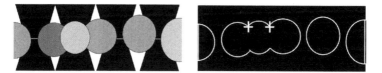

FIGURE 2.1.16 Another form of defective performance of the singulation system (left). The machine vision system must detect two contact points between adjacent fruit to detect this malfunctioning (right).

FIGURE 2.1.17 Vibration table can be used to align soft products, like satsuma segments.

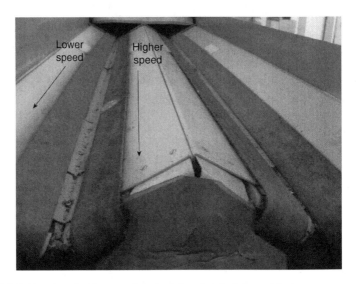

FIGURE 2.1.18 Set of rollers employed to singulate fruit in packinghouses.

utilized for fresh fruit. They are designed to receive fruits in bulk and separate them into individual pieces. They consist of two inclined belts forming a V-shape lane. The belts move forward at different speeds making the fruits rotate slowly as they move forward. If two fruits move parallel to each other, the one coming into contact with the fastest belt will be forced to move ahead of the other. Both systems can be used in combination with a conveyor belt that receives the produce at a relatively higher speed, thereby increasing the separation between close objects.

2.1.2.1.2 Transport
The individualized product travels on carrying elements, which can be conveyor belts, rollers, or a chain of basket-like receptacles made of plastic. The size of the carrying elements and the separation between them are determined by the product. Hence, there are different versions of these elements, ranging from those destined for transporting "cherry" tomatoes to those used for melons.

Rollers formed as truncated cones (Figure 2.1.19) allow the product to be rotated on its major axis of inertia, enabling most of its surface to be inspected as it advances (Leemans et al. 2004).

The transport systems are complemented with damping elements (foam rollers, strips of soft materials, etc.) to avoid product collisions and to prevent damage when the product falls to a lower level (Figure 2.1.20).

2.1.2.1.3 Sensors
The mission of the sensors is to provide information on the quality attributes of the product. The biological, physical, and chemical properties related to quality are not uniformly distributed around the whole fruit, depend on external factors (i.e., temperature and humidity), and evolve with time. For these reasons, sensors often give different responses depending on the moment of the measurement and the

Vision Systems

FIGURE 2.1.19 Conveyor shaped as truncated cones transport fruit. When they rotate, the machine vision system can inspect different parts of the fruit surface.

particular zone of the product being observed. Moreover, sensor signals are affected by temperature changes (both in the fruit and in the sensor), relative humidity, electromagnetic noise, etc. These factors complicate the assessment of the repeatability and reproducibility of the sensor readings and modify their accuracy. The target is to avoid deviations out of the range of tolerance permitted by the quality standards.

FIGURE 2.1.20 Foam roller for preventing damage at the transfer points of the machine.

Size and weight are primary quality attributes that most often require inspection. Machine vision systems complement this information by providing morphological and appearance parameters. More developed machines also include devices for measuring internal quality parameters, like spectrophotometers, x-rays, or nuclear magnetic resonance imaging (MRI), although these last two techniques are currently used only under laboratory conditions.

The most common and simple sensors for estimating weight are loading cells (Figure 2.1.21), and sometimes size is inferred from this measurement. Weight can also be estimated from the object's estimated size through machine vision. Nevertheless, the error in the estimation of weight from size can be considerable, due to the fact that the apparent density of each piece observed is not constant. For this reason, many current electronic calibrators use load cells. These are devices that convert the force exerted by the fruit and the transport unit into the deformation of strain gauges, which induces changes in their resistance and which is then converted into an electric signal. Normally, if the fruit travels in receptacles when the weight is measured, the weight of all of them is recorded off-line and is later discounted from the total weight during on-line work.

The appearance of the fruit is evaluated by means of machine vision systems. The video cameras are normally located above the devices that transport the fruits in such a way that one camera inspects several objects simultaneously. Normally, several images are obtained as the product is moved and rotates, with the aim of maximizing the inspected surface. In this case, it is very important to control the rotation speed and adapt it to the size of the product and to the advance speed.

This method does not permit analysis of the region of the object situated at both extremes of its rotation axis. A solution employed by some manufacturers consists of placing mirrors that reflect these areas to the cameras' field of vision. However, it is

FIGURE 2.1.21 Load cell used for on-line weighing of fruit.

then essential to maintain the mirrors clean, which is difficult in an industrial environment. Commercial electronic calibrators also employ several cameras to obtain a great number of images, acquired from different angles, allowing inspection of most of the fruit's surface. Nowadays, cameras sensitive to near infrared are increasingly included in the inspection chambers. In this case, conventional color cameras are used to estimate the color of the object and to detect the presence of defects on the skin, while near-infrared sensitive cameras are used to estimate the size and shape of the product units, since fruits have greater NIR reflectance than the background; furthermore, dirt on the transport devices is usually invisible in this part of the spectrum.

The demand for products with guaranteed organoleptic qualities has boosted the development of devices capable of estimating internal quality attributes, like sweetness. In general, these devices measure the absorption of the near-infrared radiation and are becoming common in current packinghouses. Early devices worked in contact with the fruit, which slowed down the machines. Contact-free sensors are currently being commercialized and give good results with thin-skinned fruits, but have lower accuracy in thick-skinned fruits, as the sugar content of the skin interferes with the measurement.

2.1.2.1.4 Decision about the Quality of the Product

The final objective of an electronic sorter is to separate the produce depending on a combination of the quality attributes that is estimated automatically with the given sorting line. For this purpose, a computing system evaluates the measurements from the sensors and, according to its preprogramming, decides the category that should be assigned to each inspected object.

The classification techniques normally applied in commercial machines require supervised learning. Before on-line work, a representative sample from each product category is selected, from which the quality parameters that permit the generation of a classification model are obtained. During on-line operation, every new inspected object is assigned to the corresponding class according to this model.

The most widely used classification models are based on statistical methods. One of the commonest is the method of minimum distance, which assigns objects to the class whose reference sample is nearest in the n-dimensional space defined by the parameters that are measured. Other methods are based on discriminant analysis derived from Bayes theorem. In addition to statistical classifiers, NNs have also been introduced as described in Section 2.1.1.3.

Some systems use a decision tree to determine quality. For example, first of all, the objects are separated according to the presence of external defects. Then, within each defect cluster they are classified by size and, finally, by color.

Automatic calibration machines have different outlets to discharge the product depending on its assigned category. All electronic calibrators possess an interface where the operator can define the attributes to be measured; how to combine these attributes to assign each object to a category; and, finally, how the product will be grouped, relating each of the combinations of selected attributes to specific outlets.

2.1.2.1.5 Systems for Separating the Units of Product

The systems used for separating the units into categories are currently one of the major differences between commercial sorting machines. It should be borne in mind that the product is discharged while advancing at a speed that is often greater than 1 m s^{-1}. Furthermore, some height transfers are made between transport systems at two different heights, which causes collisions and impacts that may damage the product.

From the moment when the product is inspected until it is discharged, it travels a known distance in a certain time and the control system must know its exact position. For this purpose, electronic encoders are located directly or indirectly over the motor rollers of the transport system. The encoder generates a number of electronic pulses proportional to the angle that motor has rotated. These pulses are received by the calibrator's electronic control and feed the tracking system, which calculates the precise position of every inspected object at any time. The tracking system also activates the device that discharges the product unit when it reaches the corresponding outlet.

Product separation devices depend on the kind of receptacle used for transport, but, in general, they are activated by electromagnets. For example, transport systems based on truncated cone rollers incorporate a pin in their lower position upon which electromagnets situated at the outlet act. When the electromagnet is in the standby position, the pin passes below an unloading ramp. When the tracking system detects an object passing through the assigned outlet, it excites the electromagnet and a lever makes the pin pass over the ramp, which makes the truncated cones discharge.

In the case of small products, pneumatic expulsion systems are commonly used (Figure 2.1.22). Here, instead of electromagnets, the tracking system activates electrovalves, which produce the ejection of pressurized air, pushing the product off the transport system (Blasco et al. 2008).

FIGURE 2.1.22 Pneumatic systems are used for removing soft objects from the conveyor belt. This picture shows a pneumatic nozzle.

Vision Systems

2.1.2.2 Major Constraints of Machine Vision Systems in Commercial Applications

The automation of the inspection of agricultural products shows certain particularities with regard to other industrial sectors, such as the great variability of the objects inspected, due to differences between species and varieties as much as to individual differences in shape, color, size, etc. Likewise, the physiological evolution that agricultural products continue to experience after harvesting continuously modifies their properties. Moreover, the fragility of the product should also be considered as it conditions the kind of machines and techniques that are used. Finally, the great amount of samples to be inspected per time unit (i.e., in citrus packinghouses machines have to inspect up to 20 fruits/s) require very fast responses, which makes reduction of computing times an essential task. Major constraints of machine vision systems in current commercial applications are related to

- Lighting: In the installations, there are always external light sources that interfere with lighting in the inspection chamber. The lighting system must be sufficiently powerful to adequately illuminate the whole scene, which often produces certain reflections on the surface of the fruit or the line's rollers, causing the machine vision system to give erroneous results.
- Image background: Appropriate color selection of the background, which should contrast with the color of the product, simplifies fruit inspection, but it should be borne in mind that the rollers, filters, and lenses can be obscured by dust and product remains, which produce unwanted effects in the image, negatively affecting its subsequent analysis.
- Image resolution: The resolution of the images affects the measurement of size and geometric and morphological characteristics. But it also affects defect detection: there are defects that are impossible to detect at low resolutions. On the other hand, the lower the resolution, the greater the image-processing speed and the transference time from the camera and the computer. The current machinery works with relative low resolutions (if a scene of 400×300 mm is investigated on a picture of 380×240 pixel size, the resolution will be ~ 1.053 mm^2/pixel) to cover the requirements of real time, which complicates the detection of certain defects. A situation of compromise should be reached between processing speed and resolution.
- Objects are not sufficiently singularized: Some current systems do not check these aspects because of the requirements of real-time processing, which produces errors in product classification.
- Electromagnetic noise: Sorters are equipped with cameras, which transmit the video signal via cables that are closely located to the machinery's wiring, electrical engines, and electric boards. These create electromagnetic fields that can negatively affect the signal, creating distortion and introducing noise. Additionally, the vibration of the machinery may cause a progressive loss of focus in the camera lenses.
- Need for very high-speed processing: The inspection algorithms are usually very time consuming, and depend on the number of parameters to be

calculated, the number of images processed for each object, and their size. To achieve fast, precise, and reliable systems, specific acceleration by hardware is sometimes required.

At present, inspection machines based on machine vision systems provide satisfactory measurements of quality attributes related to size, shape, color, and appearance. The qualification of external color is necessary not only to satisfy consumers' preferences but also to determine the product's destination, whether this can be degreening, storage in cold chambers, or immediate transport to the final destination. There is still a lack in feasible grading methods with regard to the evaluation of external defects. The most challenging tasks are related not to detection, which is solved in many cases, but rather to the identification of the individual defects, since some of them (i.e., those produced by fungal infestations or those that can favor these infestations) may affect the sanitary situation of whole batches that may lead to further microbial growth during transport to the consumer. Consequently, new machine vision techniques using invisible parts of the spectrum are being developed for early identification.

2.1.3 FUTURE OF MACHINE VISION APPLICATIONS FOR RAW MATERIAL INSPECTION

2.1.3.1 Exploitation of Fluorescent Properties

Ultraviolet light has been used to induce fluorescence, which is the result of the excitation of a molecule by high energy light (short wavelength range) and subsequent instantaneous relaxation with the emission of lower energy light (longer wavelength range).

Fluorescence allows certain types of external damage to be observed. Ultraviolet (UV) sources (350–380 nm) that may be either fluorescent tubes or mercury vapor lamps can induce visible fluorescence (500–600 nm) of essential oils present on the skin of citrus produced by cell breakage. Fluorescence then augments the contrast of damage caused by fungal infestations (Figure 2.1.23). Mechanical damage caused to tangerines has been also detected by using fluorescence to check for the presence of essential oil in the skin as a result of the lesions produced (Uozumi et al. 1987).

The chlorophyll molecule produces fluorescence, when it is excited with wavelengths of around 420 or 680 nm (Gibbons and Smillie 1980). This property is used to locate incipient peel injury and is associated with skin damage, infections, or fecal contaminations (Lefcourt and Kim 2006).

Some researchers have observed differences in the product's fluorescent response when it has been damaged during excessive cold storage. These changes have been studied in bananas and mangos and associated to a reduction of chlorophyll in the skin of these fruits (Smillie et al. 1987). The ease of inducing fluorescence in agricultural products and the fact that this can be captured by standard cameras make this technique feasible to be implemented and hence could shortly be employed on a commercial basis. Also, chlorophyll fluorescence kinetic analysis provides a tool for receiving data on the plant response to biotic and abiotic stress (see Sections 4.2 and 4.3).

Vision Systems 115

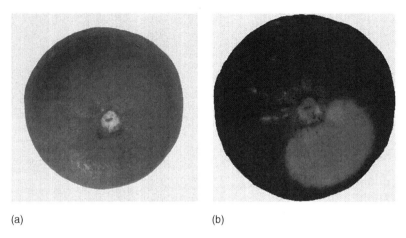

(a) (b)

FIGURE 2.1.23 Fungal infestation cannot be seen in the visible image (a). However, it contrasts clearly with the rest of the skin in a fluorescence image (b).

2.1.3.2 Hyperspectral Images

The vision systems commonly employed in automatic inspection tasks, mimicking the human eye, use electromagnetic radiation in three bands of the visible spectrum, centered on what are perceived as the colors red, green, and blue. Combining images acquired in these three relatively wide bands fails to take advantage of the fact that some defects on the fruit skin are better appreciated at specific wavelengths (Gómez et al. 2006) (Figure 2.1.24), or in combinations of images acquired at the sensitive narrow band passes (Kim et al. 2004).

The first approaches to select sensitive wavelengths were carried out in spectrophotometric studies (see Chapter 3). An advance in the use of narrowband information is the use of hyperspectral cameras, which acquire a large number of monochromatic images of the same scene at preprogrammed wavelengths. The set of monochromatic images acquired constitutes a hyperspectral image. The hyperspectral image systems arise from the field of remote satellite sensing research. Their use is expanding rapidly to other scientific fields, such as medicine, food technology, and precision agriculture, or to the inspection of agricultural products (Yang et al. 2002; Xing et al. 2005).

Hyperspectral vision systems consist of two core compounds: an image sensor (often a CCD) and a filter responsible for selecting the wavelength of the radiation that reaches the sensor. There are various kinds of filters that perform this task, the acoustic optical tunable filter (AOTF) and the liquid crystal tunable filter (LCTF) being the most significant (Poger and Angelopoulou 2001).

The AOTFs base their mode of action on the piezoelectric properties of materials and are built by joining piezoelectric transducers of an appropriate crystalline material. By exciting the transducers with the appropriate radio frequency, perturbations are produced in the material that interacts with the photons, thereby providing the capacity of frequency selection (Bei et al. 2004).

LCTF (Figure 2.1.25) devices are based on the combination of Lyot filters. A Lyot filter is made from birefringent plates usually made of quartz. Because the

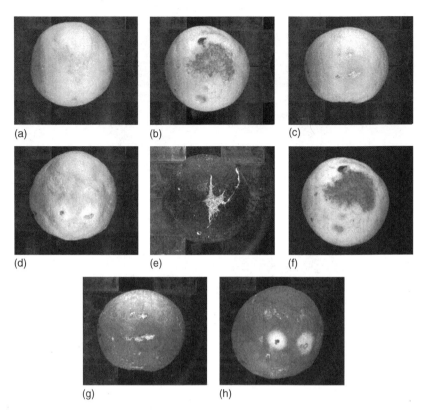

FIGURE 2.1.24 Captured images of different external damages in the same oranges using a standard monochromatic camera were acquired at particular wavelengths (from left to right: 480, 630, 530, and 540 nm). These wavelengths enhance the contrast between the defects and the sound skin.

plates are birefringent, the ordinary and extraordinary polarization components of a light beam experience a different refractive index and thus have a different phase velocity. By rotating the plates, one can shift the wavelengths of the transmitted photons (Hecht 2001).

The hyperspectral images constitute a much broader source of information than a conventional color image. The hyperspectral image is composed of a set of monochromatic images, which combined form a three-dimensional structure with two spatial dimensions x and y and a third component that contains the spectral information. On the other hand, this advantageous fact can become a problem without suitable data preprocessing, since in general these images present a lot of redundant information (Shaw and Burke 2003). Hyperspectral images have been successfully employed to predict firmness and soluble solids content of apples (Lu 2004), although care must be taken to reduce noise (Peng and Lu 2006).

Another detail to bear in mind is that when raw hyperspectral images are analyzed it is the radiance of the scene rather than its reflectance that is being analyzed. When a hyperspectral image is acquired, it is necessary to carry out the

Vision Systems

FIGURE 2.1.25 LCTF coupled to a CCD camera used for the acquisition of hyperspectral images of fruits.

opportune compensations to separate the reflectance of the scene from the radiance, and to apply techniques to reduce the amount of information obtained.

The demanding restrictions of working in real time often require reduction of the dimensionality of the problem and selection of the greatest amount of nonredundant information from a reduced number of wavelengths sensitive to the problem. Unsupervised methods such as principal components analysis or supervised ones, such as linear discriminant analysis are commonly employed.

2.1.3.3 Look at the Internal Quality

At present, the evaluation of an agricultural product's internal quality is carried out by batch sampling and destructive techniques, so that not all of the production can be inspected. For this reason, nondestructive techniques are being studied, which are still limited in extent because of their high cost or the technological difficulties involved. This part of the chapter is devoted to a brief description of those techniques that are more explored or that may be included in sorting machines for fresh and processed fruit in the near future. A more extensive description can be found in other chapters of this book.

2.1.3.3.1 X-Rays

Two x-ray techniques have been described for inspecting the fruits' interior, which associate changes in the intensity of the image with damage to tissues. The first consists of placing the fruit between an x-ray emitter and a detector, which measures the energy absorbed by the fruit. Such systems have been employed for detecting internal defects in different kind of fruits and vegetables (Schatzki et al. 1997).

The cameras used to capture this type of x-ray image are linear: they acquire the image line by line as the object advances. This produces gaps in the image if the object advances very quickly, and overlapping if it goes too slowly, for which the image normally needs previous restoration treatment before being processed (Haff and Schatzki 1997).

Damage that is particularly difficult to detect by standard methods is that caused by the dehydration of the pulp when a citrus crop is affected by frost. This change in the pulp affects fruit density and can be detected by an x-ray system (Johnson 1985). However, these methods have not been effective for the detection of seeds in mandarins. X-rays can enhance the contrast between tissues of different densities and mandarin seeds have densities that are close to the rest of the pulp (Figure 2.1.26).

In a second technique, known as computerized tomography (CT), the fruit is bombarded with x-rays from different positions on the same plane around the axis of rotation of the sample. The energy traverses the object and is measured by a detector resulting in an image of x-ray absorption of a single plane inside the object. With the information obtained, a three-dimensional image of the interior of the object is constructed. This technique has been investigated for estimating the internal quality of different fruits and vegetables (Morishima et al. 1987).

2.1.3.3.2 Magnetic Resonance Imaging

The MRI most frequently relies on the relaxation properties of excited hydrogen nuclei in water and lipids. It is based on the fact that protons in the atomic nuclei are positively charged and their movement induces a magnetic field, which makes them behave like little magnets. In the absence of an external magnetic field, the protons in a tissue sample are oriented randomly in all directions. When excited by the influence of an external magnetic field, such as that produced by a powerful magnet, they then align themselves with the magnetic field. The frequency at which the protons spin is called the precession frequency. If a short pulse of radio waves (RF) is emitted with the same frequency as the precession frequency, the protons absorb energy from that pulse. When the RF pulse stops, the protons release this energy while returning to their

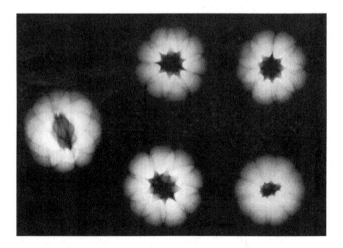

FIGURE 2.1.26 X-rays enhance the contrast between tissues of different densities but mandarin seeds have densities that are close to the rest of the pulp and cannot be seen in this image.

original orientation. The relaxation time of protons in different tissues varies according to their molecular structure. By combining gradient magnet fields in an appropriate way, gray level bidimensional images of thin slices of fruit can be obtained. Specific sequences of pulses may increase the differences in the relaxation time between tissues that can be used, for instance, to differentiate between those with great content in lipids (i.e., seeds) and those with greater free water content (i.e., pulp, rotten tissues). MRI can also be used to detect the presence of air cavities inside the fruit caused by the activities of pests or by physiological disorders.

The MRI has been applied for the detection of seeds in static images of mandarins (Moltó and Blasco 2000). In these images, seeds contrast very well with the pulp of the fruit and the images can therefore be segmented using simple thresholds (Figure 2.1.27). The problem arises when the object moves, as it is the case on a packing line. In this case, preprocessing is important and required

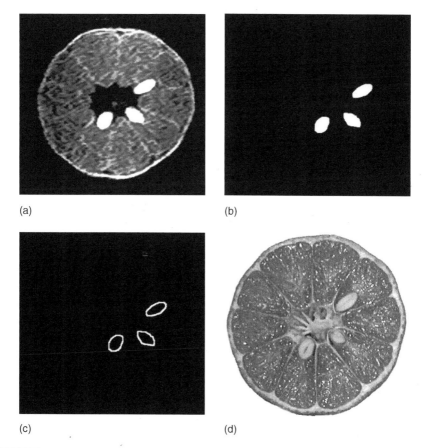

FIGURE 2.1.27 (a) Magnetic resonance image of a static mandarin. (b) Application of segmentation algorithms to detect seeds. (c) Extraction of the borders of the objects. (d) Fruit is opened to ensure that the detected objects were seeds.

(Hernández et al. 2005). MRI has also been applied to find internal damages, such as that caused by freezing (Gambhir et al. 2005).

Nowadays, MRI equipment is still far too expensive, but even so application of this technology to agro-food inspection has a very promising future. A more detailed description of these techniques can be found in Section 5.3.

2.1.3.4 Combining Several Sensors for Complete Quality Assessment

When carrying out any inspection task, we make use of the combination of our senses, as well as the previously acquired knowledge (Amerine et al. 1965). As sensors often measure only a single constituent or quality property, combined techniques have to be optimized to measure overall quality. Commercial application of these techniques is beneficial for the consumer as well as for the producer, whose products have to comply with market demands of uniform high quality (Butz et al. 2005).

A fruit quality inspection system is sounder and more reliable when it possesses different sensors for obtaining information. The combination of this information is carried out by means of techniques called sensor fusion, which work using both complementary and redundant sensors. A sensor fusion approach combined machine vision to estimate fruit color, with infrared spectrometry to estimate sugar content for improving the estimation of ripeness (Steinmetz et al. 1999). Multisensor systems for grading tomatoes by color and firmness are also reported; some of them combined information as heterogeneous as color, electrical impedance, mechanical impedance, and sugar content to predict tomato ripeness (Schotte and De Baerdemaeker 1998).

The advantages of redundant fusion techniques are clear if we bear in mind that the decision is based on global knowledge and that, furthermore, the failure of some of the sensors could be mitigated by the information extracted from the rest. This leads to much sounder and more reliable decisions than those taken through the information deriving from separate sensors. As an example, redundant sensor fusion has been employed for estimating the firmness of different fruits (Steinmetz et al. 1996). Fully automated machines that use fusion techniques to improve quality assessment are also found in the literature. For instance, a system capable of using machine vision, firmness sensors, and NIR sensors to assess the quality of fresh fruit and that consequently packs these fruits in boxes using a robot (Moltó et al. 1997).

The increase in accuracy of new sensors for quality assessment and the increase in computer speed boost the development of automated machines based on fusion techniques.

REFERENCES

Affeldt, H.A. and R.D. Heck. 1993. Optics for produce quality evaluation: Laser diffusion for orange peel thickness measurement. In *Proceedings of the SPIE 1836*, pp. 252–260.

Aleixos, N., J. Blasco, F. Navarrón, and E. Moltó. 2002. Multispectral inspection of citrus in real-time using machine vision and digital signal processors. *Computers and Electronics in Agriculture* 33:121–137.

Amerine, M.A., R.M. Pangborn, and E.B. Roessler. 1965. *Principles of Sensory Evaluation of Foods*. Academic Press, New York, ISBN: 0120561506.

Bayer, B.E. 1976. Color imaging array. US Patent No. 3971065.

Bei, L., G. Dennis, H. Miller, T. Spaine, and J. Carnahan. 2004. Acousto-optic tunable filters: Fundamentals and applications as applied to chemical analysis techniques. *Progress in Quantum Electronics* 28:67–87.
Bellon, V., G. Rabatel, and C. Guizard. 1992. Automatic sorting of fruit: Sensors for the future. *Food Control* 3:49–54.
Bennedsen, B.S., D.L. Peterson, and A. Tabb. 2005. Identifying defects in images of rotating apples. *Computers and Electronics in Agriculture* 48:92–102.
Berns, R.S. 2000. *Billmeyer and Saltzman's Principles of Colour Technology.* 3rd edn. Wiley-Interscience, pp. 120–132. ISBN: 978-0-471-19459-0.
Binenko, V.I., N.V. Voronov, and D.V. Nedotsukova. 1989. Automation of quality control of some species of vegetables and citrus fruits by optical methods. *Soviet Agricultural Sciences* 1:33–37.
Blasco, J., N. Aleixos, J.M. Roger, G. Rabatel, and E. Molto. 2002. Robotic weed control using machine vision. *Biosystems Engineering* 83:149–157.
Blasco, J., N. Aleixos, and E. Moltó. 2003. Machine vision system for automatic quality grading of fruit. *Biosystems Engineering* 85:415–423.
Blasco, J., S. Cubero-García, S. Alegre-Sosa, J. Goméz-Sanchís, V. López-Rubira, and E. Moltó. 2008. Short communication. Automatic inspection of the pomegranate (*Punica granatum* L.) arils quality by means of computer vision. *Spanish Journal of Agricultural Engineering* 6(1):12–16.
Blasco, J., N. Aleixos, J. Gómez, and E. Moltó. 2007a. Citrus sorting by identification of the most common defects using multispectral computer vision. *Journal of Food Engineering* 83:384–393.
Blasco, J., N. Aleixos, and E. Moltó. 2007b. Computer vision detection of peel defects in citrus by means of a region oriented segmentation algorithm. *Journal of Food Engineering* 81:535–543.
Blasco, J., S. Cubero, R. Arias, J. Gómez, F. Juste, and E. Moltó. 2007c. Development of a computer vision system for the automatic quality grading of mandarin segments. *Lecture Notes in Computer Science* 4478:460–466.
Brosnan, T. and D.W. Sun. 2004. Improving quality inspection of food products by computer vision—a review. *Journal of Food Engineering* 61:3–16.
Brown, G.K., D.E. Marshall, and E.J. Timm. 1993. Lighting for fruit and vegetable sorting. ASAE Paper No. 93-6069.
Butz, P., C. Hofmann, and B. Tauscher. 2005. Recent developments in non-invasive techniques for fresh fruit and vegetable internal quality analysis. *Journal of Food Science* 70:131–141.
Cerruto, E., S. Failla, and G. Schillaci. 1996. Identification of blemishes on oranges. EurAgEng Paper No. 96F-017.
Chen, Y.R., K. Chao, and M.S. Kim. 2002. Machine vision technology for agricultural applications. *Computers and Electronics in Agriculture* 36:173–191.
CIE Standard S 014-1/E:2006. 2006. Colorimetry—Part 1: CIE standard colorimetric observers. CIE-Commission Internationale de l'Eclairage. Kegelgasse 27 A-1030 Wien, Austria.
CIE Standard S 014-2/E:2006. 2006. Colorimetry—Part 2: CIE standard illuminants. CIE-CIE-Commission Internationale de l'Eclairage. Kegelgasse 27 A-1030 Wien, Austria.
Delwiche, M.J., S. Tang, and J.F. Thompson. 1990. Prune defect detection by line-scan imaging. *Transactions of the ASAE* 33:950–954.
Díaz, R., L. Gil, C. Serrano, M. Blasco, E. Moltó, and J. Blasco. 2004. Comparison of three algorithms in the classification of table olives by means of computer vision. *Journal of Food Engineering* 61:101–107.
Du, C.J. and D.W. Sun. 2004. Shape extraction and classification of pizza base using computer vision. *Journal of Food Engineering* 64:489–496.
Du, C.J. and D.W. Sun. 2006a. Estimating the surface area and volume of ellipsoidal ham using computer vision. *Journal of Food Engineering* 73:260–268.

Du, C.J. and D.W. Sun. 2006b. Learning techniques used in computer vision for food quality evaluation: A review. *Journal of Food Engineering* 72:39–55.

Duda, R.O., P.E. Hart, and D.G. Stork. 2000. *Pattern Classification*, 2nd edn., Wiley-Interscience, pp. 107–111. ISBN: 978-0-471-05669-0.

Duprat, F., H. Chen, M. Grotte, D. Loonis, and E. Pietri. 1995. Laser light based machine vision system for nondestructive ripeness sensing of Golden apples. In *Proceedings of the First Workshop on Control Applications in Post-Harvest Processing Technology*, pp. 58–93.

Felföldi, J., A. Fekete, and E. Györi. 1996. Fruit colour assessment by image processing. EurAgEng Paper No. 96F-031.

Fernández, L., C. Castillero, and J.M. Aguilera. 2005. An application of image analysis to dehydration of apple discs. *Journal of Food Engineering* 67:185–193.

Fisher, Y. 1998. *Fractal Image Encoding and Analysis* Springer, pp. 3–17. ISBN 978-3-540-63196-5.

Fu, K.S. and J.K. Mui. 1981. A survey on image segmentation. *Pattern recognition* 13:3–16.

Fukunaga, K. 1972. *Introduction to Statistical Pattern Recognition*. Academic Press, New York, ISBN: 0122698517.

Gaffney, J.J. 1973. Reflectance properties of citrus fruit. *Transactions of the ASAE* 16:310–314.

Gambhir, P.N., Y.J. Choi, D.C. Slaughter, J.F. Thompson, and M.J. McCarthy. 2005. Proton spin–spin relaxation time of peel and flesh of navel orange varieties exposed to freezing temperature. *Journal of the Science of Food and Agriculture* 85:2482–2486.

Ghazanfari, A., J. Irudayaraj, A. Kusalik, and M. Romaniuk. 1996. Machine vision grading of pistachio nuts using Fourier descriptors. *Journal Agriculture Engineering Research* 68:247–252.

Gibbons, G.C. and R.M. Smillie. 1980. Chlorophyll fluorescence photography to detect mutants chilling injury and heat stress. *Carlsberg Research Communications* 45:269–282.

Gómez, J., J. Blasco, N. Aleixos, F. Juste, and E. Moltó. 2006. Hyperspectral computer vision system for early detection of *Penicillium digitatum* in citrus fruits. In *XVI CIGR world Congress: Agricultural Engineering for a Better World*, Bonn, Germany, pp. 241–242.

Gómez, J., E. Moltó, G. Camps, L. Gómez, N. Aleixos, and J. Blasco. 2008. Automatic correction of the effects of the light source on spherical objects. An application to the analysis of hyperspectral images of citrus fruits. *Journal of Food Engineering* 85:191–200.

Guedalia, D. and Y. Edan. 1994. A dynamic artificial neural network for coding and classification of multisensor quality information. *ASAE Paper 94-3053*.

Gui, J., X. Rao, and Y. Ying. 2007. Fruit shape classification using support vector machine. In *Proceedings of SPIE 6764*:39.

Gunasekaran, S. 1987. Image processing for stress cracks in corn kernels. *Transactions of the ASAE* 30:266–270.

Haff, R.P. and T.F. Schatzki. 1997. Image restoration of line-scanned x-ray images. *Optical Engineering* 36:3288–3296.

Haralick, R.M., K. Shanmugam, and I. Dinstein. 1973. Textural features for image classification. *IEEE Transactions on Systems Man and Cybernetics* 6:610–621.

Harrel, R.C. 1991. Processing of color images with Bayesian discriminate analysis. In *I International Seminar on Use of Machine Vision Systems for the Agricultural and Bio-Industries*, Montpellier, pp. 11–20.

Hecht, E. 2001. *Optics*, 4th edn., Addison Wesley. ISBN: 978-0805385663.

Heinemann, P.H., H.J. Sommer, C.T. Morrow, R. Beelman, C. Kao, and R. Hughes. 1991. An automated mushroom inspection system using artificial intelligence and machine vision. *ASAE Paper No 91-7001*.

Heinemann, P.H., Z.A. Varghese, C.T. Morrow, C.T. Morrow, H.J. Sommer III, and R.M. Crassweller. 1995. Machine vision inspection of Golden Delicious apples. *Applied Engineering in Agriculture Transactions of the ASAE* 11:901–906.

Hernández, N., P. Barreiro, M. Ruiz-Altisent, J. Ruiz-Cabello, and M.E. Fernandez-Valle. 2005. Detection of seeds in citrus using MRI under motion conditions and improvement with motion correction. *Concepts in Magnetic Resonance Part b-Magnetic Resonance Engineering* 26B:81–92.

Hough, P.V.C. 1962. Method and means for recognizing complex patterns. U.S. Patent 3069-654.

Howarth, M.S., J.R. Brandon, S.W. Searcy, and N. Kehtarnavaz. 1992. Estimation of tip shape for carrot classification by machine vision. *Journal of Agricultural Engineering Research* 53:123–139.

Hwang, H., C.H. Lee, and S.C. Kim. 1998. On-line sorting for mushroom via neuro-image processing. *SENSORAL 98. International Workshop on Sensing Quality of Agricultural Products 1*, pp. 199–210.

Johnson, M. 1985. Automation in citrus sorting and packing. In *Proceedings of Agrimation, I Conference and Exposition*, Chicago, USA, pp. 63–68.

Kim, M.S., A.M. Lefcourt, Y.R. Chen, and S. Kang. 2004. Uses of hyperspectral and multispectral laser induced fluorescence imaging techniques for food safety inspection. *Key Engineering Materials* 270–273:1055–1063.

Kleynen, O., V. Leemans, and M.F. Destain. 2005. Development of a multi-spectral vision system for the detection of defects on apples. *Journal of Food Engineering* 69:41–49.

Leemans, V., H. Magein, and M.F. Destain. 2004. A real-time grading method of apples based on features. *Journal of Food Engineering* 61:83–89.

Leemans, V. and M.F. Destain. 2006. Application of the hough transform for seed row localisation using machine vision. *Biosystems Engineering* 94:325–336.

Lefcourt, A.M. and M.S. Kim. 2006. Technique for normalizing intensity histograms of images when the approximate size of the target is known: Detection of feces on apples using fluorescence imaging. *Computers and Electronics in Agriculture* 50:135–147.

Lev, Z.H. 1998. Machine vision in the packing house-putting neural networks to work. *Proceedings of SPIE 3543*:82–90.

Li, Q., M. Wang, and W. Gu. 2002. Computer vision based system for apple surface defect detection. *Computers and Electronics in Agriculture* 36:215–223.

Liao, K., M.R. Paulsen, and J.F. Reid. 1994. Real-time detection of colour and surface defects of maize kernels using machine vision. *Journal of Agricultural Engineering Research* 59:263–271.

Lin, W.C., J.W. Hall, and A. Klieber. 1993. Video imaging for quantifying cucumber fruit color. *Hort Technology* 3:436–439.

Lu, R. 2004. Multispectral imaging for predicting firmness and soluble solids content of apple fruit. *Postharvest Biology and Technology* 31:147–157.

Marchant, J.A. and C.M. Onyango. 2003. Comparison of a Bayesian classifier with a multilayer feed-forward neural network using the example of plant/weed/soil discrimination. *Computers and Electronics in Agriculture* 39:3–22.

Miller, B.K. and M.J. Delwiche. 1989. Peach defect detection with machine vision. *ASAE Paper No. 89-6019*.

Moltó, E., J. Blasco, V. Steinmetz, A. Bourely, F. Navarrón, and G. Perotto. 1997. SHIVA: A robotics solution for automatic handling, inspection and packing of fruit and vegetables. In *Bio-Robotics 97: International Workshop on Robotics and Automated Machinery for Bio-Productions*, Valencia, Spain, pp. 65–70.

Moltó, E., J. Blasco, V. Escuderos, J. García, R. Díaz, and M. Blasco. 1998. Automatic inspection of olives using computer vision. *SENSORAL 98. International Workshop on Sensing Quality of Agricultural Products 1*, pp. 221–229.

Moltó, E. and J. Blasco. 2000. Detection of seeds in mandarins using magnetic resonance imaging. In *Proceedings of the International Society of Citriculture*. Orlando, FL, pp. 1129–1130.

Moltó, E., N. Aleixos, J. Blasco, and F. Navarrón. 2000a. Low-cost real-time inspection of oranges using machine vision. In *Processing AgriControl 2000. International Conference on Modelling and Control in Agriculture Horticulture and Post-harvested*, Wageningen, The Netherlands, pp. 309–314.

Moltó, E., N. Aleixos, J. Blasco, and F. Navarrón. 2000b. Assessment of citrus fruit quality using a real-time machine vision system. In *Proceedings of the 15th International Conference on Pattern Recognition 1*, Barcelona, pp. 482–485.

Morishima, H., Y. Seo, Y. Sagara, Y. Yamaki, and S. Matsuura. 1987. Non-destructive internal quality detection of fresh fruits and vegetables by CT-scanner. In *Proceedings of International Symposium on Agricultural Mechanization and International Cooperation in High Technology Era*, Tokyo, Japan, pp. 342–349.

Okamura, N.K., M.J. Delwiche, and J.F. Thompson. 1991. Raising grading by machine vision. *ASAE Paper No. 91-7011*.

Ozer, N., B. Engel, and J. Simon. 1995. Fusion classification techniques for fruit quality. *Transactions of the ASAE* 38:1927–1934.

Patil, R.T. and S. Sokhansanj. 1992. Particle size characterization in alfalfa cubes using machine vision and NIR. *ASAE Paper No. 92-6541*.

Paulsen, M.R. and W.F. McClure. 1986. Illumination for computer vision systems. *Transactions of the ASAE* 29:1398–1404.

Paulus, I., R. De Busscher, and E. Schrevens. 1997. Use of image analysis to investigate human quality classification of apples. *Journal Agricultural Engineering Research* 68:341–353.

Pearson, T.C. and T.F. Schatzki. 1998. Machine vision system for automated detection of aflatoxin-contaminated pistachios. *Journal of Agricultural and Food Chemistry* 4:2248–2252.

Peng, Y.K. and R.F. Lu. 2006. Improving apple fruit firmness predictions by effective correction of multispectral scattering images. *Postharvest Biology and Technology* 41:266–274.

Peterson, C. 2001. How it works: The charged-coupled device or CCD. *Journal of Young Investigators*, http://www.jyi.org/volumes/volume3/issue1/features/peterson.html.

Poger, S. and E. Angelopoulou. 2001. Multispectral sensors in computer vision. Stevens Institute of Technology Technical Report, CS-2001-3.http://www.cs.stevens.edu/~elli/tech-report.multi.pdf.

Qing, Z.S., B.P. Ji, and M. Zude. 2007. Predicting soluble solids content and firmness in apple fruit by means of laser light backscattering image analysis. *Journal of Food Engineering* 82:58–67.

Recce, M., J. Taylor, A. Plebe, and G. Tropiano. 1996. High speed vision-based quality grading of oranges. In *Proceedings of the International Workshop on Neural Networks for Identification, Control, Robotics and Signal/Image Processing (NICROSP96)*, Venice, Italy, pp. 136–144.

Rigney, M.P., G.H. Brusewitz, and G.A. Krauzler. 1992. Asparagus defect inspection with machine vision. *Transactions of the ASAE* 35:1873–1878.

Ruiz, L.A., E. Moltó, F. Juste, F. Plá, and R. Valiente. 1996. Location and characterization of the stem-calyx area on oranges by computer vision. *Journal of Agricultural Engineering Research* 64:165–172.

Schatzki, T.F., R.P. Haff, R. Young, I. Can, L.C. Le, and N. Toyofuku. 1997. Defect detection in apples by means of X-ray imaging. *Transactions of the ASAE* 40:1407–1415.

Schotte, S. and J. De Baerdemaeker. 1998. Use of sensor fusion to detect green picked and chilled tomatoes. *SENSORAL'98. International Workshop on Sensing Quality of Agricultural Products*. Vol. 2, Montpellier, France, pp. 609–617.

Shaw, G. and H. Burke. 2003. Spectral imaging for remote sensing. *Lincoln Laboratory Journal* 14:3–28.

Singh, N., M.J. Delwiche, R.S. Johnson, and J. Thompson. 1992. Peach maturity grading with color computer vision. *ASAE Paper No. 92-3029*.

Singh, N., M.J. Delwiche, and R.S. Johnson. 1993. Image analysis methods for real-time color grading of stonefruit. *Computers and Electronics in Agriculture* 9:71–84.

Smillie, R.M., S.E. Hetherington, R. Nott, G.R. Chaplin, and N.L. Wade 1987. Applications of chlorophyll fluorescence to the postharvest physiology and storage of mango and banana fruit and the chilling tolerance of mango cultivars. *ASEAN Food Journal* 3:55–59.

Steinmetz, V., M. Crochon, V. Bellon-Maurel, J.L. Garcia, and P. Barreiro. 1996. Sensors for fruit firmness assessment: Comparison and fusion. *Journal of Agricultural Engineering Research* 64:15–28.

Steinmetz, V., J.M. Roger, E. Moltó, and J. Blasco. 1999. On-line Fusion color camera and spectrophotometer for sugar content prediction of apples. *Journal of Agricultural Engineering Research* 73:207–216.

Studman, C. and L. Ouyang. 1997. Bruise measurement by image analysis. In *Proceedings of V Symposium on Fruit, Nut and Vegetable Production Engineering*. Davis, CA, pp. 1–7.

Sun, D.-W. 2007. *Computer Vision Technology for Food Quality Evaluation*. Elsevier Academic Press, San Diego, CA, ISBN: 978-0-12-373642-0.

Tao, Y., P.H. Heinemann, Z. Varghese, C.T. Morrow, and H.J. Sommer III. 1995a. Machine vision for color inspection of potatoes and apples. *Transactions of the ASAE* 38:1555–1561.

Tao, Y., C.T. Morrow, P.H. Heinemann, and J.H. Sommer III. 1995b. Fourier-based separation technique for shape grading of potatoes using machine vision. *Transactions of the ASAE* 38:949–957.

Tao, Y. and Z. Wen. 1999. An adaptive spherical image transform for high-speed fruit defect detection. *Transactions of the ASAE* 42:241–246.

Throop, J.A., D.J. Aneshansley, W.C. Anger, and D.L. Peterson. 2005. Quality evaluation of apples based on surface defects: Development of an automated inspection system. *Postharvest Biology and Technology* 36:281–290.

Tillett, R.D. and B.G. Batchelor. 1991. An algorithm for locating mushrooms in a growing bed. *Computers and Electronics in Agriculture* 6:191–200.

Unay, D. and B. Gosselin. 2006. Automatic defect segmentation of 'Jonagold' apples on multispectral images: A comparative study. *Postharvest Biology and Technology* 42:271–279.

Uozumi, J.L., S. Kawano, M. Iwamoto, and K. Nishinari. 1987. Spectrophotometric system for quality evaluation of unevenly colored food. *Journal of the Japanese Society for Food Science and Technology* 34:163–170.

Upchurch, B.L., J.A. Throop, and D.J. Aneshansley. 1994. Influence of time, bruise-type and severity on near-infrared reflectance from apple surfaces for automatic bruise detection. *Transactions of the ASAE* 37:1571–1575.

Van Eck, J.W., G.W.A.M. Van der Heijden, and G. Polder. 1998. Accurate measurement of size and shape of cucumber fruits image analysis. *Journal of Agricultural Engineering Research* 7:335–343.

Vapnik, V.N. 1999. *The Nature of Statistical Learning Theory*. 2nd edn., Springer-Verlag, New York, USA. ISBN: 9780387987804.

Varghese, Z., C.T. Morrow, P.H. Heinemann, J.H. Sommer III, Y. Tao, and R.M. Crassweller. 1991. Automated inspection of golden delicious apples using color computer vision. *ASAE Paper No. 91-7002*.

Wright, J.N., F.E. Sistler, and R.M. Watson. 1984. Measuring sweet potato size and shape with a computer. *Louisiana Agriculture* 28:12–13.

Xing, J., C. Bravo, P.T. Jancsók, H. Ramon, and J. De Baerdemaeker. 2005. Detecting bruise on 'Golden Delicious' apples using hyperspectral imaging with multiple wavebands. *Biosystems Engineering* 90:27.

Yang, C., S. Prasher, J. Whalen, and P. Goel. 2002. Use of hyperspectral imagery for identification of different fertilisation methods with decision-tree technology. *Biosystems Engineering* 83:291–298.
Yang, Q. 1993. Finding stalk and calyx of apples using structured lighting. *Computers and Electronics in Agriculture* 8:31–42.
Zheng, C., D.W. Sun, and L. Zheng. 2006a. Correlating colour to moisture content of large cooked beef joints by computer vision. *Journal of Food Engineering* 77:858–863.
Zheng, C., D.W. Sun, and L. Zheng. 2006b. Recent developments and applications of image features for food quality evaluation and inspection—a review. *Trends in Food Science & Technology* 17:642–655.
Zheng, C., D.W. Sun, and L. Zheng. 2006c. Recent applications of image texture for evaluation of food qualities—a review. *Trends in Food Science & Technology* 17:113–128.

2.2 COMPUTER VISION FOR QUALITY CONTROL

CHAOXIN ZHENG AND DA-WEN SUN

2.2.1 INTRODUCTION

Quality control of agro-food products is an important issue in the food industry. As the demand of food and food products is increasing in a worldwide scale, it is significant for food manufacturers to ensure that the products they deliver to customers meet at least the quality level requested by authority regulations for food safety. Nevertheless, customers are willing, happy, and usually desired to pay a higher price for products with higher quality. Regarding this, quantifying and grading food products into different quality classes are important for both manufacturer profit and safety issues, and therefore techniques are brought into the food industry for quality evaluation.

The most common technique would be human inspection and still is in some manual processing lines. Quality standards are established by the industry or qualified institutes, who also provide some gold standards for humans to judge and grade food products. Human inspectors are trained, based on these standards, to inspect food products. Because of the slow inspecting procedure, only a few products are sampled from a large number of products. This certainly cannot ensure that all products are at the same standard level. Furthermore, human inspectors are subjective and, thus, evaluation results may vary even for the products with very similar qualities. The inspecting procedure is tedious and unrepeatable because products are destroyed while being measured. To overcome the subjective nature of human evaluation, instruments have been developed simulating the human evaluation process.

Manufacturers are seeking alternative inspecting techniques, and so far one of the most potential and promising techniques is computer vision. First introduced about a few decades ago, computer vision has established substantial applications in food quality evaluation. Applications can be found for fruit, vegetable, meat, fish, and many other food products. Early computer vision techniques are quite simple, that is, simulating human vision in judging food qualities. As research goes on, the quality parameters that can be evaluated by computer vision are not only restricted to those that can be seen, but also to those that are determined after being consumed such as meat texture evaluated by adapted instruments.

In this chapter, we present a brief review of the current applications of computer vision for the quality control of the processed food, not only of the products that are "seen" but also of the products that are "felt." It is focused on the scope of applications of computer vision. However, first of all, a brief general introduction of image analysis and machine learning techniques is presented, which gives readers some ideas in the core of computer vision. Subsequently, we demonstrate the applications of computer vision for quality control of processed food by examples respectively in fruit and vegetable, grain, meat, and many other products, and discuss and summarize the applications of different techniques based on the applications. This chapter is aimed at providing some fundamentals and guidelines for researchers who intend to get involved in the research of computer vision for quality control of food products.

2.2.2 Image Analysis and Machine Learning

2.2.2.1 Image Analysis

The typical objective at the image analysis step in a quality evaluation vision system is to label the food objects in images and extract their essential visual properties, called image features. The labeling process is called image segmentation while the later is called image feature extraction. In this section, the current techniques used for food image segmentation and feature extraction will be introduced.

2.2.2.1.1 Image Preprocessing and Segmentation

Images captured from CCD, MRI, x-ray imaging, and infrared imaging (IRI) are subject to defects such as low contrast and high noise. In low-contrast images, the details of objects or the boundary between objects are not clearly visible, while in images corrupted by high-level of noise, the feature extracted might not actually reflect or represent the actual properties of the images. Therefore, defects need to be removed or reduced such that the subsequent results will not be affected. To increase contrast, one may use the histogram scaling and histogram equalization techniques. The adaptive histogram equalization, a modified histogram equalization technique, is used by which the contrast can be enhanced without changing the scale of the histogram. Filters can be used to remove noise. Because of the fact that it does not shift the intensity like linear or other filters do, median filter with the kernel of 3×3 is performed. Note that this size of the kernel can be increased depending on the quantity of noise.

Although computers do not have the capability of recognizing food objects like human beings do, programs can be used to instruct computers for doing some complex tasks, for example, image segmentation. The objective of image segmentation for food images is to label or to partition the food images into two non-overlapping parts, food objects and background. The former will be further used for analysis, while the latter will be discarded. Image segmentation is very important for a vision-based task because all subsequent processes rely on the segmentation results. If the images cannot be properly segmented, it is unlikely that the feature extracted can correctly describe the objects and it is thus impossible to use these features for measuring food qualities. Much more effort has been spent on developing an automatic and powerful segmentation technique with the self-learning ability. Unfortunately, such a technique is still unavailable. Current image segmentation techniques can only be used for processing a limited number of images with some common properties (Wu et al. 1998). In food quality evaluation, segmentation quality is usually compromised by the speed because a large number of food products are to be evaluated in a very short time. Among all segmentation techniques, histogram thresholding is the most efficient one.

In histogram thresholding, a limiting value is automatically selected by an objective or cost function developed based on information theories. Variance-based (Otsu 1979), peak-and-valley, and entropy-based functions are three of the most widely used objective functions. In our research group, we have developed a very powerful segmentation technique based on the variance objective function (Zheng et al. 2006a). Two thresholds, one for intensity and the other for the average, are subsequently selected by using a fast recurring algorithm. As an example, the techniques were tested on a laptop with 256 MB RAM and 1400 MHz CPU clock (TOSHIBA Tecra S1, TOSHIBA Corp., Japan), and the average time for segmentation was 0.12 s. Figure 2.2.1 shows the scheme of the two-dimensional thresholding.

2.2.2.1.2 Image Feature Extraction
Image features are the low-level features used to represent the visual and meaningful properties of images or objects within the images. Color, shape, size, and texture (pattern) are four of the most important image features that have been used extensively for food quality evaluation. In our previous review (Zheng et al. 2006b), we have summarized the different visual features to be used for food quality evaluation. Therefore, only a brief introduction is presented here. Readers may refer to our original review.

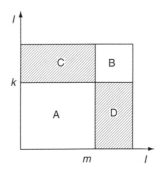

FIGURE 2.2.1 Scheme of the 2D thresholding: k and m are the thresholds used to segment regions with intensities (I) that fall in A. Region B is the background, and C, D corresponding to the noise presented in the image.

To analyze color, one may use different color spaces, such as HSI and LUV space, converted from the RGB space that is generally used for image storage. Shape can be measured in two different ways: either from the object boundary or from the object region. The most famous method used in the former is the Fourier transform. Object boundary is encoded into a set of radius and the Fourier transform is applied. Shape features can be extracted from the magnitude and frequency. The features based on object region are mostly spatial moments, which are the statistical measurements of the spatial distribution of the objects pixels. Size features are quite simple to measure in digital images, simply by counting the number of pixels.

Texture is perhaps the most complicated feature, and so far, there is no formal and scientific definition for texture. However, researchers have developed two different rough definitions: one is that the texture is the variance of intensities across pixels and the other is that texture is the spatial dependency between pixels. A great number of methods have been developed for texture analysis, naming the COM (Haralick et al. 1973), autoregressive models (Kartikeyan and Sarkar 1991), wavelet transform, and Gabor filters. Texture analysis techniques are detailed in the review by Zheng and coworkers (2006c). In our working group, we have developed a novel texture analysis technique called region-primitive technique (Zheng et al. 2006d). This region-primitive technique originated from the discoveries of vision research and uses the concepts of texton proposed by Julesz (1979). An image is partitioned into small and non-overlapping parts corresponding to the concept of texton; within each part, the intensity variation is small so it is assumed that no texture is perceptible (Figure 2.2.2). Texture features can thus be obtained from the spatial properties and the placement of the textons. The region-primitive technique had been used to

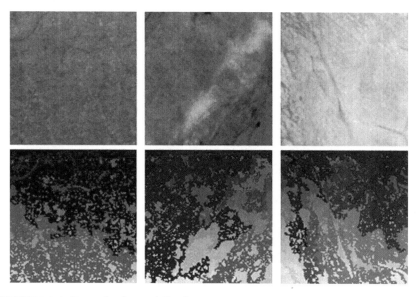

FIGURE 2.2.2 (See color insert following page 376.) Beef, pork, and lamb image with their region primitives constructed (bottom).

classify meat images where it was found that it outperformed the pixel-run technique (Galloway 1975).

2.2.2.2 Machine Learning

After extracting the visual and meaningful features from images, the problem encountered is how to establish the relationship between the features and the quality of agricultural products from a set of known samples so that the quality of products can be predicted by simply using their image features. With the fast development of both theory and applications in artificial intelligence, this process can be possibly conducted by using some mathematical models. These models have the ability of learning the pattern of the input signals as humans do, and thus are called machine learning. In quality evaluation issues, the learning models can be used to learn the behaviors or patterns of quality while the image features change.

2.2.2.2.1 Statistical Classification
The statistical classification (SC) extracts the statistical properties of the training data set using explicit underlying probability models, for instance Bayesian theory (Du and Sun 2006). With these statistical properties, the resultant values of the testing data can be generated. There are generally three types of SC techniques, naming the Bayesian learning, discriminant analysis, and nearest neighbor. The first technique generates the probability distribution of desired output based on the input features, and then reasons the probability together to produce the final output. Discriminant analysis considers the different input features and creates a new feature that is supposed to be the best combination of the original features as the differences between the predefined groups are maximized. The nearest neighbor classification rule is a non-parametric classification procedure that assigns a random vector to one of the resultant populations. In SC, one has to make an assumption of the distribution of the input features, and results are strongly dependent of the assumptions made. However, this is a nontrivial task since little knowledge can be gained for the image features and thus SC does not produce very good classification if the assumption of the distribution is not close to the actual one.

2.2.2.2.2 Neural Networks
NN, first proposed in the 1950s, are originally built to simulate the behaviors of neurons in human brains. The fundamental theory of NN is the winner-take-all rule, that is, only the neuron that first exceeds the threshold and becomes the winner is fired, while the others are suppressed. A typical NN consists of a set of neurons, usually divided into layers, and the weights that connect the neurons together. The organization decides the architecture of the network while the weights are adjusted during training to fit or simulate the input. There are two types of networks, feed-forward and feed-backward networks. In the former, weights of the networks are adjusted according to the feed forwarded from the beginning of the network, while in the latter, feedback is retrieved from the output layer of the network, and propagated back to change the weights in the middle layers.

Vision Systems

One of the main advantages of the NN over SC is that NN can learn the distribution of the input data and reveal the topology of the input while in SC the user has to assume or generalize the distribution to use the SC. However, the initialization of the network is based on random weights; the results of the final weights of the network do not remain always the same. In other words, the results from the NN are not repeatable if the network is not trained properly. Undertraining and overtraining are two main problems. It is also quite difficult to determine the number of samples for training. Nevertheless, because of the powerful function approximation and prediction ability, NN is widely used in food quality evaluation. Features extracted from images and the quality measurements are the input and output of the network, respectively, while network is trained to produce a desired output with the input presented. The weight back-propagation network (Figure 2.2.3) is a powerful network for such a prediction or classification purpose, and has been widely used in food quality evaluation.

2.2.2.2.3 Other Learning Techniques

Two other main learning techniques used in agricultural products include the fuzzy logic and the decision tree. Fuzzy logic, related to fuzzy sets and probability theory, is introduced by Lotfi Zadeh from University of California, Berkeley, in the 1960s. The general idea is that instead of claiming an entity as true or false, the true or false is fuzzed into a percentage of true or false. For example, a glass of 100 mL containing 30 mL water is described as 70% empty and 30% full. The fuzzy logic is quite suitable for measuring agricultural products. For instance, the tenderness of beef or pork can be expressed as 75% tender and 25% tough, which is more useful

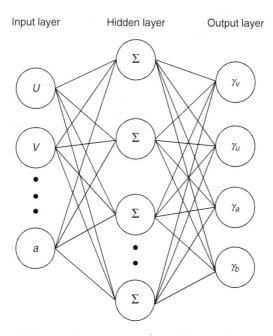

FIGURE 2.2.3 Architecture of the back-propagation NN.

rather than simply tender or tough. Decision tree is to build a tree structure for classifying the input signal into different leaves. Starting from the root, the input is cut into half, with each being a child of the root, by using a threshold. This process is repeated until every leaf of the tree is examined and no more division can be made.

2.2.3 APPLICATIONS

2.2.3.1 Quality Control of Fruits and Vegetables

2.2.3.1.1 Defection Detection

Color, shape, and texture are used frequently in quality control of fruits and vegetables. Color and texture reflects the appearance and surface properties of a product. On products with even surface, the color will remain mostly the same or change in a very regular pattern, which means texture will become even on the surface. When the product surface becomes uneven, for example, if defects or contaminants appear on the surface, there will be a sharp change of color or texture. On the basis of this assumption, various defects and contaminants on fruit or vegetable surface can be detected automatically using machine vision. The problem of defect detection is then converted into a two-class classification problem: how to use pattern recognition technique to classify the two sets of pixels—one set defect and the other non-defect. The input of the classifier are the image features consisting of different combinations of color, texture, and shape depending on the applications, while the output of the classifier is in the binary scale—1 being non-defect pixel and 0 for defect pixels or vice versa. The input and output are then presented to train a classifier, which can be linear discriminate functions or single-layer perceptron if they are linear separable, and multilayer perceptron or SVM if they are not linear separable. Owing to the complex nature of the problem and the huge number of features being used, it is suggested here that several experiments are conducted to compare and select the classifier with the best performance.

A large number of such applications can be found for apples. Leemans and coworkers (1999) used color features extracted from pixels in the RGB (red, green, blue) spaces to segment defects on 'Jonagold' apples. However, it was further found that the accuracy could still be improved by using features extracted from images generated by using multispectral imaging technique (Kleynen et al. 2003). Nevertheless, the satisfactory segmentation results ($>70\%$) relied on the use of an SC technique, that is, the Bayesian theory (Leemans and Destain 2004).

One problem here is the optimal selection of feature combinations. If too few features are selected, the accuracy will be affected, as the important information for the judgment of defect/non-defect is not included in the feature set. On the other hand, if too many features are selected, the information reduction might cause under-performance of the classification model since the features are not independent. Moreover, the increase of features will consume more computing power and memory, which slows down the classification. To solve the problem, it is usually plausible to feed the input into a data dimension reduction techniques such as principle component analysis (PCA), independent component analysis (ICA), singular vector decomposition (SVD), and nonnegative matrix factorization (NMF). These techniques are quite popular in the areas of computer vision and pattern recognition and have been applied in many other areas especially in biology and medicine, but not in the agriculture except the PCA.

2.2.3.1.2 Sorting and Grading

Other than the detection of defects on surface, evaluation of product maturity and quality changes during storage is another important issue for quality inspection of fruits and vegetables especially in tomatoes. In early 1985, Sarkar and Wolfe (1985) used gray intensities of images to classify green and red tomatoes. The classification rate in the study was very high considering the computing technique at that time. However, the evaluation process of classifying tomatoes into green and red classes still did not meet the requirement of the U.S. standards in which six grades, that is, green, breakers, turning, pink, light red, and red, are used. Further study was conducted by Choi and coworkers (1995), who used a color CCD camera instead of black-and-white camera, extracted features from the HSI color space, and achieved a much better classification rate of 77.5% for a total of 120 tomatoes based on the six grades of U.S. standards. Significant correlation between color features, obtained from the CIE $L^*a^*b^*$ color space, and tomato maturity was found (Jahns et al. 2001). A strong relationship was also found between color and the storage temperature, and between color and storage time, and a model was developed to describe the relationship (Lana et al. 2005). The model worked with 84%, 94%, and 89% accuracy on the red, green, and blue color channels, respectively. Texture features were used to classify the grade of apples after dehydration, and the accuracy of the classification was up to 95% (Fernández et al. 2005).

The fundamental that underlies this grading and sorting is that human beings judge the grading and sorting based on their observation of the surface color of tomatoes. Recall from the earlier text that the fundamentals and principles of computer vision are to simulate human visions and make decisions in a similar way to humans. However, the nature of computer vision and human vision differs in a fundamental way. Computer vision is highly quantitative while human vision, on the other hand, is quite fuzzy. Therefore, in the grading and sorting process, computer vision might perform better than human vision because of the quantitative and objective measurement of color features of tomatoes. In other words, computer vision is not easily fooled as human being.

2.2.3.1.3 Prediction

With MRI technique (see Section 5.3) the internal structure of apples during ripening and storage, and by texture analysis the properties of apples during ripening and storage can be evaluated (Létal et al. 2003). Texture features can also be used to predict sugar content of orange with the correlation coefficient of 0.83 (Kondo et al. 2000). Thybo and coworkers (2004) used MRI to acquire images of the internal structure of raw potatoes and obtained texture features by using the following four different techniques: first-order statistics (FOS), convolution masks (CM), COM, and run-length matrix (RLM), which were used to predict the sensory qualities including hardness, adhesiveness, mealiness, graininess, moistness, and specific gravity after the potatoes were cooked.

The underlying question here is that how much information about the other quality measurement of the fruits and vegetables is carried on the image features. Sensory scores/qualities made on the basis of the visual inspection are based on human vision, which is more or less similar to computer vision. Consequently, it is quite natural to assume that the image features obtained are highly correlated to the visual sensory scores. How about the qualities, such as chemical components like sugar content? It seems that the correlation between sugar content and image features are not as related as it is between sensory quality and image features. Ignoring their internal connections might lead to good results that are produced randomly and are not reproducible, which is not usable, or falling into the bottleneck of performance since the relationship between the two components are quite small, the regression results would not be good.

2.2.3.2 Quality Control of Grain

2.2.3.2.1 Sorting and Grading

The quality control of grain hardly depends on the use of ordinary CCD camera; instead, x-ray, near-infrared (NIR) spectroscopy, and thermal imaging techniques are used. Various studies (Karunakaran et al. 2004a,b) were conducted to classify grains infested and uninfested by different stages of insects. The accuracy was over 96%. Using NIR spectroscopy, it is possible to measure the hardness and vitreosity of grain kernels, to perform the color classification, to identify damaged kernel, to detect insect and mite infestation, and to inspect mycotoxins (Singh et al. 2006). Thermal imaging can be used to detect insect-infested kernels and the different classes of wheat. In these applications, the general procedure is used: images are captured using soft x-ray, NIR spectroscopy, or thermal imaging; a preprocessing and segmentation were performed to identify grains in the images; features were measured from color, histogram, shape, and texture; and finally either statistical classifier or NNs are conducted linking image features and grain qualities or grades together.

However, this method still has some disadvantages. Firstly, the inspection is dependent on the assumption that the network or classifier has learnt or seen all the behavior between image features and their qualities or grades. If the network faces input with image features that have never been presented to the network before, the predicted output will be much less accurate. Therefore, the sample should cover a wide enough range of qualities or grades and image features so that the network can learn all the possible behaviors of image features when the grades and qualities changes. In reality, this is very difficult to achieve due to unpredictable future. Fortunately, this problem can be solved by monitoring the performance of the system. When the accuracy of the system is too low, one can prepare a new set of samples to retrain the system so that the new relationships between image features and grades/qualities can be learnt and "remembered" by the system.

2.2.3.2.2 Segmentation

The challenge here remains in the segmentation process and there were a large number of features to be classified while the time allowed is quite limited. Because of the small size of a single grain, images captured generally contain a large number of grains together. These grains can be isolating, touching, or overlapping each other. For touching and overlapping kernels, the image segmentation needs to break the touching and overlapping kernels apart. Otherwise, the features extracted cannot correctly represent the kernels and thus the classification results will be strongly affected. Although image erosion and dilation is one solution to this problem, these two operations require that the size of the overlapping part is pre-known, which is not usually available. Therefore, other techniques were used and the most promising ones were the watershed algorithm (Vincent and Soille 1991) and the active contour models (Wang and Chou 2004).

Watershed algorithm consists of two stages: seed detection and seed growing. In the seed detection procedure, markers or seeds inside a grain are detected based on the assumption that the intensity of the marker is at the lowest or highest, depending on the characteristics of the image. The seed is then grown by merging neighboring pixels into it. Unlike the region-growing-and-merging image segmentation techniques, the growing in watershed algorithm operates at the same time for every seed such that the boundary between two connecting seeds can be automatically and objectively detected. The watershed algorithm indeed performs quite well even with strongly overlapping grains. However, it usually leads to over-segmentation while facing with noisy image and much effort has to be spent in overcoming the over-segmentation problem.

The active contour model initially generates a coarse contour of the object and deforms the contour based on two quantitative measurements, one from the external of the contour and one from the internal of the contour. The contour is reformed until the combination of the two measurements reaches the minimum. Unlike the watershed algorithm, the active contour only works for individual grain and for each grain, a global constrain of the contour has to be defined to deal with the overlapping problem.

The advantages of the watershed algorithm over the active contour model are that the watershed algorithm can effectively solve the problem of overlapping and touching grains, which cannot be easily solved by the active contour model. However, the active contour model can generate much smoother and very close contour to the actual grain boundary. In other words, the accuracy of the active contour model would be higher than the watershed algorithm for individual grain. Nevertheless, for image consisting of a large number of grains, the computation load of active contour model is too high for real-time application. Therefore, it is recommended here to use watershed algorithm in grain image segmentation for large-scale applications.

2.2.3.3 Quality Control of Other Foods

In our food refrigeration and computerized food technology (FRCFT) research laboratory, we have conducted a series of research work on pizza quality evaluation. A lot of effort has been spent on and we have achieved some success in evaluating the acceptance of pizza base and toppings. Depending on the different ways that customers observe the quality of products, different image features were used to evaluate the qualities of bases and toppings. Size and shape were mostly used for pizza bases, while color is preferred for toppings.

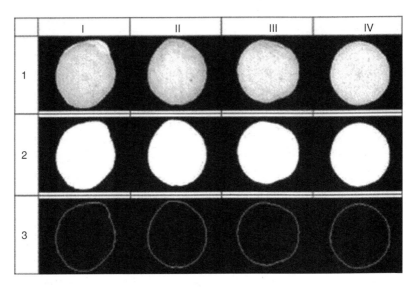

FIGURE 2.2.4 Images of four different classes of pizza base and their segmented images.

In the early study (Sun and Brosnan 2003), the size and shape of pizza base are measured using the ratio between area and perimeter. On the basis of the size and shape measurements, the quality of pizza bases can be divided into four different classes: flowing base, poor alignment, poor pressing, and standard. Later on, Du and Sun (2004) used a different shape descriptor, that is, Fourier transform, to extract shape features from pizza base segmented using Canny edge detection techniques. The features extracted were then fed into an SVM model and the quality of pizza bases was quantified into four different classes (Figure 2.2.4).

For the toppings (Du and Sun 2005), color of pizza sauce was quantified into only 256 different values, which were then again fed into an SVM classification model where pizza sauce was discriminated into five classes. The overall accuracy of the classification result was 0.90 out of a total of 120 samples.

In quality control of fish, color is one of the most frequently used features. In these studies, color features are extracted from the $L^*a^*b^*$ space using statistical measurements such as mean and standard deviation. These features are then used to correlate the lipid rates or sorting fishes (Louka et al. 2004). It was also found that the sorting rate by using computer vision techniques was better than the sensory panels. Size features such as the measurements of area can also be used for sorting fishes. The sorting lays on the fact that fishes can be discriminated by thresholding on the size (Parsonage and Petrell 2003). However, this fact is not always true when facing with the huge varieties of fish. Therefore, some other features such as shape are also used to sort fish (Zion et al. 2000).

2.2.4 Image Features for Quality Evaluation: Usefulness and Limitation

As stated previously, size, shape, color, and texture are the main image features often used in evaluating the quality of agricultural food products. Color is the first factor of consumer impression for food products, and thus can be used to determine the acceptability of the product for customers. During processing, the chemical or physical components of food products are usually reconstructed; color, which is strongly affected by these components, can reflect the reconstruction of these components. Although the changes of chemical and physical components are difficult to measure or quantify, color of digital images are very easy to measure using computer vision techniques (Yam and Papadakis 2004). Processed food products are usually not uniform but constituted of different parts with different properties and qualities such as pizza toppings consisting of cheese, vegetables, and meat. These constitutions usually have different colors, and thus a significant variation of color between two pixels in images generally indicates that the two pixels belong to different objects or different parts in an object with different characteristics. In this respect, color features might also be used to detect defects in food products, such as those on the surface of apples, or to classify products having different qualities (Leemans et al. 1999).

When images of products are captured under the same scale, the size of products in digital images becomes comparable. It is therefore possible to use size feature in digital images to determine the changes of product size during processes and the final product yields (Tu et al. 2000). One of the main reasons for this is that measurement of products size performed by human or by instruments is tedious and labor intensive, while measurements by computer vision technique are quite efficient and effortless (Zheng et al. 2006a). It can usually be accomplished by simply counting the number of pixels at the location where the size is measured.

Shape is another significant factor that affects the decision of consumers on purchasing, and thus, similar to color, shape features measured by computer vision techniques have also played a nontrivial role in grading processed food products (Leemans and Destain 2004). The typical applications of shape features are characterizing the shape of products during processing to study the changes and monitor the processing, evaluate the acceptance of products by consumers in respect of shape, additionally discriminating different kinds of products or the same kind of products but with different qualities (Ghazanfari and Irudayaraj 1996).

As discussed earlier, texture is generally referred to as the dependencies among pixels or the variations across pixels. Such dependencies or variations can be expressed in more detail as some surface properties such as coarseness, fineness, granulation, randomness, lineation, and hummocky. All these properties reflect how the food product surfaces are composed or structured in a similar way in which humans can see and observe texture (Amadasun and King 1989). Texture, therefore, is generally used to correlate the sensory properties of food products. Apart from this, research also found that texture is quite useful in determining the chemical or physical properties of beef, much more important than color and size (Huang et al. 1997). However, similar research has not been introduced for other products, and it is hoped that such research work can be started in the near future to find out the importance of texture in measuring the chemical and physical properties of pork, vegetables, and fruits.

2.2.5 Conclusions

In this chapter, we briefly reviewed the current computer vision techniques used for food quality evaluation in each step including image preprocessing and segmentation, image analysis, and machine learning. The applications of such techniques were then demonstrated by reviewing their use in the quality control of fruits and vegetables, grain, meat, and other food products. In most of these applications, it was shown that quality evaluation results obtained by computer vision techniques are quite comparable to those from the human inspectors. However, the cost using computer vision is much lower, the speed much higher, and safety is no longer a concern as human inspectors are free from the evaluation processes. Therefore, computer vision techniques should have substantial potential in their future use in the food industry to replace the current manual evaluation techniques.

REFERENCES

Amadasun, M. and R. King. 1989. Textural features corresponding to textural properties. *IEEE Transactions on Systems, Man, and Cybernetics* 19:1264–1274.

Choi, K., G. Lee, Y.J. Han, and J.M. Bunn. 1995. Tomato maturity evaluation using colour image analysis. *Transactions of the ASAE* 38:171–176.

Du, C.J. and D.W. Sun. 2004. Shape extraction and classification of pizza base using computer vision. *Journal of Food Engineering* 64:489–496.

Du, C.J. and D.W. Sun. 2005. Pizza sauce spread classification using colour vision and support vector machines. *Journal of Food Engineering* 6:137–145.

Du, C.J. and D.W. Sun. 2006. Learning techniques used in computer vision for food quality evaluation: a review. *Journal of Food Engineering* 72:294–302.

Fernández, L., C. Castillero, and J.M. Aguilera. 2005. An application of image analysis to dehydration of apple discs. *Journal of Food Engineering* 67:185–193.

Galloway, M.M. 1975. Textural analysis using grey level run lengths. *Computer Graphic and Image Processing* 4:172–179.

Ghazanfari, A. and J. Irudayaraj. 1996. Classification of pistachio nuts using a string matching technique. *Transactions of the ASAE* 39:1197–1202.

Haralick, R.M., K. Shanmugam, and I. Dinstein. 1973. Textural features for image classification. *IEEE Transactions on Systems, Man, and Cybernetics* 6:610–621.

Huang, Y., R.E. Lacey, L.L. Moore, R.K. Miller, A.D. Whittaker, and J. Ophir. 1997. Wavelet textural features from ultrasonic elastograms for meat quality prediction. *Transactions of the ASAE* 40:1741–1748.

Jahns, G., H.M. Nielsen, and W. Paul. 2001. Measuring image analysis attributes and modelling fuzzy consumer aspects for tomato quality grading. *Computers and Electronics in Agriculture* 31:17–29.

Julesz, B. 1979. Textons, the elements of texture perception and their interaction. *Nature* 290:91–97.

Kartikeyan, B. and A. Sarkar. 1991. An identification approach for 2-D autoregressive models in describing texture. *Graphical Models and Image Processing* 53:121–131.

Karunakaran, C., D.S. Jayas, and N.D.G. White. 2004a. Detection of internal wheat seed infestation by Rhyzopertha dominica using X-ray imaging. *Journal of Stored Products Research* 40:507–516.

Karunakaran, C., D.S. Jayas, and N.D.G. White. 2004b. Identification of wheat kernels damaged by the red flour beetle using X-ray images. *Biosystems Engineering* 87:267–274.

Kleynen, O., V. Leemans, and M.F. Destain. 2003. Selection of the most efficient wavelength bands for 'Jonagold' apple sorting. *Postharvest Biology and Technology* 30:221–232.

Kondo, N., U. Ahmad, M. Monta, and H. Murase. 2000. Machine vision based quality evaluation of Iyokan orange fruit using neural network. *Computers and Electronics in Agriculture* 29:135–147.

Lana, M.M., L.M.M. Tijskens, and O. van Kooten. 2005. Effects of storage temperature and fruit ripening on firmness of fresh cut potatoes. *Postharvest Biology and Technology* 35:87–95.

Leemans, V. and M.F. Destain. 2004. A real-time grading method of apple based on features extracted from defects. *Journal of Food Engineering* 61:83–89.

Leemans, V., H. Magein, and M.F. Destain. 1999. Defects segmentation on 'Golden Delicious' apples by using colour machine vision. *Computers and Electronics in Agriculture* 20:117–130.

Létal, J., D. Jirák, L. Šuderlová, and M. Hájek. 2003. MRI "texture" analysis of MR images of apples during ripening and storage. *Lebensmittel-Wissenschaft und-Technologie* 36:719–727.

Louka, N., F. Juhel, V. Fazilleau, and P. Loonis. 2004. A novel colorimetry analysis used to compare different fish drying processes. *Food Control* 15:327–334.

Otsu, N. 1979. A threshold selection method from gray-level histograms. *IEEE Transactions on Systems, Man, and Cybernetics* 9:62–66.

Parsonage, K.D. and R.J. Petrell. 2003. Accuracy of a machine-vision pellet detection system. *Aquaculture Engineering* 29:109–123.

Sarkar, N. and R.R. Wolfe. 1985. Feature extraction techniques for sorting tomatoes by computer vision. *Transactions of the ASAE* 28:970–979.

Singh, C.B., J. Paliwal, D.S. Jayas, and N.D.G. White. 2006. Near-infrared spectroscopy: Applications in the grain industry. CSBE Paper No. 06-189, Canadian Society for Bioengineers, Winnipeg, MB.

Sun, D.W. and T. Brosnan. 2003. Pizza quality evaluation using computer vision—part 1: Pizza base and sauce spread. *Journal of Food Engineering* 57:81–89.

Thybo, A.K., P.M. Szczypiński, A.H. Karlsson, S. Dønstrup, H.S. Stødkilde-Jørgensen, and H.J. Andersen. 2004. Prediction of sensory texture quality attributes of cooked potatoes by NMR-imaging (MRI) of raw potatoes in combination with different imaging analysis methods. *Journal of Food Engineering* 61:91–100.

Tu, K., P. Jancsók, B. Nicolaï, and J. De Baerdemaeker. 2000. Use of laser-scattering imaging to study tomato-fruit quality in relation to acoustic and compression measurements. *International Journal of Food Science and Technology* 35:503–510.

Vincent, L. and P. Soille. 1991. Watersheds in digital spaces: An efficient algorithm based on immersion simulations. *IEEE Transactions on Pattern Analysis and Machine Intelligence* 13:583–598.

Wang, Y.C. and J.J. Chou. 2004. Automatic segmentation of touching rice kernels with an active contour model. *Transaction of the ASAE* 47:1803–1811.

Wu, H.S., J. Gil, and J. Barba. 1998. Optimal segmentation of cell images. *IEEE Proceedings of Vision, Image and Signal Processing* 145:50–56.

Yam, K.L. and S.E. Papadakis. 2004. A simple digital imaging method for measuring and analyzing colour of food surfaces. *Journal of Food Engineering* 61:137–142.

Zheng, C., D.W. Sun, and C.J. Du. 2006a. Estimating shrinkage of large cooked beef joints during air-blast cooling by computer vision. *Journal of Food Engineering* 72:56–62.

Zheng, C., D.W. Sun, and L. Zheng. 2006b. Recent applications of image texture for evaluation of food qualities—a review. *Trends in Food Science & Technology* 17:113–128.

Zheng, C., D.W. Sun, and L. Zheng. 2006c. Recent developments and applications of image features for food quality evaluation and inspection—a review. *Trends in Food Science & Technology* 17:642–665.

Zheng, C., D.W. Sun, and L. Zheng. 2006d. Segmentation of beef joint images using histogram thresholding. *Journal of Food Process Engineering* 29:574–591.

Zheng, C., D.W. Sun, and L. Zheng. 2007. A new region-primitive method for classification of colour meat texture based on size, orientation, and contrast. *Meat Science* 76:620–627.

Zion, B., A. Shklyar, and I. Karplus. 2000. In-vivo fish sorting by computer vision. *Aquacultural Engineering* 22:165–179.

3 VIS/NIR Spectroscopy

Bernd Herold, Sumio Kawano, Bernd Sumpf, Peter Tillmann, and Kerry B. Walsh

CONTENTS

3.1 Spectrophotometer Technology 143
    3.1.1 Light Sources 143
        3.1.1.1 Basics 143
        3.1.1.2 Thermal Light Sources—Blackbody Radiation 145
        3.1.1.3 Spectral Lamps 146
        3.1.1.4 Light Emitting Diodes 146
        3.1.1.5 Laser 147
    3.1.2 Spectral Selective Detection 152
        3.1.2.1 Prism Monochromator 152
        3.1.2.2 Grating Monochromator 152
        3.1.2.3 Interference Filters 152
        3.1.2.4 Fourier-Transform Spectrophotometer 153
        3.1.2.5 Laser Spectrophotometer 153
    3.1.3 Detection 154
        3.1.3.1 Detectors 154
        3.1.3.2 Excitation Modes 156
Further Reading 157
3.2 Monitoring and Mapping of Fresh Fruit and Vegetable
Using VIS Spectroscopy 157
    3.2.1 Product Pigment Contents, Spectral Signature,
and Spectral Indices 157
    3.2.2 Spectroscopy in the VIS Wavelength Range 160
    3.2.3 Mapping Technique, GPS Technique, and Data Management 170
        3.2.3.1 Design of the Portable Spectrophotometer Equipment 170
        3.2.3.2 Fruit Analyses Using the Portable Spectrophotometer
Equipment with Probe for Light Interaction 174
        3.2.3.3 GPS Technique 175
        3.2.3.4 Data Management 175
    3.2.4 Application Examples 175
        3.2.4.1 Apple Fruit 175
        3.2.4.2 Sweet Cherry 181
        3.2.4.3 Carrot Root 181
        3.2.4.4 Potato Tuber 184

3.2.5 Mapping of Spectral Indices of Apple Fruit
Development in Orchard 185
    3.2.5.1 Fruit Maturity in Dependence on Geographical Site 185
    3.2.5.2 Effect of Fruit Yield per Tree on Maturity Progress......... 186
    3.2.5.3 Effect of Climatic Condition on Maturity Progress 186
    3.2.5.4 Mapping... 187
3.2.6 Conclusions 189
References 191
3.3 Near-Infrared Spectroscopy 192
    3.3.1 Introduction 192
        3.3.1.1 Setting the Task 192
        3.3.1.2 Going Further 194
    3.3.2 Factors for Success.. 195
        3.3.2.1 Attribute... 195
        3.3.2.2 Sample 196
        3.3.2.3 Instrument 198
    3.3.3 Spectroscopy and Chemometrics Theory 207
        3.3.3.1 Data Preprocessing 207
        3.3.3.2 Spectroscopy of Water 209
        3.3.3.3 Sugar in Water.. 212
        3.3.3.4 Calibration Based on Multivariate Regression............. 214
        3.3.3.5 Chemometric Terms 215
        3.3.3.6 Chemometrics: Calibration Techniques........................ 217
        3.3.3.7 Chemometrics: PLSR Wavelength Selection 219
    3.3.4 Application in Food 222
        3.3.4.1 Fresh Fruit 222
        3.3.4.2 Grains and Grain Products 225
    3.3.5 On-Line Quality Monitoring 229
        3.3.5.1 Definition 229
        3.3.5.2 Issues 229
        3.3.5.3 Examples... 229
Acknowledgment 234
References 234
3.4 Network of NIRS Instruments 239
    3.4.1 Need for Calibration Transfer 239
    3.4.2 Strategies to Remedy the Situation 240
        3.4.2.1 On the Level of the Predicted Results............................ 241
        3.4.2.2 On the Spectral Level 242
        3.4.2.3 In the Calibration Model 243
        3.4.2.4 Advantages and Disadvantages 245
    3.4.3 Examples 246
        3.4.3.1 Grain Analysis Based on Filter Instrument..................... 246
        3.4.3.2 FOSS Infratec Grain Network... 246
        3.4.3.3 VDLUFA Forage Maize Network.................................... 247
References 248

3.1 SPECTROPHOTOMETER TECHNOLOGY

BERND SUMPF

3.1.1 LIGHT SOURCES

3.1.1.1 Basics

Optical radiation is defined as emission in the spectral range from 100 nm up to 1000 μm (Figure 3.1.1). It includes the ultraviolet (UV) radiation from 100 up to 380 nm, the visible (VIS) region from 380 till 780 nm, and the infrared (IR) radiation with longer wavelength than 780 nm. The range from 780 nm up to 2.5 μm is called near-infrared (NIR) region, the region from 2.5 to 25 μm is the mid-infrared (MIR) region, and the range above is the far-infrared (FIR) region.

Beside the wavelength (λ) for the position of an emission or an absorption line also other physical values are often used: frequency, $\nu[1/s] = c/\lambda$ (c = speed of light, 299,792,458 m/s); energy, $E[J] = h \cdot c/\lambda$ (h = Planck's constant, $6.62606876 \times 10^{-34}$ W s^2) or $E[eV] = 1.24/\lambda$ [μm] (please note that lambda is in micrometer here); wave number, $\tilde{\nu}[\text{cm}^{-1}] = 1/\lambda$.

The different physical values can be converted to each other using

$$\lambda = \frac{c}{\nu}, \quad \nu = \frac{c}{\lambda}, \quad \tilde{\nu} = \frac{\nu}{c}, \quad \nu = \tilde{\nu} \cdot c \tag{3.1.1}$$

The emission and absorption of optical radiation are described by three basic processes, which were suggested by Einstein in 1917. Assuming a two-level atom system as shown in Figure 3.1.2, photons can be emitted in a process called spontaneous emission (Sp_Emi). From an excited state (S_i), the relaxation to the ground state (S_0) occurs. This process depends on the number of atoms in the excited state and the lifetime in the excited state. From the ground state, atoms can be excited into the upper state because of the absorption of radiation (Abs). Here the strength of the absorption depends on the transition itself, the number of atoms in the ground state, and the intensity of the light. In addition to these obvious processes, Einstein introduces the process of stimulated emission (St_Emi). Here the emission of photons occurs in the presence of light. This process depends on the transition, the number of atoms in the excited state, and the intensity of the light.

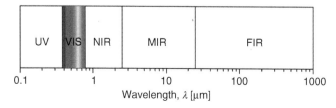

FIGURE 3.1.1 Optical spectrum between 0.1 and 1000 μm.

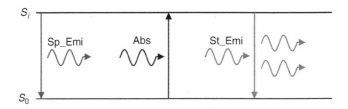

FIGURE 3.1.2 Basic emission and absorption processes.

Solids, fluids, and gases have characteristic emission and absorption spectra. There are three different physical processes:

- Electronic transitions: During the emission or absorption of light, the state of the electrons changes. Because of their high mobility and herewith the high energy, the observed spectra are in the UV or VIS range.
- Vibrational transitions: The interaction with light changes the vibrational state of the molecule. The vibrational energy is smaller compared to electronics transitions. The typical wavelength range is the NIR and MIR.
- Rotational transitions: Here only the rotational state of a molecule is changed. The rotational energy is even smaller and herewith the absorption or emission occurs in the MIR and FIR range.

Despite their different nature, all transitions deliver information on substances and the composition of mixtures.

In the case of absorption, a measured spectrum of i substances can be analyzed by using Beer–Lambert law:

$$I(\tilde{\nu}) = I_0(\tilde{\nu}) \cdot \exp\left[-\sum_i \mu_{a_i}(\tilde{\nu}) \cdot L\right] \quad (3.1.2)$$

where
$I(\tilde{\nu})$ is the measured light intensity at the wave number $\tilde{\nu}$
$I_0(\tilde{\nu})$ is the emitted light intensity at the wave number $\tilde{\nu}$
i is the index that indicates the different substances
$\mu_{a_i}(\tilde{\nu})$ is the absorption coefficient of the substance i at $\tilde{\nu}$
L is the length of the absorption path

A characteristic absorption, for example, an absorption line (Figure 3.1.3) is characterized by their position $\tilde{\nu}_0$, their line strength S, and the line shape $\Phi(\tilde{\nu} - \tilde{\nu}_0, \Delta\tilde{\nu})$. The width $\Delta\tilde{\nu}$ of the line is measured as full width at half maximum (FWHM).

For rotational–vibrational lines of gases in the MIR, at low pressures, the line width is typically determined by the Doppler width $\Delta\tilde{\nu}_D$ and at normal pressure, by the collisional line width $\Delta\tilde{\nu}_C$. Typical values are $\Delta\tilde{\nu}_D = 10^{-3}$ cm^{-1} and $\Delta\tilde{\nu}_C = 0.1$ cm^{-1}.

VIS/NIR Spectroscopy

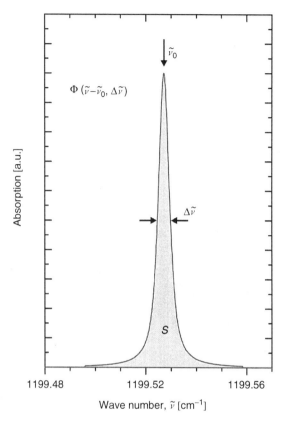

FIGURE 3.1.3 Absorption line with their characteristic parameters (shown for the H_2S absorption line from the ν_2 band at 1199.5271 cm^{-1}): Line position, $\tilde{\nu}_0$; line width, $\Delta\tilde{\nu}$; line strength, S; line profile, Φ.

In the VIS and UV regions, the measured structures are a superposition of electronic, vibrational, and rotational transitions. Therefore, the structure is much broader and can reach values up to 10 cm^{-1}.

According to these typical widths, the emission width of the excitation light source and the resolution of the spectral system have to be smaller compared to the measured structure width. The properties of different light sources and of various spectrophotometer systems will be discussed in the following sections.

3.1.1.2 Thermal Light Sources—Blackbody Radiation

Using the three processes suggested by Einstein, it is possible to deduce Planck's law for the emission of a blackbody source. This law describes the spectral distribution of a body in thermal equilibrium. The optical energy density (ρ[J s m^{-3}]) is given by

$$\rho_\nu = \frac{8 \cdot \pi \cdot h \cdot \nu^3}{c^3} \cdot \frac{1}{\exp\left(\dfrac{h \cdot \nu}{k \cdot T}\right) - 1} \quad (3.1.3)$$

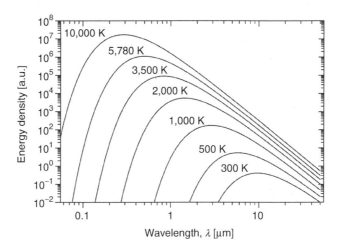

FIGURE 3.1.4 Planck's law for different temperatures 300 K $\leq T \leq$ 10,000 K.

with Boltzmanns constant $k = 1.3806503 \times 10^{-23}$ W s/K and $\pi = 3.14159$. Examples for Planck's law are given in Figure 3.1.4. The different curves illustrate the temperature dependence of the radiation. The curve for $T = 5780$ K, which is the temperature of the solar surface, has their maximum at 500 nm. When changing the temperature, the maximum wavelength changes according to Wien's displacement law:

$$\lambda_{max} \cdot T = 2897.790 \; \mu m \cdot K \qquad (3.1.4)$$

All thermal light sources can be described by this law. A typical example is the incandescent lamp. With a typical temperature $T = 1500$ K of the spiral-wound filament, the maximum of the emission is at 1.9 µm in the NIR.

For a wavelength selective excitation, parts of the spectra had to be selected accompanied by a strong reduction of the output power. Therefore, often light sources with characteristic emission lines are used in spectroscopic applications.

3.1.1.3 Spectral Lamps

Spectral lamps are mostly based on an electric discharge. The lamp is filled with the respective gas, which emits its typical emission spectrum. A selection of available light sources together with the positions of the strongest emission lines in the VIS are compiled in Table 3.1.1.

These lamps require a relatively sophisticated setup and have rather poor efficiencies. Therefore, nowadays, light emitting diodes (LEDs) based on semiconductors are often used.

3.1.1.4 Light Emitting Diodes

An alternative to the above-discussed spectral lamps are semiconductor-based LEDs. The devices consist of different semiconductor materials, for example, AlGaAs in the red to NIR range, GaP in the green to yellow region, and GaN in the UV and the

TABLE 3.1.1
Strong Emission Lines of Selected Spectral Lamps

Gas	Wavelength [nm]
Cadmium (Cd)	467.8, 480.0, 508.6, 515.5, 643.8
Helium (He)	501.6, 587.6, 656.0
Mercury (Hg)	404.6, 435.8, 491.6, 546.0, 577.0, 579.1
Krypton (Kr)	441.3, 467.8, 480.0, 508.6, 515.5, 643.8
Sodium (Na)	498.2, 568.8, 589.0, 589.6, 616.0
Neon (Ne)	534.1, 540.0, 585.2, 594.4, 614.3, 621.7, 638.3, 640.2, 650.6, 653.7
Rubidium (Rb)	420.1, 421.5, 620.6, 629.8, 780.0, 794.8
Zinc (Zn)	468.0, 472.2, 481.0, 518.0, 636.3

blue range. Emission wavelengths of, for example, 405, 470, 525, 570, 590, 605, 624, and 660 nm are commercially available. The emission peak width of the sources is typically larger than 20 nm measured at FWHM.

A wavelength adjustment can be performed by illuminating specific luminescent materials with the light of an UV-LED. Herewith, it is possible to manufacture white light LEDs and customized wavelength.

The advantage of LEDs is their small excitation current and the resulting possible low-cost operation.

For several applications, the spectral width and the output power of the discussed light sources based on Sp-Emi are not sufficient. Here the use of laser is necessary.

3.1.1.5 Laser

Laser is an acronym for **l**ight **a**mplification by **s**timulated **e**mission of **r**adiation. A laser consists of optical active medium, an optical resonator, and an external excitation (Figure 3.1.5).

Laser operation occurs if the following conditions are fulfilled:

- First laser condition "Inversion": The number of electrons in the excited state had to be larger than in the ground state. If this condition is fulfilled, the intensity of the St-Emi emission is larger compared to the spontaneous

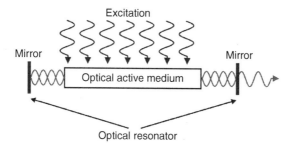

FIGURE 3.1.5 Principle scheme of a laser.

emission. A further amplification of the light intensity is possible by optical feedback in a resonator at its characteristic modes.
- Second laser condition: Within the resonator the optical gain had to be larger than all losses, that is, absorption losses within the resonator and losses due to the emitted light.

If both conditions are fulfilled, the so-called laser threshold is reached and laser operation occurs.

Based on the active medium, laser can be distinguished into the following:

- Solid state laser (e.g., ruby-laser, Nd:YAG, Nd:glass, Ti:sapphire)
- Gas laser (e.g., He:Ne, Ar-ion, CO_2, CO)
- Semiconductor laser (e.g., $A_{III}B_V$–GaAs, InP, GaN, GaSb and $A_{IV}B_{VI}$–PbS, PnSe, PbTe)
- Dye laser (e.g., coumarin, oxacine, rhodamin)

The excitation of laser can be performed in different ways:

- Optical pumping with flash lamps or other laser
- Collisional excitation in a gas discharge
- Chemical pumping
- Current through a *p–n* junction in case of semiconductor laser

With respect to the application, one additional feature is important: tuneability of the laser. Some of the laser, for example, the He:Ne laser, the ruby laser, the Nd:YAG laser, or the Ar-ion laser, operate at one or more fixed emission wavelengths. It is possible to switch between the different wavelengths, but it is not possible to tune the wavelength. The tuning range is limited by the maximal gain width of the optically active medium. In the case of the He:Ne laser, the Doppler width of the Neon with 0.04 cm^{-1} (0.002 nm) allows only a very small tuning.

Laser with a wider gain region, for example, Ti:sapphire, dye laser, or semiconductor laser, allows a tuning over several 10 nm. The tuning is performed by the implementation of gratings, prisms, and filters in the laser resonator. In the following, a small number of laser used in spectroscopic applications will be presented:

3.1.1.5.1 Nitrogen Laser
The nitrogen laser had the strongest emission lines in the UV spectral range at a wavelength of 337.1 nm. The laser emits short laser pulses with a pulse length shorter than 20 ns. The maximal repetition rate is about 100 Hz. With these short pulses, the laser is well suited for fluorescence spectroscopy and also as pump source for other laser, laser photochemistry, and in material processing.

3.1.1.5.2 Argon-Ion Laser
The laser emission is caused by transitions between excited states of ionized argon ions. The emission wavelengths are 457.9, 476.5, 496.5, 488.0, and 514.5 nm. Output powers up to 20 W are possible.

Applications of the laser are fluorescence spectroscopy, pump source for other laser, diagnostics, medicine, and material treatment.

3.1.1.5.3 Helium:Neon Laser

The He:Ne laser can operate in the VIS and NIR range with an output power up to several milliwatt. The most intense emission lines are the well-known red emission at 632.8 nm, the NIR line at 1.1523 μm, and the MIR line at 3.3913 μm.

The laser can operate in continuous-wave (cw) operation and has a large coherence length, a good beam quality and is suitable as reference laser for calibration purposes.

3.1.1.5.4 Nd:YAG

This laser is based on yttrium aluminum garnet $Y_3Al_5O_{12}$ host crystal with Nd as lasing material. The primarily used lasing transition is at 1064 nm, but there are also laser lines at 946, 1338, and 1834 nm.

The laser can operate in cw mode and also in pulsed mode. The laser is used also for material treatment, as pump sources for nonlinear frequency conversion, as pump source for other laser, in length measurements, for data communication, in medicine, etc.

There are high-power Nd:YAG laser available with several J pulse energy suitable for the pumping of x-ray laser or even for fusion reactors.

3.1.1.5.5 Ti:Sapphire Laser

The Ti:sapphire laser is a widely tunable laser with a tuning range from 685 up to 1080 nm. There are narrow line-width laser available suitable for spectroscopy, high-power femtosecond laser, and widely tunable laser sources, for example, for light detection and ranging (LIDAR) applications. Output powers up to several Watts are known. Beside the direct application of the laser for spectroscopic applications, the laser is often used as pump source for nonlinear frequency conversion.

3.1.1.5.6 Diode Laser

Diode laser is commercially available at different wavelengths (Figure 3.1.6). Lasers are available from 340 nm in the UV up to 33 μm in the MIR. Unfortunately, especially in the UV and VIS range, which is interesting for several biological and medical applications, there are some gaps in the available diode laser.

A coarse selection of the laser wavelength can be performed by the composition of the material. It is possible to mix x-parts of GaAs and $(1-x)$-parts of InAs to obtain $Ga_xIn_{1-x}As$. If $x = 1$, that is, GaAs, the emission wavelength is about 870 nm. A smaller Ga-content leads to longer wavelength. Besides this coarse tuning, a fine-tuning by using temperature and excitation current is possible. Moreover, by using an external cavity, laser arrangement tuning ranges up to 100 nm are possible.

Laser diodes have excellent conversion efficiency with values larger than 70%. They are easily excitable by electrical current, which can be moreover used for an easy modulation. A scheme of a diode laser and mounted devices are shown in

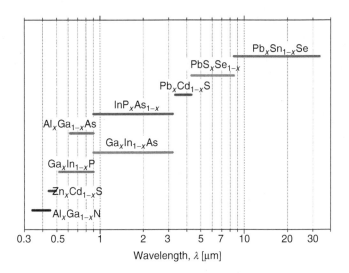

FIGURE 3.1.6 Emission ranges of different compound semiconductor laser materials.

Figures 3.1.7 and 3.1.8. The thickness of the laser is about 120 μm, the length typically between 2 and 4 mm, and the width of the chip is about 600 μm. In the case of broad-area laser, the stripe width is between 60 and 200 μm; for ridge waveguide laser, the ridge has a width of about 3 μm.

Diode laser is mechanically robust and a reliable operation up to several thousands hours is possible.

With internal grating diode laser, an emission line width smaller than 2 MHz (at $\lambda = 1$ μm this is line width of 0.01 pm) is available. This allows their application even in high-resolution spectroscopy. Diode laser can also be used in material processing, medicine (e.g. photodynamic therapy), fluorescence diagnostic, process control, as pump source for other laser, and also for nonlinear frequency conversion.

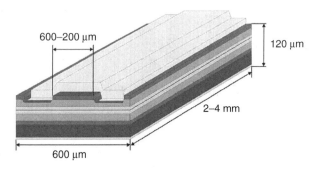

FIGURE 3.1.7 Scheme of a semiconductor laser.

VIS/NIR Spectroscopy

FIGURE 3.1.8 Different package forms of semiconductor laser in comparison to the size of paperclip. (Copyright FBH/schurian.com.)

3.1.1.5.7 Light Sources Based on Frequency Conversion

To achieve wavelengths, which were not directly accessible, nonlinear frequency conversion processes can be used. Certain materials without inversion symmetry can exhibit nonlinear effects. When exciting these materials with high-intensity light, for example, a laser, not only the incident wavelength is propagating, but also other wavelengths can be generated. These effects are proportional to the square of the intensity of the incoming light and to the nonlinear optical coefficients of the material. The so-called phase matching condition, that is, the proper phase relationship between the interacting waves, and between the pump light and the generated light, had to be fulfilled. Often used nonlinear processes are the doubling of the frequency of the exciting laser beam (second harmonic generation, SHG), the formation of the sum or the difference of two incoming waves (the sum frequency generation, SFG, and difference frequency generation, DFG, respectively). The combination of SHG and SFG leads to the generation of the third harmonic (third harmonic generation, THG). Performing two SHG processes in series also the generation of the fourth harmonic is possible (fourth harmonic generation, FHG).

For the Nd:YAG laser with the fundamental wavelength at 1064 nm, the SHG to 532 nm, the THG to 355 nm, and the SHG to 266 nm are often used. The combination of the widely tuneable Ti:sapphire laser with nonlinear frequency conversion offers a light source from 225 up to 3800 nm.

Materials for nonlinear frequency conversion are, for example, lithium niobate ($LiNbO_3$), potassium titanyle arsenate ($KTiOAsO_4$, KTA), potassium titanyl phosphate ($KTiOPO_4$, KTP), lithium triborate (LiB_3O_5, LBO), rubidium titanyle arsenate ($RbTiOAsO_4$, RTA), rubidium titanyle phosphate ($RbTiOPO_4$, RTP), β-barium borate (β-BaB_2O_4, BBO), and bismuth triborat (BiB_3O_6, BiBo).

3.1.2 SPECTRAL SELECTIVE DETECTION

The design of the spectral selective detection depends on the requested resolution. Resolution can be defined with the Rayleigh criterion, which describes the smallest distance, where two spectral lines with a distance $\Delta\tilde{\nu}$ can be separated. Assuming that both lines have the maximal intensity 1, they can be resolved, if the minimum between the two lines is smaller than $8/\pi^2 \approx 0.81$.

The spectral resolution, R_s, is defined as

$$R_s = \frac{\tilde{\nu}}{\Delta\tilde{\nu}} = \frac{\nu}{\Delta\nu} = \frac{\lambda}{\Delta\lambda} \tag{3.1.5}$$

3.1.2.1 Prism Monochromator

A classical method for a spectral selection is the application of a prism. The spectral resolution can be calculated using

$$R_s = B \cdot \frac{dn}{d\lambda} \tag{3.1.6}$$

where
  B is the basis length of the prism
  $dn/d\lambda$ is the dispersion of the material, that is, the dependence of the refractive index n on the wavelength λ

Assuming a basis length of $B = 5$ cm and a dispersion $dn/d\lambda$ of about 130 cm^{-1}, for example, fused silica or CaF_2, the resolution is about 650, that is, a width of 1.5 nm could be resolved at 1 μm wavelength. This is only sufficient for very broad spectral features.

3.1.2.2 Grating Monochromator

A higher resolution could be obtained by using a grating monochromator. Here the resolution depends on the number of illuminated lines in the grating N and the diffraction order m.

$$R_s = m \cdot N \tag{3.1.7}$$

Gratings in the VIS typically have 1200 lines/mm. Working in the first order, that is, $m = 1$, and illuminating 25 mm, a resolution of 30,000 is obtained. At 1 μm this means that structures of about 0.03 nm can be resolved.

3.1.2.3 Interference Filters

One way for doing spectral selection is the use of interference filters, for example, Fabry–Perot filters. This filter is based on two parallel mirrors, which were exactly parallel orientated to each other.

VIS/NIR Spectroscopy

The free spectral range (fsr) of a Fabry–Perot interferometer is

$$\text{fsr} = \frac{1}{2 \cdot n \cdot L} \tag{3.1.8}$$

where
 n is the refractive index
 L is the distance between the two mirrors

If the reflectivity of the mirrors, R, is high, the transmission curve (described by an Airy function) has narrow peaks and the filter is sharp. The half width, $\Delta\tilde{\nu}$, is

$$\Delta\tilde{\nu} = \frac{1-R}{\pi \cdot \sqrt{R}} \cdot \text{fsr} = \frac{1-R}{\pi \cdot \sqrt{R}} \cdot \frac{1}{2 \cdot n \cdot L} \tag{3.1.9}$$

The spectral resolution, R_s, is

$$R_s = \frac{\pi \cdot \sqrt{R}}{1-R} \cdot \frac{\tilde{\nu}}{\text{fsr}} \tag{3.1.10}$$

Using this formula, an interference filter with an fsr of 0.33 cm^{-1} and a mirror reflectivity of 0.99 could reach at 10,000 cm^{-1} (1 μm) a resolution of 10^7. Such filter with its equidistant transmission peaks can be used as relative frequency normal (étalon).

3.1.2.4 Fourier-Transform Spectrophotometer

A Fourier-transform spectrophotometer (FTS) is based on a Michelson interferometer. The spectral resolution depends on the scanning length, L, of the interferometer.

$$R_s = 2 \cdot L \cdot \tilde{\nu} \tag{3.1.11}$$

An FTS with a scanning length of 50 cm can reach a resolution of 10^7 at 1 μm, that is, 0.0001 nm. Because of the necessary mathematical data treatment, this theoretical resolution is not achieved in the experiment. FTS systems are typically used to obtain wide-range overview spectra, whereas laser spectrophotometers are used as magnifier to investigate in more detail the structures.

3.1.2.5 Laser Spectrophotometer

The decisive component in a laser spectrophotometer (Figure 3.1.9) is the tuneable laser source. An easy tuning over large spectral ranges together with a narrow-spectral line width is requested. Beside the optical channel for the substance under study, a laser spectrophotometer should have an option for measuring the intensity of the emission from the source and moreover reference channels. Depending on the application, this could be a reference for the concentration of the substance or a reference concerning the relative and absolute line positions.

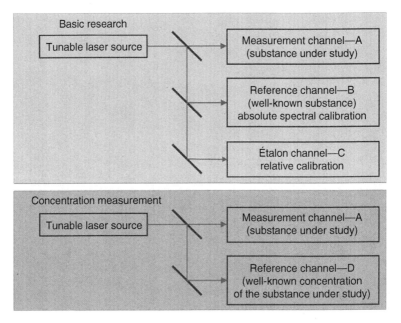

FIGURE 3.1.9 Scheme of laser spectrophotometers for basic research and concentration measurements.

In a spectrophotometer suitable for basic research investigations, the light from the laser is divided into three channels. The channel A contains the measurement cell for the substance under study. Channel B used well-known reference lines for an absolute spectral calibration. Channel C contains an étalon, which delivers the relative spectral calibration owing to the wave number equidistant peaks.

In the case of substance measurements channel, a wavelength calibration is not necessary, but a channel D with a well-known substance concentration had to be applied for calibration. In some cases, the calibration can be done by well-known reference data.

3.1.3 Detection

3.1.3.1 Detectors

For different spectral ranges and different applications, different types of detectors are available. Before discussing the detectors in detail, some more general comments should be given.

Each detector has its characteristic wavelength dependent responsivity, R_λ. If the detector signal is measured as voltage, R_λ is given in [V/W]. In the case that the signal is measured as current, the unit for the responsivity is [A/W]. For the sensitivity of the system, the detector noise has to be taken into account. Here the noise equivalent power (NEP) gives the value of optical signal on the detector equal to the noise of the detector (signal-to-noise ratio = 1). This value is given in [W].

Another characteristic parameter is the specific detectivity D^*. This value is defined as

$$D^* = \frac{\sqrt{A}\sqrt{\Delta f}}{\text{NEP}} \quad (3.1.12)$$

and can be used to compare different types of detectors. In the equation, A is the active area of the detector and Δf is the bandwidth of the electronic measurement system.

For a given detector material with a given D^*, the NEP can be calculated as

$$\text{NEP} = \frac{\sqrt{A}\sqrt{\Delta f}}{D^*} \quad (3.1.13)$$

This illustrates that a high detectivity D^* leads to a small NEP. A larger detector with the area A causes a large detector noise. And last but not least, a measuring system with a small electronic bandwidth Δf generates also a smaller NEP.

The theoretical possible maximal-specific detectivity can be described with the background limited induced power (BLIP), the signal generated by the background radiation of 300 K (see Planck's law) of an ideal detector with a certain acceptance angle. This theoretical limitation is especially relevant in the MIR for wavelengths larger than 2.5 μm.

3.1.3.1.1 Photocathode and Photomultiplier

Photomultipliers are commonly used in the UV and VIS range. Photocathode consists of compound semiconductors. Incident light generates electrons (external photoelectric effect), which were amplified by several electrodes.

The typical responsivity of a photocathode is about 0.1 A/W. The background power is below 10 nW. Photocathodes are available in the spectral range between 150 and 1600 nm. Their quantum efficiencies can reach values up to 25%. Reaction times in the ns-range are possible.

3.1.3.1.2 Photodiodes and Photoresistors

Semiconductor-based detectors are photodiodes and photoresistors. Both can consist of different materials like silicon, which is sensitive in the range from 190 up to 1060 nm. Silicon-based photodiodes can reach sensitivities up to 0.7 A/W and NEP in the 1 nW range.

Germanium can be used in the spectral range from 800 up to 1650 nm, whereas InGaAs is used as photodetector material in the spectral range from 0.9 up to 2.6 μm.

Light in the MIR range can be detected by applying InSb (up to 5.5 μm) and HgCdTe (up to 20 μm) as detector material.

3.1.3.1.3 Detector Arrays

Beside single-element detector, a combination of several detectors on one chip is suitable for parallel measurement at different wavelength. These arrays are available based on silicon, InGaAs, and also HgCdTe for different wavelength ranges.

Linear sensors with more than 2048 photodiodes are now widely used in standard spectrophotometer and allow an optical multichannel measurement, which is especially required in Raman and fluorescence spectroscopies.

Additional to the noise of each emitter, in the case of the charge-coupled device (CCD) linear sensors, the different responsivity of each element had to be taken into account. This effect is known as fixed pattern noise.

3.1.3.2 Excitation Modes

3.1.3.2.1 Continuous Wave and Modulation

The intensity of a laser operating in cw mode can be measured directly by measuring the direct current (DC) at the photodetector. Such a measurement suffers often from the limited stability of DC measurements. Therefore, the modulation of the laser is a commonly used method.

The easiest way to modulate the laser beam is the direct modulation of the laser intensity. This can be done by using a mechanical chopper, which periodically interrupts a light beam. Often they consist of a motor and a metal disk with slots. Typical frequencies for mechanical choppers are in the kHz-range. Using a reference signal, the chopper provides information on the frequency and moreover the phase between the chopper signal and the measurement signal. Both signals allow the use of frequency-selective and phase-sensitive detections. For this purpose, lock-in voltmeters are used. Because of the small electronic bandwidth of these devices, the signal-to-noise ratio can be significantly improved compared to the DC measurement. The relatively small modulation frequency limits the temporal resolution of such systems. A higher modulation frequency can be achieved by using optical modulators, for example, Pockels cells.

In the case of semiconductor laser, a direct modulation of the excitation current is possible. Because of the different injection current, a different heating of the laser occurs. This leads to a change in the emission wavelength of the laser. This offers the opportunity to scan over absorption features. Combining this with the measurement at the modulation frequency or even on higher overtones, extremely small detection limits are possible. This well-established technique is often used for concentration measurements of trace species, especially in atmosphere. Also in this case the measurement time is rather large.

3.1.3.2.2 Pulsed Excitation

Beside the excitation of laser in cw mode, it is also possible to excite laser in pulse mode. Lasers with pulses the μs-range down to the fs-range are available. For pulses down to the ns-range, the direct measurement is possible using fast analog–digital converter, fast digital storage oscilloscopes, or boxcar integrators. To detect the shape of a 1 μs pulse properly, a spectral bandwidth in the MHz-range is necessary. According to Equation 3.1.13, this large bandwidth leads to a higher noise level on the detector and typically smaller sensitivities in the signal measurement. The advantage of this method is the relatively short measurement time. This allows a fast online monitoring of reaction processes.

For the measurement of even faster processes, often the method of the single-photon counting is applied. By measuring the time, which individual photons need to emit, the temporal behavior of fluorescence processes can be detected.

Other techniques to detect ultrafast processes used the so-called streak cameras or optical autocorrelators.

FURTHER READING

Bachmann, F., P. Loosen, and R. Porpawe. *High Power Diode Lasers—Technology and Applications*. Springer-Verlag GmbH.
Bergmann, L., C. Schaefer, and H. Niedrig. 2004. *Lehrbuch der Experimentalphysik: Optik. Wellen- und Teilchenoptik*. Auflage 10. Gruyter, Berlin, Germany.
Bleicher, M. 1990. *Halbleiter—Optoelektronik*. Hüthig, Heidelberg, Germany.
Demtröder, W. 2002. *Laser Spectroscopy—Basic Concepts and Instrumentation*. Springer-Verlag GmbH, Berlin, Germany.
Diehl, R. 2000. *High Power Diode Lasers—Fundamentals, Technology, Applications*. Springer-Verlag GmbH, Berlin, Germany.
Drexhage, K. 1977. *Dye Lasers*. 2nd rev. ed., Vol. 1. Springer-Verlag, Berlin/New York.
Duarte, F.J. and L.W. Hillman. 1990. *Dye Laser Principles*. Academic, New York.
Ebeling, K.J. 2008. *Integrated Optoelectronics—Waveguide Optics, Photonics, Semiconductors*. Springer-Verlag GmbH, Berlin, Germany.
Schubert, E.F. 2006. *Light-Emitting Diodes*. Auflage 2. Cambridge University Press.
Träger, F. 2006. *Springer Handbook of Lasers and Optics*. Springer-Verlag, Berlin, Germany.

3.2 MONITORING AND MAPPING OF FRESH FRUITS AND VEGETABLES USING VIS SPECTROSCOPY

BERND HEROLD

3.2.1 Product Pigment Contents, Spectral Signature, and Spectral Indices

A rapid nondestructive sensing technique could be very helpful to monitor and to manage different working steps directly in horticultural production, at harvest as well as in fruit and vegetable postharvest processing. In particular, optical measurements have potential to operate on-line during the produce growth and development on the plant, for monitoring the developmental stage and for recognizing deviations from expected produce quality. Because of the fast response of measurable pigmentation and structural tissue changes in the plant organs, the influence of different external factors like geographical site, weather and soil conditions, nutrition status, etc. can be accessed at the time the plant responds, and allow the producer to react at an earlier date than in conventional production.

Visual (VIS) spectroscopy covers the wavelength range from 400 to 750 nm. Within this wavelength range, several important plant pigment groups (chlorophyll, anthocyanins, and carotenoids) can be detected. Each pigment absorbs light within characteristic wavelength bands. The degree of light absorption in each of these wavelength bands correlates with the relative pigment content of the cell tissue included in the measurement. The Beer–Lambert law (Equation 3.2.1) shows the relation between wavelength-specific light absorption (A) and the concentration of absorbing molecules in a sample. However, Beer–Lambert law assumes highly diluted molecules in transparent (low scattering) isotropic medium (Workman and Springsteen 1998):

$$A = \log \frac{I_0}{I_T} = \log\left(\frac{1}{T}\right) = \varepsilon_i \cdot c_i \cdot L \qquad (3.2.1)$$

where
I_0 is the intensity of incident light
I_T is the intensity of transmitted light
T is the transmittance, I_0/I_T
ε_i is the molar extinction coefficient
c_i is the concentration
L is the thickness of material/path length

The cell tissue of the real produce shows deviations from all three assumptions. Furthermore, geometrical and surface characteristics influence the optical measurement. Therefore, this formula is not exactly valid for an intact produce. Nevertheless, it allows qualitatively interpreting changes in cell tissue properties.

Spectroscopic measurements can be carried out on parts of, or on the whole, produce. Accordingly, the spectral signature recorded by the measurement indicates partially or completely the pigmentation in the produce cell tissue. Because of the biological nature, considerable differences of the pigment concentration exist between the individual products as well as within an individual product. Additionally, certain local changes of cell tissue structure (cell size, cell wall properties) can occur due to exogenous and endogenous factors. Therefore, precise monitoring of product development requires performing repeated measurements with identical configuration. Moreover, it is important to assure that optical measurements are carried out over a certain time period under equal environmental conditions for monitoring the development of living product.

Quality changes are related to changes in the chemical composition that can be acquired by spectroscopic measurements. This technology is widely used, for example, in remote sensing of earth surface to investigate changes of vegetation growth and development. The spectral signature of plant canopy in VIS and NIR wavelength range indicates sensitively the seasonal development and particularly the effect of stress conditions on chlorophyll and water content (Figure 3.2.1). Reflectance from plant foliage is largely controlled by plant pigments in the spectral region between 0.4 and 0.7 µm, by internal leaf structure between 0.75 and 1.0 µm, and by foliar moisture content between 1.2 and 2.4 µm. Any physiological or environmental factor resulting in a difference in the amount or composition of pigments present in the leaves, in the amount of cell-wall/air interface inside the leaves, or in the moisture content of the leaves can produce a change in canopy or leaf reflectance. Many investigators have identified differences between species in response to

VIS/NIR Spectroscopy

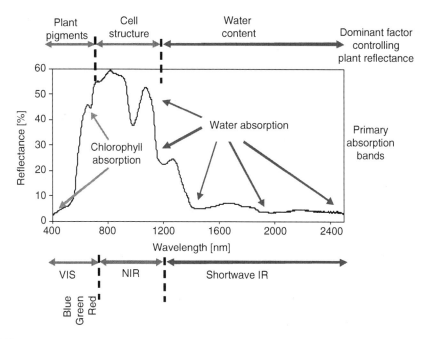

FIGURE 3.2.1 General example of a reflectance plot for apple fruit. (From Short, N.M., Sr., Technical and historical perspectives of remote sensing, NASA publication, 2006.)

various biological and environmental factors. Responses of plant canopy to environmental factors often result from complex interactions of several physiological and morphological changes (Olson 1986).

Horticultural products show similar physiological behavior, and related changes in spectral signature can be observed. For instance, during maturation of apple fruits on the tree, typical physiological processes occur: ethylene synthesis, starch conversion to sugar, increasing of respiration rate, chlorophyll content decrease, and anthocyanins accumulation (Knee 2002). It is difficult to monitor these processes completely in nondestructive way. However, monitoring of chlorophyll content decrease and of anthocyanins accumulation can be done nondestructively by spectroscopic measurement and enables to indicate the fruit maturity and quality progress (Zude 2003). Resulting spectroscopic data can provide useful information to predict the maturity stage and related optimum harvest date.

Also, ripening and quality changes of fruit after harvest and during storage cause characteristic changes of spectral signature that can be detected by spectroscopy. It is possible to identify apple fruits with internal quality defects like water core or cell tissue discoloration and to remove them by optical sorting before marketing.

The VIS spectral signature indicates very sensitively the changes in the degree of light absorption by pigments within selected wavelength bands. On the other hand, there exist wavelength bands with relatively almost zero light absorption. These wavebands are influenced mainly by the scattering properties of the tissue. In this regard, it is a feasible approach to build spectral indices derived from the relationship between the light intensities measured at two or more wavelengths. These indices are used immediately as criteria to describe produce quality changes in various applications. For example, in remote sensing of plant canopies, the normalized difference vegetation index (NDVI), according to Equation 3.2.2, is frequently used to describe changes in plant vitality

$$\text{NDVI} = (I_{\text{NIR}} - I_{\text{RED}})/(I_{\text{NIR}} + I_{\text{RED}}) \quad (3.2.2)$$

where
I_{NIR} is the light intensity measured in the NIR channel (0.76–0.90 μm)
I_{RED} is the light intensity measured in the RED channel (0.63–0.68 μm)

the data are acquired by satellites (e.g., Landsat TM channels 3 and 4).

For estimating the plant chlorophyll content, several indices were tested. Lichtenthaler and coworkers (1996) found that the red-edge index derived from spectral reflectance measurements could be a useful indicator of chlorophyll content in leaves of different plants. If the chlorophyll content decreases, then a shift of the red-edge toward shorter wavelengths is observed. The red-edge is defined as the inflection point at the red flank of spectral signature, that is, around 700 nm, and provides more reliable measurements of chlorophyll content than other indices.

3.2.2 Spectroscopy in the VIS Wavelength Range

A spectrophotometer consists of light source, means to disperse the light into spectral components, means to conduct the light from the source to the sample and from the sample to the detector, and light detector (see Section 3.1).

Depending on the sample properties that are to be analyzed, different modes of sample presentation can be used. According to contributions in the textbook of Ozaki and coworkers (2007), four modes of sample presentation have to be considered: transmission, reflection, transflection, and interaction (see Section 3.3).

In the case of the transmission mode, the incident light illuminates perpendicular to one side of the sample and the transmitted light is detected from the opposite side. This mode of presentation is widely applicable for liquids without scattering or in low-scattering conditions, where a cuvette is used. So the Beer–Lambert law is almost applicable; it is possible to consider only the absorption properties. Obviously, this condition is not feasible for intact fruits and vegetables. Mostly, they have an inhomogeneous structure with different light scattering properties (see Section 1.3). Nevertheless, the transmission mode is in practical use to get suitable spectra of some horticultural products with thick skin, where in the interaction mode the light might not provide sufficient information on inner portion of the product. However, very high intensity of illumination is needed.

In the case of reflection mode, the incident light also illuminates perpendicular to the sample surface. The light propagates in the sample with a series of absorption, scattering, diffraction, and transmission. Finally, diffuse-reflected light radiates from the sample surface. In this case, the sample should be opaque. The Kubelka–Munk

VIS/NIR Spectroscopy

theory, according to Equation 3.2.3 (I'Anson 2007), is applicable for this mode, where both absorption and scattering properties are important factors to explain the variation of measured spectra

$$R_\infty = 1 + \frac{\mu_a}{\mu_s} \sqrt{\frac{\mu_a^2}{\mu_s^2} + 2\frac{\mu_a}{\mu_s}} \approx 1 - \sqrt{\frac{2\mu_a}{\mu_s}} \qquad (3.2.3)$$

where
  R_∞ is the diffuse reflectance from sample with infinite thickness
  μ_a is the absorption coefficient
  μ_s is the scattering coefficient

The incident light cannot reach a deep position in the sample because of high absorption or multiple scattering. If the sample has sufficient thickness, the optical sample thickness should be regarded as infinite, in which case, according to Ozaki and coworkers (2007), attention has to be paid to only the absorption coefficient in the Kubelka–Munk equation. Such a situation is useful for analyzing the spectra of powdered and solid samples with thickness more than 1 cm. Approximately, this theory may be applicable for fruits and vegetables with high density, that is, if the light cannot penetrate deep into the cell tissue. Because of the small penetration depth, only data of substances located just under the produce surface can be obtained.

Attention is also to be paid to the optical geometry of the instrument. Standard viewing geometries for reflectance measurements are the D/8 sphere geometry and the 45/0 geometry. The first one means diffuse illumination of the sample produced by an integrating sphere and viewing at 8° to the sample's normal (Figure 3.2.2). Usually, this configuration includes gloss in the measurement. In the 45/0 geometry, the illumination is presented at 45° to the normal and the sample is viewed on the normal (Figure 3.2.3), that is, gloss is excluded from the measurement. The decision on the viewing configuration to be selected depends on the sample properties as well as on the requirements of measurement. In many cases, intact fruits and vegetables have an uneven sample surface. Because the D/8 sphere geometry does not use

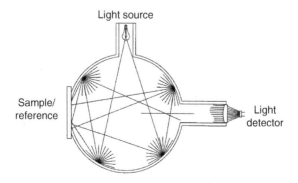

FIGURE 3.2.2 D/8 sphere geometry of optical reflectance measurement. (Courtesy of Accuracy Microsensors, Inc., Pittsford, New York, USA.)

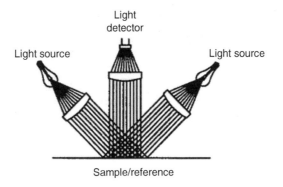

FIGURE 3.2.3 45/0 geometry of optical reflectance measurement. (Courtesy of Accuracy Microsensors, Inc., Pittsford, New York, USA.)

directional light, it seems to be more advantageous for reflectance measurements in horticulture.

In the case of the transflection mode, there is a combination of transmission and reflection. Incident light is transmitted through the sample and then scattered back from a reflector, which is made of ceramic or aluminum to be compatible with the diffuse reflection characteristics of the instrument. The measured spectra of not only liquid but also turbid material can be detected by the integrating sphere or the detector closely settled to the sample. Particularly, spectra from small volumes of the sample can be sensitively measured. This mode requires a specific sample preparation and, therefore, it seems not to be suited for intact horticultural products.

In the case of the interaction mode, an interaction probe with a concentric outer ring of illuminator and an inner portion of receptor is usually used. The end of the probe is in contact with the surface of the sample. For turbid material, the light propagates in the sample, similar to the reflection mode, with a series of absorption, scattering, diffraction, and transmission. However, in difference to the reflection mode, only the light transmitted through the sample can be detected. This mode is termed remission in physics. The effective path length of light transmission through the sample depends on the distance between illuminator and receptor. A theoretical approach to describe the light propagation in the interaction mode could be the point spread function according to Equation 3.2.4 (Praast et al. 2003)

$$\mu_s(r) = \frac{\mu_s(1 - R_\infty^2)^2 \cdot e^{-\mu_s 2br}}{(1 - R_\infty^2 \cdot e^{-\mu_s 2br})^2} \quad (3.2.4)$$

where
 $\mu_s(r)$ is the point spread function
 μ_s is the light scattering coefficient
 r is the radial distance from incidence point

$$2b = \frac{1}{R_\infty} - R_\infty$$

Equation 3.2.4 is used in the printing process to explain the lateral scattering of light in paper, that is, some part of light scattered in the paper is absorbed by the ink, and therefore the actually reflected quantity of light is smaller than expected. The area covered by ink can be compared with the distance between illuminator and receptor in the interaction mode. For a fixed distance r, a low value of the point spread function results on the one side from a black area (small reflectance and small light scattering coefficient), and on the other side from a white area (high value of R_∞ and large light scattering coefficient). In the first case, there is a strong light absorption, and in the other case, there is a small light penetration depth because of the high scattering coefficient. Based on the interrelations described by Equation 3.2.4, it could be possible to optimize the design of interaction probes.

The interaction mode is widely used in VIS and NIR (Section 3.3) spectroscopies to determine valuable compounds in fruits or vegetables, in particular with thin skin.

The monitoring of fruit and vegetable production makes great demands on spectroscopic technique. Obviously, nondestructive and rapid operation is necessary. Further demands on the instrumentation are compactness, robustness in particular including minimum influence of environmental conditions on the measuring results, easy usability, and low costs.

In many applications, small size of the sensor is more important than high spectral resolution and high sensitivity. To maximize flexibility, spectral range should be large and restricted only by detector sensitivity. Resolution should be widely and quickly adaptable to each task, to optimize signal-to-noise ratio. The device should be easy to align permanently and to integrate with other components. For persistent surveillance or long-term monitoring, actuator power usage and data handling (processing, storing, transmitting, etc.) power usage should be minimized (Bhalotra 2004). Under these restrictions, the currently preferred spectrophotometer instrument for monitoring should be based on diode array technology.

Mainly two different types of diode arrays are described (Ozaki et al. 2007): (1) devices with emitting diode array (EDA) and (2) devices with photodiode array (PDA) or CCD.

A limited number of devices with EDA have been developed. These very compact spectrometers are used for specialized applications requiring only a few wavelengths. For determining chlorophyll and moisture, the TW meter is used. It was developed by McClure and associates (McClure 2002), and is a handheld device with three sequentially operated diode emitters (700, 880, and 940 nm).

In contrast, devices with PDA or CCD detector contain a miniaturized fixed grating and acquire the full spectrum. The resolution is limited by the spectral resolution (Rayleigh criterion) and the number of pixels in the array.

A large number of PDA- and CCD-based spectrometer modules (Figure 3.2.4) for different application areas are commercially available (Ocean Optics, Dunedin, Florida; Carl Zeiss MicroImaging GmbH, Jena, Germany; Boehringer Ingelheim Pharma GmbH & Co. KG, Ingelheim, Germany; HORIBA Jobin Yvon Inc., Edison, New Jersey). For industrial color measurements, the Pausch Messtechnik GmbH offers the handheld device type color 5d for diffuse reflectance measurement that uses several LEDs for illumination covering the VIS range from 400 to 700 nm,

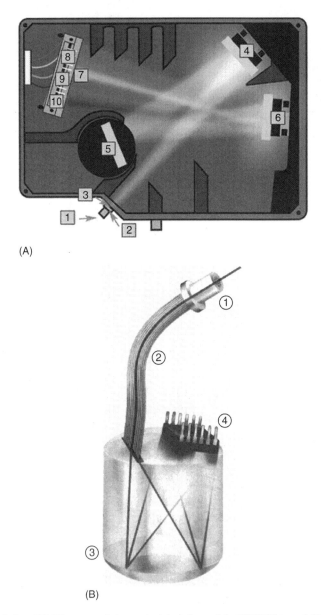

FIGURE 3.2.4 (A) Diagram of the optical bench used in HR2000+ and HR4000 high-resolution spectrometer: 1, SMA connector; 2, entrance slit; 3, long-pass absorbing filter; 4, collimating mirror; 5, grating and wavelength ranges; 6, focusing mirror; 7, detector collection lens; 8, detector (CCD array); 9, variable long-pass order-sorting filter; 10, UV detector upgrades. (Courtesy of Ocean Optics, Inc. Dunedin, Florida, USA.) (B) Design principle of the spectral sensor MMS1: 1, SMA connector; 2, fiber cross-section converter as optical input; 3, holographically recorded and blazed imaging grating; 4, photodiode readout. (Courtesy of Carl Zeiss MicroImaging. Jena, Germany.)

VIS/NIR Spectroscopy

FIGURE 3.2.5 Handheld color meter type color 5d. (Courtesy of Pausch Messtechnik GmbH, Haan, Germany.)

and acquires spectral data with a resolution of 20 nm (internally 5 nm) (Figure 3.2.5) (Pausch Messtechnik GmbH, Haan, Germany).

Other types of portable industrial spectrophotometers include a full-spectral illumination by flash and cw e.g. by xenon flash and tungsten lamps, miniaturized spectrometer module, and two standard viewing geometries (Konica Minolta Business Solutions Europe GmbH, Langenhagen, Germany; X-Rite, Incorporated, Grand Rapids, Michigan). One of the most comfortable portable spectrophotometers for industrial color measurements uses diffuse illumination with flash lamps, a spectral sensor with silicon-based PDA for the wavelength range from 360 to 740 nm with resolution of 10 nm, and allows to measure with and without gloss as well as with UV control (Figure 3.2.6) (Konica Minolta Business Solutions Europe GmbH, Langenhagen, Germany). Silicon-based PDA or CCD devices cover the wavelength range from 200 to 1000 nm (Table 3.2.1), and InGaAs-based PDA devices cover the range from 1000 to 2000 nm.

The signal of a spectral measurement can be referenced by using Equation 3.2.5

$$\text{Spectrum} = \frac{\text{sample} - \text{dark}}{\text{reference} - \text{dark}} \qquad (3.2.5)$$

where
sample is the signal measured from the sample
dark is the signal measured under dark reference (0%)
reference is the signal measured from a white reference (100%)

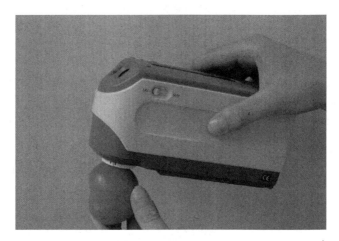

FIGURE 3.2.6 View of the portable spectrophotometer type CM-2600d measuring the light diffusely reflected from tomato fruit.

The difference between reference and dark signals should be as large as possible to obtain a reasonable signal-to-noise ratio. Therefore, suited conditions for dark and white reference measurements have to be selected.

A comparison of three different types of miniaturized spectral sensing devices in connection with D/8 sphere geometry shows that they have similar performance within the VIS wavelength range (Figure 3.2.7). It should be noted that each of these devices is equipped with different illumination source: CM-503i with xenon flash lamp, MMS1 with halogen lamp, and color 5d with several LEDs.

As discussed above, an important aspect of spectrometer measurement on intact horticultural products is the sample presentation. Most suitable modes of sample presentation seem to be the common reflection geometry used also in other application fields, and also the interaction (equals remission) mode.

The reflection mode is preferred, if only information on or close to the product skin is required. Advantageously, low intensity of illumination is necessary to obtain reliable spectral data. Mainly, measurements of reflection are used to determine the color of product surface.

TABLE 3.2.1
Specification of Two Widely used Miniaturized Spectral Sensors

Sensor Type (Manufacturer)	MMS1 NIR Enhanced (Carl Zeiss MicroImaging GmbH, Jena, Germany)	USB4000 (Ocean Optics, Dunedin, Florida)
Detector	Photodiode detector array	Linear CCD array
Number of pixels	256	3,648
Wavelength range [nm]	400–1,100	300–1,100
Stray light [%]	0.1	0.1
Dynamic range	65,000:1	1,300:1

VIS/NIR Spectroscopy

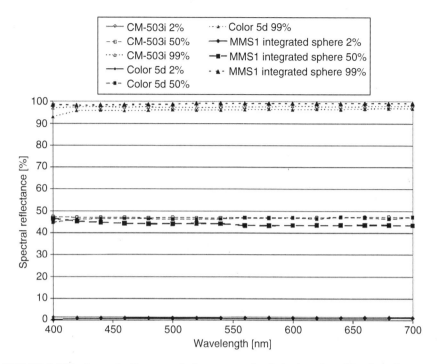

FIGURE 3.2.7 Spectral reflectance of reference samples (Labsphere, Inc., Wessling, Germany) measured with three spectrophotometer instruments: CM-503i (Konica Minolta Business Solutions Europe GmbH, Langenhagen, Germany) MMS1 + D/8 sphere (Carl Zeiss MicroImaging GmbH, Jena, Germany), and color 5d (Pausch Messtechnik GmbH, Haan, Germany). For this comparison, data were acquired from each 20 nm within the range from 400 to 700 nm.

In the interaction mode, the light penetrates through the skin and scatters inside of product tissue. A portion of light reflected from inside the fruit is recovered and measured. As the optical path length through the product depends on the geometric distance between illuminator and receptor (and scattering coefficient, Section 1.3), larger distances provide more information on inner portion of the product. However, higher intensity of illumination is required to obtain reliable data with this additional information. For use of portable spectrophotometer, because of restriction of power supply, the usable optical path length is confined. The optimum distance between illuminator and receptor can be explored in dependence on the optical density of product tissue. Basic studies of light transmission through the product tissue showed that the penetration depth is approximately equal to the distance between illuminator and receptor.

Regarding these restrictions, recently a commercial handheld spectrometer with LEDs for illumination and sample presentation in interactance mode has been developed (Zude 2006). The hardware of the handheld instrument consists of optical components (integrated light cup, spectrophotometer, as well as IR-thermometer), electronic circuits, 128 kB memory, microprocessors, display, and accumulators (Figure 3.2.8).

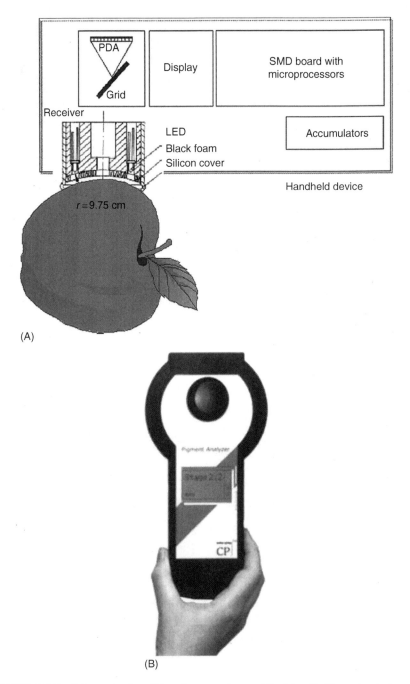

FIGURE 3.2.8 Schematic view (A) and camera image (B) of handheld spectrophotometer with PDA-grid (Pigment Analyzer PA1101, CP, Germany), with electronic circuits (surface-mounted devices, SMD), and integrated light cup for measuring in interaction mode at the fruit surface. (Courtesy of Control in Applied Physiology [CP], Falkensee, Germany.)

VIS/NIR Spectroscopy

Incident light is provided by a combination of four white LEDs (peak wavelength at 450 ± 60 nm), two red LEDs (peak wavelength at 660 ± 20 nm), two far red LEDs (peak wavelength at 690 ± 45 nm) and a reference LED (peak wavelength at 780 ± 20 nm) arranged circularly around the receiving fiber. LEDs emitting diffuse radiation at narrow angles, frequently, for example, 30° according to CIE standards (Section 2.1), were selected due to improved light distribution into the plant tissue. The light source and the fruit were kept in close contact during the measurement supported by black foam to avoid measuring reflectance instead of remittance. The radiation from the light source enters the sample matter and interacts with the texture leading to scattering as well as with absorbing compounds leading to extinction of the radiation. Backward scattered radiation is received by the fiber cross section that is placed in the center of the light cup. A miniaturized PDA spectrophotometer (450–1100 nm, 3.3 nm resolution) reads the incoming photons by means of a sample remission (equals interaction) spectrum (Chen and Nattuvetty 1980; Zude 2003) at automatically adjusted integration time (50–400 ms). The spectrophotometer circuits of PDA developed and current-to-voltage conversion method enables high-precision signal detection even at low output levels that are necessary for fruit interaction readings. The PDA operates with charge integration mode (Hamamatsu, Hamamatsu City, Japan).

The output is linear to the sum of light exposure during integration time (light intensity × integration time). Such linearity is approved in the range from

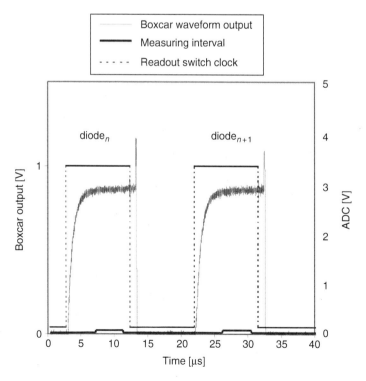

FIGURE 3.2.9 Readings of discrete photodiodes related to two specific wavelength recorded with digital oscilloscope.

0% to 90% of full-scale output. The readout of PDA is an iterative process, where single photodiodes are sequentially discharged. Differentially, charges are amplified and a switched integrator provides the current to voltage conversion. The switched integrator uses analogue electronics, supported by digital control, monitoring one discrete diode in time. The boxcar averaging applied is an analogue measurement, when the signal is averaged over a short time gate. In Figure 3.2.9, the boxcar output of two photodiodes is visualized by means of oscilloscope (TDS210, Tektronix, Beaverton, Oregon) recordings. Measurements were carried out in the steady phase of the signal, marked as measuring interval in Figure 3.2.9. Nonlinear properties of wavelength representation have been calibrated with a third-order polynomial function (Zeiss, Jena, Germany). For obtaining intensity values intermediately appearing between discrete wavelengths, the curves were spline interpolated after smoothing by means of Savitzky–Golay algorithm.

3.2.3 Mapping Technique, GPS Technique, and Data Management

The mapping of spectral data of fruits or vegetables during the growth period has to include also site-specific data of environment and of climatic conditions. Beside the most recently commercially available devices (Pigment Analyzer, Fantec, etc.), in former research work, a portable spectrophotometer equipment has been developed to collect data on the tree (Truppel et al. 2005).

3.2.3.1 Design of the Portable Spectrophotometer Equipment

The spectrophotometer was designed and built up for acquisition of spectral signature together with surface temperature of horticultural objects in VIS and NIR wavelength range from 500 to 1000 nm during development in situ. The block diagram of the spectrophotometer system is shown in Figure 3.2.10.

This system was based on commercial original equipment manufacturer (OEM) components and consisted of two main parts (in Figure 3.2.10 separated by the dashed line):

1. Portable part with spectrophotometer device, headset for audio communication, wireless communication system with two transceivers to transmit measuring data and audio signals, and power supply
2. Stationary part with notebook personal computer (PC), wireless communication system to transmit measuring and audio data, and power supply

The wireless data transfer was used to overcome restrictions that result from selection of less powerful handheld computer systems as well as from additional weight and reduced freedom of movement. The equipment worked over distances up to 80 m under open-field conditions between operator and PC.

The spectrophotometer device consisted of the spectral sensor Type MMS1 NIR enhanced (Carl Zeiss MicroImaging GmbH, Jena, Germany) (Table 3.2.2), the electronic controlling unit Type LOE1 (tec5 AG, Oberursel, Germany), and the sensing probe.

VIS/NIR Spectroscopy

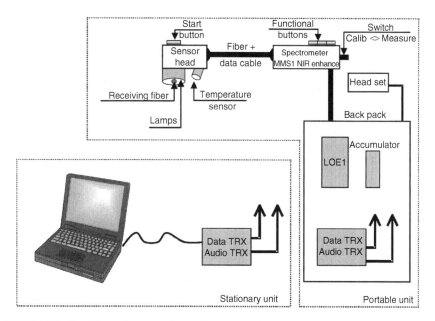

FIGURE 3.2.10 Block diagram of portable spectrophotometer system: TRX, transceiver; LOE1, electronic control unit; MMS1 NIR enhanced, spectrophotometer module.

The spectral sensor was integrated into metallic box with permanent cable connection to the electronic controlling unit. The box was provided with a switch to change between two modes of operation. Additionally, three buttons were placed at this box as user interface to perform operation and communication. The 240 mm

TABLE 3.2.2
Technical Data of Spectral Sensor MMS1 NIR Enhanced

Parameter	Data
Optical entrance	SMA connector, fiber cross-section converter from circular to rectangular shape (entrance slit of spectral sensor)
Grating	flat-field, 366 1/mm (in center) blazed for 600 nm
Spectral range	400–1100 nm
Wavelength accuracy	Absolutely 0.3 nm
Temperature drift	<0.02 nm/°C
Spectral pixel distance	$\Delta\lambda$ pixel 3.3 nm
Spectral resolution	$\Delta\lambda$ Rayleigh 10 nm
Sensitivity	1013 counts/W s (with 14-bit conversion)
Total dimensions	70 × 60 × 40 mm (with housing), length of cross-section transformer fiber: 240 mm
Diode array sensor	Hamamatsu Type S 4874-256 Q
Pixel number	256
Pixel size	25 × 2500 μm

FIGURE 3.2.11 Sensing probe for measurement of light interaction on fruit, the forefinger operates the start button, and the sensor box is affixed to forearm.

long glass fiber of the spectral sensor was attached to the handheld sensing probe. Different sensing probes were developed:

- Small cylindrical probe for light interaction on fruits and vegetables (Figure 3.2.11)
- Cylindrical probe (integrating sphere) for diffuse reflection (Figure 3.2.12)
- Gripper probe for complete transmission through plant leaves (Figure 3.2.13)

FIGURE 3.2.12 Sensing probe for measurement of diffuse light reflection. The cylindrical body contains an integrating sphere with 60 mm diameter. Below the lamp holder can be seen. The measuring aperture is located at the visible front side.

VIS/NIR Spectroscopy

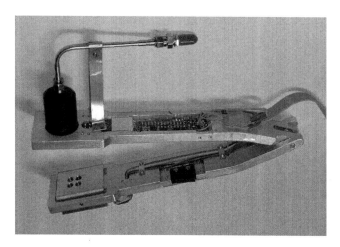

FIGURE 3.2.13 Sensing probe for measurement of light transmission through a leaf. Four lamps are placed in the holes of one blade (below), while a lens is integrated in the opposite blade (above) to focus the transmitted light on the fiber entrance.

Probes can be exchanged to adapt the system suitably to the measuring object, for example, to fruit or leaf. The outstanding attribute of each sensing probe is the single-handed operation. The probes were equipped with miniaturized light bulbs to illuminate the sample. The power supply of light bulbs was completed with particular electronic control circuitry to ensure for each measurement more than 1 s start-up phase and thereafter stable lamp operation. The measurement was started by means of a button placed on the sensing probe.

Particularly, the sensing probe for light interactance geometry was designed to detect spectral signature of skin and adjacent tissue of apple fruit. For this purpose, six light bulbs and the light detecting cross-section glass fiber were concentrically grouped on the front side of the probe. Soft foam rings were placed around every lamp and the glass fiber to fit probe and fruit surface. The start button for measurement and white balance, respectively, was installed on the backside of the probe (Figure 3.2.12). Additionally, a thermopile sensor was positioned at the probe to detect the fruit surface temperature. The temperature was measured each time before lamp operation, and data transfer was carried out during the start-up phase of illumination.

The electronic controlling unit LOE1 was necessary to set the measuring conditions and to read out the measured data. One address of the unit interface was used to read the handling of the four functional buttons and the switch for operation mode, to control the switching of the light bulbs, and to read the data of the temperature sensor.

The communication between this controlling unit and PC was performed via standard serial port. This communication path was cut off and a wireless modem pair both on the portable and the stationary part was placed in between. A high-frequency (HF) modem was used to transmit measured temperature and spectral optical data between portable part and stationary part with PC. Additionally, an analogous audio communication channel was installed between portable and stationary parts consisting of headset, additional electronic circuitry to control the audio signal processing, and ultra-high-frequency (UHF) transceiver pair. The complete wireless communication system was housed in a separate box with two antennas. Two functions were carried out: (1) transmission of annotations on measuring object from portable part to stationary part, spoken by the operator and (2) transmission of information on data from stationary part to portable part, audible for the operator.

To supply the entire portable part for several hours, the origin accumulators of LOE1 with 1.2 V/1.3 Ah were not sufficient. They were replaced by NiMH accumulators (Type NH GP 370AFH-1Z) with 1.2 V/3.7 Ah (Reichelt Elektronik, Sande, Germany). The complete portable part was placed in a backpack with a total weight of 4 kg.

The PC used on stationary part had the following parameters: Pentium III, 700 MHz, 128 MB RAM, 6.4 GB HD, serial COM port, sound card with input and output, and Win98 with Windows sound system. It was connected to transceivers paired to them of the portable part. Accordingly, additional electronic circuitry was built up to control the audio-signal processing, and the complete system was housed in a separate box with antennas. The power supply of stationary part was provided by car battery.

To operate the spectrophotometer, in-house developed software was used. The communication between PC and spectrophotometer device as well as data handling on PC was controlled by DELPHI program based on the DLL available from manufacturer of LOE1. Additional software functions were installed for handling of stored audio data. Further software was developed to control the temperature measurement and the operation of light bulbs and UHF transceivers. The spectrophotometer software allowed fully automated operation including automated calibration and adaptation of sensitivity by controlling of integration time.

3.2.3.2 Fruit Analyses Using the Portable Spectrophotometer Equipment with Probe for Light Interaction

First tests of the device were carried out in an apple orchard during preharvest and harvest period. The stationary part was placed in opened car boot, while the car was parked between the tree rows. The operator carried the backpack with portable part. The cable with sensor box and sensing probe as well as the headset were placed according to Figure 3.2.11. Before measurements, the calibration mode was used to calibrate the spectrophotometer. For dark balance, the front of sensing probe was completely shaded. The white balance was carried out by attaching the probe to a white Teflon block. Then, in the measuring mode, the sensing probe was attached to the skin of selected fruits and measurements were carried out. The completion of each measurement was automatically confirmed by transmitting an audible comment from PC. This comment allowed the operator to assess the usability of measurement.

Because the spectrophotometer equipment is suited for single-handed operation, the operator was free to use the other hand to hold the fruit to provide for sufficient contact between fruit and sensing probe.

3.2.3.3 GPS Technique

The application of global positioning system (GPS) is proved to be valuable for localization and navigation in agriculture, particularly for site-specific farming (precision farming). However, for cultivation of perennial plants like trees, the reliability is restricted due to shadowing effects and reflection (multipath effects). An alternative technology to identify the individual tree could be provided by radio frequency identification (RFID). In the latter case, the individual object is equipped with a transponder with unique identification number for wireless data acquisition.

3.2.3.4 Data Management

A Web-based system for analyzing the data gathered was developed in the frame of a bilateral project of Germany and Slovenia. The aim of data analysis is to monitor site-specific pigment changes in apple. Such data analysis would enable the real-time evaluation of the fruit maturity in different growing regions.

With this concept, data can be recorded in different growing regions (or production systems) and transferred from the orchard to the marketing office. The data input will be supported by means of a Web site (client) based on Internet browser technology, programmed in C sharp. The data will be stored in a protected SQL database. An additional application server (middle part in Figure 3.2.14) will be used to analyze the data by means of algorithms developed and continuously improved for this purpose (Figure 3.2.14).

3.2.4 Application Examples

3.2.4.1 Apple Fruit

The conventional determination of the optimum harvest date for apple fruit is based on several parameters of physiological development and quality (Osterloh 1980; Streif 1983). Depending on the properties of cultivar, the mature fruit should not only have typical size, shape, and color, but also high internal quality parameters. The start of apple fruit maturation is indicated by a distinct increase of ethylene production. However, the acquisition of this parameter is sophisticated and expensive (Wilcke 2002). Furthermore, starch conversion, firmness, soluble solids, acidity, color, and chlorophyll content indicate the fruit maturity stage.

Frequently used methods to determine the fruit maturation are the starch iodine test, Magness–Taylor firmness test, and refractometry of juice. Weekly sampling is

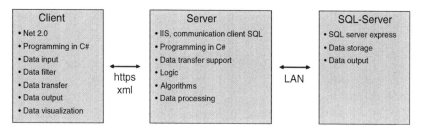

FIGURE 3.2.14 Schematic structure of data input and data visualization via Internet.

necessary to allow precise prediction of optimum harvest date. The use of these destructive methods requires sufficient experience to obtain correct interpretation of results, because the apple cultivars exhibit different rate and variability of maturity progress (Höhn et al. 1999; Wilcke 2002).

Therefore, a nondestructive method could be very useful to improve the performance of determination of fruit maturity. Spectral measurements on the fruit allow sensitive detecting the pigment changes occurring during fruit development on the tree (Osterloh 1980; Herold et al. 2005).

The spectrophotometer with sensing probe for light interaction readings (Truppel et al. 2005) was tested first time in commercial orchard during the season 2003 to acquire the spectral signatures of identical apple fruits on the tree during maturity progress in preharvest and harvest period. The measurement was carried out on the fruit cheek. The spectral signatures of 'Elstar' apples showed typical changes in the range from 550 to 700 nm owing to anthocyanins accumulation and decrease of chlorophyll content, respectively (Figure 3.2.15). In the range above 750 nm mainly, absorption owing to water content was indicated.

In particular, the decrease of chlorophyll content was represented by the increase of the light intensity around the absorption peak near 680 nm, and by the synchronous shift of the red-edge toward shorter wavelengths (Figures 3.2.15 and 3.2.16). The decline of the red-edge curve was gradually reduced during the period of measurement. The data of red-edge and of skin chlorophyll content of the same fruit analyzed in laboratory were sufficiently highly correlated (Figure 3.2.17).

However, the time course of red-edge data acquired from sequentially harvested fruit was affected by random variability because of measurements on different fruits. The acquisition of red-edge by repeated spectral measurements on the same fruit during development on the tree allowed to reduce this variability, and simultaneously to increase the number of fruits to enhance the reliability of results. The gradually

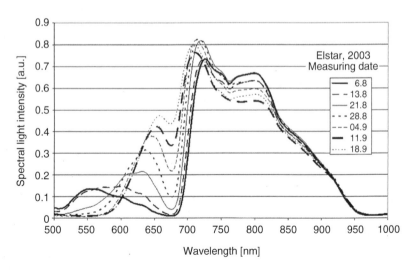

FIGURE 3.2.15 Change of average spectral signature of $n = 73$ 'Elstar' fruits during the period from August to September 2003.

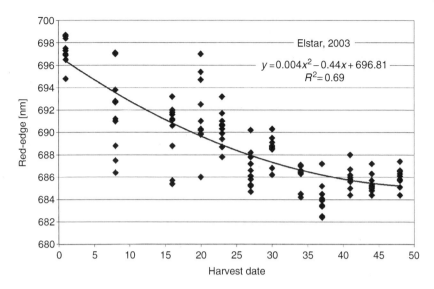

FIGURE 3.2.16 Change of the single red-edge values measured during the season 2003 on each 10 sequentially harvested 'Elstar' fruits (day 1 = August 6).

lowered decline of red-edge curve was confirmed, but no clear criterion of harvest maturity level could be identified (Figure 3.2.18).

Instead of the calculation of red-edge wavelength, an approximately equivalent index of chlorophyll decrease could be determined by applying Equation 3.2.6 to the original spectral data

$$I_{\text{ChlD}} = \frac{I_{750} - I_{700}}{I_{750} + I_{700}} \tag{3.2.6}$$

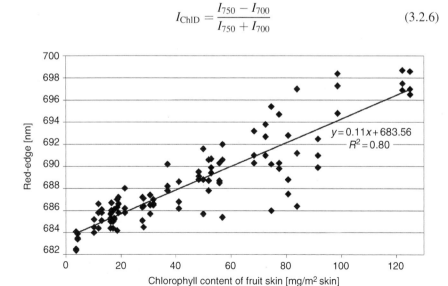

FIGURE 3.2.17 Relation between red-edge values and chlorophyll content acquired on 'Elstar' fruits during the season 2003.

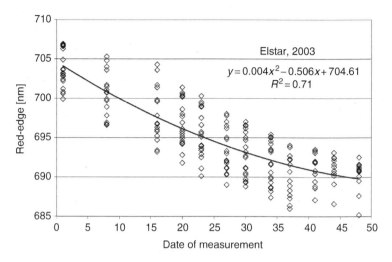

FIGURE 3.2.18 Change of the single red-edge values repeatedly measured on the same 20 fruits during the season 2003 (day 1 = August 6).

where
I_{ChlD} is the index of chlorophyll decrease
I_{750} is the reflectance at the wavelength 750 nm
I_{700} is the reflectance at the wavelength 700 nm

The correlation between data of red-edge and index of chlorophyll decrease (I_{ChlD}) seemed to be high and was confirmed during further seasons for several cultivars (Figure 3.2.19) also with high correlation to the chemically analyzed chlorophyll content (Figure 3.2.20).

On the other hand, the anthocyanins accumulation was represented by the decrease of measured light intensity around the absorption peak near 550 nm. Typically, during this process, the curve of spectral signature between 630 and 570 nm inclined. Although the spectral signature in this wavelength range is also affected by other pigments such as carotenoids and chlorophyll, it seemed to be useful to describe the anthocyanins accumulation by an index (Equation 3.2.7)

$$I_{AnI} = \frac{I_{630} - I_{570}}{I_{630} + I_{570}} \qquad (3.2.7)$$

where
I_{AnI} is the index of anthocyanins changes
I_{630} is the signal at the wavelength 630 nm
I_{570} is the signal at the wavelength 570 nm

The indices of anthocyanins increase and of chlorophyll decrease could be presented simultaneously on the same scale; hence they offered a quick and precise insight on pigment changes of apple fruit during maturity development on the tree (Figure 3.2.21).

VIS/NIR Spectroscopy

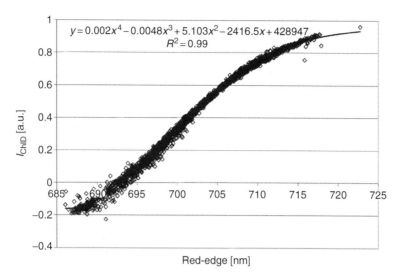

FIGURE 3.2.19 Correlation between data of red-edge and index of chlorophyll decrease measured on the same fruits of 'Elstar', 'Pinova', and 'Topaz' during development on the tree, during the season 2006.

Apple cultivars show appreciable differences in fruit maturity progress. These differences are caused by genetic properties and can be observed by measuring the change of physical and chemical fruit parameters during maturity development. During maturity development from August to October, the three cultivars 'Elstar',

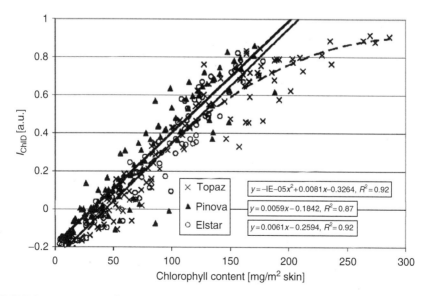

FIGURE 3.2.20 Relation between spectral index of chlorophyll decrease and chlorophyll content analyzed in laboratory and during the maturity development of apple fruit of three apple cultivars from August 7 to October 9, 2006.

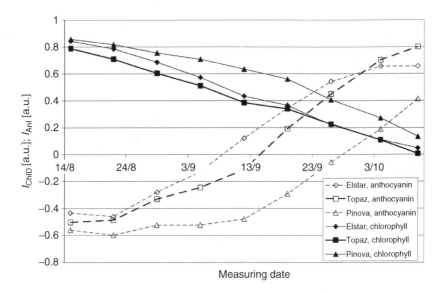

FIGURE 3.2.21 Average course of indices of chlorophyll decrease and anthocyanins increase of each 40 same fruits of three apple cultivars during the maturity development on the tree, during the season 2005.

'Pinova,' and 'Topaz' exhibited specific differences in the spectral signature. The spectral signatures of each group of 40 fruits per cultivar were measured on the tree between August 15 and October 4, 2005. At both dates, the spectral curves of 'Elstar' and 'Topaz' were similar, but 'Pinova' showed lower values for anthocyanins index and higher values for chlorophyll index. On October 4, the shape of spectral curve of 'Pinova' differed significantly from the shape of both the other cultivars, particularly in the wavelength range from 500 to 650 nm. Supposable, that could be caused by typical differences of pigment composition (data not shown). Nevertheless, the index of anthocyanins increase appeared to be very useful parameter for rapid and nondestructive determination of anthocyanins accumulation.

Comparing the course of these parameters during a single season, clear differences between the individual cultivars were found (Figure 3.2.21). However, these differences have not been confirmed for several seasons. This fact pointed out the complexity of fruit maturation. Therefore, for reasonable identification of fruit maturity stage, it will be necessary to acquire several independent parameters.

Steady curves of average index values of chlorophyll decrease and of anthocyanins increase were observed on 'Elstar' in 2003. The index of anthocyanins increase arrived at a boundary value of 0.9 on September 9, while the index of chlorophyll decrease continued to reduce even though with lower decline. Because the velocity of anthocyanins increase commonly depends on highly variable factors, the arrival date at the boundary value was not estimated to provide a sufficient criterion on harvest maturity level. The standard deviation (SD) of both indices indicated considerable variation between the individual fruits of the tree during fruit development. This high variation was typical for 'Elstar' and confirmed the well-known fact that several picking dates would be optimal in this case.

3.2.4.2 Sweet Cherry

The maturity development of sweet cherries can be monitored by spectral measurements in a similar way as shown for apples. The spectrophotometer was equipped with the miniaturized module MMS1 and a modified sensing probe for light interaction consisting of a handle with two light guides with lateral distance of 10 mm, one for incident light and another one for light detection. For this study, each 40 fruits of German sweet cherry. 'Spansche Knorpel' were harvested at five dates during the season 2001. Immediately after harvest, the fruits were transported to laboratory and measured with spectrophotometer.

The sensing probe was carefully attached to the fruit cheek, and possible influence of ambient light was restricted by covering sensing probe and fruit with dark plastic film. Similarly as on apple, the spectral signature of sweet cherry showed typical changes because of a distinct decrease of chlorophyll content that was VIS at wavelengths around 680 nm (Figure 3.2.22). Additionally, an increasing absorption because of anthocyanins accumulation was found in the wavelength range near 550 nm. The decrease of chlorophyll content could be described by an approximately linear decline of the red-edge introduced above (Figure 3.2.23). An appreciable variation of the red-edge data of single fruits seemed to exist. This variation corresponded to those of other cherry fruit parameters determined in laboratory (Table 3.2.3).

3.2.4.3 Carrot Root

In carrots, the main pigments are the carotenoids, of which β-carotene dominates, also providing the main nutritional value of carrots as a source for provitamin A.

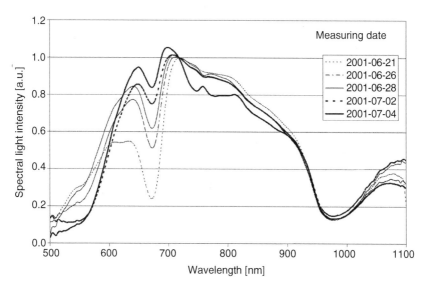

FIGURE 3.2.22 Average spectral signatures of each 40 sweet cherries from 5 different harvest dates during the season 2001 (the spectral curves were normalized by division of the data by the data at the wavelength 717.5 nm).

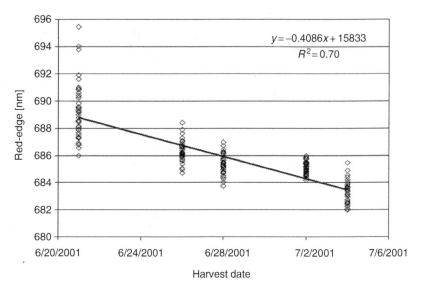

FIGURE 3.2.23 Decline of red-edge during the maturity development of sweet cherries from June 21 to July 04, 2001 (single data of each 40 fruits per harvest date).

Color descriptors are often used in practice for characterizing pigment changes in fresh horticultural products. However, in the present study, the sums of carotenes and color measurements using the $L^*a^*b^*$ color space were weakly correlated, showing maximum coefficients of determination of 0.26, 0.30, and 0.57, respectively. Such results were expected from earlier studies on apples (Merész et al. 1994) and could be explained by the nonlinear changes of product pigment contents, masking effects, as well as tissue scattering properties resulting in various effects on the product's color appearance.

In contrast, the nondestructive determination of pigments by means of spectral spectroscopy could provide more clear results. Experiments were carried out on fresh carrots. All samples were analyzed nondestructively, taking two readings per sample

TABLE 3.2.3
Average and SD of Cherry Fruits Harvested at Five different Dates during the Season 2001

Harvest Date	Fresh Mass [g/fruit]		Soluble Solid Content [°Brix]		Red-Edge [nm]	
	Average	SD	Average	SD	Average	SD
June 21	4.86	0.68	12.6	1.4	689.2	2.1
June 26	6.02	0.70	12.3	1.8	686.3	0.8
June 28	5.86	0.75	14.0	1.5	685.3	0.7
July 2	6.23	0.88	14.4	1.6	685.1	0.4
July 4	7.44	0.88	12.5	1.7	683.3	0.8

at a product temperature of 18 ± 0.5°C, and subsequently samples were analyzed for pigments.

A handheld PDA spectrophotometer device (Pigment Analyzer 1101, CP, Germany, wavelength range from 350 to 1100 nm) was applied for recording spectral signature of carrots. A light cup with LEDs (peak wavelengths at 460, 550, 660, 690 nm, FWHM = 30–60 nm) served as light source, while alumina (Ferro-Ceramic Grinding Inc., Wakefield, Massachusetts) was used as white reference providing a feasible standard for the light interaction mode. Spectra were collected with a resolution of 3.3 nm.

To analyze the data of spectral measurements (I_R), the red-edge (Lichtenthaler et al. 1996) was calculated on the second derivation, $I_R''(680 \text{ to } 720 \text{ nm}) = 0$. For determining carotenoids in carrots, a similar index named the car index representing the inflection point at the long wave flank of the cumulative carotenoids absorption peak was applied. This car index was calculated in the wavelength range from 510 to 610 nm, $I_R''(\lambda 510 \text{ to } 610 \text{ nm}) = 0$, using the firmware (SpectroMeter.exe, CP, Falkensee, Germany) of the handheld spectrophotometer.

As expected, the discrete detection of pigments taking into account their absorption at specific wavelength ranges was higher correlated than color data. The peak widths of carrot chlorophyll and carotenoids absorption particular showed characteristic changes according to the tissue pigment contents (Figure 3.2.24).

In remote-sensing applications and single-product analysis, the phenomenon of peak widening with increased contents could be used for quantitative analyzing the fruit pigment content. Accordingly, the red-edge indicated the changes of chlorophyll content in the spectral signature. When comparing individual carrots from both cultivars, low correlation of the red-edge and the carrot chlorophyll content were found due to low contents found in the samples. However, variation in the red-edge values (Figure 3.2.25) pointed to at least minor chlorophyll contents in the carrot samples.

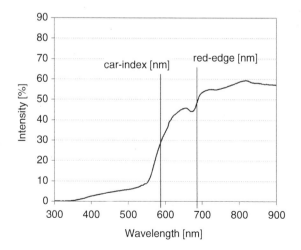

FIGURE 3.2.24 Carrot spectrum with marked inflection points of pigment absorption is indicated for red-edge, $I_R''(680 \text{ to } 720 \text{ nm}) = 0$, and car index, $I_R''(510 \text{ to } 610 \text{ nm}) = 0$.

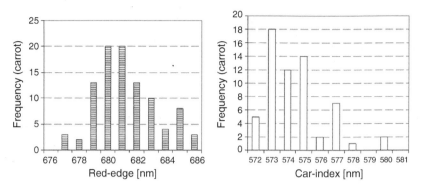

FIGURE 3.2.25 Distribution of red-edge (left) and car index (right) in carrots cv. 'Bolero.' (Modified from Zude, M., Birlouez-Aragon, I., Paschold, J., and Rutledge, D.N., *Postharvest Biol. Technol.*, 45, 30, 2007.)

Higher carotenoid contents led to peak widening and a resulting shift of the car index to higher wavelengths. The coefficient of determination for the sum of the carotenes was $R^2 = 0.80$. The distribution of the car index suggested a large variation in the individual carotene contents (Figure 3.2.25). Such findings were confirmed by means of chromatographic analyses. Summarizing, the application of these indices appeared to be more reasonable than working with color data, since they are related directly to the molecule group absorption at specific wavelength ranges.

3.2.4.4 Potato Tuber

The occurrence of black spots in potato tubers is a serious quality defect. The black-spot susceptibility can be studied by drop tests. For this purpose, potato tubers 'Karlena' were taken from a commercial storage room. Before the test, the tubers were stored under cold storage condition at 4°C. The experiments have been carried out under room temperature condition, so that the resulting product temperature during the tests was about 10°C. The tubers were placed on a fall apparatus in well-defined position, and then dropped onto a fixed metallic plate. To detect a priori existence of black spots within the tuber flesh, a nondestructive method was necessary.

Therefore, the spectrophotometer with sensing probe for light interaction (Truppel et al. 2005) was applied to the tuber part striking against the plate. Already the visual monitoring of spectral signature (Figure 3.2.26) measured on the potato was useful to detect the existence of black spots. The spectral signature of potato tuber without any defect showed a characteristic peak at 720 nm. After significant mechanical impact and several days under cold storage condition, the potato tubers developed black spots. The spectral signature of tubers with black spots changed in characteristic way so that the peak at 720 nm gradually disappeared. This change could be signalized by means of an adapted normalized spectral index according to Equation 3.2.8

$$I_{blsp} = \frac{I_{720} - I_{740}}{I_{720} + I_{740}} \tag{3.2.8}$$

VIS/NIR Spectroscopy

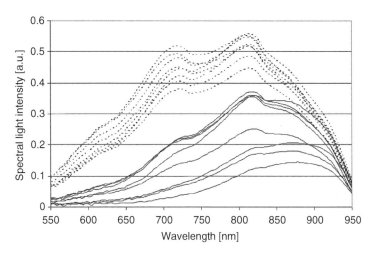

FIGURE 3.2.26 Spectral signature of typical examples of potato tubers 'Karlena' with (solid line) and without (dashed line) black spots.

where

I_{blsp} is the spectral black-spot index
I_{720} is the signal at the wavelength 720 nm
I_{740} is the signal at the wavelength 740 nm

If the spectral black-spot index value decreases below zero, increasing black spots are found in the potato tissue. The spectral black-spot index allowed to separate tuber parts with and without black spots with sufficiently high accuracy.

3.2.5 MAPPING OF SPECTRAL INDICES OF APPLE FRUIT DEVELOPMENT IN ORCHARD

3.2.5.1 Fruit Maturity in Dependence on Geographical Site

Spectral measurements in apple orchards showed significant differences in the development of apple trees and related fruit development depending on the geographical site. Of course, the fruit development differs between locations with significantly different geographical altitudes or latitudes because of climatic differences. However, potential causes of differences in fruit development could also be different soil conditions and, particularly, different availability of water. For instance, a commercial orchard showed significant differences of the availability of water in upper site in comparison with lower site at distances of a few 10 m. The trees of the upper site grown under slight drought stress were smaller and held less branches, leaves, and fruits. The indices of chlorophyll decrease and anthocyanins increase measured on 'Elstar' fruit in the upper site changed more rapidly indicating a forced rate of maturity development (Figure 3.2.27).

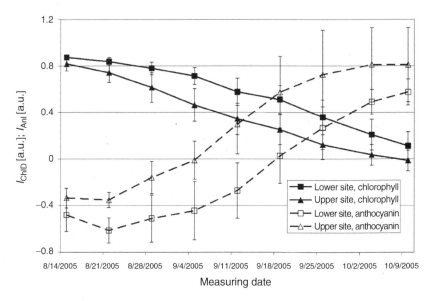

FIGURE 3.2.27 Average course and SD of indices of chlorophyll decrease and anthocyanins increase, respectively, during maturity development of each 20 same fruits. 'Elstar' on the tree during the season 2005. Data of trees from sites with drought stress (upper site) and well irrigated (lower site).

3.2.5.2 Effect of Fruit Yield per Tree on Maturity Progress

The fruit yield of an apple tree is an essential cause of fruit maturity progress. A study has been carried out on two neighboring trees with similar habit but different number of fruits. As expected, the fruit maturity development was more rapid on the tree with lower fruit yield. This fact could be confirmed by evaluation of the indices of chlorophyll decrease and anthocyanins increase measured on the same fruits of 'Elstar' during the season 2003 and other parameters finally analyzed in laboratory (Figure 3.2.28, Table 3.2.4).

3.2.5.3 Effect of Climatic Condition on Maturity Progress

Spectral measurements of 'Elstar' fruits have been carried out with the same spectrophotometer equipment during several seasons in the same orchard. Typical differences in the course of the indices of chlorophyll decrease indicated different rates of fruit maturity development (Figure 3.2.29).

During the seasons 2003 and 2006, stronger declines of the indices of chlorophyll decrease were observed indicating a higher rate of maturity development. When compared with the course of the temperature sums during the according seasons, then the differences could be explained (Figure 3.2.30). The reduced temperature sum is defined as sum of the daytime temperatures reduced by the physiological threshold of 5°C from beginning of May to harvest date in September or October. The highest values of reduced temperature sum were recorded during the seasons 2003 and 2006, while the lowest values occurred during the seasons 2004

VIS/NIR Spectroscopy

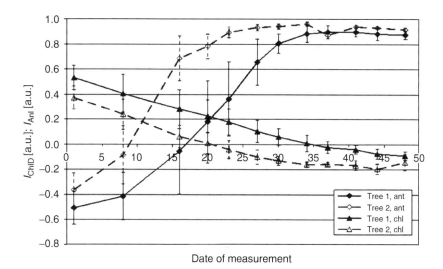

FIGURE 3.2.28 Average course and SD of indices of chlorophyll decrease (chl) and anthocyanins increase (ant), respectively, of 10 same 'Elstar' fruits during development on the tree (season 2003, day 1 = August 6); each 10 fruits were measured on two trees with different fruit yield (tree 1: 79 fruits, tree 2: 22 fruits).

and 2005. According to the given experience, the temperature has most important influence on the fruit maturity progress. However, additional influencing factors have to be considered such as relative humidity, precipitation, wind, and solar radiation. Models were developed to predict the optimum harvest date based on climatic conditions; however, currently, these models do not sufficiently correspond with the real fruit maturity development. Therefore, the spectral data could be helpful for future development of predicting models.

3.2.5.4 Mapping

Spectroscopy in optical geometry for remittance readings was applied for monitoring the maturity-related fruit chlorophyll and quality-related anthocyanins contents on tree. Apples (*Malus x domestica* 'Elstar'/M26) were monitored in a marked area of

TABLE 3.2.4
Average Quality Parameters of 'Elstar' Fruits in 2003 of the Two Trees (see Figure 3.2.28) at Harvest Date

Parameter	Tree 1	Tree 2
Fruit volume [cm³/fruit]	175.0	246.0
Fresh mass [g/fruit]	145.1	197.0
Starch index (1–10)	9.9	9.6
Fruit flesh firmness [N]	61.3	48.5
SSC [°Brix]	14.0	15.6

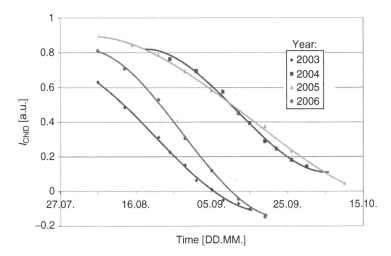

FIGURE 3.2.29 Comparison of change of index of chlorophyll decrease measured on 'Elstar' fruits of the same orchard during the seasons from 2003 to 2006.

four rows each with 150 trees in a commercial orchard. Two times a week 200 fruit spectra were recorded over a period of 5 weeks around the optimum harvest date. Conventional destructive analyses were carried out on a subsample of 20 fruits per measuring date. Geoelectrical measurements were performed with a resistivity meter (4-Point light, LGM Lippmann, Schaufling, Germany). The instrument is based on the four-point method, in the present study with Wenner array geometry: two electrodes were injecting low-frequency alternating current into the soil, while another pair of electrodes served as potential probes measuring the voltage drop. Raw data were temperature corrected and calibrated on the soil–water content using gravimetrical readings as references.

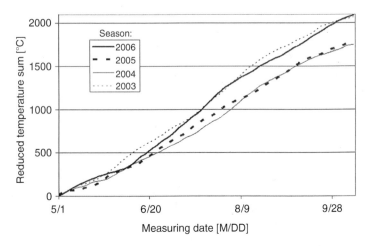

FIGURE 3.2.30 Reduced temperature sums recorded during the seasons from 2003 to 2006 in Potsdam, Germany.

VIS/NIR Spectroscopy

As shown earlier, the calibration results of nondestructive pigment analyses show high coefficients of determination $R^2 > 0.90$ and low values of (root mean square) RMSE < 7%. The maturity-related chlorophyll content decreased by 13.7% in the measuring period, while the quality-related anthocyanins value increased by 6.0% in the measuring period. The calibration of geoelectrical readings on the soil–water content resulted in $R^2 = 0.66$ and RMSE = 21%, representing an absolute value of soil–water content = 1.1%.

Spatial distribution in the soil–water content varied in absolute values from 3.2% to 6.8%. The minimum values forced a drought stress for the trees resulting in a reduced leaf area of 2.05 m^2/tree (mean leaf area = 13.9 cm^2), while trees with higher water availability showed a ratio of 4.46 m^2/tree (mean leaf area = 26.9 cm^2). The fruit yield was reduced to one-third in the drought zone. Nondestructive fruit pigment monitoring and soil–water content showed inverse correlations for the maturity-related chlorophyll and the quality-related anthocyanin. In drought stress zones, the maturity was accelerated, while the fruit quality in term of anthocyanins accumulation was enhanced. However, the better external quality was not correlated with internal quality parameters.

Concluding, site-specific data obtained by means of geoelectrical methods and optical fruit sensing provide sensitive data for monitoring spatial variation (Figure 3.2.31) of the crop properties because of water supply (Figure 3.2.32).

3.2.6 Conclusions

Spectral measurements on fruit and vegetable can be carried out rapidly and nondestructively, and provide useful data on produce quality and physiological development. Particularly, spectral measurements allow to simultaneously detect several independent parameters within the visible as well as in the NIR wavelength ranges.

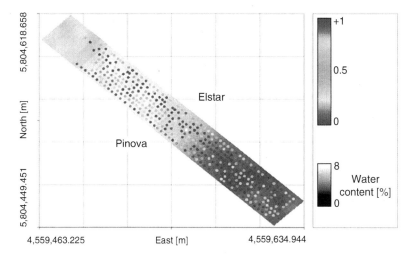

FIGURE 3.2.31 (See color insert following page 376.) Mapping of soil–water content (gray scale) and the I_{ChlD} values (false color scale) measured nondestructively on September 13, 2007 in two cultivars, each four rows. (Zude and Gebbers, pers. communication, 2007.)

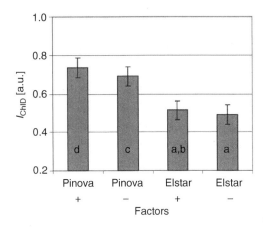

FIGURE 3.2.32 Group-wise comparison of I_{ChlD} values from drought zones (−) and zones with sufficient water supply (+) for two cultivars on September 13, 2007. (Zude, pers. communication, 2007.)

It is expected that the use of robust miniaturized spectrophotometers at relatively low costs will offer advantages for production of high-quality fruit and vegetable.

However, for reasonable use of spectral measurements, numerous factors have to be considered. Besides the optical properties of the produce, the environmental conditions affect the measuring results. Experimental tests are inevitable to elaborate the optimum configuration for every product under the specific conditions.

For fresh intact products under variable environmental conditions, the measurement of light interaction could be recommended. Wavelength range, geometrical configuration, and timing of measurement have to be adapted to the product characteristics. Not only properties of product surface, skin, cell structure, and chemical composition, but also external influences by weather and production technique could be important for use in spectrophotometer technique.

For instance, apple fruit development and maturation could be monitored by repeated measurements on the same product. The visible wavelength range allows to sensitively detect changes in content of pigments such as chlorophyll, anthocyanins, and carotenoids. The pigmentation could be precisely described by use of different indices derived from spectral signature and highly correlated with actual pigment content according to reference analyses in laboratory. Results of spectral measurements showed the potential to detect the influence of geographical site (climatic and soil conditions), of fruit yield per tree, of seasonal differences (temperature sum), and of specific properties of the cultivar.

However, only spectral data would not enable the user to describe satisfactorily the fruit maturity stage. According to existing knowledge, the user of conventional parameters for determination of fruit maturity needs always several independent parameters as well as specific experience to obtain a usable result. This means that sufficiently long experiences also in spectral sensing will be required for practical use in quality monitoring.

REFERENCES

Bhalotra, S.R. 2004. Adaptive optical microspectrometers and spectra-selective sensing. Dissertation, Department of Applied Physics, Stanford University Palo Alto, California, USA.
Chen, P. and V.R. Nattuvetty. 1980. Light transmittance through a region of an intact fruit. *Transaction of the ASAE* 23: 519–522.
Herold, B., I. Truppel, M. Zude, and M. Geyer. 2005. Spectral measurements on 'Elstar' apples during fruit development on the tree. *Biosystems Engineering* 91: 173–182.
Höhn, E., D. Datwyler, F. Gasser, and M. Jampen. 1999. Streifindex und optimaler Pflückzeitpunkt von Tafelkernobst. *Schweizerische Zeitschrift für Obst- und Weinbau* 18: 443–446.
Knee, M. 2002. Fruit quality and its biological basis. *Fruit Technology and Quality Handbook*. C.H.I.P.S., Weinheim, Germany.
I'Anson, S.J. 2007. Optical properties of paper and board. Lectures, University of Manchester, UK. Available at http://www.ppfrs.org.uk/ianson/paper_physics/Kubelka-Munk.html, August.
Lichtenthaler, H.K., A. Gitelson, and M. Lang. 1996. Non-destructive determination of chlorophyll content of leaves of a green and an aurea mutant of tobacco by reflectance measurements. *Journal of Plant Physiology* 148: 483–493.
McClure, W.F. 2002. Hand-held near-infrared spectrometry: Status, trends, and futuristic concepts. In: *Near Infrared Spectroscopy*. Proceedings of the 10th International Conference on Near Infrared Spectroscopy, Kyongjgu, Korea, eds. A.M.C. Davies, 131–136. NIR Publications, Chichester, United Kingdom.
Merész, P., T. Lovász, and P. Sass. 1994. Postharvest physiological changes of apple in respect of picking date. In: *The Post-Harvest Treatment of Fruit and Vegetables—Current Status and Future Prospects*. Proceedings of the 6th International Symposium of the European Concerted Action Program COST 94, October 19–22. Oosterbeek, Netherlands, 251–257.
Olson, C.E. 1986. Evaluation of the airborne imaging spectrometer for remote sensing of forest stand conditions. Final Technical Report, University of Michigan, Ann Arbor, Michigan, USA. Available at http://ntrs.nasa.gov/archive/nasa/casi.ntrs.nasa.gov/19870012863_1987012863.pdf, August 2007.
Osterloh, A. 1980. *Obstlagerung*. Deutscher Landwirtschaftverlag, Berlin Germany.
Ozaki, Yu, W.F. McClure, and A.A. Christy. 2007. *Near-Infrared Spectroscopy in Food Science and Technology*. Wiley Interscience, Hoboken, New Jersey, USA.
Praast, H., T. Ziegenbein, and L. Göttsching. 2003. Der Einfluss von Lichtfang und Druckübertragungsprozess auf die Druckqualität. *Das Papier* 5: 43–48. Available at http://www.ipwonline.de/download/zellchem/2003/dp050304.pdf - German, August 2007.
Short, N.M., Sr. 2006. Technical and historical perspectives of remote sensing. NASA publication. Available at http://rst.gsfc.nasa.gov/Intro/Part2_5.html, August 2007.
Streif, J. 1983. Der optimale Erntetermin beim Apfel. I. Qualitätsentwicklung und Reife. *Gartenbauwissenschaft* 48: 154–159.
Truppel, I., A. Jacobs, B. Herold, and M. Geyer. 2005. Mobile spectrometer equipment for apple fruit quality detection. Technical Report, Leibniz-Institute for Agricultural Engineering Potsdam-Bornim, Department of Horticultural Engineering, Potsdam, Germany.
Wilcke, C. 2002. Ernteterminbestimmung und Qualitätsvorhersage bei Äpfeln. *Sächsiche Landesanstalt für Landwirtschaft, Schriftenreihe* 7: 24–36.
Workman, J., Jr. and A.W. Springsteen. 1998. *Applied Spectroscopy: A Compact Reference for Practitioners*. Academic Press, San Diego/London/Boston/New York/Sydney/Tokyo/Toronto.
Zude, M. 2003. Comparison of indices and multivariate models to non-destructively predict the fruit chlorophyll by means of visible spectrometry in apple fruit. *Analytica Chimica Acta* 481: 119–126.

Zude, M. 2006. *Manual Pigment Analyzer PA1101*, Vol 3. Control in Applied Physiology GbR, Falkensee, Germany, 45pp.

Zude, M., I. Birlouez-Aragon, J. Paschold, and D.N. Rutledge. 2007. Nondestructive spectral-optical sensing of carrot quality during storage. *Postharvest Biology and Technology* 45: 30–37.

3.3 NEAR-INFRARED SPECTROSCOPY

KERRY B. WALSH AND SUMIO KAWANO

3.3.1 INTRODUCTION

3.3.1.1 Setting the Task

Can you remember when you settled into a steady job and the pay checks became regular? Do you recall agonizing over the decision to purchase a new sound system, or a car? I imagine you debating whether you really needed the new item, weighing up the benefits in terms of finances and in terms of quality of life. Having decided that the new item was justified, the debate shifts to the price and features offered. Do you need a sports car, a family wagon, a people mover, or a utility vehicle? Do you want a reel-to-reel player or an iPod? There can be a steep learning curve here, as you attempt to learn from the salespeople, enough to know which features are worth asking for. This process can be iterative, when you learn about the available features, you can rationalize spending more funds.

Now I imagine you in a quality control role, working with fresh fruit, grain, or oil. You have a complement of instruments that you already use to assess the quality of your product. However, you have heard about near-infrared spectroscopy (NIRS). It sounds wonderful, with no sample preparation and a near instantaneous measurement. It can even be implemented on the process line and you can close down the quality control laboratory. Can this be true? You search for suppliers of such equipment and are surprised by the number of providers, and then you are overwhelmed by the descriptions of the various competitor units. How can you rationally assess these units on the basis of fitness for purpose and value for money?

This section will not allow you to become an expert in NIRS instrumentation, NIRS theory, or chemometrics; however, this section should provide sufficient background on the technology and its application to guide you in an evaluation of instrumentation for a given task.

So, briefly, what is NIRS and what advantages and limitations does it offer? We will spend the rest of the section expanding this summary, but the following paragraphs should be sufficient for you to judge whether the technique is appropriate to your application. NIRS is a spectroscopy involving the NIR region of the spectrum (Figure 3.1.1), that is, the wavelength region from approximately 780 to 2500 nm. That part of the NIR spectrum from 780 to 1100 nm is termed short-wave NIR (SWNIR), near NIR, or the Herschel region. The use of UV wavelengths (100–380 nm) and VIS–FIR spectroscopy (380–780 nm, see Section 3.2) is discussed elsewhere in this book. In general, compounds absorb in the VIS and UV regions of the spectrum because of

VIS/NIR Spectroscopy 193

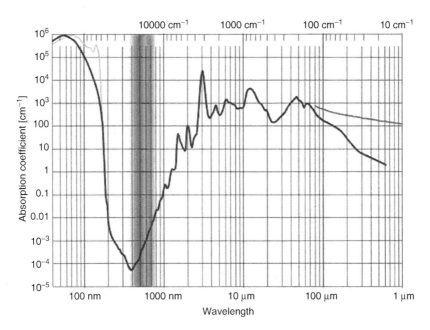

FIGURE 3.3.1. Absorption spectrum of water. (From Chaplin, M., Water structure and science. Available at www.lsbu.ac.uk/water/vibrat.html, December 19, 2007. With permission.)

electronic transitions. Absorption in the IR and NIR is principally due to the vibration (stretching, rotation) of chemical bonds, principally C—H, O—H, N—H, and S—H.

IR spectroscopy (2,500–25,000 nm) is an established technique for fingerprinting organic chemicals. However, the absorption coefficient for biological material is orders of magnitude higher in the IR than in the NIR (Figure 3.3.1). In consequence, IR spectroscopy is of value only as a surface measurement technique to depths in the order of micrometers. The use of NIR, and in particular SWNIR, with lower absorptivity, allows for higher penetration depths (see Sections 1.3 and 5.2). This is essential for use with nonhomogenous biological material. However, the typical penetration depth even for SWNIRS does not exceed 10 cm, and is usually less than 10 mm. Further, there is a penalty for working with the NIR–SWNIR, in that the sharp absorption peaks observed in the IR region for a given chemical bond in a given chemical environment are broad and overlapped with absorption peaks of other absorbers in the NIR–SWNIR. NIRS–SWNIRS is therefore less useful than IR for chemical identification. Further, this overlapping of peaks means that interpretation of NIR spectra is not as simple as that of the IR. Actually, it is rarely possible to base an assessment on a single wavelength. This limitation is even more marked for the SWNIR. Thus, NIRS is intimately linked to the use of multivariate statistics, involving either nonlinear techniques such as support vector machines and neural networks, or linear techniques such as multiple linear regression (MLR) and partial least squares (PLS) regression.

The apparent absorption of a sample is also a function of its optical properties, and this can be defined by its scattering and absorption coefficients (see Section 1.3).

The scattering behavior of a sample is a measurable quantity that is of value for some applications, for example, the attribute of wheat grain hardness is related to the cell size, which affects the light scattering properties of the grain.

In short, NIRS is "the most practicable and exciting analytical technique to hit the agricultural and food industries since Johann Kjeldahl introduced the Kjeldahl technique" because of its "speed in testing, its flexibility in sample size and presentation methods [to the instruments], and its relative freedom from the need for sample preparation" (Williams and Norris 2001a). The theory of this field is maturing, and has been well reviewed, so our task in this section is to provide an overview from which you can direct your ongoing education.

3.3.1.2 Going Further

Normally, this section would be at the end of a chapter or book. But in this case, I believe it is useful for you to have a feeling of the depth of material and support that exists for this technique.

The International Council for Near Infrared Spectroscopy (ICNIRS, see www.icnirs.org) maintains a bibliography of NIRS-related research papers, and hosts an international meeting every second year.

National or regional societies exist in many areas. These societies are involved in the organization of conferences and workshops, and offer professional support. Typically, these groups meet every second year, in sequence to the international meeting. The U.S. group, which predates ICNIRS, hosts the international diffuse reflectance conference (www.idrc-chambersburg.org). This conference is well known for its calibration shoot-out, in which participants are provided with spectral data and reference values before the event, and submit their solution to the set problem.

Some other examples of regional groups include the following:

Asia: Asia is one of the tertiary poles of the NIRS community. The Asian consortium of NIRS was established as a consensus among participants from China, Korea, Thailand, and Japan when the Japan–Korea Joint Symposium was held in June 2006 at Seoul, Korea. A specific mission of the consortium is to promote the transfer of knowledge and know-how on NIRS among Asian countries. The Asian NIRS symposium will be held every two years from 2008. A domestic NIR meeting is also held in Japan, Thailand, and China. In Japan, a forum has been held every year in Tsukuba, while an NIR workshop has been held almost every year in Kasetsart University, Bangkok, Thailand. In China, the first Chinese conference on NIRS was held in August 2006 and the conference will be held every two years.

Australia: The Australian Near-Infrared Spectroscopy Users Group is a section of the Royal Australian Chemical Institute (www.anisg.com.au). A meeting with workshops is held every second year, in the off year to the international conference.

Europe: The instrument manufacturers form an active support group. For example, the free platform http://www.spectroscopyeurope.com/ provides advice on applications, instruments, as well as hints for the proper data processing method. However, scientific exchange on the use of NIRS in food analyses is found scattered in specific conferences and workshops of each technical community, for example, the European symposium on polymer spectroscopy is an annual event.

There is a published refereed journal devoted to the topic, the *Journal of Near Infrared Spectroscopy* (*JNIRS*, www.impublications.com/nir/page/nir), although of course NIRS-related articles are published into a wide range of journals related to the application area, and also to discipline areas (e.g., *Applied Spectroscopy*, *Measurement Science and Technology*, *Chemometrics*). The JNIRS Web site offers a discussion forum, and also offers a range of key texts.

Recommended further reading includes the textbooks of Williams and Norris (2001a) for a comprehensive overview of the use of NIRS in agricultural and food applications, and Mark and Workman (2003) for coverage of statistics relevant to this field.

3.3.2 FACTORS FOR SUCCESS

3.3.2.1 Attribute

Successful implementation of any analytical technique requires sound knowledge of the attribute and the sample. However, it is often the case that an industry can forget the uncertainties associated with the measure of a particular attribute. Often, a relatively simple technique has been widely adopted. This technique, with its own flaws in terms of indexing the attribute of interest, may become the basis against which a new method is assessed. The introduction of a new analytical approach, such as NIRS, can bring issues relating to the measure of the attribute of interest into sharp focus.

Consider that if we wished to use NIRS to measure the sucrose concentration of a solution containing only sucrose and water, our task would be relatively straightforward. However, if we are asked to use NIRS to assess the sweetness of an intact fruit, we must begin to question the definition of the attribute. What is sweetness? The level of sugars in the fruit? The industry currently typically uses a method based on a refractometric measure of soluble solids content (%SSC [°Brix]) in mechanically squeezed juice. But consider that a peach with a high organic acid content will not taste as sweet as a peach with the same SSC reading but with a lower acid content. Also, the SSC (refractometric) reading will assess both sugars and organic acids in juice. For peach juice, up to 10% of the SSC reading will be due to organic acids. Further, the taste of the fruit will also be related to the amount of juice released as the fruit tissue is crushed in the mouth and to the SSC of fruit tissue, which is not uniform (during crushing, the SSC value of the first release of fluid can be different to later release).

So an application to use NIRS to assess the SSC of intact fruit must proceed with caution, given knowledge that the SSC value is influenced by the extractability of juice from the sample and that the refractometer reading is of all soluble compounds. In other words, the NIRS and reference methods may be measuring different attributes.

However, SSC at least represents a measure of a chemical attribute. There is often temptation to use NIRS to assess nonchemical attributes. For example, another eating quality criterion for fresh fruit is mouthfeel, described as the crunchiness of an apple. The standard industry measure of this attribute is to take a penetrometer reading. In such a measurement, the skin of the fruit is shaved off, and the force required to plunge a cylinder of known diameter a certain distance into the fruit is measured. Is it sensible to attempt to use NIRS to predict a fruit penetrometer reading? No, not in terms of absorption. However, it is possible that change in the cell walls of the tissue may result in a change in the scattering of light. Such a measurement will require different instrumentation (e.g., to measure the time of flight of a photon or backscattering images) to that used for absorption spectroscopy, and this area is not further considered here (but see Sections 1.3.2 and 5.2).

3.3.2.2 Sample

As noted earlier, successful implementation of any analytical technique requires sound knowledge of the sample. However, such knowledge is often not in place.

There are several classic pitfalls to be avoided in this respect, as follows:

3.3.2.2.1 Sampling Strategy

Sampling is a major bugbear to any analytical technique. How do you obtain a representative sample from a truck containing 10 t of grain? Actually, as a rapid, noninvasive technique, NIRS offers a potential solution if plant infrastructure allows for it. For instance, the NIRS equipment can be mounted to view a stream of grain during the unloading process. Any lot sampling procedure is fraught with problems, and it is essential to demonstrate that the lot is representative of the population.

A commonly asked question is that of how many samples (n) are required to gauge the mean and variation of a population. This will be dependent on the SD of the population, the required uncertainty, and the required probability level (Equation 3.3.1).

$$n = (y \cdot SD/\varepsilon)^2 \qquad (3.3.1)$$

where

y is the student's t-statistic (e.g., $y = 1.96$, given a 95% probability level and $v = $ infinity)
ε is the desired uncertainty of the measurement
SD is the standard deviation of the population (Valero and Ruiz-Altisent 2000)

If SD is unknown, an iterative process must be implemented to gain an estimate of this variable.

One example: a tomato processing plant received fruit from growers in large open trucks. At the entrance to the plant, the trucks crossed a weighbridge and a grab sampler took a 10 kg sample from the top of the truck. The sample was immediately blended and SSC assessed. The truck proceeded to a water dump, and its contents were processed to tomato paste. Plant management was mystified by an apparent loss of solids in that the calculation of the entering trucks exceeded that of the waste streams and the final product. It eventuated that growers had learnt that certain areas of their field produced fruit of higher SSC and were using these fruit in a top layer when loading trucks.

VIS/NIR Spectroscopy

Sampling strategy will make or break an application. This short section can only remind you of the importance of this issue.

3.3.2.2.2 Sample Homogeneity and Optical and Reference Sampling Volumes

Biological products are nonhomogenous; however, we have a tendency to regard them as homogenous. For example, a standard of 12% SSC might be set on the SSC of a pineapple fruit. However, there can be over a 4% SSC difference between the top of a pineapple fruit and the bottom, and between the sunny side (the northern side in the southern hemisphere) and the shady side (the northern side in the northern hemisphere). So you will offer the bottom, northern side of the pineapple to your guests if you like them and if your fruit is from South Africa.

But with such a variation of SSC within the fruit, how is a standard on fruit SSC to be implemented? Obviously, there must be clear guidelines on the sampling procedure. Further, the optical sampling volume of a spectroscopic system must be matched to the sampling volume of the reference method (Long et al. 2002).

As an example, can you foresee any issue should I seek to develop a model of the temperature of an object based on SWNIR (750–1100 nm) spectra from a transmission optical geometry and a reference measurement using a noncontact IR thermometer that uses a reflectance geometry? The transmission geometry means that the NIRS measurement will represent an average of the sample, while the reference technique relies on an IR wavelength that has a high-absorption coefficient, and thus a low-effective penetration depth, that is, it is very much a surface measurement. When ambient temperatures are changing, surface and internal temperatures may be quite different, leading to a mismatch between the reference and SWNIRS measurements.

3.3.2.2.3 Sample Composition (Matrix) and Variation Over Time

The complexity of sample composition (sample matrix) is an issue to any analytical technique. For NIRS, this issue is manifested in two ways: (1) the absorption features recorded in NIRS and especially SWNIRS are broad and overlapped (see Section 3.3.3), such that it is difficult to differentiate the spectral features owing to a particular attribute and (2) this sample matrix may change between lots or over time.

For example, the analysis of starch and sucrose using NIRS relies on absorption features related to O—H and C—H vibration. The chemical environment of those bonds is slightly different in starch and sucrose, but, at least for the SWNIR region and in an aqueous environment, the spectral features of the two compounds are difficult to distinguish. Thus, it is not unexpected that Subedi and coworkers (2007) report that it is not possible to assess the SSC of ripening mango fruit, in which there is both starch and sugar present.

This issue assumes even greater importance if either the chemical composition or aspects of the physical composition of the sample that affect the passage of light are changed between populations. An NIRS calibration model is built using spectral and reference data from a given set. The validity (accuracy and precision) of this model in predicting attribute level in other populations will depend on how well the chemical and physical variations present in the new sets were represented in the calibration set.

An analogy may be helpful here. You know that N deficiency is manifest in a plant as general chlorosis (yellowing) of lower leaves. The degree of yellowing can be used to gauge the N content of the leaf. In your mind, you have calibrated the leaf N-level using the spectral information received by your eye. However, you know that you cannot use this calibration with other plant species, and you also know that the calibration can be upset by other factors. For example, Fe deficiency also causes chlorosis. So it is with NIRS; we are not measuring the desired attribute directly, but rather making a correlation between the spectra of the sample and the attribute of interest. The more complex the sample, and the more change in this composition between samples, the less robust this calibration model in terms of use with new populations.

A related issue is the range of values of the attribute of interest. The bottom line here is that it is important to have a range of values in the calibration set that is representative of the range of values to be assessed in future populations.

Thus, an NIRS model developed for the oil content of canola seed using data from one season in which the oil content was higher than normal cannot be expected to predict subsequent seasons well.

3.3.2.2.4 Captured Variance

In summary, we have alerted you to the need to choose a sampling method that adequately represents the sample and the population, and to match to this an optical geometry. The success of the NIRS method is then dependent on how well variance, in terms of the attribute of interest, and other, interfering, attributes are captured in the calibration set. Interfering attributes may be chemical constituents or physical factors (e.g., change in the optical properties of the sample). These concepts are discussed further in Section 3.3.3.

3.3.2.3 Instrument

Spectrophotometer technology has been discussed earlier (Section 3.1). In this section, we will expand on this earlier discussion, with focus on aspects relevant to NIRS.

The design of an NIR spectrophotometer requires optimization of both individual and integrated components for the required application. In comparing and choosing between instruments for your quality control laboratory, you will need to obtain technical specifications on each instrument and consider these features in context of your application.

Of particular importance are the choices of

- Source–sample–detector optical geometry
- Method and frequency of referencing
- Light source
- Spectrograph wavelength range and resolution
- Spectrograph speed
- Dynamic range and signal-to-noise ratio of the detector system

A useful summary of these issues is provided by Williams and Norris (2001b).

VIS/NIR Spectroscopy

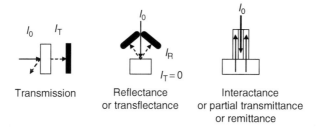

FIGURE 3.3.2 Sample presentation geometry (I_0: incident light, I_T: transmitted light, I_R: reflected light).

In the following section, an introduction is provided to each item, and then a case study is made of the choice of instrumentation for a particular application.

3.3.2.3.1 Source–Sample–Detector Optical Geometry
There are several categories of optical geometry possible (Figure 3.3.2). Some of these categories are not totally exclusive, so it is wise to look for further detail for a given instrument. These categories include the following:

- Transflectance
- Reflectance
- Partial transmittance or remittance
- Interactance

NIR transmittance spectroscopy is sometimes abbreviated as NIRT, while NIR reflectance spectroscopy can be referred to as NIRR.

In a transmittance optical geometry, the light source–sample–detector is aligned at 180° (i.e., in a line, such that sample path length is known). The amount of light reaching the detector will be a function of the absorption and scattering coefficients of the sample. The familiar Beer–Lambert law applies to measurements made using a transmittance optical geometry and for a non-scattering, highly diluted sample.

In transflectance, light is passed through the sample to a reflector, and is then reflected back through the sample to the detector. Thus, the effective path length is twice the sample thickness. However, if the sample is highly scattering, much of the detected light will not have passed to the reflector. Thus, a transflectance probe used with a scattering sample is really making a reflectance or interactance measurement.

In a reflectance optical geometry, the detector views an area of the sample that is illuminated, and thus receives both radiation that has been specularly reflected from the sample, that is, contains no information about the internal composition of sample, and radiation that has interacted with the sample, termed diffuse reflection.

In a partial transmittance geometry, sometimes also referred to as remittance, the detector is shielded from receiving specular reflections, however the light source–sample–detector angle is less than 180°. For example, the sample might sit in a cup with a detector viewing the sample from below, and be illuminated on a horizontal plane above the cup (i.e., light source–sample–detector angle of 90°).

An interactance geometry is perhaps best defined as a short path length partial transmittance geometry, in which the illuminated and detected areas are nearby. This term is commonly used for fiber optic sampling probes in which the illuminating fibers are arranged adjacent to, or separated by a short distance, the receiving fibers.

3.3.2.3.2 Method and Frequency of Referencing

To calculate absorption, the intensity of the light beams reaching the sample and transmitted through the sample must be known. The intensity of the incident beam will vary with changes in the lamp or other optics within the instrument, and can change due to external influences such as humidity (with gaseous water causing absorption of certain frequencies) or additional ambient light levels. Further, lamp and detector behavior may change with ambient temperature or age. To accommodate such issues, most instruments have a referencing procedure.

The referencing procedure may involve two detectors, with one detector viewing radiation that has interacted with the sample, and the other detector viewing radiation from the lamp. Such a dual detector strategy involves increased equipment cost and does not control for changes between the detectors. However, this strategy is very appropriate for precise readings essential in pulsed, time-resolved analyses. In NIRS instruments, however, it is more common to have a single detector (or detector array) with sequential measurement of radiation from the sample and the lamp. The frequency of referencing must be determined in context of the measuring conditions, for example, variation in humidity, ambient light, stability of lamp and detector.

Usually, the reference measurement does not involve a direct path between the lamp and the detector, but rather the replacement of the sample with a highly reflective material. In this way, any change in the optics of the instrument is monitored.

A manual re-referencing between each sample is often reported with use of the scanning grating Foss 6500 (e.g., Guthrie and Walsh 1997). For the application of in-line fruit grading, Rei and coworkers (1994) developed a system involving a rotating chopper operating on a 50% duty cycle and at a predetermined frequency for continual dark referencing between samples, here fruit. Hashimoto and coworkers (2002) also described the use of black-coated shutter device, driven by a rotary solenoid, to allow dark referencing between every sample. Further, a material coated in white color was applied to the cover lid of the instrument to allow for a manually initiated white-referencing procedure. However, these procedures do not allow for a measure of ambient stray light levels. The Perten DA7000 employs a 30 Hz chopper, facilitating continual referencing to lamp output and also monitoring of the sample during light and dark intervals (Perten DA7000 Operation Manual). This procedure should allow for a measure of ambient stray light.

The choice of reference material is another variable. Gold is very reflective in the IR and in the NIR; however, it does have a VIS absorption spectrum (it is yellowish). Teflon and white ceramic are alternatives with a fairly flat and high reflectance through the VIS and NIR range. Polished aluminum has good reflectivity through the VIS and NIR; however, this is sensitive to temperature (Figure 3.3.3), with a broadband feature peaking around 820 nm (Oriel Instruments 2003).

Wavelength calibration of the spectrophotometer is most often done by the manufacturer. However, in a scanning spectrophotometer regular recalibration is

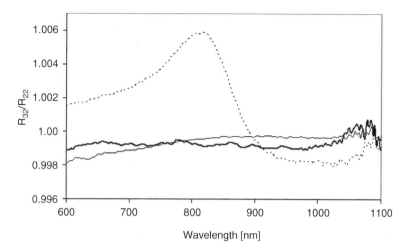

FIGURE 3.3.3 Temperature sensitivity ($\Delta T = 10°C$) for different referencing materials, expressed as the ratio of the detector response at the elevated temperature (at 32°C, denoted as R_{32}) to that at room temperature (at 22°C, denoted as R_{22}), for Teflon (thin solid line), gold (thick solid line), and aluminium (dotted line).

required. This is typically achieved with reference to narrow absorption peaks of a material such as polystyrene. For monolithic PDA spectrophotometers, an initial calibration to set the wavelengths is generally considered sufficient; however, this very much depends on the stability of optical bench employed in the instrument as components differentially expand with change in temperature (Walsh et al. 2000). Some instruments, such as the Perkin Elmer Lambda 950, carry out an automatic calibration using the known lines of the xenon lamp in the VIS wavelength range. Small mercury-vapor lamps (available with sub-miniature version A [SMA] fiber optic coupling) can be used for wavelength calibration in the SWNIR.

Photometric response correction is rarely considered by the user of an instrument, at least until it is desired to transfer a calibration model between instruments. However, this aspect should be firmly in the mind of the manufacturer. Each instrument is slightly unique in terms of its optical properties, although good quality control during manufacture can reduce this variation. Each detector is also unique in terms of its spectral sensitivity, varying with the lot of manufacture. Further, sometimes the manufacturer of a spectrometer can change the specification of the detector used. For example, around the year 2000, Zeiss changed the specification of the Hamamatsu PDA used in their MMS1 NIR enhanced spectrometer to an array that eliminated signal carryover between subsequent readings but that was less sensitive in the SWNIR.

Transfer of a calibration model between instruments typically requires the development of a transfer function that accommodates photometric response differences between the instruments. This function is developed using spectra acquired from a set of samples using all instruments (e.g., Greensill and Walsh 2001; see also Section 3.4).

3.3.2.3.3 Light Source

There is a range of NIR and SWNIR light sources available (see also Section 3.1). Fluorescent lamps do not emit SWNIR or NIR. Thus, it is possible to ignore fluorescent room light when acquiring NIR spectra. Diode lasers emit radiation with an FWHM of 10–20 nm. Lasers can offer more intensity at a single wavelength than other sources. However, consider that if a 100 W tungsten halogen bulb emitted radiation at equally all wavelengths between 400 and 2400 nm, it would emit 500 mW over each 10 nm segment. Also, as SWNIR and NIR are non-VIS, the use of such a 500 mW laser involves significant safety hazards. Further, maintaining a stable level of output from a diode laser is not without problems, for example, output peak wavelength and intensity of output are sensitive to temperature.

An LED may have slightly wider emission ranges than a laser diode, for example, 30 nm FWHM for an LED with peak intensity at 980 nm and up to 35 nm for fluorescent LED with peak wavelengths in the VIS spectrum (Miller and Zude 2004). An increasing range of these devices exist for SWNIR applications, in terms of the peak wavelength of emission. Several different LED units can be bonded together on one carrier to be able to address the specific wavelengths needed (Zude 2006). A series of LED units can be activated in series to illuminate the sample, with employ of a single detector, rather than a detector array. Such an arrangement increases acquisition time but decreases the cost of instrumentation. LED units are characterized by long life, the use of much less power than tungsten halogen lamps, and a square output that eliminates the need for a warm-up phase. As such these light sources have potential for portable applications.

The quartz tungsten halogen (QTH) lamp remains "king" within SWNIRS and NIRS applications. This type of lamp is a broadband emitter in the VIS and NIR regions. QTH lamps are commonly used for VIS light applications (e.g., domestic lighting, car headlights), but they actually emit more radiation beyond the VIS, in the SWNIR and NIR. The technology is low cost, readily available, and the lamps have a reasonable life (ca. 1000 h). The lamp filaments are also reasonably compact, which is useful in designing optical systems that focus and collimate light. This is often required in SWNIRS and NIRS systems to maximize the intensity of radiation incident on the sample.

Given their ubiquitous use, it is worth learning a little more about this type of lamp. In the operation of a QTH lamp, tungsten evaporates from the filament, migrates to, and deposits onto, the lamp wall, then reacts with the halogen gas (usually bromine), and migrates back to the filament as tungsten bromide. The tungsten bromide reacts with the lamp filament, depositing tungsten back onto the filament. The process is then repeated. Because of this cleaning action, the radiant output of a QTH lamp is more stable than that of other lamps (e.g., incandescent lamps), and a QTH lamp has an extended service life. However, because of the chemical processes occurring within the QTH lamp, the luminescent output of the lamp can have an interesting temporal profile during start-up. Of course, the intensity and spectral profile of output will also be dependent on the supply voltage; therefore, a regulated power supply is required to eliminate this variable.

It is well recognized that there is a need for system warm-up to achieve stability in detector response and tungsten halogen lamp output (Figure 3.3.4). Most equipment manufacturers recommend a warm-up period of up to 1 h. For example, Walsh and coworkers (2000) noted a change in spectral output from a Philips halotone lamp over the first 30 min of operation, with a 5% decrease in luminosity at 650 nm, and a 2% increase at 830 nm. Hashimoto and coworkers (2002) proposed a system for

VIS/NIR Spectroscopy

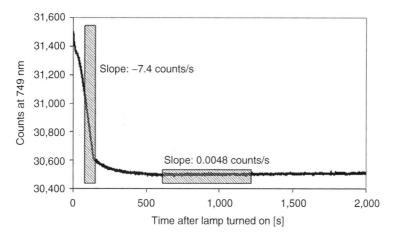

FIGURE 3.3.4 Temporal profile of lamp output (at 749 nm) from power-up. Count refers to the analogue to digital output of a 15 bit device.

monitoring lamp temperature to assess when the system had achieved stability, thereby reducing the warm-up waiting period.

Each brand of lamp, and potentially each batch of manufacture, will vary in its spectral output and its warm-up behavior (Figure 3.3.5). In using a particular brand of instrument, you assign responsibility for quality control over this issue to the manufacturer, and you will need to buy their expensive replacement lamp rather than a generic lamp!

The lamp itself is also one component of the light source. Usually, the lamp filament is positioned at one focal point of a parabolic reflector, with the intent of producing a collimated beam of light. However, each lamp assembly has its peculiarities in terms of the spatial profile of the light output. Again, in using commercially available equipment, you trust that the optics of the system has been optimized.

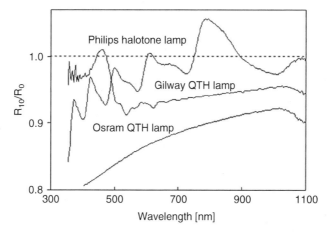

FIGURE 3.3.5 Lamp warm-up behavior, expressed as the ratio of detector response after 10 min to the initial spectrum (at power-up) of three commercially available lamps (from Philips, Gilway, and Osram).

Obviously, however, caution must be employed if you seek to customize the lamp-sample optical geometry.

3.3.2.3.4 Ambient Light

Ambient light is a potential source of error in NIRS and SWNIRS, particularly where the ambient light source is spectrally distinct from the output of the spectrophotometer lamp, and where ambient light conditions are changing. Roger and coworkers (2003) reported that external lighting had no impact on model performance; however, in this case, the external light level was varied by adding additional lamps of the same type, and thus the same spectral output, as used in the spectrophotometer assembly.

Most bench-top systems feature a light-tight sample compartment, but this is not true for all in-line or field portable applications. Outdoor operation of a spectrophotometer imposes a harsh test on instrumentation. The sun is a strong emitter of SWNIR and NIR, and both the quantity and quality of this radiation varies (Figure 3.3.6). Saranwong and colleagues (2003) reported the use of a portable NIR unit developed by Fantec on field mangoes through the expediency of enclosing the fruit in a light-tight silver bag to shield it from sunlight. Obviously, a handheld spectrophotometer intended for field use should directly address the issue of sunlight interference (e.g., Zude et al. 2008).

3.3.2.3.5 Wavelength Range

Do you need an instrument that acquires over a spectral range from 400 to 1050 nm, 800 to 2000 nm, 1100 to 2500 nm, or another range? There are two pressures in this decision: the first is a spectroscopic issue related to your application and the second is a cost-available instrumentation issue.

There will be an optimum wavelength range for your application, determined by the thickness of the sample you wish to work with (thicker samples requiring shorter wavelengths) and the attribute of interest. However, instrument manufacturers are constrained by the availability of detectors. For SWNIRS and NIRS, there are three major classes of detectors: silicon (Si) detectors that operate over the range 400–1050 nm, indium gallium arsenide (InGaAs) detectors that operate over the range 800–1700 nm (or, with doping, this range is shifted up to 2200 nm or higher), and lead sulfide (PbS) detectors that operate over the range 1100–2500 nm.

3.3.2.3.6 Spectrograph Dispersive System—Speed and Stability

The main options to produce monochromatic light are interference filters, acousto-optical tunable filters (AOTF), and gratings. Interference filters are a relatively cheap option, although this equipment is locked into the use of discrete wavelengths. An AOTF can scan through a range of wavelengths; however, this imposes a time constraint on the acquisition of a spectrum, and the technology is not cheap. Grating-based systems currently dominate in the market. In the future, micromechanical systems (MEMS) that act as variable wavelength filters may become cheaply available.

PbS detectors are typically available only as single detectors. Thus, the instrument design must incorporate a mechanism to deliver monochromatic light to the detector. This might involve a rotating filter set, an AOTF, or a scanning grating. The

VIS/NIR Spectroscopy

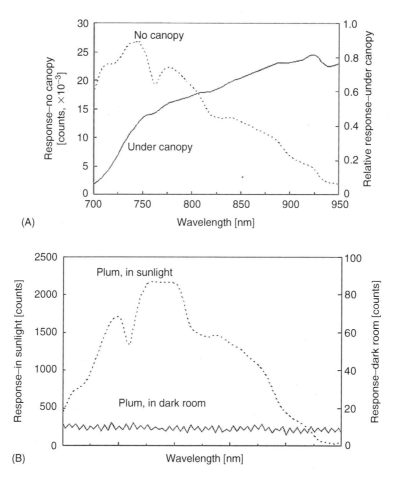

FIGURE 3.3.6 (A) Solar spectrum, analogue to digital counts (dotted line) and spectrum of light under a tree canopy, as a ratio to the solar spectrum (solid line); and (B) spectrum of light passing through a plum fruit, for a fruit illuminated by sunlight.

monochromator can be placed in the light path before the sample (a predispersive system) or after the sample (a postdispersive system). Scanning monochromator systems involve use of an encoder to match the angle of the grating and the peak wavelength reaching the detector. Such a mechanical system can be sensitive to vibration. The scanning systems also require a relatively longer time to acquire a spectrum, typically >1 s for one scan of the spectrum. This can be a limiting factor for application to on-line systems (i.e., for assessment of more than one item per second).

InGaAs and Si detectors are available in arrays. This allows for a postdispersive system in which the grating (dispersive element) is fixed. That is, light from the sample is dispersed and then falls across the array. As a fixed system, such units can be relatively insensitive to mechanical shock. A typical minimum integration time for a unit with 256 pixels is 4 ms.

3.3.2.3.7 Wavelength Resolution

First, let us define some terms that should not be confused with wavelength resolution. An instrument vendor of a scanning grating system may specify the readout interval, for example, 2 nm. This interval represents how frequently readings were taken as the dispersed radiation was scanned across the detector. It is independent of wavelength resolution. An instrument vendor of an array system may specify pixel dispersion, for example, ~3.2 nm. This interval represents the wavelength range received by each detector (pixel) in the array. However, this figure will vary across the array. Another figure that may be quoted is the interpolation interval, for example, 1 nm. This is merely an interpolation of the pixel data and, again, it is not a measure of wavelength resolution.

Wavelength resolution is determined by the optics of the system, by the width of the entrance slit, and by the quality of the dispersive element. If a line spectra (a spectral feature that is less than 1 nm in width) is viewed by the spectrophotometer, a typical dispersive system will see a peak of measurable width. This is typically characterized by measuring the FWHM of line spectra from a mercury argon lamp. This wavelength resolution of a grating spectrograph will vary across the wavelength range, and should therefore be specified at a specific wavelength, for example, 13 nm at 910 nm.

To utilize the available resolution of the dispersive system, a detector array should have a pixel dispersion of at most one-third of the resolution of the system (e.g., 3 nm pixel dispersion for optics achieving 10 nm resolution).

If higher resolution is required, then the use of a Fourier transform near-infrared (FTNIR) spectroscopy system is warranted. The resolution of FTNIR systems can be varied at the time of spectral acquisition. Note that such systems present spectral data in terms of wave number [cm^{-1}] rather than wavelength [nm].

The required wavelength resolution is determined by application. It is generally considered that a resolution of 10 nm is adequate for most SWNIRS and NIRS applications (e.g., Greensill and Walsh 2000a), although there is indication that the higher wavelength resolution offered by FTNIR units assists in successful calibration transfer between instruments. Also, any application in which narrow spectral features are present would benefit from the use of a system with a higher wavelength resolution. One (nonfood) example is offered by Barton and coworkers (2006), who reported that cotton fiber contaminated relatively with low levels of sugars (sticky cotton) could not be distinguished from normal cotton using standard NIRS. However, contaminated cotton was distinguished from normal cotton using higher resolution FTNIR spectra.

3.3.2.3.8 Signal-to-Noise Ratio of the Detector System and the Repeatability of the Instrument

The signal-to-noise ratio is a characteristic of the detector and associated electronics. Obviously, a higher value is desirable. But perhaps a more meaningful measure to the instrument user is a repeatability value. This can be characterized as the SD of repeated measures of a sample giving a spectra at close to the saturation value for the detector. For example, the SD of 50 repeat scans of the white reference can be calculated. An useful way of expressing this characterization is to calculate the absorbance spectrum for replicate scans, using the first scan as the reference, and plot the SD of absorbance against wavelength. For this parameter, a lower value is preferred. High-end units such as the FOSS NIR Systems 6500 or Perkin Elmer Lambda 950 will achieve a minimum value around 20 μA. A typical value for the Zeiss MMS1 unit referred to in the previous section is 500 μA. In comparing such values, however, it is important to confirm the level of detector saturation and the time interval over which the replicate spectra were collected: the longer the interval, the harsher the test.

Obviously, there will be a trade-off between a lower repeatability value and the cost of the instrument. Thus, once again, the acceptable level for these characteristics must be set for the intended application. If the absorption features under consideration are gross, then a cheaper system is acceptable.

3.3.3 SPECTROSCOPY AND CHEMOMETRICS THEORY

An understanding of the theory of spectroscopy in terms of spectral features and band assignments of both the attribute of interest and of interfering factors is essential for the development and optimization of robust chemometric calibration models. Such an understanding will inform the choice of instrumentation, for example, in terms of optical geometry and wavelength range. Similarly, an understanding of chemometrics is required in the selection of appropriate data preprocessing and regression analysis techniques to establish an application. However, during the course of regular operation of an established application, the number of variables is, fortunately, constrained. For example, while regular calibration maintenance (also termed model updating or recalibration) may be required to deal with changes in the instrument or change in the samples, an in-depth understanding of chemometric theory is not required of the practitioner. Nonetheless, an appreciation of these areas will assist a practitioner to deal with problems in operation.

In this section, we aim to present a primer for the practitioner, covering basic aspects of spectroscopic and chemometric theories. We will focus on water and sugar as major constituents of biological materials. If your application area involves other materials, such as protein, oil, or others, then you are advised to seek a similar primer on the spectroscopy of these materials.

3.3.3.1 Data Preprocessing

In a spectrophotometer, the detector records an energy spectrum as a function of the number of photons received and the detectors efficiency at converting those photons to an electrical signal. Both the energy spectrum of light received from the sample (S) and from a reference (W) are typically recorded. Further, as a measure of electrical noise in the system, the energy output of the detector is also recorded in the absence of any light input (D).

Traditionally, in applications involving a transmission optical geometry, light attenuation by the sample is recorded at each wavelength (pixel) in terms of transmittance (T), where T is calculated as $(S-D)/(W-D)$. Absorbance (A) at each wavelength (or pixel in an array spectrometer) is then calculated using Equation 3.3.2:

$$A = -\log\left(\frac{\text{sample spectrum} - \text{dark spectrum}, 0\%}{\text{white spectrum}, 100\% - \text{dark spectrum}, 0\%}\right) = \log\frac{1}{T} \quad (3.3.2)$$

where

$$\left(\frac{\text{sample spectrum} - \text{dark spectrum}}{\text{white spectrum} - \text{dark spectrum}}\right)$$

In applications involving a reflectance optical geometry, light attenuation by the sample, relative to the reference, is referred to as reflectance (R). A term analogous to absorbance is then calculated. This term is usually presented in terms of the calculation log $1/R$.

However, in most applications of NIRS and SWNIRS, spectra contain high levels of scattered light. This condition invalidates the use of the Beer–Lambert law. Therefore it is not given that calculation of absorbance is required. For example, McGlone and coworkers (2003) report use of transmittance values. Nonetheless, most studies use log $1/T$ or log $1/R$ data in chemometric analyses. Given that the samples employed in NIRS studies typically scatter as well as absorb light, it is apparent absorbance or the attenuation coefficient (see Section 1.3) that is calculated.

If material used for the reference is used as a sample, the calculated absorption spectrum should be a flat line around zero. The deviation around zero will be higher at both ends of the operative wavelength range of a given detector, compared to the middle region. This is because detector photon conversion efficiency is lower toward the edge of its operative range, such that the recorded energy level of the dark measurement becomes significant relative to that of the sample and lamp reference.

Note that absorbance is a logarithmic function. Thus for an absorbance of 1.0, the sample returns only 10% of the light returned by the reference. For an absorbance of 3.0, the sample returns only 0.1% of the light returned by the reference. At this point, noise becomes significant to the low sample energy spectrum, so only for a spectrometer with a very low signal-to-noise ratio could an absorbance value above 2.5 be trusted.

Further, most studies employ a second derivative of absorbance data, as this procedure eliminates baseline effects (removing part of the scattering effects). Derivatives are also useful in spectroscopy to identify peak positions even if VIS only as a shoulder in the absorption spectrum, but have the disadvantage of magnifying noise in the spectra. Derivatives can be calculated using user-specified intervals, and so this pretreatment can also act as a smoothing function. However, there is more than one method of calculation of a derivative (e.g., gap, Fourier transform, Savitsky–Golay; Hruschka 2001), and it is important to be aware of such differences if moving data between software packages.

Various other pretreatment options such as multiplicative scatter correction, standard normal variate, and detrending are available (for an useful summary, see Ozaki et al. 2001). Wavelet filtering is a relatively new technique that has been reported to offer some benefit (e.g., Coomans 2006). Other techniques that aim to remove factors from the data set that are orthogonal to the target will continue to become available, and should be trialed (e.g., Zude et al. 2007).

3.3.3.2 Spectroscopy of Water

Water is an extremely strong absorber for most of the UV and MIR. Thus, for liquid water, an absorbance value of 1.0 is achieved by an absorption path length of only ~0.01 mm for 7 μm MIR radiation. For 1 μm NIR radiation, however, an absorbance value of 1.0 is achieved by an absorption path length of ~100 mm. This long path length has obvious advantages for the assessment of intact biological samples, and thus, this neglected NIR spectral window has been prominently used in a plethora of applications including fruit quality analysis (e.g., Saranwong et al. 2003; Walsh et al. 2004).

As a major constituent of biological materials and contaminant of dry materials, it is worth understanding the spectroscopy of water further. Consider that variation in the relative humidity of the air may introduce additional water absorption features into an acquired spectrum.

Absorption of UV-VIS radiation (200–780 nm) causes transitions or excitation of valence electrons into higher electronic states. Absorption in the NIR (780–2500 nm) and MIR (2.5–25 μm) is due to molecular vibrations and to a lesser extent rotations.

Water, H—O—H, is a relatively simple molecule and thus should have relatively simple vibrational spectra. The O—H bond has three fundamental vibrations: a symmetric stretch mode (ν_1), a bending mode (ν_2), and an asymmetric stretch mode (ν_3) (Table 3.3.1). Each of these vibrational modes gives rise to overtones or combination bands in the NIR, with the intensity and position of these bands depending on the absorption cross section and molecular anharmonicity, respectively. The ratio of the absorption intensities of $\nu_1:\nu_2:\nu_3$ is 0.87:0.33:1.00 for liquid water (Chaplin 2007). Relative to that of water vapor, the fundamental O—H

TABLE 3.3.1
Assignment of O—H Vibrational Bands of Liquid Water

	Assignment	Peak Wave Number [cm^{-1}]	Peak Wavelength
Fundamental (MIR)	ν_1 stretch mode	3,277	3.05 μm
	ν_2 bend mode	1,645	6.08 μm
	ν_3 stretch mode	3,490	2.87 μm
First overtone	$a\nu_1 + b\nu_3, a+b=2$	6,800	1.47 μm
Second overtone	$a\nu_1 + b\nu_3, a+b=3$	10,310	970 nm
Third overtone	$a\nu_1 + b\nu_3, a+b=4$	13,530	739 nm
Fourth overtone	$a\nu_1 + b\nu_3, a+b=5$	16,500	606 nm
First combination	$a\nu_1 + \nu_2 + b\nu_3, a+b=1$	5,260	1.90 μm
Second combination	$a\nu_1 + \nu_2 + b\nu_3, a+b=2$	8,330	1.20 μm
Third combination	$a\nu_1 + \nu_2 + b\nu_3, a+b=3$	11,960	836 nm
Fourth combination	$a\nu_1 + \nu_2 + b\nu_3, a+b=4$	15,150	660 nm

Source: From Maeda, H., Ozaki, Y., Tanaka, M., Hayashi, N., and Kojima, T., *J. Near Infrared Spectrosc*, 3, 191, 1995; Chaplin, M. Water Structure and Science, available at www.lsbu.ac.uk/water/vibrat.html, December 19, 2007.

Note: a and b are integers, ≥ 0.

stretching modes of liquid water are shifted to a lower frequency, while the bending mode is shifted to an increased frequency (Bernath 2002). These frequency shifts are further enhanced in solid water (ice). As the O—H stretching frequency increases with increasing temperature, the absorbance band associated with this vibration decreases (i.e., shifts to shorter wavelengths).

Because of its negative charge, each oxygen atom in a water molecule may associate with up to four other hydrogen atoms in what is known as H-bonding. In liquid water, water molecules can thus be H-bonded to up to four other water molecules, so determining the structure of water in the mixture model (Abe et al. 1995). Thus, water has five potential states: S_0 (water with no hydrogen bonding), S_1, S_2, S_3, and S_4 (corresponding to the number of H-bonds per water molecule).

The extent of H-bonding will influence the energy of vibration of the covalent O—H bond. Thus, Maeda and coworkers (1995), using second-derivative absorbance spectra acquired using an FTNIR, observed at least five distinct spectral features superimposed on the first-overtone O—H ($\nu_1 + \nu_3$) band and assigned this to the S_0, S_1, S_2, S_3, and S_4 water species.

This theory can be used to explain why the apparent absorption peaks of water display a shift to shorter wavelengths as temperature increases (at an average rate of ~0.5 nm/°C). For example, the first-overtone ($\nu_1 + \nu_3$) band for liquid water was observed to shift from 1461 to 1418 nm as the temperature changed from 5°C to 85°C (Maeda et al. 1995), 1460 to 1424 nm for the temperature range from 6°C to 80°C (Segtnan et al. 2001), and 1450 to 1420 nm as the temperature varied from 20°C to 80°C (McCabe et al. 1970). Performing a spectral decomposition of the water absorption band, Maeda and coworkers (1995) interpret this absorbance band shift as because of a change in the proportion of each S species of water. Thus, an absorption feature at 1412 nm was noted to become stronger, and a feature at 1491 nm to become weaker, with increased solution temperature. The number of S_0 species was calculated to increase sharply with temperature, the number of the S_1 species to remain fairly constant, and the number of S_2, S_3, and S_4 species to decrease with increasing water temperature.

To illustrate this point in the SWNIR (see also Golic et al. 2003), spectra were acquired of pure water at a range of temperatures (from 5°C to 80°C, at temperature intervals of about 5°C). Spectra were also acquired of sucrose solutions at 0%, 4%, 8%, 12%, and 16% w/w (the approximate range of sugar in most fruit), at each five different temperatures over the range 10°C–50°C. As expected, the absorbance and second-derivative absorbance plot of water contain features consistent with the second-overtone O—H band at 970 nm, the third-overtone O—H band at 740 nm, and the third-combination O—H band around 840 nm (Figure 3.3.7A).

The shift of the second-overtone O—H band at 970 nm with temperature is obvious, while the third-overtone band and third-combination band at 740 and 840 nm, respectively, demonstrate very little wavelength shift (Figure 3.3.7B). Why is the temperature-induced wavelength shift for the second- and third-overtone bands lower? This can be explained by considering the high-molecular anharmonicity of symmetric and asymmetric O—H vibrations, leaving not only the vibrational energy spacing between higher order overtones closer to each other, but also the spacing between the vibrational levels of S_0, S_1, S_2, S_3, and S_4 species.

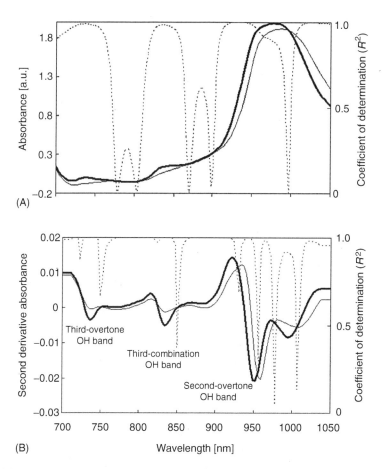

FIGURE 3.3.7 Absorbance (A) and second derivative of absorbance spectra (B) of water at 80°C (thick solid line) and 5°C (thin solid line) and the correlation coefficient of determination between raw and second derivative of absorbance at each wavelength and temperature (dotted line).

A plot of the coefficient of determination (R^2) of a regression between absorbance (Figure 3.3.7A) or second-derivative of absorbance (Figure 3.3.7B) and temperature (the temperature correlation spectrum) illustrates a high correlation at the peak of the third-overtone (740 nm) and third-combination (840 nm) O—H bands, and a much lower correlation at the peak of the second-overtone (970 nm) band. Of course, a poor correlation existed wherever the sample spectra at the different temperatures intersect each other. At these crossover points, there is minimal (or no) information on the constituent of interest (in this case, temperature). Such areas should be excluded from a calibration model.

Interpretation of the correlation plot for second-derivative spectra is difficult, given that the calculation of the derivative also entails a smoothing function that acts to blur information across a wavelength range.

However, the average correlation with temperature, across all wavelengths, for the absorbance spectra is significantly lower than that for second-derivative absorbance spectra (0.76 compared to 0.92). Second-derivative absorbance spectra, therefore, besides its usual benefits of slope and baseline removal, and spectral smoothing, also has the added benefit of producing more stable spectral regions and generally better correlation with the constituent, for example temperature, of interest.

3.3.3.3 Sugar in Water

Water is an excellent solvent for ionic (i.e., salts) and nonelectrolytic (e.g., sugar) compounds because of its polarity, small size, and high dielectric constant. Sugars not only consist of $C-H_n$ groups, but also contain $O-H$ groups that can H-bond with water, creating a number of different configurations. This results in broad absorption bands, even at the fundamental frequency. The extent of this H-bonding will depend on the concentration of the sugar. Thus, sugar $O-H$ vibrations are sensitive to water and sugar concentrations, as well as temperature, but relatively insensitive to pH and ionic strength (Reeves 1994). Therefore, chemometric calibration models based on sugar $C-H$ vibrations (910–930 nm) should be more robust with respect to temperature than models based on the sugar $O-H$ vibrations.

A spectrum of a sucrose solution is dominated by features related to water. Increasing sucrose concentration is associated with a shift of absorbance band due to second-overtone $O-H$ vibration (at around 960 nm) to shorter wavelengths (Figure 3.3.8). This behavior is consistent with an increase in the extent of H-bonding, with the $C-H_n$ and $O-H$ groups of sugar serving as H-donors for water molecules, thereby creating structures of water that have higher numbers of H-bonded

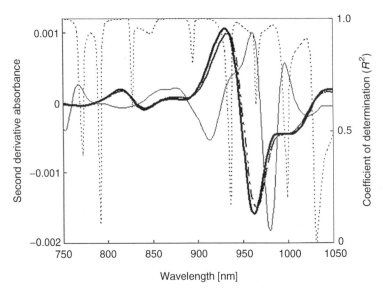

FIGURE 3.3.8 Second derivative absorbance spectra for water only (thick solid line), 16% w/v sucrose (dash–dot line) and crystalline sugar (thin solid line), and the correlation coefficient of determination between second derivative of absorbance at each wavelength and solution sucrose concentration (dot–dot line). All samples at 20°C.

species (i.e., S_2, S_3, S_4). Effectively, adding sugar to water has a similar effect to the structure of water species as a decrease in temperature. As noted for temperature change, the third-overtone band experiences very little, if any, wavelength shift with increasing sucrose concentration.

In the second derivative of absorbance spectrum for dry crystalline sugar, the broad absorption band around 920 nm is interpreted as a convolution of the third-overtone CH-vibration (910 nm) and the third-overtone CH_2-vibration (930 nm). The well-defined peak around 986 nm is ascribed to a sugar O—H vibration, which is less prominent for amorphous sucrose (Kawano et al. 1992), the prevalent state in aqueous solution.

Generally, with samples at constant temperature, high correlation between sucrose concentration and the second derivative of absorbance was noted at wavelengths corresponding to sugar C—H and sugar O—H vibrations, but it certainly was not limited to those wavelengths. Regions of high-sucrose correlation, for instance, are seen across the interval 800–900 nm, where the third O—H combination band is situated. However, when temperature was varied, the R^2 between absorbance at single wavelengths and sucrose concentration of the samples decreased markedly (Figure 3.3.9).

The peak wavelength of the second-overtone O—H vibration decreased with increasing temperature, and increased with increasing sucrose concentration (Figure 3.3.10). The linearity of the shift with temperature across a range of sucrose concentrations is quite evident, suggesting that sugar molecules in a water matrix cause a set perturbation of the water structure (i.e., the extent of H-bonding). Indeed, it is possible to assess the concentration of solutes such as NaCl, which have no NIR spectra of themselves, by their influence on the water spectrum (if temperature is held constant).

How does our spectroscopic knowledge apply to a specific task, such as the assessment of intact fruit for sucrose content? First, the choice of SWNIRS is logical, given the penetration depth capable with these wavelengths. Second, we know to expect the fruit spectra to be dominated by water features. Indeed, it is not necessarily expected that the wavelengths of peak-absorbance features of the attribute of interest, sucrose, will be heavily weighted in a multivariate model, given that information on

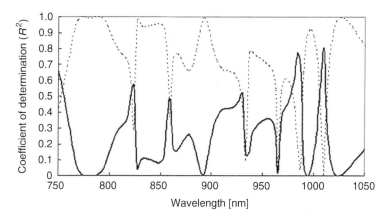

FIGURE 3.3.9 Correlation coefficient of determination between raw absorbance at each wavelength and solution sucrose concentration (solid line) and solution temperature (dotted line). Samples of varying sucrose concentration and temperature.

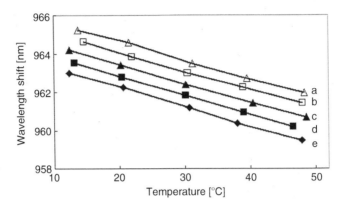

FIGURE 3.3.10 Peak wavelength of the second overtone OH vibration, for solutions of (A) 16% sucrose, (B) 12% sucrose, (C) 8% sucrose, (D) 4% sucrose, and (E) water only, at varied temperatures.

the shoulders of the peaks may be less influenced by interferences such as water. Third, we expect temperature to influence the water spectra, and so this variable must be accommodated in any modeling exercise.

3.3.3.4 Calibration Based on Multivariate Regression

A given application firstly requires good instrumentation (e.g., high-repeatability value) and an appropriate optical geometry. No amount of chemometrics can compensate for poor choices or procedures in these areas. However, given a good set of spectral data, there are several available options for model development.

While earlier NIRS and SWNIRS systems were very site specific, requiring a high level of expertise on site, the responsibility for calibration is shifting off site, to a firms' centralized site, to the instrument vendor, or to a third-party provider. This is consistent with the prediction of Williams and Norris (2001a) that "Calibration methods will continue to improve as software becomes [...] more comprehensive [...] instruments purchased for operating in the field will [...] serve as a medium for scanning samples and relaying the spectral data to the calibration center by email."

Nevertheless, it remains useful for the user of the technology to have an appreciation of the available chemometric techniques, for in purchasing an instrument package the hardware features (as reviewed above) are only half the story. The capabilities and features of the calibration system are every bit as relevant as those of the hardware in making a purchase decision.

As we have noted earlier, this text can only serve a primer, alerting you to standard practices and things that you should know about. For greater depth of understanding, we recommend the formal training offered through the various regional and international NIRS societies, by the equipment and chemometric software package suppliers and by consultants. For those who learn from texts, the chapter by Martens and Naes in Williams and Norris (2001a) is recommended for study. Chapter 6 of this text offers also further detail on this topic, which we assume you have mastered in the following discussion.

3.3.3.5 Chemometric Terms

The first priority in any discussion is to establish a common language, here a definition of chemometric terms is provided (as used throughout this section) (Table 3.3.2). Unfortunately, there is no one term that can be used to indicate the performance of a model, so we need to define a few! Also unfortunate is the reality

TABLE 3.3.2
Definition of Common Chemometric Terms

Term	Abbreviation	Equation	Comments
Standard deviation	SD	$SD = \sqrt{\frac{\sum_{i=1}^{n}(x_i - \bar{x})^2}{n-1}}$	Measure of the dispersion of the reference values for the population
Bias	Bias	$Bias = \sum_{i=1}^{n} \frac{(x_i - y_i)}{n}$	Measure of the difference between reference and predicted values
Root mean square of residual errors of calibration	RMSEC	$RMSEC = \sqrt{\frac{\Sigma(x_i - y_i)^2}{n-1}}$	Estimate of the variation of the reference and predicted values of the calibration set
Root mean square of residual errors of cross-validation	RMSECV	$RMSECV = \sqrt{\frac{\Sigma(x_i - y_i)^2}{n}}$	Where x_i and y_i are samples left out of the calibration set during cross-validation
Root mean square of residual errors of prediction	RMSEP	$RMSEP = \sqrt{\frac{\sum_{i=1}^{m}(c_i - \hat{c})^2}{m}}$	Estimate of the variation of the reference and predicted values of a validation set
Standard error of prediction corrected for bias	SEP	$SEP = \sqrt{\sum_{i=1}^{m} \frac{(c_i - \hat{c}_i - Bias)^2}{m}}$	Estimate of the variation of the reference and predicted values of the validation set, corrected for bias
Standard deviation of validation differences	SDVD	$SDVD = \sqrt{\sum_{i=1}^{m} \frac{(c_i - \bar{c}_i)^2}{m-1}}$	
Coefficient of determination	R^2	$R^2 = \frac{Regression\ SS}{Total\ SS}$ $= \frac{\sum_{i=1}^{n}(x_i - \bar{x})(y_i - \bar{y})}{\sum_{i=1}^{n}(x_i - \bar{x})^2 \sum_{i=1}^{n}(y_i - \bar{y})^2}$ $= 1 - \left(\frac{RMSEC}{SD}\right)^2$	The fraction of the variance within the predicted values explained by the calibration model

(continued)

TABLE 3.3.2 (continued)
Definition of Common Chemometric Terms

Term	Abbreviation	Equation	Comments
Variance ratio	1 − VR	$1 - \left(\dfrac{RMSECV^2}{SD^2}\right)$	
Standard deviation ratio	SDR	$SDR = \dfrac{SD}{RMSECV \text{ or } RMSEP \text{ or } RMSEC}$	Standardizing RMSEP, RMSEC, or RMSECV between populations
Ratio performance deviation	RPD	$RPD = \dfrac{SD}{RMSECV \text{ or } RMSEP \text{ or } RMSEC}$	As above but bias corrected

Note: The following notation is used:
  n = number of samples in the calibration set
  x_i = actual or reference attribute value of samples in the calibration set i, $i = 1, 2, \ldots, n$
  $\bar{x}$ = mean reference attribute value of samples for population $= \dfrac{1}{n}\sum_{i=1}^{n} x_i$
  $\lambda_{i,j}$ = absorbance value at wavelength position (pixel) j for sample i; $i = 1, 2, \ldots, n$ and $j = 118\text{--}224$
  y_i = predicted sample attribute value I, $I = 1, 2, \ldots, n$, for samples in the calibration set
  c_i = actual or reference attribute value of samples in prediction/validation set i, $i = 1, 2, \ldots, m$
  $\hat{c}_i$ = predicted attribute value of samples in prediction/validation set i, $i = 1, 2, \ldots, m$
  m = number of samples in the prediction/validation set

---

that these definitions are not universal. Different authors, and different manufacturers and software vendors, will favor the use of different, if related, terms, or, perhaps more frustratingly, will use different abbreviations. For example, the chemometric software package "The Unscrambler" (Camo) has used the abbreviation RMSEP for RMSECV (root mean square of residual errors of cross-validation) as defined here. Note that it is important to be clear as to whether the statistics quoted apply to a calibration, a cross-validation, or a prediction exercise.

Bias is the average difference between the predicted and actual values. Another parameter of interest is how well the predicted and actual values fit to a line. This can be defined in terms of the regression coefficient of determination (R^2) and the root mean square error of prediction (RMSEP). Note that R^2 and bias-corrected RMSEP (SEP) are linked by the SD of the population, as

$$R^2 = 1 - \left(\frac{SEP}{SD}\right)^2 \qquad (3.3.3)$$

Imagine the case where actual and predicted values fit a line with a slope of unity exactly (i.e., $R^2 = 1.0$), but the line has a nonzero intercept. In this case, a bias exists (a special case, with bias equal to intercept). Alternatively, the mean of the predicted values could equal that of the reference values (i.e., zero bias), but R^2

could be poor. Consider a technique that has an SEP of 1.0 unit. If used with a population of SD 1.0 unit, the R^2 of the prediction will be 0, yet if used with a population of SD 10.0, the R^2 of the prediction is 0.99. Thus, all terms, bias, R^2, and RMSEP, have value in describing the results of a model prediction.

Unfortunately, comparison of literature reports is usually complicated by variation in population size and structure with respect to the attribute of interest. It is therefore critical that publications detail the population SD for the attribute of interest. Indeed, reporting of a range of statistics (including R, RMSECV, SD, number of samples used, number of outliers removed, and number of principal components used) is required to interpret calibration model performance

3.3.3.6 Chemometrics: Calibration Techniques

For some applications, a simple relationship between absorbance at one or a few wavelengths and the attribute of interest exists. Many applications, however, depend on the use of multivariate statistics, utilizing information in broad regions of the spectrum.

These multivariate techniques fall into two broad camps: linear and nonlinear techniques. The linear techniques of MLR and partial least squares regression (PLSR) are commonly employed. For the application of fruit SSC assessment, Kawano and coworkers (1992) have favored stepwise MLR, manually selecting wavelengths related to the attribute under consideration (e.g., for sucrose, 1057 and 906 nm, being second and third overtones, respectively, of C—H stretching vibration). This approach ensures an empirical basis to the developed models. However, PLSR modeling has a greater following, and is presented as the method of choice in many commercially available chemometric software packages.

Given the popularity of the PLSR method, it would be inappropriate not to offer at least a short description of foibles of the method (see Chapter 6 for a description of the technique). In PLSR, the spectra are described by a set of factors or components. The more the factors used, the better the calibration set will be modeled. However, at some point, the data set is over-fitted, for example, modeling noise in the spectra. In this case, the prediction performance of the model with independent data is decreased. Commonly, a cross-validation procedure is used to optimize the number of factors, wherein sets of samples are left out of the calibration set to act as a validation group. Nonetheless, the choice of the number of factors to be used remains problematic. As a rule of thumb, a model with very few factors (e.g., <4) or many factors (e.g., >10) for a sample type with a complex matrix should be viewed with suspicion. For the sample water–sucrose mixtures discussed previously, PLS regression on water temperature required only two factors to capture more than 99% of the variance of the data set, but for water–sucrose mixtures at different temperatures, three factors were needed.

A factor consists of a vector of values (loadings) by wavelength. In PLSR, the factors are regressed onto the values of the attribute of interest, with each factor contributing to a final set of b coefficients (a vector of values by wavelength) that represent the regression model. The loadings of the individual factors or the model b coefficients can be inspected for spectroscopic significance; however, it is the nature of the technique that the information related to a given attribute is not contained in a single factor, but is spread over several factors. There are other techniques (e.g., Independent Components Analysis) that attempt to quarantine the information owing to a given constituent to a single factor. Finally, although PLSR is a linear regression technique, its use of multiple factors does mean that it is surprisingly able to model a degree of nonlinearity.

A number of nonlinear modeling techniques are also available, including neural networks and support vector machines. FOSS (www.foss.de) employs neural network models with a range of their NIRS instrumentation. Such techniques have promise, but particular care must be taken to avoid over-fitting of the data. Also, interpretation of the models is often impossible, so the operator must ensure the quality of the input data. For example, for a fruit SSC application, Peiris and coworkers (1998) reported better prediction results ($R_p^2 = 0.48$, SEP $= 0.52\%$ SSC) using a neural-network calibration model than achieved using a PLSR calibration based on the wavelength range 780–980 nm ($R_p^2 = 0.32$, SEP $= 0.74\%$). However, a (seven times) larger population was used for the neural-network calibration exercise and thus the results are not directly comparable. A small gain in performance was reported for use of a least squares support vector machine algorithm for modeling of a grape spectra and acidity data set (Chauchard et al. 2004).

3.3.3.6.1 Fitting a Model

Some issues, such as the selection of the number of PLS factors, are specific to the regression technique. The following issues of outlier removal and validation are relatively generic across the methods.

An outlier is a sample that does not fit the model. The outlier is commonly identified as either an X outlier, different in terms of its spectrum, or a Y outlier, identified on the basis of a high residual (predicted value minus actual value). Obviously, removal of outliers will improve the apparent fit, but such a result can represent a false sense of security, and may actually represent the removal of useful variance from the model, such that the model will be poorer in prediction of new sets. Some regression approaches allow for removal of outliers to some arbitrary limit, for example, residuals greater than a given limit, or up to 10% of the population. However, in the strict sense, no sample should be removed unless there is a rationale for its removal, at least in terms of Y outliers.

When a model is used in prediction of a new set for which there are no reference values available, obviously it is impossible to calculate a Y residual. However, as the X residual is based only on the spectrum, this character can be calculated. As such, it is reasonable to remove samples from the calibration set based on the X residual, and apply the same rule in use of the model in prediction of new samples. There are various methods for calculating a term that can be used as a measure of the X residual, with leverage and Mahalanobois distance being commonly employed in conjunction with PLSR.

As mentioned earlier, it is important to differentiate between calibration, validation, and prediction results. In reading any report of an NIRS application, you should take care to understand the authors' trial design in this respect. All too often either only calibration model statistics (e.g., RMSEC [root mean square of residual errors of calibration]) or cross-validation statistics (e.g., RMSECV) are presented. Where prediction statistics (e.g., RMSEP) are presented, the prediction is often based on a subset of the population used for calibration. The latter case is really a glorified cross-validation activity. The model should be used in prediction of a truly independent group of samples (e.g., harvested from different fields or seasons). Many reports are quite vague on how independent the prediction samples are from the calibration set.

Another issue is in application to a quality control situation, in terms of the level of classification success (Miller and Zude 2004). Such application depends on the level and spread of attribute levels in a population, indexed by mean and SD, the prediction error, indexed by RMSEP, and the desired control point. For example, if it is desired to maintain an attribute level at 10 units, but if the population mean is 10 units and SD 1.0 unit, then the use of a sorting technology with RMSEP of 1.0 is futile (fruit will be randomly assigned as above and below the criterion point).

Finally, there is the issue of how to compare model performance. Is a model resulting in an RMSEP of 0.55 really any better than another delivering a result of 0.61? Where the same set of samples (same reference values) have been used with the two methods, it is possible to compare the variation in the residuals of the two methods. This allows for a test of the statistical significance of the difference between the two results (Fearn 1996).

3.3.3.7 Chemometrics: PLSR Wavelength Selection

The use of non-informative regions of the spectrum can only add noise to a modeling exercise. This task, then, is to identify the informative regions in a spectrum with overlapping peaks (e.g., Hruschka 2001). The general adage, no prediction without interpretation, no interpretation without prediction, gives the general principle that the wavelength region selected should have some spectroscopic or chemical basis in terms of relation to the attribute of interest. Admittedly, however, other workers have attempted a strictly statistical approach to reduce colinearity in spectral data sets (e.g., Rossi et al. 2006).

Correlation spectra and MLRs are helpful tools to identify spectral regions or wavelengths that are associated with the constituents of interest. Yet, when performing PLS regression models, very often non-optimized wavelength ranges or spectral regions are used for model development. The optimal wavelength range to be used is very much dependent on the physical and chemical state of the sample, and also on limitations of the instrumentation being used.

For example, for fruit quality SSC assessment, a patent by Dull and coworkers (1992) stipulates a wavelength range of 800–1050 nm, while Bellon and coworkers (1993) used the loading factors from a PLS model to recommend the region 847–977 nm for peach SSC assessment. Walsh and coworkers (2004) commonly utilize the region ~735–930 nm, which includes the important third-overtone O—H and C—H functional groups of the water–sugar matrix of fruit.

Ozaki and coworkers (2001) have developed a moving window PLS regression method to obtain informative spectral regions, for example, for the determination of cholesterol and glucose in serum solutions (Kang et al. 2005). This moving windows approach has the advantage of mapping out all possible combinations of PLS intervals. The challenge is to combine this with outlier removal and selection of optimum number of factors for each interval, and to add a level of interpretation as to the spectroscopic significance of the chosen window.

For example, a variant of the moving windows approach was used, based on code developed in MatLab (MathWorks Inc., United States) and using PLS Toolbox 3.5 (Eigenvector Research Inc., United States), to perform PLS regressions on every possible spectral interval in a spectral data set, plotting a map of RMSEC values to allow an easy visualization of optimum wavelength ranges for the analysis (Guthrie et al. 2005a). This map consisted of a two-dimensional color plot, with the wavelength start (WS) position on one axis, the end wavelength (WS + window width) on the second axis, and the respective RMSEC value for every WS/end combination shown in the third dimension as a color plot. The position of this window, and its size, was moved across the entire wavelength range in 1 nm intervals. Approximately, 60,000 individual PLS regressions made up a single PLS spectral map (700–1050 nm) in this exercise. With the number of latent variables constant, model performance (i.e., RMSEC values) was directly comparable for the different spectral regions.

This procedure was applied to the sucrose–water–temperature data set considered above (i.e., Figure 3.3.10). The best PLS spectral interval for modeling sucrose concentrations was from 770 to 910 nm (RMSEC = 0.35°C, $R^2 = 0.999$), while the worst interval was from 920 to 970 nm (RMSEC = 2.9°C, $R^2 = 0.75$) (Figure 3.3.11). Inclusion of wavelengths encompassing the second-overtone band of water and sugar O—H stretch resulted in degraded model performance. Inclusion of wavelengths relevant to the third-combination O—H band and the third-overtone C—H stretch band was beneficial to the model.

3.3.3.7.1 Assessing a Calibration Model

In this example, spectra were collected with a prototype handheld NIR unit (iQ, Integrated Spectronics, Australia) using a transflectance optical geometry. The

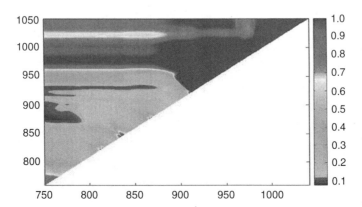

FIGURE 3.3.11 PLS–RMSEC map for various wavelength intervals, in terms of the start (*x*-axis) and finish (*y*-axis) wavelengths for the interval. All PLS models used three factors, with no outlier removal. Data set of sucrose–water mixtures, with temperature varying from 12°C to 48°C (see Figure 3.3.9). Bar to right of graph provides a gray scale index for RMSEC value.

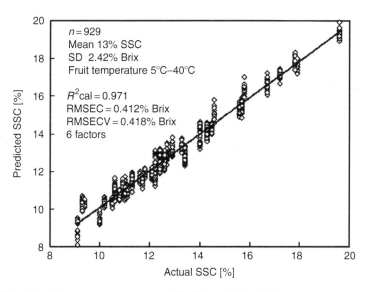

FIGURE 3.3.12 Temperature compensated PLSR model for SSC of peaches and nectarines.

wavelength region was optimized using the procedure outlined above, and a temperature compensated six factor PLSR model was developed on the SSC of peaches and nectarines. Calibration model statistics are summarized in terms of R^2, RMSEC, and RMSECV (Figure 3.3.12, Table 3.3.3). Note that the reporting of statistical results on the performance of a model (Table 3.3.3) is not a complete substitute for inclusion of a scatter graph of predicted values against actual values.

TABLE 3.3.3
Calibration and Prediction Statistics for a Stone Fruit SSC Model

	R^2	SEC/V/P	Bias	Slope	RPD
Model development					
Calibration model	0.97	0.41	0	1	5.87
Cross-validation	0.97	0.42	0	1	5.79
Prediction					
Instrument 22°C	0.87	0.62a	1.39	1.02	2.55
Instrument 40°C	0.90	0.55a	1.37	1.04	2.87
Sunlight	0.38	2.88b	−0.83	0.30	0.55
Sunlight corrected	0.89	0.58a	1.10	1.12	2.72

Note: The calibration set consisted of $n = 929$, mean $= 13.0\%$, SD $= 2.4\%$ SSC. The model used six factors. A leave one out cross-validation procedure was used. The prediction set consisted of $n = 60$, mean $= 8.4\%$, SD $= 1.6\%$ SSC, with predictions made under four conditions: (i) room lighting, instrument at 22°C; (ii) room lighting, instrument at 40°C; (iii) sunlight, instrument at ambient temperature, but operated without a referencing correction; and (iv) as for (iii), but instrument operated with referencing correction. RMSEP values followed by the same letter are not significantly different at a 95% confidence level.

The model was validated using an independent prediction set acquired some 20 months after the model was developed, and under different conditions (Table 3.3.3). The mean SSC for the prediction set was relatively low (8.4% compared to 13.0% SSC for the calibration set). Model prediction statistics were assessed with the instrument temperature kept at a constant 22°C (laboratory temperature), increased to 40°C, and operated in full sunlight. The prediction was considered acceptable in terms of the bias-corrected RMSEP and ratio performance deviation (RPD) values; however, the high bias value indicates a need for either bias correction of results or model updating. No significant difference (Fearn 1996) was observed in the RMSEP value between the acquisition conditions. However, model performance was significantly degraded when a referencing feature was disabled.

3.3.4 Application in Food

NIRS has many established applications in the food and feed industries (Kawano 200X). Typical applications include beer, biofuels, flour, grain, sugar, oils and fats, and wine.

So let us take on some practical examples! What features are you looking for in your application to allow you to choose between instrument vendors? While the theory of the NIRS field has been maturing, the available instrumentation is still progressing through a developmental phase (e.g., McClure 2003), so it is still a period of buyer beware!

You now know to assess the available units on the basis of source–sample–detector optical geometry, light source, spectrograph wavelength range and resolution, dynamic range and signal-to-noise ratio of the detector system, and the method and frequency of referencing, in context of your specific application.

3.3.4.1 Fresh Fruit

You wish to acquire the capacity to grade your intact fresh stone fruit for SSC using SWNIRS. You are interested in grading fruit on a pack-line at up to 10 items per second, and also in assessing fruit on the tree. What are the design criteria for such instruments? Relative to the pharmaceutical or petrochemical industries, or even the processed food industry, horticulture is an extensive operation dealing in units of low value. Thus, cost is a major criterion to equipment purchase. But what technical specifications should be met?

Design criteria for this task were set as early as 1958 (Birth and Norris 1958), although commercial instrumentation was not available until the 1990s. The appropriate wavelength range was defined as 700–1100 nm, taking advantage of the second- and third-overtone O—H and C—H bands in this region that have relatively low absorptivity in water. This feature allows for collection of spectral information from some depth of the fruit sample. Subsequent work has refined these concepts, for example, the required wavelength resolution and signal-to-noise limits (e.g., Long et al. 2005).

For assessment of intact fruit, a reflectance measurement would render the unit sensitive to changes in surface characteristics (McGlone et al. 2003). However, given the high optical density and inhomogeneous nature of fruit, such as central seed in peach, a 180° transmittance geometry may be impractical. Partial transmittance or interactance geometries could be considered. Geometries that allow for a noncontact optical configuration are particularly relevant to on-line grading (Greensill and Walsh 2000b).

Any equipment acquisition should also allow an upgrade path. One of the strengths of NIRS technology is that it can be used in prediction of more than one attribute. Therefore, while the current aim may be for SSC grading, it is also desirable to factor in requirements (e.g., optical geometry) for other applications, for example, assessment of dry matter in fruits that store starch or oil, or detection of internal defects such as brown heart and translucency in apples.

Of the various spectrophotometer technologies available, the state of current technology favors the use of an Si photodiode detector array, a dispersive grating, and a tungsten halogen lamp for this application (see Section 3.3.2.3). A scanning grating with single detector is too slow for the in-line application, and too bulky and vibration sensitive for the portable application. An AOTF can be used instead of a grating, allowing operation with a single, rather than an array, detector, but these systems are expensive, relative to grating–PDA based systems. LED units hold promise for portable applications, given their power efficiency instantaneous warm-up period.

Each technology has its limitations. For example, while PDA spectrophotometers are relatively insensitive to vibration, these units are sensitive to temperature, in terms of thermal noise, altered detector sensitivity, and changes in wavelength calibration because of difference in the thermal expansion coefficients of the components of the optical bench of the instrument (Walsh et al. 2000). Such effects may be minimized by control of detector temperature. This solution incurs an extra power requirement, of no significance to a bench-top unit, but problematic to a portable unit. Another potential solution involves frequent re-referencing.

However, you know that hardware specifications are only half the answer. Instrument offers must also be assessed in context of their software, allowing for calibration model transfer between units and for model updating.

You quickly come up with a list of commercially available instrumentation. Integration of Si PDAs and halogen lamps into fruit pack-lines occurred first in Japan in 1988 (Mitsui, Fantec, and Saika P/L), then in Australasia (Colour Vision Systems and Taste Technologies P/L) and Europe (Sacmi P/L) in 2000. Other providers in Europe (MAF-Roda, Greefa, and Aweta P/L) started doing so since 2004. A laser diode based technology was available from Sumitomo P/L (Japan) during the late 1990s, but this unit is no longer commercially available.

Your task is now to obtain technical specifications on each of these units and compare them in light of your knowledge of the requirements of the operation.

Specifications and criteria for instrument comparison for this application could include the following:

- Optical geometry to suit on-line application and extension of application to other attributes (e.g., transmittance optics may be required if it is desired to assess internal browning as well as SSC)
- Compatibility with existing conveyor technology (to initiate spectral acquisition and to associate predicted SSC data with other information, e.g., fruit weight, for grading purposes)
- Wavelength range of 400–1050 nm (enabling VIS and SWNIR spectroscopic applications)
- Wavelength resolution of <10 nm
- Repeatability value of <1 mA (SD of 50 repeated measures of a standard) from 600 to 900 nm
- Low instrument sensitivity to vibration, ambient temperature, and ambient light level in terms of repeatability value or repeatability of prediction for SSC in fruit
- Characterized lamp warm-up period
- RMSECV for a model of SSC in apple fruit of <0.7% SSC
- Outlier detection strategy (i.e., a means to avoid prediction of samples that are not described by the model)
- Chemometric modeling procedure (MLR, PLSR, and others)
- Ease of modeling
- Stability of models, frequency of model updating required

An useful practical test would be to provide each vendor with the same sets of fruit for calibration of their units, with your judgment based on the performance of the units in prediction on independent fruit sets, in terms of R and RMSEP. A harsher test would be to provide fruits for calibration of one unit, and ask the vendor to transfer the model to another unit, with that unit assessed in prediction of an independent set of fruits.

Similarly, there are a number of portable units available specifically for the fruit market. Temma and coworkers (2002) described the development of a handheld unit that was subsequently commercially released (Amamir Taster, Amamir P/L, Japan). A handheld unit has also been developed by Fantec P/L (Japan) and described by Saranwong and coworkers (2003). More recently, Miller and Zude (2004) described the development of a unit released by Control in Applied Physiology GbR (unit PA1101, Germany). Also, the portable sister (iQ) to the inline Colour Vision Systems unit was released by Integrated Spectronics (Australia), and described by Golding and associates (2006) and Subedi and coworkers (2007). Brimrose (United States) has also released a luggable version of their AOTF-based spectrophotometer. An MEMS-based handheld is also available (Polychromatix P/L, United States), but this unit uses a detector technology that operates outside the SWNIR.

Excepting the Brimrose and Polychromatix units, all of the units mentioned are based on a dispersive grating in Si photodiode detector array system. Excepting the

VIS/NIR Spectroscopy

German system, which utilizes LEDs for illumination, all of the units are based on a tungsten halogen lamp system. The same criterion for equipment comparison will apply as for the in-line system, with the additional considerations of

- Sensitivity to ambient light
- Sensitivity to instrument temperature
- Ergonomics of portability
- Power supply (operational life, number of spectra recorded per battery charge)
- Ease of information exchange from field unit to base station (e.g., model updating)

To reiterate, however, SWNIRS technology involves an indirect measurement of the attribute of interest, that is, the spectra are measured, and then related to the attribute. As a result, the ultimate measure of the performance of a unit, and its ability to do the task, lies not in a one-off assessment, for example, a comparison of technical specifications, but in the performance of the instrument and calibration model in prediction of subsequent populations, for example, across multiple populations of stone fruit (Golic and Walsh 2006), mandarin (Guthrie et al. 2005b), melon (Guthrie et al. 2005c), and mango (Subedi et al. 2007).

3.3.4.2 Grains and Grain Products

Grains have been the object of studies using NIRS for a long period. Many studies on this application have been undertaken, with NIRS now used as a standard method for routine analyses. Initially, NIR spectra were measured of ground samples, but measurement of whole kernels has superseded this practice, as grinding is laborious and time consuming.

According to the *NIRT Handbook*, published by USDA-FGIS, United States, NIRT using whole grains was first employed in 1996 in place of NIRR using ground samples. NIRT is now used as the standard method for determining official protein and oil contents in soybeans, and official protein content in wheat (USDA 1996).

Williams and coworkers (1985) reported on the accuracy and precision of NIRT for determining protein and moisture contents in wheat and barely. The SEP for NIRT was slightly higher than that for NIRR. However, the elimination of grinding and cell-loading errors and large sample size, with consequent reduction in sampling error, tends to compensate for the difference in accuracy. Comparison of commercial NIRT and NIRR instruments for analysis of whole grains was performed. Both approaches were comparable in accuracy and reproducibility (Williams and Sobering 1993; Delwiche et al. 1998).

When grains are utilized as raw materials for food industry, there are processing characteristics as well as constituents that should be evaluated. In case of wheat, its kernel texture (degree of hardness or softness) is the most important factor affecting wheat flour functionality. This character exerts a significant influence on the yield of flour and on flour-damaged starch incurred during milling. Methods including NIRS defining hardness by measuring a property of the wheat after it is ground are usually measuring some aspect of the resulting particle size distribution because harder wheats have a larger mean particle size than softer wheats. In case of NIRR using ground samples, phenomena that spectral baseline is shifted by particle size variations between samples are used for measuring hardness (Norris et al. 1989). By NIRT using whole kernels, Williams (1991) tried to measure hardness or softness of wheat. The investigation has shown that an NIRT method (whole grains) is capable of predicting wheat hardness with precision equal to that of the reference method (grinding–sieving method) and that it is slightly superior to the NIRR method (ground samples).

In a rice milling plant, whiteness and taste are the most important factors for quality control. Whiteness, affected by degree of milling (DOM), is related to nutritive value of rice. It has been reported that whiteness can be measured by NIRS (Delwiche et al. 1996). However, whiteness seems not to be measured directly by NIRS but to be measured by determining the amount of bran remaining in the surface of milled rice kernel in this case. NIR determination of surface lipid content (SLC) of milled rice has also been investigated because there was a strong correction between (DOM) and SLC (Chen et al. 1997; Li and Shaw 1997).

Rice taste is generally assessed by sensory evaluation using steamed rice. Numerous studies on the relationship between rice taste and physicochemical properties have also been performed. It is well known that rice taste is a function of chemical constituents such as protein, moisture, amylose, fatty acid (fat acidity), and minerals (Hosaka 1987). The lipids in stored rice are hydrolyzed and oxidized to free fatty acids or peroxides, causing acidity to increase and significantly deteriorating the taste, as well as producing off-flavor. NIR determination of fat acidity of whole kernel and ground rough rice has been made, and a good calibration equation for ground sample ($R^2 = 0.95$, SEP = 0.73 mg of KOH/100 g of dry matter) has been obtained (Li and Shaw 1997). Amylose is related to cooked rice texture quality. NIRT (800–1050 nm) of unground brown rice or milled rice was demonstrated to adequately screen for apparent amylose content (AAC) (Villareal et al. 1994). A method for determining AAC, based on the NIRR spectrum (1100–2498 nm) of ground milled rice, was also investigated (Delwiche et al. 1995). The SEPs were 0.79% for NIRT and 1.04% for NIRR, respectively. In addition, prediction of cooked rice texture quality using NIRR analysis of whole-grain milled sample has been done (Windham et al. 1997).

Taste-related formulas that derived taste score from the constituents have been developed. However, it is impractical for quality control to use the results of time-consuming chemical analyses. To overcome this problem, rice taste analyzers based on NIRS have been developed.

The first rice taste analyzer developed by Satake Engineering Co., Ltd. Tokyo, Japan, consists of NIR instrument provided by Bran + Luebbe GmbH. The analyzer is based on the experimental proved result that rice taste is fixed by the balance of moisture, protein, amylose, and fatty acid. In a practical procedure, milled rice is ground, and the ground sample is kept at a constant temperature oven for more than 1 h, and then the NIR measurement is performed to determine each constituent. From the constituents, the taste score can be calculated using taste-related formula that relates the constituents to taste score. A taste score can be generated in only a few minutes. The rice taste analyzer includes software that calculates the blending ratio to perform lowest price at the same taste, or to perform best taste at the same price (Hosaka 1987).

As for wheat products in processed foods, the NIR method is widely used for quality control of raw materials, intermediate products, and final products. In case of wheat flour, NIRR method has been used successfully to determine the degree of starch damage of commercially milled flour (Osborne and Douglas 1981; Osborne et al. 1982; Morgan and Williams 1995), which is important in relation to the baking properties of the flour, because the degree of starch damage has a strong influence on the water absorption. If the starch has much damage during flour milling, hydration of starch progresses, and then the water absorption properties of the flour are affected. It was reported that the degree of starch damage could be measured at wavelengths corresponding to overtones and combinations of H-bonded O—H vibration.

In relation to quality of biscuits and biscuit doughs, the possible use of NIRR method to monitor their fat, sucrose, dry flour, and water has been reported (Osborne et al. 1984a), and the simultaneous analysis of ascorbic acid, azodicarbonamide, and L-cysteine in bread improver mix has been achieved (Osborne 1983). As for bread, NIR analyses of sliced white bread and air-dried bread were done, and very good results have been obtained (Osborne et al. 1984b).

In the analyses of wheat products, NIR method is widely used, say, for determination of fat and sucrose in dry cake mixes (Osborne et al. 1983), and fiber in processed cereal foods (Baker 1983; Kays et al. 1996, 1997); while no sufficient accuracy has been obtained in NIR determination of α-amylase activity of wheat (Osborne 1984), and loaf volume at bread baking (Starr et al. 1983).

In soy sauce manufacturing, quality control of intermediate and final products by many components such as total nitrogen, sodium chloride, alcohol, reducing sugar, lactic acid, glutamic acid, and glucose is demanded. Wet chemistry analytical method needs much labor and time for these analyses. To compensate this problem, the Kikkoman Co., a soy sauce manufacturer, developed an automatic chemical composition analyzer for soy sauce (Kobayashi et al. 1990). The analyzer consists of an NIR spectrometer (InfraAlyzer 400 or Infralyzer 500), temperature controller, automatic sampler, and pumps. A certain amount of soy sauce, collected by the automatic sampler, is sent to the NIR spectrometer at a constant flow rate controlled by the pump through the temperature controller maintained at 20°C. The NIR measurement is made automatically. After the measurement, the sample cell and tube are washed with cleansing liquid. It takes about 3 min to analyze one sample including the washing process. Raw soy sauce in each fermentation vessel had a different chemical composition; therefore, to maintain the preferred quality of soy sauce, certain amounts of sauce from different fermentation vessels are blended in the bottling process. Components of each lot of soy sauce, such as sodium chloride (SEP = 0.09%), total nitrogen (0.01%), alcohol (0.12%), lactic acid (0.12%), and glutamic acid (0.05%), are analyzed automatically by NIRS. In 1993, the NIR method has been adopted to screen the soy sauce samples as the JAS (Japanese Agricultural Standard) quality inspection method in Japan. In assay of soy sauce, high accuracy could be provided in determination of sodium chloride. The first wavelength of 1445 nm in the calibration for sodium chloride was assigned to water band.

Single-kernel NIRS is a promising technique of use to grain breeding programs. Delwiche (1993) used an adapted designed clip to measure hardness of wheat using SWNIR in transmittance mode. Attempts to use the more informative long-wavelength region to measure a single-kernel spectrum with a typical reflectance mode were made (Delwiche and Massie 1996; Delwiche 1998). The sample presentation developed was, theoretically, excellent but hardly used for high-speed commercial applications, such as sorting machines. To solve this subject, a special sample holder for single kernel using long-wavelength region in transmittance mode was developed for measuring protein and moisture contents in brown rice and milled rice (Rittiron et al. 2004, 2005). Using the method developed, a calibration model for brown rice with an SEP of 0.24% w/w dry base for moisture and 0.4% w/w dry base for protein could be obtained. In case of milled rice, the DOM is one of the factors affecting the accuracy of a calibration model, and then the model with the DOM compensation was developed. By examining the normal distribution of the protein content measured by the single-kernel spectroscopy, a blended bulk could be separated from a pure bulk as it did not exhibit a normal protein distribution while a pure one did. By doing the collaborative research between National Food Research Institute Tsukba, Japan and Shizuko Seiki Co., Ltd. Shizuoka-Ken, Japan, a prototype of sorting machine for single kernels of brown rice has been developed as shown in Figure 3.3.13.

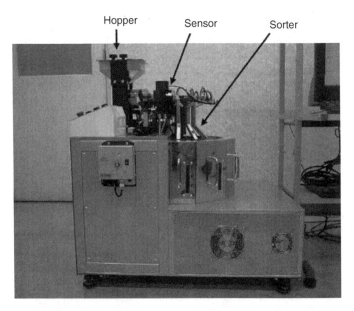

FIGURE 3.3.13 Prototype of a sorting machine for single kernels of brown rice, developed by Shizuoka Seiki Co. Ltd.

VIS/NIR Spectroscopy

3.3.5 ON-LINE QUALITY MONITORING

3.3.5.1 Definition

Most technologies develop in a laboratory and begin their operational life in the form of bench-top equipment. Where this is in demand, the technology will be reengineered for portable application, or for application in association with a process line. In the latter case, the unit may sit adjacent to, but independent of, the process line (at-line), may analyze a representative substream of the main process line (in-line), or may be implemented in the main process line itself (on-line). Be cautious however, these definitions (at-line, on-line, and in-line) are not universal, and are sometimes used interchangeably.

3.3.5.2 Issues

There are no fundamental issues of spectroscopy or chemometrics that separate an at-line application from an on-line application. Rather, all of the considerations mentioned above come into play. The geometry of the process line may limit the positioning of the spectrometer, so the relevance of the optical geometry and sampling volume should be questioned. The process line is moving at speed, so the protocol by which a sample is gathered for a reference determination should be questioned. Another issue related to the available sample variance is that in a controlled process, it is desired to keep parameter levels within relatively tight limits, with these limits exceeded only in case of a breakdown. Calibration of an analytical method in such a scenario is difficult, in that updating of the model regularly during periods of normal operation will decrease the variance of samples used for calibration.

3.3.5.3 Examples

To illustrate these points, the operation of the SWNIR on-line fruit quality assessment unit (InSight) offered by Colour Vision Systems (www.cvs.com.au) is profiled.

This unit utilizes a partial transmittance or interactance optical arrangement (Figure 3.3.14; Greensill and Walsh 2000b). As such, only one side of the fruit is assessed, unless used on a pack line that employs rollers to spin the fruit as it moves under the spectrometer. The unit is capable of assessing up to 10 items of fruit per second (i.e., up to a belt speed of 1 m/s), depending on fruit size. Operating with a 100 W QTH lamp, an integration time of around 20 ms is required, depending on fruit commodity (Walsh et al. 2004). At a belt speed of 1 m/s, there is movement of 20 mm during the period of spectral acquisition. Reference methods must sample an equivalent volume of fruit. A PLSR calibration model is used, with a Mahalanobois distance based outlier detection routine employed, such that samples that appear unlike the calibration set are not predicted. In quality control, the attribute level is predicted and assessed destructively in test groups (ca. 20 fruits). If the prediction performance is unacceptable, spectral data and reference values are e-mailed to a vendor site for model updating.

FIGURE 3.3.14 On-line sorting machine for fruit, utilizing interactance optics, as developed by Color Vision Systems P/L.

The following example is drawn from the work of Golic and Walsh (2006) in relation to the use of a prototype InSight unit with stone fruit. The SWNIR spectra of moving fruit are no different to that of stationary fruit, and are dominated by water features, for example, at 840 and 960 nm (Figure 3.3.15). Indeed, it is a source of marvel that such featureless spectra can be used to determine SSC levels, but there is the strength of spectroscopy and chemometrics! Calibration models on aqueous sample are very sensitive to a difference in temperature between the calibration and validation sets, because of the influence of temperature on the O—H features; however, the simple expediency of adding fruit of a range of different temperatures to the calibration set avoids this issue (Figure 3.3.16). Other, more sophisticated, approaches can also be used (e.g., external parameter orthogonalization; Roger et al. 2003). In the stone fruit orchard where the work was undertaken, a different variety was harvested every few weeks through the season. Model robustness using this instrumentation was apparently influenced not so much by variety as by population mean for the attribute of interest (SSC) (Figure 3.3.17). Model predictions were affected not so much in terms of precision (e.g., prediction R^2 values were relatively stable) as in terms of accuracy, as reflected in bias error. Better predictive performance was achieved for assessment of dry matter in intact mango using a later InSight model (Figure 3.3.18; Subedi et al. 2007).

In this example, once the technology was in use within the fruit pack house, the variation in fruit SSC over time was obvious (Figure 3.3.17D). Consider that a

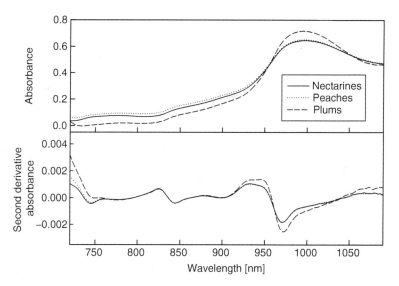

FIGURE 3.3.15 Spectra of nectarines, peaches, and plums collected using an in-line SWNIRS system (average of 20 fruit spectra for each fruit type). (From Golic, M. and Walsh, K. B., *Anal. Chim. Acta*, 555, 286, 2006. With permission.)

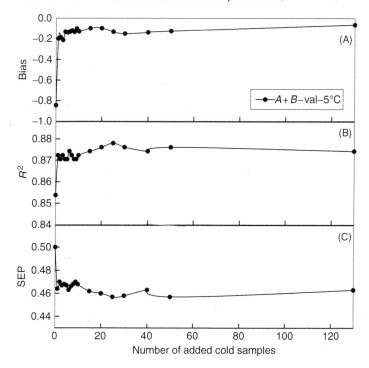

FIGURE 3.3.16 Influence of the addition of cold (5°C) samples into a calibration data set based on samples at 20°C on the prediction of a validation set of cold (5°C) samples, in terms of model (A) bias, (B) R^2, and (C) SEP. (From Golic, M. and Walsh, K. B., *Anal. Chim. Acta*, 555, 286, 2006. With permission.)

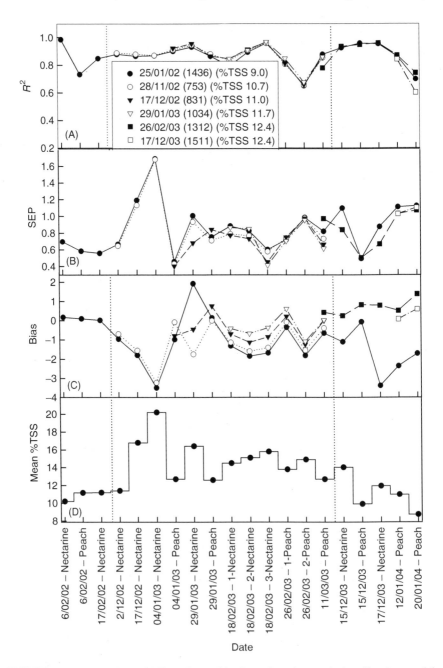

FIGURE 3.3.17 Predictive ability of (combined peach and nectarine) calibration models across new fruit populations. The legend in panel (A) indicates the date of model creation, the number of calibration samples, and the mean SSC of the calibration data sets. Prediction statistics are reported as follows: (A) R^2, (B) SEP, and (C) bias. Mean SSC for each predicted population is shown in (D). Vertical dotted lines represent boundary lines between the three fruiting seasons. (From Golic, M. and Walsh, K. B., *Anal. Chim. Acta*, 555, 286, 2006. With permission.)

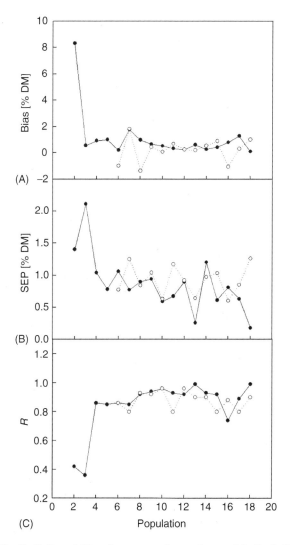

FIGURE 3.3.18 Predictive ability of a mango dry matter model. Prediction statistics are reported as follows: (A) bias, (B) SEP, and (C) R^2. Open circles: populations 1–4 used to develop a model, used in prediction of population 5 onwards; closed circles: calibration model continually updated (i.e., model on population 1 used to predict population 2, model on population 1 + 2 used to predict population 3, etc.). The 18 populations vary in mean dry matter content from 14.7% to 21.4%, and involve four mango varieties, harvested from September 28, 2005 to January 25, 2006. (From Subedi, P., Walsh, K. B., and Owens, G., *Postharvest Biol. Technol.*, 43, 326, 2007. With permission.)

minimum acceptable standard of 11% SSC is suggested for stone fruit. In this case, farm management attention was then devoted to changing agronomic factors to improve fruit SSC. The on-line grader was useful for rapid assessment of fruit harvest from different agronomic treatments (Walsh et al. 2006).

The fresh fruit application is a very small application area, as a relative "johnny-come-lately" entrant. Also, while capable of separating fruit into two grades, the accuracy and precision of predictions in this application (e.g., $R^2 = 0.8$) is not nearly as impressive as that for some other application areas. For example, NDC (http://www.ndc.com/NDC/) has a long-established pedigree in this area, offering process line solutions based on interference filter based technology, primarily for applications involving determination of water and fat contents. Another notable, international group offering SWNIR and NIRS solutions for the food and agricultural industries is FOSS P/L (www.foss.com.dk). This group supplies scanning–grating and diode array technologies, backed up by neural network and PLSR calibration support.

ACKNOWLEDGMENT

The support of Christo Leibenberg is acknowledged, particularly with respect to Figures 3.3.3 through 3.3.12.

REFERENCES

Abe, H., T. Kusama, S. Kawano, and M. Iwamoto. 1995. Analysis of hydrogen bonds in water using spectrum decomposition technique applied to near-infrared absorption spectra. *Journal of the Spectroscopical Society of Japan* 44: 247–253.

Baker, D. 1983. The determination of fiber in processed cereal foods by near-infrared reflectance spectroscopy. *Cereal Chemistry* 60: 217–219.

Barton, F.E., II, J.A. de Haseth, and D.S. Himmelsbach. 2006. The use of two-dimensional correlation spectroscopy to characterize instrumental differences. *Journal of Molecular Structure* 799: 221–225.

Bellon, V., J.L. Vigneau, and M. Leclercq. 1993. Feasibility and performances of a new, multiplexed, fast and low-cost fiber-optic NIR spectrometer for the on-line measurement of sugar in fruits. *Applied Spectroscopy* 47: 1079–1083.

Bernath, P.F. 2002. The spectroscopy of water vapour: Experiment, theory and applications. *Physical Chemistry Chemical Physics* 4: 1501–1509.

Birth, G.S. and K.H. Norris. 1958. An instrument using light transmittance for nondestructive measurement of fruit maturity. *Eighteenth Annual Meeting of the Institute of Food Technologists*. Chicago, IL, pp. 592–595.

Chaplin, M. 2007. Water Structure and Science. Available at www.lsbu.ac.uk/water/vibrat.html, December 19, 2007.

Chauchard, F., R. Cogdill, S. Roussel, J.M. Roger, and V. Bellon-Maurel. 2004. Application of LS-SVM to non-linear phenomena in NIR spectroscopy: Development of a robust and portable sensor for acidity prediction in grapes. *Chemometrics and Intelligent Laboratory Systems* 71: 141–150.

Chen, H., B.P. Marks, and T.J. Siebenmorgen. 1997. Quantifying surface lipid content of milled rice via visible/near-infrared spectroscopy. *Cereal Chemistry* 74: 826–831.

Coomans, D. 2006. Adaptive wavelet modeling of NIR data. *12th Australian Near Infrared Spectroscopy Conference: NIR a Fruitful Science*. Rockhampton, Australia, p. 41.

Delwiche, S.R. 1993. Measurement of single-kernel wheat hardness using near-infrared transmittance. *Transactions of the American Society of Agricultural Engineers* 36: 1431–1437.

Delwiche, S.R., M.M. Bean, R.E. Miller, B.D. Webb, and P.C. Williams. 1995. Apparent amylose content of milled rice by near-infrared reflectance spectrophotometry. *Cereal Chemistry* 72: 182–187.

Delwiche, S.R. and D.R. Massie. 1996. Classification of wheat by visible and near-infrared reflectance from single kernels. *Cereal Chemistry* 73: 399–405.

Delwiche, S.R., K.S. McKenzie, and B.D. Webb. 1996. Quality characteristics in rice by near-infrared reflectance analysis of whole-grain milled sample. *Cereal Chemistry* 73: 257–263.

Delwiche, S.R. 1998. Protein content of single kernels of wheat by near-infrared reflectance spectroscopy. *Journal of Cereal Science* 27: 241–254.

Delwiche, S.R., R.O. Pierce, O.K. Chung, and B.F.W. Seabourn. 1998. Protein content of wheat by near-infrared spectroscopy of whole grain: Collaborative study. *Journal of Association of Official Analytical Chemists International* 81: 587–603.

Dull, G., R.G. Lefler, and G.S. Birth. 1992. *Nondestructive Measurement of Soluble Solids in Fruits Having a Rind or Skin*. U.S. Patent 5,089,701.

Fearn, T. 1996. Comparing standard deviations. *NIR News* 7: 5–6.

Golding, J.B., S. Satyan, W.B. McGlasson, C. Liebenberg, and K.B. Walsh. 2006. Application of portable NIR for measuring soluble solids concentration in peaches. *Acta Horticulturae* 713: 461–464.

Golic, M., K.B. Walsh, and P. Lawson. 2003. Short-wavelength near-infra-red spectra of sucrose, glucose and fructose with respect to sugar concentration and temperature. *Applied Spectroscopy* 57: 139–145.

Golic, M. and K.B. Walsh. 2006. Robustness of calibration models based on near infrared spectroscopy to the in-line grading of stone fruit for total soluble solids. *Analytica Chimica Acta* 555: 286–291.

Greensill, C.V. and K.B. Walsh. 2000a. Optimisation of instrumentation precision and wavelength resolution for the performance of NIR calibrations of sucrose in a water-cellulose matrix. *Journal of Applied Spectroscopy* 54: 1–13.

Greensill, C.V. and K.B. Walsh. 2000b. A remote acceptance probe and illumination configuration for spectral assessment of internal attributes of intact fruit. *Measurement Science and Technology* 11: 1674–1684.

Greensill, C.V. and K.B. Walsh. 2001. Calibration transfer between miniature PDA-based spectrometers in the NIR assessment of mandarin soluble solids content. *Journal of Near Infrared Spectroscopy* 10: 27–35.

Guthrie, J. and K.B. Walsh. 1997. Non-invasive assessment of pineapple and mango fruit quality using near infra-red spectroscopy. *Australian Journal of Experimental Agriculture* 37: 253–263.

Guthrie, J.A., D. Reid, and K.B. Walsh. 2005a. Assessment of internal quality attributes of mandarin fruit: 1. NIR calibration model development. *Australian Journal of Agricultural Research* 56: 405–416.

Guthrie, J.A., D. Reid, and K.B. Walsh. 2005b. Assessment of internal quality attributes of mandarin fruit: 2. NIR calibration model robustness. *Australian Journal of Agricultural Research* 56: 417–426.

Guthrie, J.A., K.B. Walsh, and C. Liebenberg. 2005c. NIR model development and robustness in prediction of melon fruit total soluble solids. *Australian Journal of Agricultural Research* 5: 411–418.

Hashimoto, H., N. Taniguchi, M. Tanaka, and Y. Nishiyama. 2002. *Device for Evaluating Internal Quality of Vegetable or Fruit, Method for Warm-Up Operation of the Device, and Method for Measuring Internal Quality*. U.S. Patent 6754600, PCT/JP01/03196.

Hosaka, Y. 1987. Evaluation of taste of rice by means of near infrared. *Proceedings, International Symposium on Agricultural Mechanization and International Cooperation in High Technology Era*. Tokyo, Japan, pp. 357–360.

Hruschka, W. 2001. Data analysis: Wavelength selection methods, Chapter 3. In Williams, P. and Norris, K. (Eds.), *Near-Infrared Technology in the Agricultural and Food Industries*, 2nd ed. American Association of Cereal Chemists, St. Paul, MN.

Kang, N., S. Kasemsumran, Y.-A. Woo, H.-Y. Kim, and Y. Ozaki. 2005. Optimization of informative spectral regions for the quantification of cholesterol, glucose and urea in control serum solutions using searching combination moving window partial least squares regression method with near infrared spectroscopy. *Chemometrics and Intelligent Laboratory Systems* 82: 90–96.

Kawano, S., H. Watanabe, and M. Iwamoto. 1992. Determination of sugar content in intact peaches by near infrared spectroscopy with fiber optics in interactance mode. *Journal of the Japanese Society of Horticultural Science* 61: 445–451.

Kawano, S. 2002. Application to agricultural products and foodstuffs, Chapter 12. In Siesler, H.W., Ozaki, Y., Kawata, S., and Heise, H.M. (Eds.), *Near-Infrared Spectroscopy*, Wiley-VCH, Morlenbach, Germany.

Kays, S.E., W.R. Windham, and F.E. Barton, II. 1996. Prediction of total dietary fiber in cereal products using near-infrared reflectance spectroscopy. *Journal of Agriculture and Food Chemistry* 44: 2266–2271.

Kays, S.E., F.E. Barton, II, W.R. Windham, and D.S. Himmelsbach. 1997. Prediction of total dietary fiber by near-infrared reflectance spectroscopy in cereal products containing high sugar and crystalline sugar. *Journal of Agriculture and Food Chemistry* 45: 3944–3951.

Kobayashi, K., K. Iizuka, T. Okada, and H. Hashimoto. 1990. Determination of chemical compositions of soy sauce by near infrared spectroscopy. In Iwamoto, M. and Kawano, S. (Eds.), *Proceedings of the 2nd International NIRS Conference*. Korin Publishing Co., Ltd., Tokyo, Japan, pp. 178–189.

Li, W.S. and J.T. Shaw. 1997. Determining the fat acidity of rough rice by near-infrared reflectance spectroscopy. *Cereal Chemistry* 74: 556–560.

Long, R., K.B. Walsh, D. Midmore, and G. Rogers. 2002. NIR estimation of rockmelon (*Cucumis melo*) fruit TDS, in relation to tissue inhomogeneity. *Acta Horticulturae* 588: 357–361.

Long, R., K.B. Walsh, and C.V. Greensill. 2005. Sugar "imaging" of fruit using a low cost CCD camera? *Journal of Near Infrared Spectroscopy* 13: 177–186.

Maeda, H., Y. Ozaki, M. Tanaka, N. Hayashi, and T. Kojima. 1995. Near infrared spectroscopy and chemometrics studies of temperature-dependent spectral variations of water: Relationship between spectral changes and hydrogen bonds. *Journal of Near Infrared Spectroscopy* 3: 191–201.

Mark, H. and J. Workman. 2003. *Statistics in Spectroscopy*, 2nd ed. Academic Press, San Diego, CA.

McCabe, W., S. Subramanian, and H.F. Fisher. 1970. A near infra-red spectroscopic investigation of the effect of temperature on the structure of water. *The Journal of Chemical Physics* 74: 4360–4369.

McClure, F. 2003. Review: 204 years of near infrared technology: 1800–2003. *Journal of Near Infrared Spectroscopy* 11: 487–518.

McGlone, V.A., D.G. Fraser, R.B. Jordan, and R. Kunnemeyer. 2003. Internal quality assessment of mandarin fruit by vis/NIR spectroscopy. *Journal of Near Infrared Spectroscopy* 11: 323–332.

Miller, W.M. and M. Zude. 2004. NIR-based sensing to identify soluble solids content of Florida citrus. *Applied Engineering in Agriculture* 20: 321–327.

Morgan, J.E. and P.C. Williams. 1995. Starch damage in wheat flours: A comparison of enzymatic, iodometric, and near-infrared reflectance techniques. *Cereal Chemistry* 72: 209–212.

Norris, K.H., W.R. Hruschka, M.M. Bean, and D.C. Slaughter. 1989. A definition of wheat hardness using near infrared reflectance spectroscopy. *Cereal Foods World* 34: 696–705.

Oriel Instruments. 2003. *Mirrors Technical Discussion*, Vol. 1, Fernandes, N. (Ed.). Oriel Instruments, Stratford, CT, pp. 10–12.

Osborne, B.G. and S. Douglas. 1981. Measurement of the degree of starch damage in flour by near infrared reflectance analysis. *Journal of the Science of Food and Agriculture* 32: 328–332.

Osborne, B.G., S. Douglas, and T. Fearn. 1982. The application of near infrared reflectance analysis to rapid flour testing. *Journal of Food Technology* 17: 355–363.

Osborne, B.G. 1983. Measurement of levels of bread improvers in concentrates by means of near infrared reflectance spectroscopy. *Journal of the Science of Food and Agriculture* 34: 1297–1301.

Osborne, B.G., T. Fearn, and P.G. Randall. 1983. Measurement of fat and sucrose in dry cake mixer by near infrared reflectance spectroscopy. *Journal of Food Technology* 18: 651–656.

Osborne, B.G. 1984. Investigations into the use of near infrared reflectance spectroscopy for the quality assessment of wheat with respect to its potential for bread baking. *Journal of the Science of Food and Agriculture* 35: 106–110.

Osborne, B.G., T. Fearn, A.R. Miller, and S. Douglas. 1984a. Application of near infrared reflectance spectroscopy to the compositional analysis of biscuits and biscuit doughs. *Journal of the Science of Food and Agriculture* 35: 99–105.

Osborne, B.G., G.M. Barrett, S.P. Cauvain, and T. Fearn. 1984b. The determination of protein, fat, and moisture in bread by near infrared reflectance spectroscopy. *Journal of the Science of Food and Agriculture* 35: 940–945.

Ozaki, Y., S. Sasic, and J.H. Jiang. 2001. How can we unravel complicated near infrared spectra?—Recent progress in spectral analysis methods for resolution enhancement and band assignments in the near infrared region. *Journal of Near Infrared Spectroscopy* 9: 63–95.

Peiris, K.H.S., G.G. Dull, R.G. Leffler, and S.J. Kays. 1998. Near infra-red spectrometric technique for nondestructive determination of soluble solids content in processing tomatoes. *Journal of the American Society for Horticultural Science* 123: 1089–1093.

Reeves, J.B., III. 1994. Effects of water on the spectra of model compounds in the short-wavelength near infrared spectral region (14,000–9091 cm^{-1} or 714–1100 nm). *Journal of Near Infrared Spectroscopy* 2: 199–212.

Rei, M., C. Tsunenori, and O. Atsushi. 1994. *Measuring Apparatus for Interior Quality of Vegetable and Fruit by Transmission Method.* Japan Patent 6300680.

Rittiron, R., S. Saranwong, and S. Kawano. 2004. Useful tips for constructing a near infrared-based quality sorting system for single brown-rice kernels. *Journal of Near Infrared Spectroscopy* 12: 133–139.

Rittiron, R., S. Saranwong, and S. Kawano. 2005. Detection of variety contamination in milled Japanese rice using single kernel near infrared technique in transmittance mode. *Journal of Near Infrared Spectroscopy* 13: 19–25.

Roger, J.-M., F. Chauchard, and V. Bellon-Maurel. 2003. EPO–PLS external parameter orthogonalisation of PLS application to temperature-independent measurement of sugar content of intact fruits. *Chemometrics and Intelligent Laboratory Systems* 66: 191–204.

Rossi, F., A. Lendasse, D. François, V. Wertz, and M. Verleysen. 2006. Mutual information for the selection of relevant variables in spectrometric nonlinear modeling. *Chemometrics and Intelligent Laboratory Systems* 80: 215–226.

Saranwong, S., J. Sornsrivichai, and S. Kawano. 2003. On-tree evaluation of harvesting quality of mango fruit using a hand-held NIR instrument. *Journal of Near Infrared Spectroscopy* 11: 283–293.

Segtnan, V.H., S. Sasic, T. Isaksson, and Y. Osaki. 2001. Studies on the structure of water using two-dimensional near-infrared correlation spectroscopy and principal component analysis. *Analytical Chemistry* 73: 3153–3161.

Starr, C., D.B. Smith, J.A. Blackman, and A.A. Gill. 1983. Application of near infrared reflectance analysis in breeding wheats for bread-making quality. *Analytical Proceedings* 20: 72–74.

Subedi, P., K.B. Walsh, and G. Owens. 2007. Prediction of mango eating quality at harvest using short wave near infrared spectroscopy. *Postharvest Biology and Technology* 43: 326–334.

Temma, T., K. Hanamatsu, and F. Shinoki. 2002. Development of a portable near infrared sugar-measuring instrument. *Journal of Near Infrared Spectroscopy* 10: 77–83.

USDA. 1996. *Near-Infrared Transmittance (NIRT) Handbook.* Federal Grain Inspection Service, USDA Washington DC; also alvailable at http://archive.gipsa.usda.gov/reference-library/handbooks/nirt/nirt-h6.pdf.

Valero, C. and M. Ruiz-Altisent. 2000. Design guidelines for a quality assessment system of fresh fruits in fruit centres and hypermarkets. *Agricultural Engineering International* II: 1–26.

Villareal, C.P., N.M.D. Cruz, and B.O. Juliano. 1994. Rice amylose analysis by near-infrared transmittance spectroscopy. *Cereal Chemistry* 71: 292–296.

Walsh, K.B., J.A. Guthrie, and J. Burney. 2000. Application of commercially available, low-cost, miniaturised NIR spectrometers to the assessment of the sugar content of intact fruit. *Australian Journal of Plant Physiology* 2: 1175–1186.

Walsh, K.B., M. Golic, and C.V. Greensill. 2004. Sorting of fruit and vegetables using near infrared spectroscopy: Application to soluble solids and dry matter content. *Journal of Near Infrared Spectroscopy* 12: 141–148.

Walsh, K.B., R. Long, and S. Middleton. 2006. Use of near infra-red spectroscopy in evaluation of source-sink manipulation to increase stonefruit soluble sugar content. *Journal of Horticultural Science and Biotechnology* 82: 316–322.

Williams, P.C., K.H. Norris, and D.C. Sobering. 1985. Determination of protein and moisture in wheat and barley by near-infrared transmission. *Journal of Agriculture and Food Chemistry* 33: 239–244.

Williams, P.C. 1991. Prediction of wheat kernel texture in whole grains by near-infrared transmittance. *Cereal Chemistry* 68: 112–114.

Williams, P.C. and D.C. Sobering. 1993. Comparison of commercial near infrared transmittance and reflectance instruments for analysis of whole grains and seeds. *Journal of Near Infrared Spectroscopy* 1: 25–32.

Williams, P. and K. Norris (Eds.). 2001a. *Near-Infrared Technology in the Agricultural and Food Industries,* 2nd ed. American Association of Cereal Chemists, St. Paul, MN.

Williams, P. and K. Norris. 2001b. Variables affecting near-infrared spectroscopic analysis, Chapter 9. In Williams, P. and Norris, K. (Eds.), *Near-Infrared Technology in the Agricultural and Food Industries,* 2nd ed. American Association of Cereal Chemists, St. Paul, MN.

Windham, W.R., B.G. Lyon, E.T. Champagne, F.E. Barton, II, B.D. Webb, A.M. McClung, K.A. Moldenhauer, S. Linscombe, K.S. McKenzie, and K.S. McKenzie. 1997. Prediction of cooked rice texture quality using near-infrared reflectance analysis of whole-grain milled samples. *Cereal Chemistry* 74: 626–632.

Zude, M. 2006. *Manual Pigment Analyzer PA1101,* Vol. 4. Control in Applied Physiology, Falkensee, Germany, 45pp.

Zude, M., I.P. Birlouez-Aragon, P.J. Paschold, and D.N. Rutledge. 2007. Nondestructive spectral-optical sensing of carrot quality during storage. *Postharvest Biology and Technology* 45: 30–37.

Zude, M., M. Pflanz, C. Kaprielian, and B.L. Aivazian. 2008. NIRS as a tool for precision horticulture in the citrus industry. *Biosystems Engineering,* 99(3): 455–459.

3.4 NETWORK OF NIRS INSTRUMENTS

PETER TILLMANN

3.4.1 Need for Calibration Transfer

Any NIR spectroscopist will one day be confronted with the question of calibration transfer. It might be a colleague asking him to share his superior calibration, it might be he has to exchange his instruments with another because of replacement or new technological developments. In rare circumstances, there might be less serious reasons to think about a calibration transfer like changes in the instrument that affect the spectral output. Almost always a direct transfer would in the first case give deviating results between the original instrument and the instrument the calibration model was transferred to. Of course, the calibration development can start from scratch but then the NIR spectroscopist will ask for alternatives, because NIR calibrations usually consist of a large number of samples, with several parameters for any sample, which were collected over a long period of time (especially for agricultural and food applications). On the other hand most calibration samples might not have been stored or belong to perishable products that are not available anymore.

The aim of all networking is to use a single calibration model on several instruments. For calibration transfer, the instrument on which the calibration model was originally developed is called the "master instrument," while the instruments on which the calibration model shall be used on are named "slave," "satellites," or "host instruments" (Figure 3.4.1). The important thing to understand is that the spectra of the calibration samples contain some information specific to the master instrument that causes trouble when doing a direct calibration transfer.

The idea of reverse standardization is to transfer the spectra of a calibration set to a new instrument instead of transferring the calibration model. Reverse standardization will always result in developing calibrations individually for the new instruments. They are no slaves in a network, each instrument will run individually afterward.

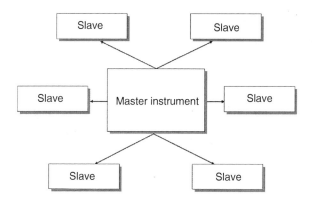

FIGURE 3.4.1 Principle of calibration transfer from the master to the slave instruments.

However, some obstacles for direct calibration transfer were first identified in practical applications. When transferring calibration models directly from one instrument to another typically the NIRS predictions from master and slave deviate. This might be as little as a bias, or a bias and slope or more serious deviations.

These deviations in the NIRS predictions can be traced back to the spectra, because all differences between predictions will not be introduced by the calibration model but by the spectra. There are four reasons for variation between spectra from two instruments that might be considered:

1. Differences in photometric response (accuracy and precision of intensities)
2. Differences in wavelength accuracy and precision
3. Differences in resolution (number and distribution of data points in spectral range)
4. Differences in the scanning range of the instruments

When a calibration transfer is been done between instruments of the same type (two individual instruments of one manufacturer with identical specifications), you would expect less problems/differences than in the case of two instruments of different manufacturers with different optical systems (i.e., diode array vs. FT instruments). In general, it is not useful to transfer calibrations between two instruments with no common scanning range. Most if not all calibration transfer is done on a common wavelength/wave number range between two instruments.

The calibration transfer can only handle differences between instruments. Any difference from the samples (temperature), from sample preparation (grinding, no grinding), or from the measurement conditions (reflexion vs. transmission) affect the absorption coefficients and cannot be treated by calibration transfer.

A strong criterion for a successful calibration transfer is that the differences between instruments are not larger than the differences between repeated measurements on a single instrument. This would result in the reproducibility being not larger than the repeatability. For practical reasons, a calibration transfer is successful, when the reproducibility for the NIRS analysis (across instruments) is equal to or better than the reproducibility of the reference method (across laboratories).

3.4.2 Strategies to Remedy the Situation

Calibration transfer came to mind in the early 1980s. At that time, some scientists wanted to exchange calibrations between laboratories and wondered how to avoid the differences in prediction. Generally, the NIRS instrument records a spectrum of a sample. Together with a calibration model this spectrum is subsequently used to calculate the NIR prediction values under question (Figure 3.4.2). When two spectra recorded on two instruments but from one sample result in different NIRS predictions, the corrections can be made at the following three levels:

1. At the predicted results
2. At the spectra
3. At the calibration model

VIS/NIR Spectroscopy

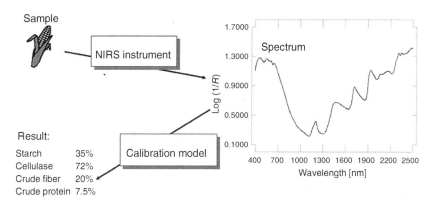

FIGURE 3.4.2 Dependent and independent variables, the latter predicted by means of the calibration model.

3.4.2.1 On the Level of the Predicted Results

The idea is to correct the NIRS predicted results, because the results recorded on the slave do not fit to those from the master. Depending on the observed deviations several proposals were made.

3.4.2.1.1 Bias Correction
Osborne and Fearn (1983) proposed the correction of the bias between NIRS predictions of different instruments. Therefore, they proposed to scan a few samples on both instruments (master and slave), calculate the systematic difference between them, and use this bias to correct the NIRS prediction of any unknown sample later (Figure 3.4.3).

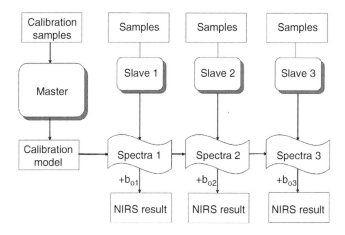

FIGURE 3.4.3 Networking of NIRS instruments using bias correction on the prediction level; b is the systematic difference between master and slave.

3.4.2.1.2 Bias/Slope Correction

Jones and coworkers (1993) proposed the correction of bias and slope. They proceeded according to the proposal by Osborne and Fearn, but would estimate and correct for a bias and a slope between the two predictions on the master and slave instruments. For a given number of samples, it is obvious that the degree of freedom to estimate bias and slope is less than predicting the bias only. Therefore, more samples are needed for the estimation of bias and slope compared to the pure bias estimation. Furthermore, the estimation of slope demands well-chosen samples with especially a good spread in the distribution of the constituent of interest.

3.4.2.2 On the Spectral Level

Because the differences between NIRS predictions must be caused by the spectra, the idea of spectral correction is straightforward. All proposals are captured by the term "standardization of instruments" or "spectral standardization" (Figure 3.4.4). Again, depending on the observed deviations between the spectra of two instruments several proposals have been made.

3.4.2.2.1 Differences Come from Differences in Photometric Response Only

If the differences between two instruments are solely due to a systematic offset in the spectra, then these might be corrected by this offset. Therefore, it would be enough to scan a single sample, calculate the difference, and correct future spectra. This proposal was made by Shenk and Westerhaus (1991) for instruments of the same type (two individual instruments of one manufacturer with identical specifications). This approach gives numerically identical results to bias correction (Osborne and Fearn 1983) when calculated on the identical sample.

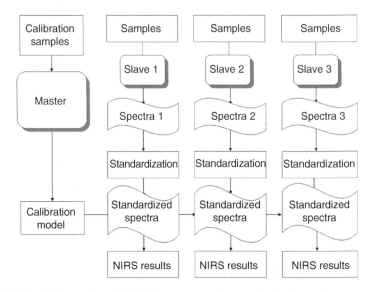

FIGURE 3.4.4 Networking of NIRS instruments using standardization of spectra.

A special version of this proposal is the assumption that the offset between spectra might not be constant at different absorption levels. For this case, Wüst and Rudzik (1990) proposed the least square fit (LSF) formula to correct a linearly increasing offset, depending on the absorption level. The correction function has to be determined by more than one sample and the same restrictions as above (Jones et al. 1993) apply. It is clear that this correction could also follow a quadratic or cubic function.

3.4.2.2.2 Differences Arise from Differences in Photometric Response and Wavelength Accuracy

For a more general solution, Shenk and Westerhaus (1989) and Shenk (1990) proposed an algorithm to determine a correction in absorption intensity and on the wavelength scale. Thereby also differences on the wavelength scale could be corrected by a window around a single wavelength. The correction on the wavelength scale is calculated as a quadratic function, where the correction on the wavelength scale is a linear function. For this algorithm, 30 samples have been proposed.

Wang and coworkers (1991) presented two proposals for the standardization of instruments, which is named direct standardization (DS) and a reduced version is named piecewise direct standardization (PDS). In the DS algorithm, the transfer function is calculated for any single wavelength from the slave instrument to each wavelength of the master instrument, while the PDS algorithm uses a predefined window on the master instrument only. In both approaches, the transfer function is calculated in a multivariate way so that nonlinear intensity changes for each wavelength can be corrected. The PDS algorithm is very flexible with respect to window size, etc. and therefore universal and powerful.

3.4.2.2.3 Standardization after Preprocessing of Spectra

Forina and coworkers (1992, 1995) proposed a two-block PLS algorithm for the spectrum transfer step. It is similar to the DS algorithm, but it preprocesses and reduces continuous monochromator spectra before calculation of the transfer function. Other proposals standardize spectra after transformation into the time or wavelet domain before standardization (e.g., Walczak et al. 1997). These approaches have not received special interest, because the advantage of transferring spectra to the time or wavelet domain (e.g., ease of filtering and storage space) is not apparent in practical applications.

3.4.2.3 In the Calibration Model

3.4.2.3.1 Creating Robust Calibration Models by Capturing All Possible Variances

The third possibility to apply corrections to minimize differences between NIRS predictions is the calibration model itself. With this approach, the calibration function shall be made less sensitive to changes in the spectra originating from the instruments and not the samples. The first choice would be to include spectra from all instruments into the calibration set (Figure 3.4.5). Thereby assuming that the calibration model will de-emphasize differences in the spectra coming from the instruments (Shenk 1992).

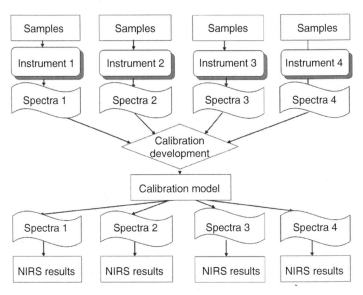

FIGURE 3.4.5 Networking of NIRS instruments using spectra from different instruments.

An alternative approach would be to include spectra from some samples collected on all instruments as the so-called stabilization samples. This concept is going back to Karl Norris. For this approach, identical samples would be scanned on different instruments and be included in the calibration sample set. A third alternative is the so-called repeatability file (Westerhaus 1990). Here some samples are scanned on all instruments and collected without the need for reference values in a repeatability file. This is included during calibration development (Figure 3.4.6).

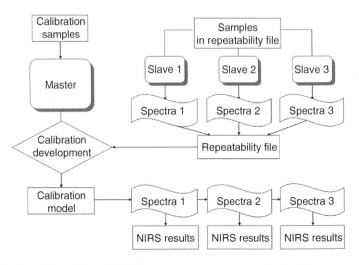

FIGURE 3.4.6 Networking of NIRS instruments using a repeatability file.

Robustness of the calibration can also be approached by selecting wavelengths with respect to their insensitivity toward wavelengths shifts appearing in different instruments (Mark and Workman 1988). Up to now the approach was designed for MLR techniques and is hardly transferable to PLS calibration models.

Another proposal for PLS calibrations is named PCA standardization (based on Roger et al. 2003; Zeaiter et al. 2005) using principle component analysis (PCA) on the spectra of identical samples collected from several instruments. It might be expected that the variation in the first PCA factor captures the variance caused by the instrument, which is subsequently removed from the calibration dataset.

For all proposals instead of all instruments, a subgroup of instruments can be used as long as the variation among instruments is represented in this subgroup. This has to be cross-checked during calibration development.

3.4.2.4 Advantages and Disadvantages

3.4.2.4.1 Bias–Slope Correction

A huge advantage of the approaches shortly described above is the feasibility. It can always be performed outside the instrument software, when the NIRS predicted results are exported. The huge disadvantage is that with any new parameter and any new calibration function bias and slope have to be recalculated for every instrument. In case of changes in the calibration this can become laborious. In addition any statistical test on the spectra (X residuals, Mahalanobis distance, etc.) will not be done correctly, because all corrections are done on the predicted NIRS results only.

3.4.2.4.2 Standardization

The standardization has to be calculated only once for any instrument and product as long as the instrument is stable over time. The disadvantage is that for any new instrument to be included into the network the standardization procedure has to be redone.

The choice of samples for standardization or bias and slope correction is crucial. Each sample or at least each product yields specific spectra. The most obvious difference is the absorption intensity and the spectral pattern. If it can be assumed that the response of the detector is linear at all absorption levels, then any single sample would be appropriate for the standardization or the calculation of the bias. Since this is usually not the case, the requirement is to use a sample of similar absorption or even better with the ideal absorption characteristics of the sample under question. This leads to choosing one or several representative samples (Dardenne et al. 1992).

3.4.2.4.3 Robust Calibrations

This approach requires to recalibrate at least once or to anticipate the need for networking during calibration development from the beginning on. But if the full diversity of variation between instruments has been collected, any new instrument will automatically be covered by the robust calibration.

The approach with a repeatability file will not yield perfect results for statistical test on the spectra, because the information in the repeatability file is not included in the calculation of the data for the test.

It should be clear to the reader that any of the above strategies can be combined. In a real setup, for example, the use of a robust calibration can be supported by an additional bias correction if needed. The choice of the given procedures is mainly limited by the instrument software, when running real-time applications. Only few software packages support the use for several of the above approaches. This limitation will typically give only the choice between bias correction and may be one of the standardization protocols (Shenk and Westerhaus 1989; Shenk 1990).

3.4.3 Examples

3.4.3.1 Grain Analysis Based on Filter Instrument

In the 1980s, grain analysis with filter instruments was done on the ground samples. Several thousand instruments were sold and run with a uniform calibration. Differences between the individual instruments were checked and corrected by the local users applying a bias correction.

3.4.3.2 FOSS Infratec Grain Network

In the whole grain network of FOSS Infratec instruments, spectra of several instruments were included during calibration development, thereby leading to a robust calibration with respect to variations between instruments. The calibration model for wheat, barley, rye, and triticale is based on several thousands of spectra collected on several instruments around the world. The calibration technique of artificial neural nets (ANN) was used, resulting in a very robust calibration based on this huge data set. Typically, an SD between instruments for protein and moisture analysis in wheat of 0.14% and 0.10%, respectively, is achieved (Figure 3.4.7). In barley, the precision

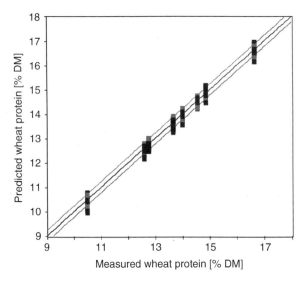

FIGURE 3.4.7 NIR predictions in an Infratec grain network of approximately 100 instruments for protein in wheat [% DM]. Lines show 95% prediction interval.

TABLE 3.4.1
Results of a Network for Oil Analysis in Rapeseed [% DM]

Sample	2701	2702	2703	2704	2705
Number of instruments	12	12	12	12	12
Mean in network	51.99	48.07	46.10	43.25	46.64
SD in network	0.76	0.73	0.51	0.52	0.41
Minimum in network	51.20	47.04	45.42	42.42	46.15
Maximum in network	53.23	48.78	46.90	44.32	47.45
Maximum difference between instruments in network	2.03	1.74	1.48	1.90	1.30
Reference value	51.44	48.24	46.34	42.42	46.38
Deviation of network mean from reference value	0.55	−0.17	−0.24	0.83	0.26

for protein and moisture analysis is slightly decreased owing to the hull content of barley resulting in 0.16% and 0.11%, respectively. This precision is in practical terms equal to that of the reference methods by means of Kjeldahl and drying oven, respectively.

3.4.3.3 VDLUFA Forage Maize Network

Based on the proposals of Shenk and Westerhaus (1989, 1991), several networks for forages and whole rapeseed analysis have been set up in the 1990s. In all cases except one, the instruments in the network are from one manufacturer. The networking is done using robust calibration with a repeatability file (Westerhaus 1990) and spectral standardization with one to six samples (Shenk and Westerhaus 1989; Shenk 1990).

The reproducibility across instruments for oil analysis using NIRS instruments in a network (Table 3.4.1) is 0.5%–0.7%, calculated as reproducibility SD. This compares to 0.6% for the Soxhlet and NMR determination of oil in oilseeds (ISO 659; ISO 10565). For glucosinolates (Table 3.4.2), the reproducibility is 1.0–1.7 μmol compared to 2.5 μmol for HPLC analysis (ISO 9167-1), calculated as reproducibility SD.

TABLE 3.4.2
Results of a Network for Glucosinolates Analysis in Rapeseed [μmol g^{-1} DM]

Sample	2701	2702	2703	2704	2705
Number of instruments	12	12	12	12	12
Mean in network	5.97	17.78	11.17	13.76	13.36
SD in network	0.72	1.70	1.11	1.28	1.14
Minimum in network	4.82	15.45	9.84	11.81	11.09
Maximum in network	7.43	20.25	12.93	15.31	15.47
Maximum difference between instruments in network	2.61	4.80	3.09	3.50	4.38
Reference value	7.13	19.17	10.40	13.42	14.31
Deviation of network mean from reference value	−1.16	−1.39	0.77	0.34	−0.95

REFERENCES

Dardenne, P., R. Biston, and G. Sinnaeve. 1992. Calibration transferability across NIR instruments. In: *Near Infra-Red Spectroscopy: Bridging the Gap between Data Analysis and NIR Applications*, eds. K. Hildrum, T. Isaksson, T. Naes, and A. Tandberg, pp. 453–458. Horwood, Chichester, United Kingdom.

Forina, M., C. Armanino, and R. Giangiacomo. 1992. A case study on the transfer of the calibration equation in NIRS. In: *Near Infra-red Spectroscopy: Bridging the Gap between Data Analysis and NIR Applications*, eds. K. Hildrum, T. Isaksson, T. Naes, and A. Tandberg, pp. 91–96. Horwood, Chichester, United Kingdom.

Forina, M., G. Darva, C. Armanino. 1995. Transfer of calibration function in near-infrared spectroscopy. *Chemometrics Intelligent Laboratory System* 27: 189–203.

ISO 659. 1996(E). *Oilseeds—Determination of Hexane Extract (or Light Petroleum Extract), Called "Oil Content"*. International Standardization Organization, Geneva, Switzerland.

ISO 9167-1. 1992(E). *Rapeseed—Determination of Glucosinolate Content, Part 1: Method Using High-Performance Liquid Chromatography*. International Standardization Organization, Geneva, Switzerland.

ISO 10565. 1998. *Oilseeds—Simultaneous Determination of Oil and Water Contents—Method Using Pulsed Nuclear Magnetic Resonance Spectrometry*. International Standardization Organization, Geneva, Switzerland.

Jones, J., I. Last, B. MacDonald, and K. Prebble. 1993. Development and transferability of near-infrared methods for determination of moisture in a freeze-dried injection product. *Journal of Pharmaceutical and Biomedical Analysis* 11: 1227–1231.

Mark, H. and J. Workman. 1988. A new approach to generating transferable calibrations for quantitative near-infrared spectroscopy. *Spectroscopy* 3: 28–36.

Osborne, B. and T. Fearn. 1983. Collaborative evaluation of universal calibrations for the measurement of protein and moisture in flour by near infrared reflectance. *Journal of Food Technology* 18: 453–460.

Roger, J.-M., F. Chauchard, and V. Bellon-Maurel. 2003. EPO–PLS external parameter orthogonalisation of PLS application to temperature-independent measurement of sugar content of intact fruits. *Chemometrics and Intelligent Laboratory Systems* 66: 191–204.

Shenk, J.S. 1990. Standardizing NIRS instruments. In: *Proceeding Third International Conference Near Infrared Spectroscopy*, eds. R. Biston and N. Bartiaux-Thill, pp. 649–654. Agricultural Research Centre Publishing, Gembloux, Belgium.

Shenk, J.S. 1992. Networking and calibration transfer. In: *Making Light Work: Advances in Near Infrared Spectroscopy*, eds. I. Murray and I. Cowe, pp. 223–226. VCH, Weinheim, Germany.

Shenk, J.S. and M.O. Westerhaus. 1989. U.S. Patent No. 4,866,644, September 12.

Shenk, J.S. and M.O. Westerhaus. 1991. New standardization and calibration procedures for NIRS analytical systems. *Crop Science* 31: 1694–1696.

Walczak, B., E. Bouveresse, and D.L. Massart. 1997. Standardization of near-infrared spectra in the wavelet domain. *Chemometrics and Intelligent Laboratory Systems* 36: 41–51.

Wang, Y., D. Veltkamp, and B. Kowalsky. 1991. Multivariate instrument standardisation. *Analytical Chemistry* 63: 23.

Westerhaus, M. 1990. Improving repeatability of NIR calibrations across instruments. In: *Proceeding Third International Conference Near Infrared Spectroscopy*, eds. R. Biston and N. Bartiaux-Thill, pp. 671–674. Agricultural Research Centre Publishing, Gembloux, Belgium.

Wüst, E. and L. Rudzik. 1990. Possibilities and problems of a network of infrared spectrometers in the dairy industry. In: *Proceeding Third International Conference Near Infrared Spectroscopy*, eds. R. Biston and N. Bartiaux-Thill, pp. 679–688. Agricultural Research Centre Publishing, Gembloux, Belgium.

Zeaiter, M., J.M. Roger, and V. Bellon-Maurel. 2005. Dynamic orthogonal projection. A new method to maintain the on-line robustness of multivariate calibrations. Application to NIR-based monitoring of wine fermentations. *Chemometrics and Intelligent Laboratory Systems* 80: 227–235.

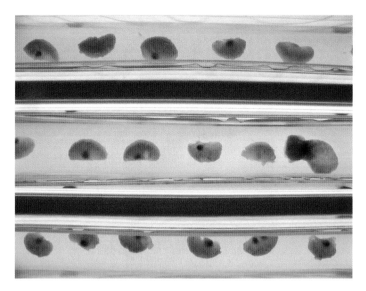

FIGURE 2.1.3 Back illumination of satsuma segments. This method provides a highly contrasted image that facilitates the detection of seeds and the analysis of size and shape.

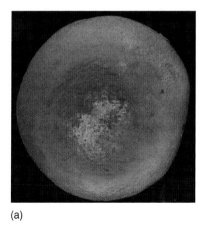

(a)

FIGURE 2.1.4 Images of the same fruit affected by green mould in visible, near-infrared, and ultraviolet bands.

(a)

FIGURE 2.1.8 Original color image decomposed into its individual components RGB and HSI.

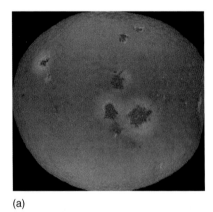

(a)

FIGURE 2.1.11 Successive steps of a region growing algorithm: from the original image to the segmented image.

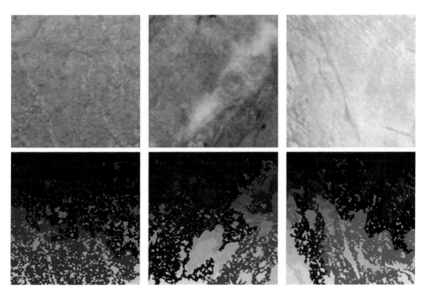

FIGURE 2.2.2 Beef, pork, and lamb image with their region primitives constructed (bottom).

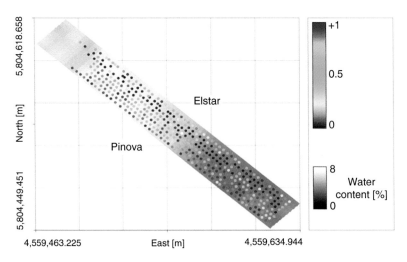

FIGURE 3.2.31 Mapping of soil–water content (gray scale) and the I_{ChlD} values (false color scale) measured nondestructively on September 13, 2007 in two cultivars, each four rows. (Zude and Gebbers, pers. communication, 2007.)

(A) (B)

FIGURE 4.2.1 Origin of the blue-green and red fluorescence of plant leaf tissues as viewed with a fluorescence microscope. (A) Fluorescence of the cross section of a green leaf of the C_4-plant maize (*Zea mays* L.). The emitted blue fluorescence is clearly seen in the cell walls of the epidermis (upper cell layer) and bundle sheet (central ring). Photo from Buschmann, C. and Lichtenthaler, H.K., *J. Plant Physiol.*, 1998. 152, 297, 1998. (B) Blue fluorescence of cell walls of an epidermis stripped off from a leaf of *Commelina communis* L. The chloroplast in the stomatal guard cells exhibits red chlorophyll fluorescence (From Lichtenthaler, H.K., Lang, M., Sowinska, M., Heisel, F., and Miehé, J.A., *J. Plant Physiol.*, 148, 599, 1996.) In both photos, fluorescence was excited in the UV (365 nm).

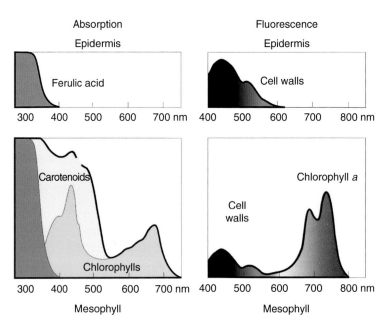

FIGURE 4.2.3 Example of absorption (left part) and UV-induced fluorescence emission (right part) in the chlorophyll-free epidermis cells (upper part) and in green mesophyll cells of a leaf (lower part). Cinnamic acids, predominantly ferulic acid, covalently bound to the cell walls, absorb in the UV, the yellow carotenoids (located in the chloroplasts) absorb in the blue, and chlorophylls absorb in the blue and in the red spectral region. The fluorescence emission spectra of leaves show a maximum in the blue (440–450 nm), a shoulder in the green (520 nm), as well as maxima in the red (690 nm) and the far-red (735 nm) spectral region. Thus, the in vivo fluorescence emission spectrum of a green leaf is essentially composed of the blue-green fluorescence of epidermis cells, and the red + far-red chlorophyll fluorescence of the green mesophyll cells.

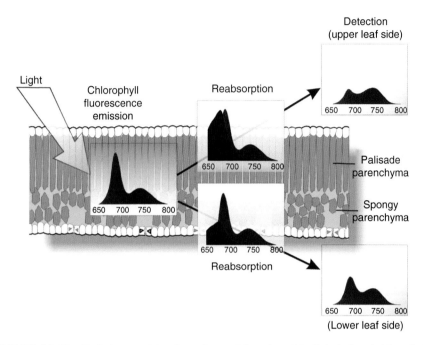

FIGURE 4.2.12 Emission, partial reabsorption, and detection of the light-induced chlorophyll fluorescence of green leaves. The scheme indicates the reabsorption primarily of the red chlorophyll fluorescence band $F690$ by the in vivo absorption bands of chlorophyll inside a leaf (shown here with a bifacial leaf structure). The chlorophyll fluorescence emission spectrum (red curve) overlaps with the chlorophyll absorption spectrum (black curve) (see also Figure 4.2.9). The amount of reabsorption depends on the chlorophyll density and is higher when detecting the chlorophyll fluorescence from the upper leaf side (high chlorophyll content in the densely packed palisade parenchyma cells) as compared to the lower leaf side (less chlorophyll in the more loosely arranged cells of the spongy parenchyma layer).

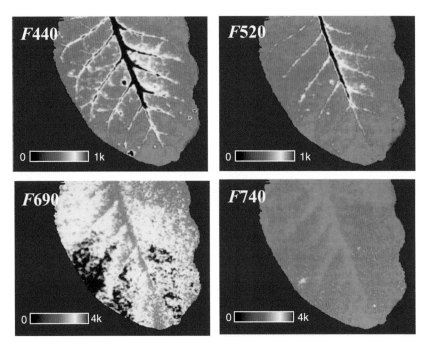

FIGURE 4.2.13 Fluorescence images of the upper leaf side of a green tobacco leaf. Shown here in false colors are the intensity of the blue ($F440$) and green ($F520$) fluorescence and the red ($F690$) + far-red ($F740$) chlorophyll fluorescence. In the images, the fluorescence yield increases from blue (no fluorescence) via green and yellow to red as the highest fluorescence. The highest blue and green fluorescence emanate from the leaf veins, whereas the highest chlorophyll fluorescence comes from the vein-free leaf regions. Note that the scales for blue and green fluorescence are different from those of the red + far-red chlorophyll fluorescence. k in the scales means kilo (=1000) counts. (Changed from Lang, M., Lichtenthaler, H.K., Sowinska, M., Summ, P., and Heisel, F., *Botanica Acta* 107, 230, 1994a; Lichtenthaler, H.K., Lang, M., Sowinska, M., Heisel, F., and Miehé, J.A., *J. Plant Physiol.*, 148, 599, 1996.)

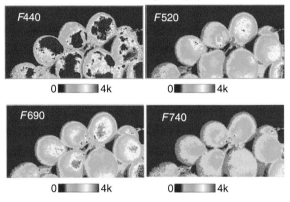
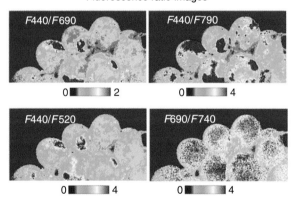

FIGURE 4.2.14 Fluorescence imaging of grapes (greenish variety) obtained from a local market. Upper part: images of blue ($F440$), green ($F520$), red ($F690$), and far-red ($F740$) fluorescence. Lower part: fluorescence ratio images blue/red, blue/far-red, blue/green, and red/far-red. The patchiness in the fluorescence ratio images of individual grapes is shown by spot-like increases in the blue fluorescence and in the fluorescence ratios blue/green, blue/red, and blue/far-red and indicates differential maturity and beginning degradation of the individual grapes. The fluorescence intensities in the images are shown in false colors; the fluorescence yield increases from blue (no fluorescence) via green and yellow to red as the highest fluorescence; k in the scales means kilo (=1000) counts. The scales in the fluorescence ratio images indicate the absolute ratio value. Note that the scales for the individual fluorescence signatures are different.

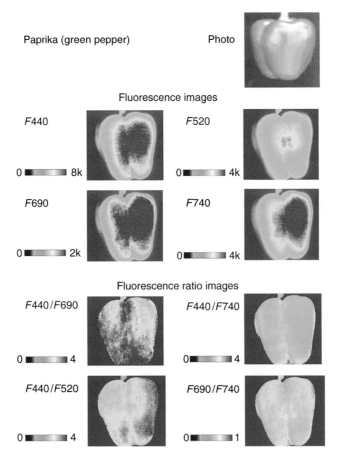

FIGURE 4.2.15 Fluorescence imaging of green bell peppers (*Capsicum annuum* L.) from a local market. Upper part: images of blue (*F*440), green (*F*520), red (*F*690), and far-red (*F*740) fluorescence. Lower part: fluorescence ratio images blue/red, blue/far-red, blue/green, and red/far-red. The fluorescence intensities are shown in the images in false colors; the fluorescence yield increases from blue (no fluorescence) via green and yellow to red as the highest fluorescence; k in the scales means kilo (=1000) counts. The scales in the fluorescence ratio images indicate the absolute ratio value. Note that the scales for the individual fluorescence signatures are different.

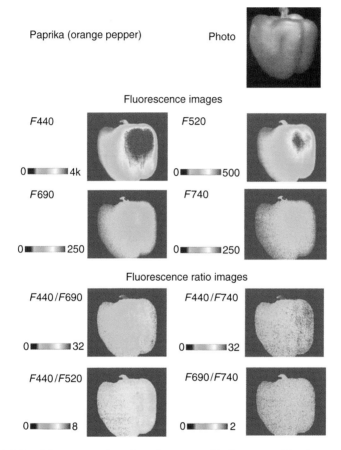

FIGURE 4.2.16 Fluorescence imaging of orange-red bell peppers (*Capsicum annuum* L.) from a local market. Upper part: images of blue (*F*440), green (*F*520), red (*F*690), and far-red (*F*740) fluorescence. Lower part: fluorescence ratio images blue/red, blue/far-red, blue/green, and red/far-red. The fluorescence intensities are shown in false colors in the images; the fluorescence yield increases from blue (no fluorescence) via green and yellow to red as the highest fluorescence; k in the scales means kilo (=1000) counts. The scales in the fluorescence ratio images indicate the absolute ratio value. Note that the scales for the individual fluorescence signatures are different.

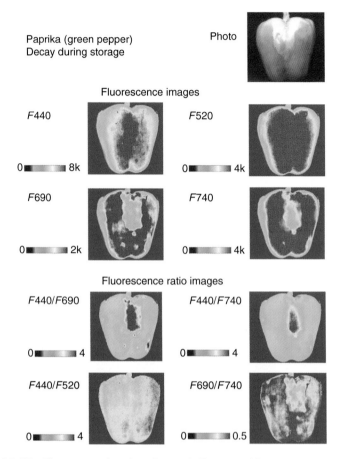

FIGURE 4.2.17 Fluorescence imaging of green bell peppers (*Capsicum annuum* L.) from a local market with decay symptoms that showed up after storage at room temperature. Upper part: images of blue ($F440$), green ($F520$), red ($F690$), and far-red ($F740$) fluorescence. Lower part: fluorescence ratio images blue/red, blue/far-red, blue/green, and red/far-red. The fluorescence images and ratio images change during decay (associated with partial chlorophyll breakdown). The fluorescence intensities are shown in false colors in the images; the fluorescence yield increases from blue (no fluorescence) via green and yellow to red as the highest fluorescence; k in the scales means kilo (=1000) counts. The scales in the fluorescence ratio images indicate the absolute ratio value. Note that the scales for the individual fluorescence signatures are different.

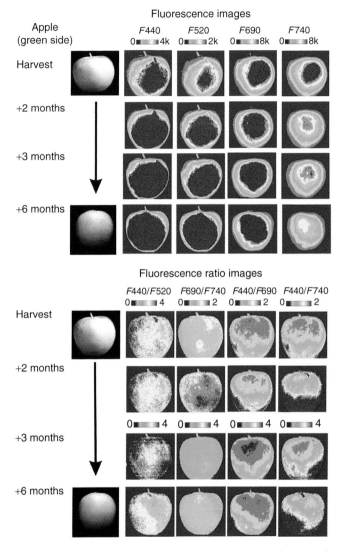

FIGURE 4.2.18 Fluorescence images of apples (*Malus x domestica* 'Braeburn') measured at room temperature after different times of storage at 4°C. Upper part: Shown are fluorescence images of blue (*F*440), green (*F*520), red (*F*690), and far-red (*F*740) fluorescence. Lower part: fluorescence ratio images blue/green (*F*440/*F*520), blue/red (*F*440/*F*690), blue/far-red (*F*440/*F*740), and red/far-red (*F*690/*F*740). The fluorescence intensities are shown in false colors in the images; the fluorescence yield increases from blue (no fluorescence) via green and yellow to red as the highest fluorescence; k in the scales means kilo (=1000) counts. The scales in the fluorescence ratio images indicate the absolute ratio value. Note that the scales for the individual fluorescence signatures are different.

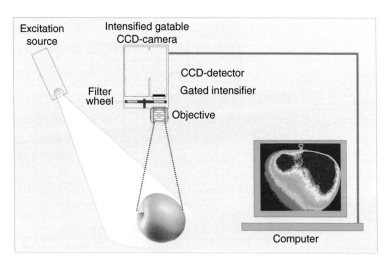

FIGURE 4.2.19 Scheme of a fluorescence imaging setup for the investigation of plant leaves and fruits. The sample (here a light-green apple) irradiated by the excitation source emits fluorescence that is taken by a camera (here an intensified gatable CCD-camera). The camera is equipped with a lens, an intensifier (here a gated intensifier with a microchannel plate synchronized to the excitation pulses), and a detector (here a CCD). The camera takes black-and-white images, i.e., the fluorescence intensities are measured in counts irrespective of the wavelength range. For choosing one of the four plant tissue fluorescence bands, for example, in the blue, green, red, and far-red wavelength region, appropriate color filters of a filter wheel are inserted in front of the intensifier. The fluorescence image is transferred to a computer that displays the fluorescence intensity in false colors with an intensity scale from dark blue (zero fluorescence intensity) via green and yellow to red (highest fluorescence intensity).

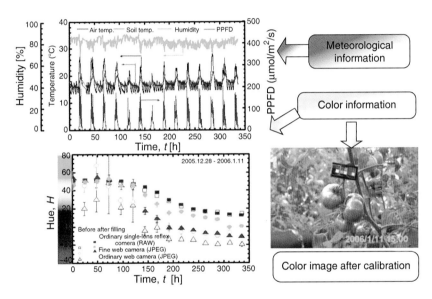

FIGURE 4.5.12 Information on remote-monitored tomatoes in greenhouse.

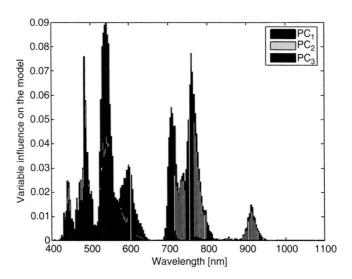

FIGURE 6.2.4 Explained variation of the variables by each PC.

4 Fluorescence

Inès Birlouez-Aragon, Claus Buschmann, Atsushi Hashimoto, Takaharu Kameoka, Michael U. Kumke, Lyes Lakhal, Gabriele Langsdorf, Hartmut K. Lichtenthaler, Hans-Gerd Löhmannsröben, Jad Rizkallaf, and Yasunori Saito

CONTENTS

4.1	Introduction	253
	4.1.1 Basics of Photoluminescence	253
	4.1.1.1 Intramolecular Deactivation	254
	4.1.1.2 Bimolecular Interaction Leading to Quenching	256
	4.1.1.3 Stokes Shift: Microscopic Interpretations and Practical Applications	258
	4.1.1.4 Anisotropy	261
	4.1.2 Instrumentation	262
	4.1.2.1 Steady-State Spectrophotometers	262
	4.1.2.2 Time-Resolved Spectrophotometers	265
	4.1.2.3 Light Sources	268
	4.1.3 Fluorescence-Based Fiber Optical Chemical Sensing	269
	4.1.4 Luminescence Probes	270
References		271
4.2	Blue, Green, Red, and Far-Red Fluorescence Signatures of Plant Tissues, Their Multicolor Fluorescence Imaging, and Application for Agrofood Assessment	272
	4.2.1 Native Fluorophores in Plants	274
	4.2.1.1 General Aspects of Plant Fluorescence	274
	4.2.1.2 Blue-Green Fluorescence of Plants	275
	4.2.1.3 Red + Far-Red Chlorophyll Fluorescence of Plants	276
	4.2.1.4 Fluorescence Excitation Spectra	280
	4.2.1.5 Fluorescence Characteristics of Plant Tissues	280
	4.2.2 Fluorescence Emission Spectra of Plant Tissues	282
	4.2.2.1 Shape of Emission Spectra in Different Plant Tissues	282
	4.2.2.2 Dependence of Fluorescence Yield on the Excitation Wavelength	284

4.2.2.3 Contrasting Wavelength Behavior of Blue-Green
and Red + Far-Red Fluorescence 285
4.2.2.4 Fluorescence Ratios as Assessment Criterion 286
4.2.3 Basis for the Variation of Fluorescence Signatures
of Plant Tissues 287
4.2.3.1 Dependence of Fluorescence on the Concentration
of Chlorophyll *a* and Ferulic Acid 287
4.2.3.2 Tissue Structure and Penetration of Excitation Light 292
4.2.3.3 Tissue Structure: Penetration of Fluorescence
from Deeper Cell Layers 294
4.2.3.4 Photosynthetic Activity 295
4.2.4 Examples for the Application of Multicolor Fluorescence
Imaging in Studies of Plant Tissues 296
4.2.4.1 Fluorescence Imaging and Fluorescence Ratio
Imaging of Plant Tissue and Fluorescence Agrofood 297
4.2.4.2 Developmental Stage 304
4.2.4.3 Strain and Stress 306
4.2.4.4 Preventive Measures in Fluorescence Imaging 308
4.2.5 Fluorescence Imaging Systems for Plant Tissues 310
4.2.6 Conclusion 314
References 314
4.3 Monitoring Raw Material by Laser-Induced Fluorescence
Spectroscopy in the Production 319
4.3.1 Basics of Laser-Induced Fluorescence Spectroscopy 319
4.3.1.1 Fluorescence Mechanism 320
4.3.1.2 Instrumentation for Fluorescence Experiments 322
4.3.2 Blue-Green LIF Spectra of Nutritional Valuable Compounds 323
4.3.3 Application in Agricultural Product Monitoring 326
4.3.3.1 Examples of BG-LIF Spectra of Agricultural Products 326
4.3.3.2 What Information Can Be Extracted from LIF Spectra? .. 327
4.3.4 Future Prospects of Laser-Induced Fluorescence
Spectroscopy in Field Measurements 331
4.3.5 Conclusion 332
Acknowledgments 333
References 333
4.4 Front-Face Fluorescence Analysis to Monitor Food Processing
and Neoformed Contamination 337
4.4.1 Introduction 337
4.4.2 Physical Approach of Process-Induced Food
Physicochemical Changes 338
4.4.2.1 Description of the Matrix-Induced Distortions 338
4.4.2.2 Assessment of the Transfer Function 342
4.4.2.3 Extraction of Pure Fluorescence Related to Native
and Neoformed Compounds 344
4.4.3 Chemometric Analysis of Front-Face Fluorescence Signal:
Assessment of Neoformed Contamination in Heat-Treated
Oils and Starch-Based Products 345

	4.4.3.1	Introduction	345
	4.4.3.2	Methodology	347
	4.4.3.3	Results	352
4.4.4	Conclusion		355

References 357
4.5   Integrated System Design 359
    4.5.1   Approaches in Multiband Spectroscopy 359
        4.5.1.1 X-Ray Fluorescence and Infrared Spectroscopy 359
        4.5.1.2 Infrared and Terahertz Spectroscopy 362
        4.5.1.3 Color Calibration for Image Analysis 364
    4.5.2   Field Server Application 365
        4.5.2.1 Concept of Field Server for Plant Monitoring 365
        4.5.2.2 Sensor Network 367
        4.5.2.3 Optical Farming 368
    4.5.3   Tasting Robot 369
        4.5.3.1 Concept of Tasting Robot 370
        4.5.3.2 Spectroscopic Data of Foods 372
        4.5.3.3 Sommelier Robot in the Future 373
References 374

4.1 INTRODUCTION

MICHAEL U. KUMKE AND HANS-GERD LÖHMANNSRÖBEN

Fluorescence spectroscopy is one of the most versatile analytical techniques available, which is also indicated by the enormous number of scientific papers published each year. In 2006, about 16,000 publications containing fluorescence can be found in the Web of Science, covering a broad spectrum from fundamental research to applied sciences, for example, in physics, life science, and geoscience. Compared with the other spectroscopic techniques, the sensitivity of fluorescence measurements is outstanding. With the latest generation of benchtop spectrophotometers, the detection of zeptomoles (typically few hundreds of molecules) is available; with the application of advanced high-end spectrophotometers, single-molecule detection is also achieved.

Fluorescence is a multidimensional method, which means the intensity and the spectral characteristics of a compound can be analyzed and additional parameters like the fluorescence decay time and the fluorescence anisotropy will yield further information, for example, on molecular kinetics.

In the following sections, a brief introduction on selected aspects of fluorescence spectroscopy is presented. Emphasis is placed upon organic molecules in solution under ambient conditions, while in the subsequent chapters the fluorescence analyses of complex samples are discussed. An in-depth explanation of basic photophysical theory and spectroscopic applications can be found in several textbooks (e.g., Turro 1978; Valeur 2002; Lakowicz 2006).

4.1.1 BASICS OF PHOTOLUMINESCENCE

Light and matter can interact in fundamental ways among which fluorescence, scattering, and absorption have been most lyrically depicted by Bohren and

Clothiaux (2006) as birth, life, and death of photons, respectively. For absorption to take place, the resonance condition ($\Delta E = h\nu$, with ΔE, energy difference between ground and excited state; h, Planck quantum; and ν, frequency of radiation) has to be fulfilled. Molecules that absorb light are called chromophores and can be of inorganic or organic character (or both in the case of transition metal complexes with organic ligands). In the following paragraph, the photophysics of organic molecules is considered in particular, but the fundamental aspects discussed are of a general concept and also valid for inorganic compounds and complexes.

In comparison to electrons, the mass of the nuclei is much higher. Therefore, the movement of the electrons can be treated separately (Born–Oppenheimer approximation), and the transition of an electron between the orbitals initiated by the absorption of a photon is considered to be vertical in a sense that the nuclei of the molecule will not change their position during the electronic transition (Franck–Condon principle). After photoexcitation, the molecule is in an electronically excited state, from which the subsequent return to the ground state occurs in a combination of several intra- and sometimes also intermolecular deactivation processes (vide infra). Photoluminescence results from radiative deactivation after photoexcitation.

Fluorescence and phosphorescence can be distinguished according to the spin (Spin $= -1/2; +1/2$) multiplicities ($= 20 \mid$ Spin $\mid + 1$) of the states involved in the emission process: An emission process that occurs without a change in the spin multiplicity is termed fluorescence, for example, singlet–singlet transitions, while an emission process with change of the spin multiplicity is called phosphorescence. The latter process is spin forbidden and consequently the corresponding rate constants are small, typically in the order of 10^3 to 10^6 s^{-1} compared to 10^7 to 10^9 s^{-1} for fluorescence.

4.1.1.1 Intramolecular Deactivation

Emission processes, which are phosphorescence (P) and fluorescence (F), and nonradiative processes, such as internal conversion (IC), intersystem crossing (ISC), and vibrational relaxation (VR), are intramolecular deactivation pathways from the electronically excited state. In addition, intermolecular deactivation (quenching) may also be operative (vide infra).

An effective way to visualize the intramolecular deactivation processes in an organic molecule is the so-called Jablonski diagram (Figure 4.1.1), in which the fundamental radiative and nonradiative intramolecular deactivation processes can be schematically summarized.

After photoexcitation, the molecule is raised from its electronic ground state (S_0) into an electronically excited state ($S_1, S_2, \ldots, S_n$). For organic molecules, the different electronic levels are consecutively numbered starting with 0, which corresponds to the electronic ground state. The same applies for the triplet state, except that here the lowest triplet state is the T_1 (Pauli principle).

Usually, the nonradiative deactivation of higher excited states ($n > 1$) occurs extremely fast with rate constants typically in the order of 10^{12} s^{-1}. In a combination of VR, in which electronic energy is transferred into vibrations and IC, in which the molecule converts from the vibrationally relaxed level of S_n to an vibrationally highly excited electronic state S_{n-1} (electronic energy is transferred into intramolecular vibrations), the molecule finally reaches the S_1 state.

Fluorescence

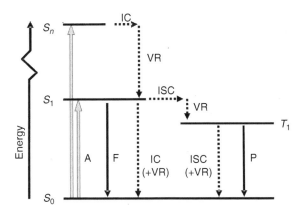

FIGURE 4.1.1 In the Jablonski diagram, the different intramolecular deactivation processes are schematically summarized. Due to light absorption (A) the molecule is in an electronically excited state S_n ($n = 1, \ldots, i$). To return to the electronic ground state S_0, the molecule converts its electronic energy in form of nonradiative processes like internal conversion (IC) and intersystem crossing (ISC) into vibrational relaxation (VR) and finally heat, or it can show fluorescence (F) or phosphorescence (P).

In addition to IC, deactivation from the S_1 state can also take place by fluorescence (F) and via ISC, in which the molecule changes its spin multiplicity, and reaches the triplet state T_n.

From these considerations, it is clear that higher excited states ($n > 1$) are extremely short lived with decay times in the picoseconds range in comparison to the S_1 state with decay times in the range of 1–100 ns. Accordingly, for the majority of organic compounds, fluorescence is almost exclusively originating from the vibrationally relaxed S_1 state (Kasha's rule).

It is also notable that the T_1 state is very long-living (decay times in the range of 1–1000 μs), thus providing ample time for photochemical reactions. This is exploited, for example, in cancer treatment with photodynamic therapy (PDT). Phosphorescence is usually only observed under special circumstances, such as low temperatures or heavy atom-perturbation.

Photophysical processes can be characterized by their quantum yields. For example, the fluorescence quantum yield is given by

$$\Phi_F = \frac{k_F}{k_F + k_{IC} + k_{ISC}} = \frac{\tau_F}{\tau_F^N} \qquad (4.1.1)$$

where

k are the rate constants for the corresponding intramolecular processes
τ_F and τ_F^N are the fluorescence decay time and the natural fluorescence lifetime, respectively, which are given by the following equation:

$$\tau_F = \frac{1}{k_F + k_{IC} + k_{ISC}} \quad \text{and} \quad \tau_F^N = \frac{1}{k_F} \qquad (4.1.2)$$

4.1.1.2 Bimolecular Interaction Leading to Quenching

In addition to intramolecular deactivation steps, intermolecular deactivation processes might play a role. Owing to the interaction of molecules, either of the same or of different kind, the fluorescence properties of a fluorophor (D) can be altered. The reduction in fluorescence quantum yield and/or fluorescence decay time is called fluorescence quenching (or in short, quenching). The molecule responsible for quenching is called the quencher (Q). In the special case where D and Q are of the same kind, this is termed self-quenching, which can be normally observed only at high fluorophor concentrations (typically $>10^{-4}$ M).

In general, two different basic interaction mechanisms of molecular quenching can be distinguished: static and dynamic quenching. These can be best discriminated by a combination of stationary and time-resolved fluorescence measurements. For a dynamic quenching, a reduction of the fluorescence intensity as well as of the fluorescence decay time is expected, while for a static quenching, only the fluorescence intensity is altered. In this case, the fluorescence decay time is not changed.

From a more general molecular view, the static quenching mechanism corresponds to a formation of a ground-state complex, which means that a new molecule is formed (Scheme 1). This might also be seen by changes in the corresponding absorption spectrum. This interaction normally already takes place in the electronic ground state of the fluorophores, and the formed complex is usually only weakly fluorescent or nonfluorescent.

$$D + Q \Leftrightarrow [DQ]$$

Scheme 1: Formation of a ground-state complex leads to static quenching in case the complex [DQ] formed is nonfluorescent.

On the other hand, in the case of dynamic quenching, a bimolecular interaction with a quencher during the decay time of the excited state of the fluorophores occurs. Such a process is diffusion controlled and thus influenced by the viscosity of the solvent. In the simplest case, the electronically excited fluorophore (D*) collides with a quencher (Q) and the electronic energy is dissipated, for example, as heat (Δ) (see Scheme 2). In this simple picture, a diffusion step is involved in the quenching, which explains the expected change of the fluorescence decay time.

$$D^* + Q \rightarrow D + Q + \Delta$$

Scheme 2: In case of dynamic fluorescence quenching, the electronically excited fluorophore (D*), interacting with a second molecule (quencher, Q), is converted into its electronic ground state (D). In the simplest case, the excess energy is transferred to the environment in form of heat (Δ).

To distinguish static from dynamic quenching, the experimental data from steady-state and time-resolved measurements are evaluated according to the so-called Stern–Volmer analysis. In the steady-state data analysis, the ratio of the fluorescence intensities I_F is analyzed in relation to the quencher concentration (c_Q). In the case of the time-resolved measurements, the data are evaluated with respect to the fluorescence decay times in the presence and absence of Q (Turro 1978; Valeur 2002; Lakowicz 2006). With static quenching, only the fluorescence intensity is decreased (because less fluorescent molecules are present in the sample), dynamic quenching will alter fluorescence intensity as well as fluorescence decay time of the sample.

From the Stern–Volmer analysis, the apparent binding constant K_{SV} (in case of pure static quenching) or the bimolecular quenching constant k_q (for dynamic quenching) is determined (Equations 4.1.3a and b). Very often, a mix of static and dynamic quenching is found. Then, a more sophisticated data analysis is required to consider both aspects in an appropriate way (Turro 1978; Valeur 2002; Lakowicz 2006).

$$\text{Static quenching:} \quad \frac{I_F^0}{I_F} = 1 + K_{SV} \cdot c_Q \quad \text{and} \quad \frac{\tau_F^0}{\tau_F} = 1 \quad (4.1.3a)$$

$$\text{Dynamic quenching:} \quad \frac{I_F^0}{I_F} = \frac{\tau_F^0}{\tau_F} = 1 + k_q \cdot \tau_F^0 \cdot c_Q \quad (4.1.3b)$$

In a more detailed view, dynamic quenching mechanisms may be categorized in processes that (a) require a close proximity of fluorophor and quencher (overlap of molecular orbitals, short range type) and (b) interactions through space (long range type). Especially, energy and electron transfer reaction have been investigated in great detail (Lakowicz 2006).

Within the context of dynamic quenching, a number of fundamental mechanisms can be distinguished:

- Energy transfer processes (Dexter type, short range; and Förster resonance type, long range)
- Electron transfer processes (intra- and intermolecular)
- Formation of excited-state complexes (exciplex) or excited-state dimers (excimer) (intra- and intermolecular)

In energy transfer processes, the interacting molecular species are normally termed donor (D) and acceptor (A) instead of fluorophor and quencher.

$$D^* + A \rightarrow D + A^*$$

Scheme 3: Energy transfer from an electronically excited donor D* to an acceptor A.

As can be deduced from Scheme 3, in this particular case the electronic energy is not dissipated but transferred to the acceptor A, which subsequently can now show luminescence itself. The distance dependence of electron and energy transfer reaction can be used for the determination of intra- and intermolecular D/A-separations. Especially, Förster or fluorescence resonance energy transfer (FRET) is a very powerful tool for the determination of molecular distances (molecular ruler) (Stryer and Haugland 1967; Turro 1978; Valeur 2002; Lakowicz 2006). In biological and medical applications, donor and acceptor molecules can be attached to biomolecules like antigens and antibodies, and the binding reaction is monitored via the appearance of the FRET process. This is the basic principle of fluorescence immunoassay (FIA) analysis, a very powerful technique in in-vitro diagnostics (Charbonnière et al. 2006; Hildebrandt et al. 2006; Hildebrandt and Löhmannsröben 2007).

4.1.1.3 Stokes Shift: Microscopic Interpretations and Practical Applications

Perylene is a prototype polycyclic aromatic hydrocarbon (PAH) with outstanding fluorescence capabilities (fluorescence quantum yield $\Phi_F \approx 1$). Derivatives of this compound, the so-called perylene dyes, constitute an important class of pigment dyes. In Figure 4.1.2, the fluorescence emission and fluorescence excitation spectrum of perylene in methanol are depicted. The spectra are shown on an energy scale expressed in wave numbers (ν) being reciprocal to wavelength (λ) with $\nu = 1/\lambda$, to better visualize the connection between the observed experimental spectra and the energy levels (electronic as well as vibronic levels) of the fluorophore. According to the Lambert–Beer law, in dilute solution (fluorophore concentration, c_F), the fluorescence intensity (detection at ν_j) obtained from excitation at ν_i is directly proportional to the molecular absorption coefficient $\varepsilon(\nu_i)$.

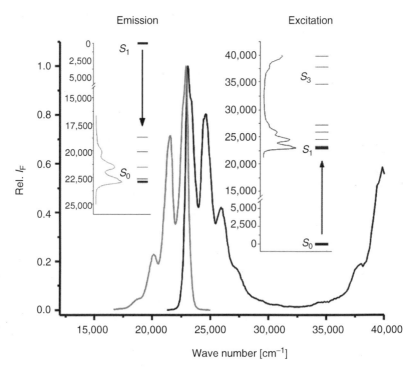

FIGURE 4.1.2 Fluorescence excitation and fluorescence emission spectra of perylene in methanol with $c_F = 1\ \mu M$, $\nu_i = 24{,}690\ cm^{-1}$ ($\lambda_{ex} = 405\ nm$), $\nu_j = 21{,}280\ cm^{-1}$ ($\lambda_{em} = 470\ nm$). Insets: Correlation between electronic selected and vibronic levels and the observed spectra.

$$I_F(\nu_i, \nu_j) = \text{const.} \cdot \varepsilon(\nu_i) \cdot \varphi_F(\nu_j) \cdot c_F \qquad (4.1.4)$$

Here, the fluorescence line shape function $\varphi_F(\nu_j)$ is introduced: $\Phi_F = \int \varphi_F(\nu_j) d\nu_j$. Equation 4.1.4 allows three important observations:

1. Ideally, fluorescence excitation spectra, obtained by scanning ν_i and keeping ν_j fixed, are qualitatively identical to absorption spectra. From Figure 4.1.2, it is then readily concluded that the fluorescence emission is shifted to lower energy relative to the corresponding absorption (in accordance with Figure 4.1.1).
2. The fluorescence emission spectra, obtained by scanning ν_j and keeping ν_i fixed, yield directly $\varphi_F(\nu_j)$.
3. Since $I_F(\nu_i, \nu_j)$ is directly proportional to c_F, Equation 4.1.4 can be used for calibration purposes and thus fluorescence spectroscopy can be applied quantitatively (vide infra).

For molecules with very similar S_0 and S_1 states in terms of electron density, dipole moment, and bond lengths, a so-called mirror symmetry between fluorescence and absorption (or fluorescence excitation) spectra is found. Many PAHs show good mirror symmetry and perylene is an almost perfect example (Figure 4.1.2). The energy difference between longest-wavelength absorption and shortest-wavelength fluorescence bands is called Stokes shift. Again, this is the case in perylene, for which the energy difference between the S_0–S_1 and the S_1–S_0 transitions is only about 200 cm^{-1} (Figure 4.1.2). The absorption and fluorescence spectra of some different PAHs are compared in Figure 4.1.3. It can be seen that for anthracene and

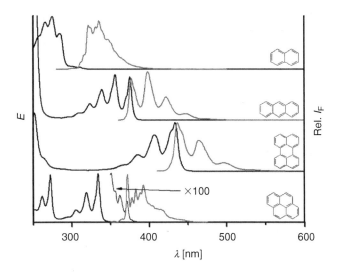

FIGURE 4.1.3 Absorption shown as the apparent extinction E and fluorescence emission spectra of some polycyclic aromatic hydrocarbons (PAHs) in methanol. For pyrene the thin-lined curve indicates the weak S_0–S_1 transition.

perylene, the mirror symmetry between absorption and fluorescence is nicely fulfilled while for naphthaline and pyrene this seems not to be the case (vide infra).

For other molecules, which are often containing functional moieties like hydroxy-, carboxy-, or amine-groups, the properties of the S_0 and S_1 states are very different. For example, the acidity of hydroxy groups of an aromatic molecule is drastically altered in the excited state. A well-known example is 2-naphthol. While the (logarithmic) acidity constant of the hydroxy group in the electronic ground state is approximately $pK_a(S_0) \approx 9.5$, it drops to about $pK_a(S_1) \approx 2.8$ after electronic excitation of the molecule (Klessinger and Michl 1989; Lawrence et al. 1991). Hence, after electronic excitation the hydroxy group of 2-naphthol will dissociate for solution pH $> pK_a(S_1)$. Depending on the solution conditions, the fluorescence spectrum can be the sum of the neutral and the anionic form of 2-naphthol (Figure 4.1.4).

Since the Franck–Condon principle can be applied for electronic transitions, the geometry (bond lengths and bond angles) of a molecule is unchanged right after the electronic transition. The molecule will be in a Franck–Condon S_1 state, which is not equal to the equilibrated S_1 state. For equilibration, intramolecular rearrangements (e.g., bond lengths and angles might be altered or rotational motions can occur) and changes in the solvation cage of the molecule may occur. These processes can happen very fast (picoseconds to nanoseconds timescale). Subsequently, the fluorescence may arise from an equilibrated S_1 state, which is lower in energy. Experimentally, this is found as a large Stokes.

The absorption and fluorescence spectrum of pyrene is shown in Figure 4.1.3. Because the S_0–S_1 transition is forbidden due to parity reasons, the corresponding extinction coefficient is very small, and hence the transition is observed in the

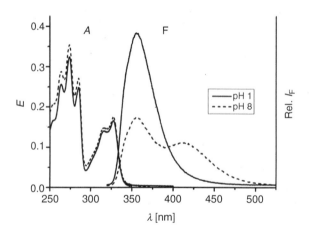

FIGURE 4.1.4 Absorption (A), recorded as apparent absorption (extinction, E) and fluorescence emission (F) spectra of 2-naphthol at pH 1 and 8. In the fluorescence spectrum at pH 8 two emission maxima can be observed at $\lambda_{em} = 355$ nm and at $\lambda_{em} = 425$ nm, which can be attributed to the 2-naphthol and to the 2-naphtholate form, respectively. The 2-naphtholate is formed in the excited state by proton dissociation of the hydroxy group as is indicated by the absorption spectra, which are almost identical for both pH.

Fluorescence

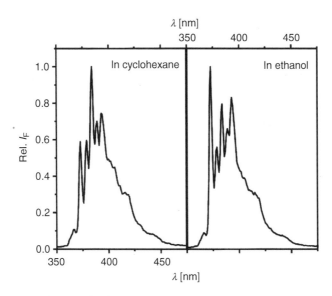

FIGURE 4.1.5 Influence of solvent polarity on the vibronic structure of the fluorescence spectrum of pyrene.

absorption spectrum only at high concentrations. Another consequence of the symmetry forbiddance is the sensitivity of some of the vibronic transitions in the fluorescence spectrum. In Figure 4.1.5, the fluorescence spectra of pyrene in cyclohexane and ethanol are shown.

In polar solvents, the intensity of the first vibronic transition ($\lambda_{em} \approx 373$ nm) is enhanced compared to nonpolar solvents. The reason for this enhancement in polar solvents is the so-called intensity stealing, which means that due to coupling of solvent vibrations the parity forbiddance of the transition is relaxed and subsequently the observed intensity is increased. This makes pyrene a valuable probe for the investigation of polarity effects on a molecular level (py scale, vide infra) (Kalyanasundaram and Thomas 1977; Valeur 2002).

4.1.1.4 Anisotropy

Excitation of a chromophore using polarized light adds another dimension to fluorescence measurements. Fluorescence anisotropy measurements can be performed stationary as well as time-resolved. The key information obtained from anisotropy measurements is on rotational motion and on the orientation between absorption and emission dipole moments. The latter represents a fundamental property of the molecule and determines the maximal (intrinsic) anisotropy of the fluorophores. The anisotropy loss due to rotational motion of the molecule is limited by its rotating volume and by the microviscosity of the solvent.

In Figure 4.1.6, the fluorescence excitation spectrum and the corresponding fluorescence anisotropy of perylene are shown in relation to the fluorescence excitation spectrum already presented in Figure 4.1.2. The observed anisotropy clearly indicates

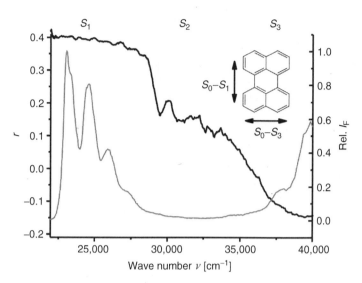

FIGURE 4.1.6 Fluorescence excitation (I_F) spectrum and anisotropy r of perylene ($\lambda_{em} = 470$ nm). The inset shows the relative orientations of the transition dipole moments of the S_0–S_1 and the S_0–S_3 transitions.

that three electronic transitions occur in the spectral range considered: Between the allowed S_0–S_1 and S_0–S_3 transitions around 25,000 cm^{-1} and 40,000 cm^{-1}, respectively, there is the forbidden S_0–S_2 transition in the spectral range of 28,000 cm$^{-1} < \nu <$ 35,000 cm^{-1}. Furthermore, from the observed anisotropy r it can be concluded that the angle between the dipole moments of absorption and fluorescence transitions is about 0° for the S_0–S_1 transition and almost 90° for the S_0–S_3 transition.

Anisotropy measurements are especially valuable for the investigation of molecular binding interactions, such as, for example, antibody reactions. In such reactions, the observed anisotropy is often greatly altered due to binding and the subsequent increase in the overall rotating volume of the complex.

4.1.2 INSTRUMENTATION

A detailed description of instrumentation for fluorescence measurements is beyond the scope of this chapter and the reader should refer to text books and special articles for detailed information (e.g., Turro 1978; Demtröder 2002; Valeur 2002). Here, a brief overview of techniques and their basic principles is given.

In general, two basic setup schemes can be distinguished: (1) steady-state measurements and (2) time-resolved measurements. More advanced methods, requiring considerable experimental efforts and expenditure, which are based on pump-and-probe principles, such as fluorescence up-conversion, are not treated here.

4.1.2.1 Steady-State Spectrophotometers

Standard steady-state fluorescence spectrometers are mostly equipped with Xe arc lamps, which are either operated in a continuous wave (cw) or in a pulsed mode

Fluorescence

(see also Section 4.1.2.3). The latter has the advantage that these instruments can also be easily employed for phosphorescence measurements. For detection, photomultiplier tubes (PMTs) are commonly used, but for special applications array detectors like charge-coupled devices (CCDs) and intensified charge-coupled devices (ICCDs) are advantageous. The detection system can be operated either in an analogous or in a single-photon-counting modus, which is by far more sensitive, but on the other hand, can bring some problems when measuring strongly fluorescing samples. In this case, sample dilution might be required to avoid saturation of the detection system. In contrast, fluorescence spectrometers operated in analogous modus are more robust, but less sensitive.

A method commonly employed to characterize the sensitivity of a fluorescence spectrometer is to measure the water Raman signal. However, when comparing information of manufacturers, it is of particular importance to ensure that the experimental conditions for the determination of the background noise levels can be different.

As shown above, the fluorescence intensity of a sample is dependent on the concentration of fluorophores excited in the sample. For practical purposes, Equation 4.1.4 will here be expressed slightly different as

$$I_F = 2.3 \cdot G \cdot I_0(\lambda_{ex}) \cdot \Phi_F \cdot \varepsilon(\lambda_{ex}) \cdot c_F \cdot L \qquad (4.1.4a)$$

From Equation 4.1.4a, it can be seen that for dilute solutions the fluorescence signal is directly proportional to an experimental factor G, to the intensity of the excitation radiation $I_0(\lambda_{ex})$, to the compound's fluorescence quantum yield Φ_F, decadic extinction coefficient $\varepsilon(\lambda_{ex})$ and concentration c_F, and, finally, to the optical path length L. The experimental factor G is specific to the apparatus and the geometry used and reflects, for example, the detector spectral sensitivity. The numerical factor 2.3 in Equation 4.1.4a derives from conversion from natural to decadic logarithm.

At higher concentrations, Equation 4.1.4a is no longer valid and a complex dependence of the fluorescence signal I_F on the concentration is operative. Also, additional parameters like inner filter effects have to be taken into account (vide infra). As both G and $I_0(\lambda_{ex})$ are usually known only in relative units, it becomes further clear from Equation 4.1.4a that the fluorescence signal intensity measured is also only useful on a relative scale. Consequently, the sample concentration can only be determined in case a proper calibration has been carried out.

The standard measurement configuration for liquid samples is a 90° arrangement between excitation and emission pathways. For this geometry it is important to bear in mind that the light absorption of the sample needs to be evaluated because this experimental setup is prone to the so-called inner filter effects. Here, first- and second-order inner filtering can be distinguished. Both effects are observed at relatively high extinctions. Typically, filter effects have to be taken into account if extinctions exceed $E \approx 0.1$ at the excitation wavelength. In the case of first-order inner filtering, the excitation light is mostly absorbed in the first layers of the sample and does not reach the center of the cuvette where the fluorescence is usually detected. Obviously, this leads to a decrease of I_F. Second-order filtering is caused by a reabsorption of fluorescence photons by the probe molecules themselves.

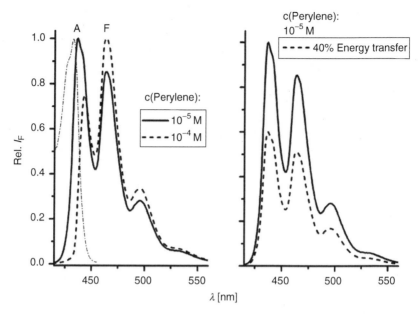

FIGURE 4.1.7 Normalized excitation (thin dashed line) and fluorescence emission spectral intensities (I_F) of perylene (in methanol, $\lambda_{ex} = 405$ nm) at two different concentrations. At high concentrations, the fluorescence spectrum (strong line) is distorted in the spectral overlap region due to reabsorption processes due to second-order inner filter effect (left). Normalized fluorescence spectra in the absence (strong line) and presence (dotted line) of energy transfer with efficiency of energy transfer is 40% (right).

Consequently, less photons are reaching the detector and, again, the fluorescence signal is decreased.

The second-order inner filtering takes place in the range of spectral overlap between absorption and emission spectra. As a consequence, a distinct spectral distortion of the fluorescence spectrum is observed in the short-wavelength part (Figure 4.1.7). To account for both the filter effects one needs to know the absorption spectra compounds in the sample, and also geometry parameters like width of the excitation and emission beams (Turro 1978; Gauthier et al. 1986; Valeur 2002).

The emission/reabsorption process is sometimes referred to as the trivial mechanism of energy transfer. However, filter effects should not be confused with true energy transfer phenomena, such as the dipolar FRET. A basic requirement for FRET is an overlap between emission spectrum of the donor and absorption spectra of the acceptor. However, in contrast to second-order inner filtering, here the signal intensity of the complete donor fluorescence spectrum is reduced, i.e., no spectral distortion is observed.

When the measurement of highly concentrated and sometimes intensely colored samples is required, a front-face detection setup can be employed to minimize the influence of inner filter effects. Here, the sample surface is mounted in an approximately 30° angle relative to the excitation beam and the fluorescence is detected on the sample surface. Such a setup is also commonly used for solid samples. The most important point is that the amount of scattered light reaching the detector has to be minimized. For solid samples, this will strongly depend on the sample properties like roughness of the surface. For such samples, it is very important to control possible experimental artifacts due to light scattering phenomena. A very helpful technique here is a time-gated detection system, in which a pulsed excitation source, for example, a short pulse laser, is used for excitation and the detection is electronically delayed relative to the excitation.

The determination of fluorescence quantum yields can be performed with various techniques. Often, manufacturers provide an instrumental correction function which accounts for the spectral sensitivity of the monochromator and PMT of the specific instrument. Correction functions are usually generated by recording the known spectral output of a standardized calibration lamp. Frequently, the correction file is limited to a certain wavelength range (often $300 < \lambda_{em} < 700$ nm), which limits its applicability.

Another method to determine Φ_F is based on quantum counters or on the measurement of samples with known fluorescence quantum yields (e.g., quinone sulfate, PAH) (Dawson and Windsor 1968; Eaton 1988; Gardecki and Marconcelli 1998). However, for such measurements, the excitation conditions as well as the spectral range of the sample and reference emission have to be identical (at best), which also limits the applicability of this simple relative method. Maybe the most promising and general approach for the determination of fluorescence quantum yields is the usage of an instrumental correction function based on a number of reference compounds with known fluorescence spectra. By measuring the fluorescence of such reference compounds it is possible to cover the spectral range $300 < \lambda_{em} < 950$ nm and may be even further. The advantage of this approach is that it is possible to renew or control the quantum correction function when components of the instrument (e.g., the PMT) are aging (Van den Zegel et al. 1986; Resch-Genger et al. 2005).

4.1.2.2 Time-Resolved Spectrophotometers

In addition to the intensity and spectral characteristics, the time dependence of the fluorescence emission can yield valuable information, for example, about dynamic molecular interaction processes. To measure the fluorescence decay time of a fluorophore it is most straightforward to use a short light pulse for excitation. The pulse length should be shorter than the decay time of the compound under investigation. As outlined above, the S_1 state deactivation of organic molecules occurs on a nanosecond timescale, whereas other luminescent materials, for example, transition metal complexes or materials containing f-elements, may have luminescence lifetimes in micro- to millisecond time ranges. Currently, pulsed lasers or light-emitting diodes (LEDs) are frequently used as excitation sources, but for some purposes flash lamps might be still useful as well.

An alternative to pulsed excitation systems are phase modulation techniques, in which the excitation radiation is modulated, for example, by Pockels cells, typically with up to megahertz to gigahertz frequency, and the phase shift and demodulation of the fluorescence is measured, and used to calculate the fluorescence decay time.

When using pulsed excitation, time-correlated single photon counting (TCSPC) using PMT, boxcar techniques using CCD detectors, or streak cameras are commercially available as turnkey systems. With a standard TCSPC setup, nanosecond decay times can easily be measured, the same is true for boxcar techniques. With streak cameras even sub-nanosecond fluorescence decays can be resolved. For faster processes, pump-and-probe techniques might be used. Streak and pump-and-probe techniques are very demanding with respect to operation conditions and costs, while boxcar and TCSPC are less expensive techniques. In TCSPC, high repetitive but low-energy laser sources are commonly used. On the other hand, for boxcar techniques lower repetition rates but higher pulse energies are required. Therefore, it is important to monitor the photostability of the sample. In Figure 4.1.8, the general timing scheme for decay time measurements using the boxcar approach is depicted.

The detector opens for a small time interval $\Delta\tau$ (detection window) and collects all luminescence photons during that time. In the next cycle, the detection window is shifted relative to the laser excitation pulse by Δt_1 and again all photons reaching the detector are collected in $\Delta\tau$. These cycles are repeated i times to measure the complete luminescence decay.

In many applications, the fluorescence of the sample is masked by background signals, for example, due to light scattering phenomena or matrix luminescence. In such cases, a gated detection like the boxcar approach might be a good option to reduce background contributions, provided that the luminescence decay of the sample signals occurs on a longer timescale. Good examples for successful

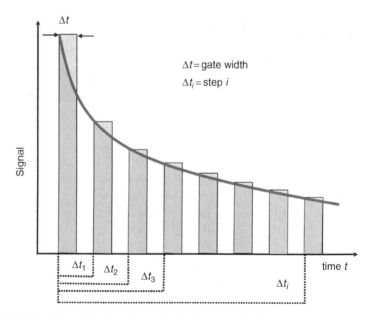

FIGURE 4.1.8 General timing scheme of a luminescence decay measurement using a boxcar technique.

Fluorescence

applications of gated detection schemes are homogeneous FIAs for in vitro diagnosis based on lanthanide complexes as luminescence probes. Herein, strong scattering and autofluorescence signals from the blood serum matrix are effectively discriminated by gated detection. Often in these assays, lanthanide ions such as europium (Eu^{3+}) or terbium (Tb^{3+}) are used as luminescence probes. The luminescence of Eu^{3+} and Tb^{3+} and its complexes is usually found in the micro- to millisecond time range, while the background signal due to light scattering and matrix components (e.g., proteins) is in the nanosecond time regime. Hence, by setting an appropriate delay in the detection channel, an essentially background-free luminescence signal is measured, which makes this type of method extremely sensitive. For a review of FIA-based in vitro diagnosis see Hildebrandt and Löhmannsröben (2007).

In combination with ICCD cameras, time-resolved emission spectra (TRES) are obtained, from which spectral as well as decay time information can be extracted. In Figure 4.1.9, TRES of Tb^{3+} in natural water is shown together with examples of a decay curve and of an emission spectrum extracted from the TRES. The emission spectrum is free of any contribution due to scattering from matrix components because of the gated detection scheme. In a standard steady-state experiment, the

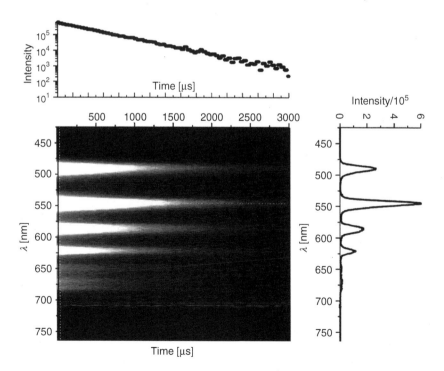

FIGURE 4.1.9 Time-resolved emission spectrum of Tb^{3+} in natural water ($\lambda_{ex} = 370$ nm, $\Delta\tau = 40$ µs). Top: Luminescence decay curve at $\lambda_{em} = 545$ nm ($\tau_{Tb} = 400 \pm 10$ µs). Left: Luminescence emission spectrum at $\Delta t = 43$ µs.

luminescence of Tb^{3+} would have been distorted by emission from dissolved organic carbon (DOC), which is always present in natural waters or other samples taken from natural sources.

4.1.2.3 Light Sources

For steady-state measurements, xenon arc lamps are commonly employed. In Figure 4.1.10, the typical emission of such a lamp is shown. It can be seen that the overall intensity as well as the spectral distribution are dependent on the operation time. For the measurement of quantum yields and excitation spectra, it is important to bear in mind that the fluorescence signal is directly proportional to the excitation intensity (Equation 4.1.4a). Some spectrometers are equipped with pulsed xenon lamps, which have a higher lifetime and allow to perform decay time measurements in the higher microsecond regime and to record phosphorescence spectra.

For time-resolved measurements, in most cases, lasers are used for excitation. For TCSPC applications, a variety of LED and diode lasers are currently available in the spectral range between 300 nm $< \lambda <$ 800 nm. However, not every wavelength is covered. Such semiconductor light sources can provide pulses in the low nanosecond down to tens of picoseconds range with repetition rates of up to the megahertz range. On the other hand, for boxcar applications laser sources with higher pulse energies are often required. With new diode pumped solid-state (DPSS) lasers in combination with optical parametric oscillators (OPOs), complete coverage of the wavelength range between 250 nm and 900 nm (and higher) is possible. For CCD-based boxcar detection, the combination of DPSS and OPO is probably the excitation source of choice. Systems with repetition rates of 10 Hz–100 Hz are commercially available. However,

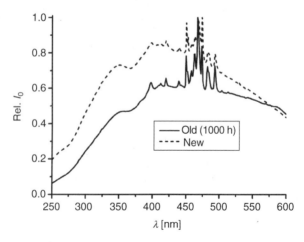

FIGURE 4.1.10 Emission spectra of a 150 W Xe arc lamp: new and after 1000 h of operation.

Fluorescence

when using such excitation sources it should be borne in mind that in some cases photodegradation of the sample under investigation might be induced.

Very recently, a completely new class of light sources, so-called white light or supercontinuum sources, have become available. They are based on nonlinear optical effects in photonic crystal fibers and represent a new class of pulsed, widely tunable lasers with TNR outstanding experimental properties, especially when combined with multichannel accusto-optic tunable filters (AOTFs). These systems combine the high output power (peak power up to 5 W) and short pulse duration (less than 100 ps) of conventional pulsed lasers with a unique spectral tunability, which spans from the violet to the near infrared. Thus, a single supercontinuum source can replace a whole array of classical lasers, which would be necessary to cover the same spectral region. In addition, a multichannel AOTF attached to a supercontinuum source allows to simultaneously couple out light of multiple wavelengths. This is essential for multicolor experiments, for example, for multiparameter detection. Although their employment is just in the beginning, it seems fair to predict that these white light sources will have revolutionary impacts in many fields of spectroscopy.

4.1.3 Fluorescence-Based Fiber Optical Chemical Sensing

The application of optical sensors in analytical chemistry and in process control is growing fast because optical techniques offer outstanding sensitivity and selectivity together with real-time data acquisition rates. Often, the sensor principle is based on a fluorescent probe, which is immobilized, for example, in a polymer matrix or bound to a surface. The fluorescence of the probe is specifically altered by certain chemical parameters. This alteration of fluorescence may be a spectral shift or a kind of fluorescence quenching. Often, optical sensors are combined with optical fibers. Optical fibers show unique properties with respect to light guiding and light manipulation capabilities. In addition to the guiding of light from the light source to the sample and from the sample to the detection system over very long distances and in sometimes extremely harsh environments, the light can also be modified by using composite materials like in graded index fibers or by the integration of microstructures like fiber Bragg gratings (FBGs).

When light is totally reflected within an optical fiber without cladding, there exists a certain probability for the light to enter into the surrounding medium and an evanescent field is formed near the fiber surface. Interaction of the evanescent field with the surrounding medium in general and, in particular, with analyte molecules close to the fiber surface will occur. Owing to the interactions, the light passing the fiber may change its properties. In a simple case, part of the light is absorbed, but also fluorescence may be induced. This has led to a relatively new spectroscopic technique, named total internal reflection fluorescence (TIRF) spectroscopy.

In fiber optical chemical sensors (FOCS), special probe molecules are bound to or are immobilized at the surface or tip of the fiber. The probe molecules can yield specific information on environmental parameters, like pH, or chemical composition, like oxygen (O_2) or ionic concentrations. In special cases, also antigen–antibody interactions can be monitored, when using antibodies attached to the fiber. FOCS can be miniaturized, for example, when the tip of an optical fiber is tapered and coated with a polymer matrix containing a luminescence probe. With such an approach, the in-vivo detection of physiological parameters within animal and plant cells is possible (Yutaka 2003; Papkovsky 2004).

During recent years, the optical determination of O_2 concentrations has witnessed a great increase in popularity. Meanwhile, so-called optodes for O_2 measurements with applications in medicine, environmental sciences, or process monitoring are commercially available. For example, the determination of oxygen is indispensable in the food production chain, since the shelf life of food is often limited by the O_2 content. Frequently, the in-line determination can only be done with noninvasive methods based on FOCS.

In succession with the development of reasonably high-powered LED, optical oxygen meters became competitive with the traditional amperometric Clark-type electrodes. In contrast to the Clark-type electrode, optical measurements do not consume oxygen, do not suffer from electrical interference, allow noninvasive measurements, and can be miniaturized for intracellular measurements. The optical oxygen measurements use a phosphorescent probe with luminescence decay times in the microseconds range. O_2 induces a dynamic quenching of the probe phosphorescence. Measuring the phosphorescence intensity or decay time allows a quantitative determination of the ambient oxygen concentration. However, for the determination a calibration step is required and a calibration curve has to be measured for the particular matrix under investigation. Embedding the probe dyes in a polymer or solgel matrix strongly increases the signal intensity and reduces cross-sensitivities. Detailed descriptions of common probe dyes, measurement techniques, and reviews of applications can be found in the literature (e.g., Yutaka 2003; Papkovsky 2004).

4.1.4 LUMINESCENCE PROBES

As outlined above, the S_1–S_0 transition in pyrene is forbidden due to parity reasons. As a consequence, the pyrene fluorescence properties strongly depend on the molecular environment. Owing to vibrational coupling, the fluorescence decay time of pyrene is increased from approximately 130 ns in highly polar environments like water to about 600 ns in nonpolar environments like hexane (Valeur 2002). Furthermore, the intensity of the fluorescence band at 373 nm is drastically altered while at the same time, the intensity of the 393 nm band is not changed. Subsequently, the ratio of these two bands is a very good estimate of the polarity of the molecular environment (py scale) (Kalyanasundaram and Thomas 1977; Valeur 2002). Pyrene has been used as a fluorescence probe to investigate the uptake of nonpolar xenobiotics by microorganisms and to investigate the location of such hydrophobic compounds in tissues. To improve the probe properties of pyrene, derivatives containing functional groups like carboxy- and hydroxyl-groups or even long alkyl chains are commercially available. Pyrene derivatives are successfully used as a pH indicator on a molecular level. Among others, dyes of the rhodamine and fluorescein family are applied to monitor the pH of Ca^{2+} ions. These dyes can be used for a well-defined labeling of proteins and DNA as well and are used to monitor these parameters on a cellular level. There is a plethora of luminescent molecular probes for chemical sensing. To mention but one example, we are currently evaluating lifetime-based pH imaging. Out of the commercial ratiometric pH sensors so far only a biscarboxy-fluorescein derivative, called BCECF, seems to meet the requirements of a reliable intracellular pH determination in living cells (Hille et al. 2008).

REFERENCES

Bohren, C.F. and E.E. Clothiaux. 2006. *Fundamentals of Atmospheric Radiation*. Wiley-VCH, Weinheim. ISBN: 3-527-40503-8l.

Charbonnière, L.J., N. Hildebrandt, R.F. Ziessel, and H.-G. Löhmannsröben. 2006. Lanthanides to quantum dots resonance energy transfer in time-resolved FluoroImmunoAssays and luminescence microscopy. *Journal of the American Chemical Society* 128: 12800–12809.

Dawson, W.R. and M.W. Windsor. 1968. Fluorescence yields of aromatic compounds. *The Journal of Physical Chemistry* 72: 3251–3260.

Demtröder, W. 2002. *Laser Spectroscopy*. Springer, Berlin. ISBN : 978-3-540-65225-0.

Eaton, D.F. 1988. Reference materials for fluorescence measurements. *Journal of Photochemistry and Photobiology B* 2: 523–531.

Gardecki, J.A. and M. Marconcelli. 1998. Set of secondary emission standards for calibration of the spectral responsivity in emission spectroscopy. *Applied Spectroscopy* 52: 1179–1189.

Gauthier, T.D., E.C. Shane, W.F. Guerin, and C.L. Grant. 1986. Fluorescence quenching method for determining equilibrium constants for polycyclic aromatic hydrocarbons binding to dissolved humic materials. *Environmental Science and Technology* 20: 1162–1166.

Hildebrandt, N. and H.-G. Löhmannsröben. 2007. Quantum dot nanocrystals and supramolecular lanthanide complexes—Energy transfer systems for sensitive in vitro diagnostics and high throughput screening in chemical biology. *Current Chemical Biology* 1: 167–186.

Hildebrandt, N., L.J. Charbonnière, R.F. Ziessel, and H.-G. Löhmannsröben. 2006. Quantum dots as resonance energy transfer acceptors for monitoring biological interactions. In *Biophotonics and New Therapy Frontiers*, R. Grzymala, and O. Haeberle (Eds.), 9 p. Proceedings SPIE, Vol. 6191, 61910W.

Hille, C., M. Berg, L. Bressel et al. 2008. In-vivo pH FLIM sensing. *Analytical and Bioanalytical Chemistry* DOI 10.1007/s00216-008-2147-O.

Kalyanasundaram, K. and J.K. Thomas. 1977. Environmental effects on vibronic band intensities in pyrene monomer fluorescence and their application in studies of micellar systems. *Journal of the American Chemical Society* 99: 2039–2044.

Klessinger, M. and J. Michl. 1989. *Lichtabsorption und Photochemie Organischer Moleküle*. VCH, Weinheim. ISBN: 3-527-26085-4.

Lakowicz, J.R. 2006. *Principles of Fluorescence Spectroscopy*. Springer, Berlin. ISBN: 0-387-31278-1.

Lawrence, M., C.J. Marzzacco, C. Morton, C. Schwab, and A.M. Halpern. 1991. Excited-state deprotonation of 2-naphthol by anions. *Journal of Physical Chemistry* 95: 10294–10299.

Papkovsky, D.B. 2004. Methods in optical oxygen sensing: Protocols and critical analyses. *Methods in Enzymology* 381: 715–735.

Resch-Genger, U., K. Hoffmann, W. Nietfeld et al. 2005. How to improve quality assurance in fluorometry: Fluorescence—Inherent sources of error and suited fluorescence standards. *Journal of Fluorescence* 15: 362–367.

Stryer, L. and R.P. Haugland. 1967. Energy transfer: A spectroscopic ruler. *Proceedings of the National Academy of Science* 58: 719–726.

Turro, N.J. 1978. *Modern Molecular Photochemistry*. The Benjamin/Cummings Publishing Company, Inc., Menlo Park (CA). ISBN: 0-8053-9354-4.

Valeur, B. 2002. *Molecular Fluorescence*. Wiley-VCH, Weinheim. ISBN: 3-527-40310-8.

Van den Zegel, M., N. Boens, D. Daems, and F.C. De Schryver. 1986. Possibilities and limitations of the time-correlated single photon counting technique: A comparative study of correction methods for the wavelength dependence of the instrument response function. *Chemical Physics* 101: 311–335.

Yutaka, A. 2003. Probes and polymers for optical sensing of oxygen. *Microchimica Acta* 143: 1–12.

4.2 BLUE, GREEN, RED, AND FAR-RED FLUORESCENCE SIGNATURES OF PLANT TISSUES, THEIR MULTICOLOR FLUORESCENCE IMAGING, AND APPLICATION FOR AGROFOOD ASSESSMENT

CLAUS BUSCHMANN, GABRIELE LANGSDORF, AND HARTMUT K. LICHTENTHALER

In this section, we present the current knowledge on the blue-green fluorescence and red + far-red chlorophyll fluorescence signals of plant tissues, give an introduction to the multicolor fluorescence imaging technique, and provide examples for its application in the assessment of agrofood.

All plant tissues show an autofluorescence in the visible spectral range (Buschmann and Lichtenthaler 1998) that is excited by UV radiation and partially also by visible light. The in vivo fluorescence, detectable during excitation by UV-A radiation, comprises the blue and green fluorescence of the cell walls, and in green tissues additionally the red + far-red chlorophyll fluorescence as shown in the fluorescence of a maize leaf cross section (Figure 4.2.1A) and the stripped off epidermis of a *Commelina communis* L. (Figure 4.2.1B). The red + far-red fluorescence, emitted by chlorophyll *a* in the chloroplasts of the green leaf or fruit mesophyll cells, have been studied extensively and applied in photosynthesis research, since they provide essential information on the functioning of the photosynthetic apparatus (see textbooks by Lichtenthaler 1988; Papageorgiou and Govindjee 2004).

Chlorophyll fluorescence can be excited not only by UV radiation but also by visible light from blue via green to orange-red. The nongreen fruit and other agrofood tissues, in turn, are characterized only by a high blue fluorescence and an often distinct but considerably lower green fluorescence, both of which are emitted by ferulic acid covalently bound to the cell walls. In the last four decades, the fluorescence measurements were performed by measuring the fluorescence yield and/or emission spectra of single leaf or fruit spots, a technique that has the disadvantage of several measurements having to be carried out at the leaf or fruit sample to gain a representative mean of the plant tissue investigated. The problem is that the fluorescence is usually not homogenously spread out across the complete leaf or fruit surface but shows a certain patchiness and heterogeneity. With the availability of color lasers of different

Fluorescence

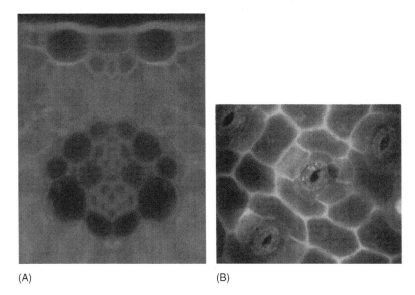

(A)   (B)

FIGURE 4.2.1 (See color insert following page 376.) Origin of the blue-green and red fluorescence of plant leaf tissues as viewed with a fluorescence microscope. (A) Fluorescence of the cross section of a green leaf of the C_4-plant maize (*Zea mays* L.). The emitted blue fluorescence is clearly seen in the cell walls of the epidermis (upper cell layer) and bundle sheet (central ring). Photo from Buschmann, C. and Lichtenthaler, H.K., *J. Plant Physiol.*, 1998. 152, 297, 1998. (B) Blue fluorescence of cell walls of an epidermis stripped off from a leaf of *Commelina communis* L. The chloroplast in the stomatal guard cells exhibits red chlorophyll fluorescence. (From Lichtenthaler, H.K., Lang, M., Sowinska, M., Heisel, F., and Miehé, J.A., *J. Plant Physiol.*, 148, 599, 1996.) In both photos, fluorescence was excited in the UV (365 nm).

wavelengths as excitation source, the fluorescence investigations were further developed to the laser-induced fluorescence (LIF) technique selecting lasers with particular wavelengths. However, the LIF-technique alone did not eliminate the great disadvantage of having to perform several measurements at different parts of the sample to obtain realistic and reliable fluorescence information.

In the last decade, the new powerful *technique of fluorescence imaging* has been developed, which is a very useful, rapid, and nondestructive tool for characterizing plant tissues, such as leaves, vegetables, and fruits. Pulsed lasers for fluorescence induction are also applied in multicolor fluorescence imaging, but pulsed flashlamps with appropriate filters are equally efficient and do not need an extreme beam expansion as lasers do (Buschmann et al. 2000; Lichtenthaler and Babani 2000). With its high spatial resolution, this extremely valuable, high-tech fluorescence imaging method allows determining the size and the developmental stages of leaves and fruits during growth, the assessment of physiological states or stress effects, as well as changes during storage and aging of fruits. The great advantage of the fluorescence imaging technique is that it simultaneously provides, for each measurement, the fluorescence information on several 10,000 pixels of the plant tissue examined, showing spatial heterogeneities of the sample as well as small local fluorescence disturbances, which may be early symptoms of stress, infections, or senescence. Thus, the results obtained by fluorescence imaging are of high statistical significance and reliability. Fluorescence imaging is best applied as multicolor fluorescence imaging by imaging all four fluorescence bands of plants. Fluorescence imaging possesses a broad span of applications and can help in screening plants and cultivars, selecting optimal properties, monitoring the action of agrochemicals, determining optimal fertilization and irrigation regimes, and controlling the quality of agrofood as well as its changes during storage. The application of the efficient multicolor fluorescence imaging technique, presently the best and superior fluorescence method, is not only of economic and ecological interest. In the future, it will become more and more important in outdoor and indoor agriculture, horticulture, and plant food production, in growth and ripening control including high-throughput postharvest sorting and packing, as well as in quality and process assessment of agrofood and the prediction of shelf life.

4.2.1 NATIVE FLUOROPHORES IN PLANTS

4.2.1.1 General Aspects of Plant Fluorescence

Several native plant molecules show after excitation with UV or visible light fluorescence, as a way of de-excitation of absorbed light energy, a process which is also known as photoluminescence. The efficiency of fluorescence emission is expressed in the fluorescence quantum yield, which is defined as the ratio of the number of photons emitted to the number of photons absorbed and results in values between 0 and 1. The fluorescence quantum yield of chlorophyll a dissolved in an organic solvent is 0.3 for a light-green diluted solution. In contrast, in an intact green plant tissue, where chlorophyll is present in highly concentrated form in the chlorophyll proteins of the chloroplast biomembranes, the quantum yield is much lower and varies between 0.01 and 0.1 (Barber et al. 1989). This indicates that only 1% (dark-green tissue) up to 10% (light-green tissue) of the absorbed light quanta are emitted as red + far-red chlorophyll fluorescence in vivo. In contrast to the stronger reflectance of plant tissues that is often indicated as percentage of the incident light, the fluorescence intensity of plant tissues is low and therefore usually not given in absolute values but almost exclusively expressed as fluorescence ratios. These fluorescence ratios exhibit a much lower variability compared to the absolute fluorescence values and are therefore more reliable than the latter. Ratios usually minimize the effect of variation due to instrumental factors (optical properties of the system, sensitivity of the detector) and sample factors (dependence on different angles and distance of excitation as well as fluorescence detection with rounded or rough surfaces). Fluorescence ratios are usually determined from the fluorescence yield measured at two different wavelengths, after two differing excitation times or under differential excitation conditions. Since the fluorescence intensity increases with the intensity of the excitation light, a good signal-to-noise ratio can only be achieved at a high intensity of excitation radiation. The latter is limited, however, by the type of light source and the energy tolerance of the plant sample. Under optimum conditions, the detection limit for fluorescent substances, also in plant tissues, is in the order of 10^{-12} mol. With conventional spectroscopic absorption measurements, substances are detectable only up to 10^{-8} mol (Valeur 2002).

Although the fluorescence of plant tissues is low, it offers the big advantage of being plant specific due to the restricted number of clearly defined fluorophores in the tissues. The endogenous (intrinsic) fluorophores of the plant tissue are the same, irrespective of the fluorescence measurement scale (i.e., from the microscopy via whole leaves or fruits to remote sensing). The major fluorophores of green plants are chlorophyll *a* that emits the red + far-red fluorescence and ferulic acid in the cell walls that emits the blue-green fluorescence.

4.2.1.2 Blue-Green Fluorescence of Plants

Blue and green fluorescence primarily derive from the blue-green fluorophore ferulic acid covalently bound to all plant cell walls as has been directly shown via a detailed chemical analysis of hydrolyzed cell walls (Harris and Hartley 1976; Lichtenthaler and Schweiger 1998) and via an indirect method using a temperature gradient (Morales et al. 1996). This genuine blue-green fluorescence is characterized by a high blue fluorescence band ranging from 420 to 430 nm and a much lower shoulder in the green wavelength region near 520 nm. The blue-green fluorescence of cell walls can easily be detected in a leaf cross section by fluorescence microscopy using UV excitation as shown in Figure 4.2.1A for the C_4-plant maize and for the stripped off leaf epidermis of *Commelina communis* L. in Figure 4.2.1B. The same blue-green fluorescence characteristics for the epidermis and green plant tissue cells are also found in the bifacial leaves of C_3-plants and in green fruits. In addition to ferulic acid bound to the cell wall, several other endogenous organic plant substances localized in other cell compartments also show blue-green fluorescence, for example, other cinnamic acids such as chlorogenic acid, caffeic acid, sinapic acid; the coumarins aesculetin or scopoletin as well as the widespread catechin; some flavonoids (e.g., quercetin) and other phenolic compounds; the alkaloid berberin; and also NAD(P)H and phylloquinone K1. The latter decomposes with a green fluorescence under UV treatment (Lichtenthaler et al. 1977). Fluorescence emission spectra of these compounds have been shown by Lang et al. (1991). General fluorescence information on such compounds is also found in other references (Goodwin 1953; Chappelle et al. 1984; Rost 1995; Morales et al. 2005). These substances, most of them secondary plant products dissolved and accumulated in the cell vacuoles (many phenolics) or functional compounds in the chloroplasts (NADPH, phylloquinone K1), may contribute slightly to the overall blue-green fluorescence emission of green and nongreen plant tissues, yet, the main signal comes from the cell walls. In fact, the intensity of the blue-green fluorescence depends on the ferulic acid content of cell walls, which is plant specific. In contrast to several assumptions of various scientists and older literature references, β-carotene does not show any blue-green fluorescence (Lang et al. 1991). Impure solutions of β-carotene, for example, isolated via chromatography from a leaf pigment extract still containing several prenylquinones and other lipophilic compounds, can exhibit, upon UV excitation, a blue-green fluorescence partially deriving from decomposing phylloquinone K1 and several other blue-green lipophilic contaminants. Purified β-carotene, however, like other carotenoids, does not show any blue-green fluorescence (Lang et al. 1991) as had routinely been investigated in our laboratory. The yellow and red carotenoids are regarded as nonfluorescent. There are reports of a very weak fluorescence with a quantum yield of 3×10^{-5} that can be noticed in picosecond fluorescence measurements, as reviewed by Truscott (1990), but this very weak fluorescence is not able to cause the large blue green fluorescence signal of plant tissues. Lignin per se apparently does not show a blue-green fluorescence. The assumption of lignin fluorescence is based on speculations, but no evidence has ever been shown. Cell walls of lignified cells are usually thicker than those of other cells, thus their ferulic acid based blue-green fluorescence is perceived to a higher degree in the fluorescence microscope. However, this phenomenon is also found in other nonlignified cells with thicker cell walls (e.g., in collenchyma cells). In this context it is noteworthy that, in contrast to intact leaves, isolated chloroplasts or thylakoid membranes show strong red chlorophyll fluorescence at 77 K, but practically no blue-green fluorescence (Stober et al. 1994). Under stress conditions, during infections or aging of plant tissues additional substances with blue or green fluorescence may show up as phytoalexins, for example, the stilbene resveratrol in vine leaves, eventually changing the shape of the fluorescence emission spectrum of the plant tissue, particularly enhancing the green fluorescence band near 520 nm. The hypersensitive reaction of tobacco leaves to infection with the tobacco mosaic virus induces the formation of the blue fluorescent scopoletin (Lenk et al. 2007a). In green leaves or fruits another essential aspect of the blue-green fluorescence of plant tissues has to be taken into consideration. The blue-green fluorescence emitted by the cell walls of the green cells is, to a large degree, reabsorbed by the blue-green light absorbing photosynthetic pigments, the carotenoids, and the chlorophylls, which are bound as pigment–protein complexes in the photochemically active thylakoids of the chloroplasts of the subepidermal leaf mesophyll cells and green cells of fruit tissue. Also, on the leaf cross section (Figure 4.2.1), the blue-green fluorescence of the mesophyll cell walls is often not detectable because it is partly reabsorbed and, in addition, superimposed by the red + far-red chlorophyll fluorescence. Because of this fact, the blue-green fluorescence of green plant tissue emanates predominantly from the cell walls of the nongreen epidermis cells that are devoid of chlorophylls and carotenoids. The typical characteristics of the plants' blue-green fluorescence and its detection are summarized in Table 4.2.1. The blue-green fluorescence is high at excitation wavelengths <330 nm, while at higher excitation wavelengths the blue-green fluorescence yield strongly declines since the responsible fluorophores hardly or no longer absorb the wavelengths >350 nm.

4.2.1.3 Red + Far-Red Chlorophyll Fluorescence of Plants

The plants' red + far-red chlorophyll fluorescence originate in the chlorophyll *a* molecules in the chloroplasts' thylakoid biomembranes of the green mesophyll cells of green leaves and fruits. Two types of chlorophylls exist in higher plants; chlorophyll *a* and the minor component chlorophyll *b* approximately in a ratio of 3:1 (see review Lichtenthaler and Babani 2004). Both chlorophylls possess characteristic absorption bands in the blue and red spectral region. As isolated pigments in an

TABLE 4.2.1
Characteristics of the Blue-Green Fluorescence Emitted by Plant Tissues

Blue-green fluorescence

Fluorescing pigments	Primarily ferulic acid covalently bound to cell walls, slight modulation by soluble cinnamic acids and flavonoids in cell vacuoles
Extraction	Cell wall bound ferulic acid: after alkaline hydrolysis
	Soluble cinnamic acids and flavonoids: with aqueous methanol
Location and origin	Cell walls (signal detected mainly from leaf epidermis)
	Vacuoles (soluble phenols, only slight interaction)
Excitation	UV radiation (e.g., N_2-laser: 337 nm; tripled frequency Nd:YAG laser: 355 nm; UV LEDs; pulsed xenon arc flashlamp with UV filter)

Fluorescence characteristics (see Figure 4.2.2)

Emission range	400–570 nm
Maxima	Near 430–450 nm (blue band $F440$)
	and a shoulder/maximum near 520–530 nm (green band $F520$)

organic solution, both chlorophylls exhibit typical fluorescence emission spectra with a high maximum in the red region and a shoulder in the far-red. However, in vivo, in the chlorophyll–protein complexes of the photosynthetic biomembranes only the fluorescence of chlorophyll *a* is emitted. This is due to the fact that in close association with chlorophyll *a*, as in the pigment–protein complexes of the photosynthetic thylakoid biomembranes, chlorophyll *b* transfers its excited state fully to chlorophyll *a*. The major characteristics of the plants' red + far-red chlorophyll fluorescence, its excitation, and detection are summarized in Table 4.2.2.

TABLE 4.2.2
Characteristics of the Red + Far-Red Chlorophyll Fluorescence Emitted by Plant Tissues

Red + far-red chlorophyll fluorescence

Fluorescing pigment	Chlorophyll *a*
Extraction	With organic solvents (e.g., acetone)
Location and origin	Chloroplasts of green leaf mesophyll cells
Excitation by	UV radiation (N_2 laser: 337 nm; tripled frequency Nd:YAG laser: 355 nm; UV LEDs, pulsed xenon arc flashlamp with UV filter); visible light from blue via green to orange-red (LEDs and lasers; He/Ne-laser: 632.8 nm; green laser: e.g., doubled frequency Nd:YAG laser: 532.5 nm; pulsed xenon arc flashlamp with appropriate filters)

Fluorescence characteristics (see Figure 4.2.2)

Emission range	650–800 nm
Maxima	In green leaves and fruits: near 690 nm (red band $F690$) and near 730–740 nm (far-red band $F740$)
	Chlorophyll *a* dissolved in organic solvent: near 680 nm

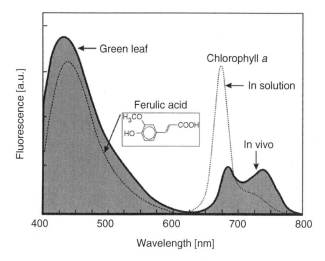

FIGURE 4.2.2 Comparison of the fluorescence emission spectrum of a whole leaf with the fluorescence spectra of the two main leaf fluorophores. In vivo fluorescence emission spectrum of an intact green tobacco leaf (*Nicotiana tabacum* L.) given as gray area below the bold line. The fluorescence emission spectra of the two main fluorophores of leaves are indicated for comparison: ferulic acid (in solvent methanol, dashed line between 400 and 600 nm) and chlorophyll *a* (in solvent acetone, dotted line between 620 and 800 nm). The spectra were taken at room temperature with UV-A excitation at 340 nm using a fluorescence spectrometer.

Similar to the blue-green fluorescence, the intensity of the chlorophyll fluorescence depends on the concentration of the fluorophore chlorophyll *a*. However, at high chlorophyll *a* levels in green leaves, vegetables, and green fruits, a partial fluorescence reabsorption takes place (see review Buschmann 2007) changing the shape of the fluorescence spectrum as outlined below (Section 4.2.3.1). Taking this knowledge into consideration when interpreting fluorescence measurements, the chlorophyll *a* fluorescence of plant tissues is a powerful trait in the assessment of the photosynthetic function and quality of green agrofood. An example of the fluorescence emission spectrum of a green leaf is given in Figure 4.2.2 where the fluorescence spectra of ferulic acid and chlorophyll *a* in solution are presented as well.

The present knowledge on the light absorption of green leaves and the origin of the blue-green fluorescence (primarily in the epidermis cells) and the red + far-red chlorophyll fluorescence (green mesophyll cells) are summarized in Figure 4.2.3. The transparent epidermis, which is free from chlorophylls and carotenoids, absorbs UV radiation with its cell wall bound ferulic acid as well as with flavonols and other plant phenolics of the cell vacuole. In contrast, the green chlorophylls and yellow carotenoids in the chloroplasts of the green subepidermal mesophyll cells are responsible for the absorption of visible light. The latter, together with the ferulic acid of the mesophyll cell walls, absorb, however, also that part of the UV radiation that has passed the epidermis without having been absorbed there. As a result of the absorption characteristics of epidermis and green cells, the epidermis emits only blue-green fluorescence, whereas the green mesophyll cells emit some blue-green

Fluorescence

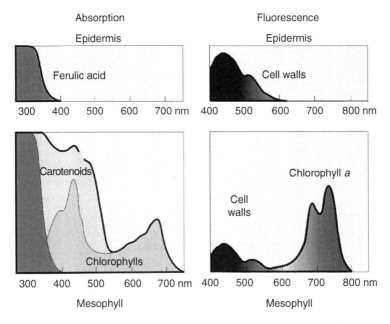

FIGURE 4.2.3 (See color insert following page 376.) Example of absorption (left part) and UV-induced fluorescence emission (right part) in the chlorophyll-free epidermis cells (upper part) and in green mesophyll cells of a leaf (lower part). Cinnamic acids, predominantly ferulic acid, covalently bound to the cell walls, absorb in the UV, the yellow carotenoids (located in the chloroplasts) absorb in the blue, and chlorophylls absorb in the blue and in the red spectral region. The fluorescence emission spectra of leaves show a maximum in the blue (440–450 nm), a shoulder in the green (520 nm), as well as maxima in the red (690 nm) and the far-red (735 nm) spectral region. Thus, the in vivo fluorescence emission spectrum of a green leaf is essentially composed of the blue-green fluorescence of epidermis cells, and the red + far-red chlorophyll fluorescence of the green mesophyll cells.

fluorescence and all of the leaf's red + far-red chlorophyll fluorescence. The overall fluorescence emission of an intact green leaf is the sum of fluorescence emission of epidermis and mesophyll cells.

This observation on the origin of the blue-green fluorescence signals from leaf and other plant tissues is further confirmed by an extensive fluorescence microscopic investigation of freshly prepared cross sections of differently pigmented wheat seedlings by Stober (1993), see also Stober and Lichtenthaler (1993b). The green primary leaves of wheat exhibited a strong blue-green fluorescence in the cell walls of epidermis cells including the epidermal leaf hairs and in the leaf bundle cells, whereas the green mesophyll cells showed the typical red chlorophyll fluorescence. Cross sections of etiolated primary leaves exhibited the blue-green fluorescence in the cell walls of all leaf cells and a faint red fluorescence that came from the low amounts of protochlorophyllide in the etioplasts of mesophyll cells (Buschmann and Lichtenthaler 1998). In contrast, the cross sections of fully white primary wheat leaves, which developed in the light in the presence of the herbicide norflurazone and were devoid of chlorophylls and carotenoids, showed only the blue-green fluorescence in all cell walls. From the cross sections of the still living cells, most of which were fully intact, the blue-green fluorescence signals came apparently exclusively from the cell walls. There was no visual indication that vacuoles and the flavonols or cinnamic acids usually dissolved in these would substantially contribute to this blue-green fluorescence emission.

4.2.1.4 Fluorescence Excitation Spectra

Isolated fluorescent substances dissolved in an organic solvent can mostly be identified via determination of their fluorescence excitation spectra because the fluorescence excitation spectra are quite similar to the respective absorption spectra of the fluorophore. For this purpose, the fluorescence yield in a fluorescence band is measured, usually the fluorescence maximum of a fluorophore, and the fluorescence yield is determined at each wavelength of a certain short wave spectral region. The blue fluorescence of a leaf or fruit is measured via the fluorescence yield at 440 nm and determined via excitation in the wavelength range from 240 to 380 nm. This yields the excitation spectrum of the blue fluorescence. For chlorophyll fluorescence, the detection is set on the red fluorescence band near 690 nm (or the far-red fluorescence band near 730–740 nm) and the fluorescence yield is registered for the excitation wavelengths from 250 to 610 nm. Examples of fluorescence excitation spectra of the red chlorophyll fluorescence (excited from 250 to 400 nm) are shown in Figure 4.2.4 indicating large differences in the excitation spectra between shade and sun leaves as well as between leaves from indoor and outdoor plants. In sun leaves and leaves from the outdoor plants, the red chlorophyll fluorescence yield around 350 nm is relatively low and increases with increasing excitation wavelength much later than in shade leaves and leaves of indoor plants.

In contrast to pure fluorophores in solution, the determination of fluorescence excitation spectra in intact plant tissues does not, however, normally allow a clear identification of the responsible fluorophore or fluorophores, in particular when several fluorophores are present, such as in the case of blue-green fluorescence. Moreover, the internal optical properties of the plant tissue can disturb both the spectral distribution of the excitation light and of the emitted fluorescence. Furthermore, the energy transfer from nonfluorescing compounds, such as from carotenoids (which also absorb exciting UV-A radiation and blue light) to chlorophylls (chlorophyll a), and from chlorophyll b (inside the light-harvesting antenna system of the chloroplast biomembranes) to the fluorophore chlorophyll a, can lead to a drastic difference between the absorption spectrum of plant tissue and its fluorescence excitation spectrum. Thus, in the investigation of plant fluorophores and the interpretation of plant excitation spectra one has to take several possibilities into consideration to arrive at a valid conclusion.

4.2.1.5 Fluorescence Characteristics of Plant Tissues

The particular characteristics of the two major fluorophores of plant tissues that determine the fluorescence emission spectra of intact leaves and fruits have been summarized above in Tables 4.2.1 and 4.2.2. The fluorescence of green leaf and fruit

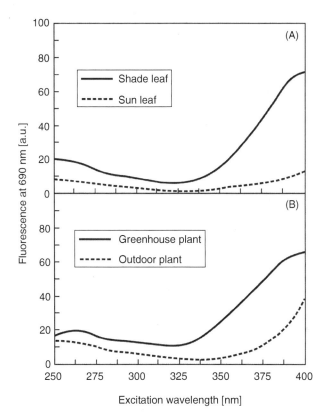

FIGURE 4.2.4 Excitation spectra of red chlorophyll fluorescence (*F*690) of plant leaves. (A) Excitation spectra of *F*690 of sun and shade leaves of the beech (*Fagus sylvatica* L.) and (B) of green maize leaves (*Zea mays* L.) from outdoor and greenhouse plants. The fluorescence yield was determined at 690 nm and at 735 nm. The excitation spectra for the *F*735 band were almost identical to the excitation spectrum of the *F*690 band shown here. (Data based on Schweiger, J., Lang, M., and Lichtenthaler, H.K., *J. Plant Physiol.*, 148, 536, 1996.)

tissue is hallmarked by blue-green fluorescence and red + far-red chlorophyll fluorescence. In contrast to green tissues, the fluorescence of yellow, orange, red, and white plant tissues of fruits, roots (e.g., carrot), or inflorescences (e.g., cauliflower) possess high blue fluorescence that decreases to a lower distinctly green fluorescence and finally levels off toward longer wavelengths. This blue-green fluorescence continuously bottoms out with increasing wavelengths toward the infrared spectral range, and it has some very low amounts in the red + far-red wavelength region. The latter, extremely low and steadily decreasing fluorescence in the red + far-red region does not, however, come from chlorophyll *a*, but might be mistaken at first glance as very low chlorophyll fluorescence, which is the case in a nongreen fruit or root tissue that does no longer possess any chlorophyll, such as an orange carrot root. In this respect, one needs to bear in mind that chlorophyll *a* fluorescence is specifically characterized, even at low chlorophyll levels, by maxima or shoulders in the red (near 690 nm) and the far-red (near $F735$ nm) and never by a low and more or less straight declining line. Another aspect to be considered in the interpretation of the red + far-red fluorescence signatures is the fact that red + far-red chlorophyll fluorescence show an induction kinetic upon illumination of the tissue that is known as Kautsky effect (see review Lichtenthaler 1992). This induction kinetic, displayed by dark-green tissue or tissue kept at low irradiance in the laboratory, is characterized by an initially high red + far-red fluorescence yield (maximum chlorophyll fluorescence) that declines with the onset of photosynthesis (photochemical light use) after a few minutes to a low steady-state level of chlorophyll fluorescence (Lichtenthaler and Miehé 1997; Lichtenthaler et al. 2005b and below in Section 4.2.3.4). In contrast, the blue-green fluorescence emission is always constant during the whole illumination period of plant tissues and does not possess any induction kinetic (Stober and Lichtenthaler 1993a,b).

4.2.2 Fluorescence Emission Spectra of Plant Tissues

4.2.2.1 Shape of Emission Spectra in Different Plant Tissues

The fluorescence emission spectra of green leaves of tobacco and maize (Figure 4.2.5) show clear maxima at 440 nm (blue), 690 nm (red), and 735–740 nm (far red). In many tissues the green fluorescence can be present as an additional, distinct band at 520–530 nm (green) that is, however, less prominent and often only visible as a shoulder in many green and nongreen plant tissues. Between dicotyledonous and monocotyledonous plants, specific differences exist in the blue-green fluorescence yield. Monocot plants (e.g., maize) exhibit a much higher blue-green fluorescence,

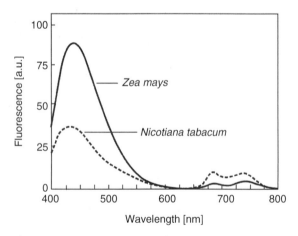

FIGURE 4.2.5 Fluorescence emission spectra of intact plant leaves. Shown are the spectra of a monocotyledonous plant (solid line, C_4-plant maize, *Zea mays* L.) and of a dicotyledonous plant (dashed line, C_3-plant tobacco, *Nicotiana tabacum* L.). Monocot plants possess in general a considerably higher blue-green fluorescence yield than dicot plants. The spectra were taken at room temperature from the upper leaf side using a fluorescence spectrometer (UV excitation at 340 nm).

Fluorescence

which is associated with a higher ferulic acid content in the cell walls as compared with dicot plants, such as tobacco. The relative blue-green fluorescence yield of tobacco shown in Figure 4.2.5 is about the highest found in dicot plants, most of them possess a much lower blue-green fluorescence. Also, in monocotyledonous plants there are gradients in the blue-green fluorescence yield, and the one shown for maize is in the medium range.

During fruit ripening (e.g., in green apples, bananas, and tomatoes) the chlorophyll is successively broken down. Thus, ripe, visually nongreen fruits possess little red + far-red chlorophyll fluorescence when excited with UV-A (Figure 4.2.6).

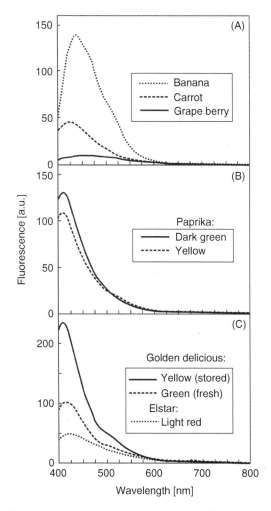

FIGURE 4.2.6 Blue-green fluorescence emission spectra of various fruits and an orange carrot obtained at a local market. The spectra were taken at room temperature using a fluorescence spectrometer (UV excitation at 340 nm). (C) Emission spectra of apples. Note that the scale of the y-axis of (C) differs from that of (A) and (B).

In some cases, the chlorophyll fluorescence can no longer be detected (e.g., in ripe tomatoes that are completely red). Then the blue fluorescence band no longer shows a maximum near 440 nm, but rather between 415 and 430 nm. In the case of ripe red tomatoes, even the blue-green fluorescence exhibits extremely low values, which are only at a level of 1%–5%, as compared with other ripe fruits so that the fluorescence emission spectrum could not be shown in Figure 4.2.6.

4.2.2.2 Dependence of Fluorescence Yield on the Excitation Wavelength

The choice of the proper wavelength range for exciting fluorescence of plant tissues is highly important:

1. Fluorescence excitation can only be carried out with wavelengths lower than the fluorescence emission (Stokes shift). Thus, the blue-green fluorescence can only be excited with UV, usually UV-A (wavelengths between 320 and 400 nm).
2. To have a high fluorescence signal (high signal-to-noise ratio) an excitation wavelength range must be chosen that is strongly absorbed by the fluorophore itself or by the pigments transferring their excitation energy to the fluorophore (e.g., carotenoids and chlorophyll *b* in case of chlorophyll *a* fluorescence of intact green plant tissue).
3. The penetration of the excitation light into the plant tissue and the path of the fluorescence radiation out of the plant tissue must be taken into account.

Thus, the UV-quanta used for excitation can strongly be absorbed already in the epidermis layer by flavonols and other plant phenolics of the cell vacuole and hardly penetrate into the green leaf mesophyll so that the fluorescence yield is rather low. This is the case in sun-exposed leaf tissues (sun leaves, high-light leaves) containing high amounts of UV-absorbing flavonols in the leaf epidermis as compared to shade leaves or low-light leaves. In addition, certain wavelength parts of the UV-induced emitted blue fluorescence can strongly be absorbed already in the epidermis layers (e.g., by the flavonols) so that it can be sensed only from cells and cell layers close to the sample surface. Owing to a rather high absorption rate of short-wave UV-quanta in the epidermis cells, especially in outdoor grown, sun-exposed leaf and needle tissue, the excitation of chlorophyll fluorescence starts only at wavelengths >340 nm and the yield of chlorophyll fluorescence then steadily increases with longer wavelengths. For this reason, the chlorophyll fluorescence excitation spectra are quite different for sun-exposed leaf and needle tissue as compared with those of shade and low-light plants (Schweiger et al. 1996). In contrast, blue excitation light (range 400–460 nm), which passes unabsorbed through the epidermis and is highly absorbed by carotenoids and chlorophylls already in the first chloroplast layer of the subepidermal mesophyll cells, provides a much higher chlorophyll fluorescence yield than UV-A excitation. Other ranges of visible light, such as green (530–550 nm) or orange light (600 nm) that are slightly absorbed by chlorophylls only and not at all by carotenoids, penetrate much deeper into the cell layers of green plant tissues than blue light. For this reason, the chlorophyll fluorescence excited by

Fluorescence

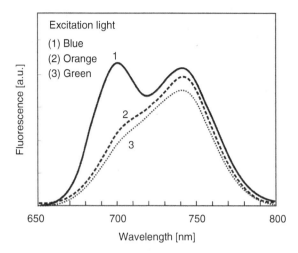

FIGURE 4.2.7 Dependence of the chlorophyll fluorescence emission spectrum on the excitation wavelength shown here for a green sun leaf of beech (*Fagus sylvatica* L.). 1, Excitation with blue light (450 nm); 2, with orange light (600 nm), and 3, green light (550 nm); the measurements were performed at room temperature. The distinct shapes of the emission spectra are due to a differential penetration depth of the excitation light into the leaf and a corresponding differential degree of reabsorption primarily of the emitted red chlorophyll fluorescence. As a consequence, the red/far-red chlorophyll fluorescence ratio $F690/F735$ is higher at blue light excitation ($\sim$1.0) as compared to orange or green light excitation ($\sim$0.6). This dependence has to be taken into consideration for the interpretation of results obtained by different authors and measuring systems. (Data based on Lichtenthaler, H.K. and Rinderle, U., *CRC Crit. Rev. Anal. Chem.* 19, S29, 1988.)

green and orange lights emanates also from deeper tissue layers and provides information on chloroplasts from deeper cell layers. In fact, the relative intensity of the red + far-red chlorophyll *a* fluorescence bands induced by visible light strongly depends on the excitation wavelength (Figure 4.2.7). Compared to excitation by orange or green light, the intensity of the red fluorescence band at 690 nm ($F690$) is higher upon blue light excitation, but distinctly lower at green or orange light excitation. As a consequence, the ratio of the two chlorophyll fluorescence bands red/far-red ($F690/F735$ or $F690/F740$) is higher at blue light excitation than at green or orange light excitation.

4.2.2.3 Contrasting Wavelength Behavior of Blue-Green and Red + Far-Red Fluorescence

Besides visible light, red + far-red chlorophyll fluorescence ($F690$ and $F735$) can also be excited by UV-A radiation, but the yield is considerably lower as compared to visible light. This is not only due to the fact that the absorption bands of chlorophyll are lower in the UV than in the blue or red region. In addition, at UV-excitation wavelengths <350 nm, the chlorophyll fluorescence can no longer be excited in plant tissues or only to a very low yield (Figure 4.2.5) as the exciting radiation is almost completely absorbed by the epidermis cells of the plant tissues. This occurs especially in sun-exposed green leaves, conifer needles, and fruits containing high amounts of UV-absorbing phenols and flavonols in their epidermal cells. As a result, the exciting UV radiation does not penetrate into the green chlorophyll-containing subepidermal cells. In contrast, the plants' blue-green fluorescence ($F440$ and $F520$) is best excited by short-wavelength UV radiation distinctly <350 nm providing high yields (Schweiger et al. 1996). At higher excitation wavelengths, however, the yield of blue-green fluorescence strongly declines. In addition, at longer UV-A wavelengths the measurement interferes with the cutoff filter applied to exclude the excitation radiation. Therefore, a simultaneous excitation of blue-green fluorescence and red + far-red chlorophyll fluorescence requires the selection of an excitation wavelength ranging from 340 to 360 nm. These wavelengths are a compromise that still provides reasonably measurable blue-green fluorescence and already distinct chlorophyll fluorescence. Thus, the whole fluorescence emission spectrum of plant tissues can be measured in one step. The same also holds true for the multicolor fluorescence imaging during the consecutive imaging of the four fluorescence bands of plants (see Section 4.2.6). One can then evaluate the relative amounts of blue-green fluorescence and red + far-red chlorophyll fluorescence, for example, in nonstressed plant tissues and compare it with the relative fluorescence yield of stressed, aged, or infected tissues of leaves, vegetables, or fruits. In most cases, this is the best and fastest approach for green or greenish plant tissues, and it is fully sufficient to obtain reliable fluorescence of the sample. If however, for example, in certain strongly sun-exposed green fruits and leaves, the UV-excited chlorophyll fluorescence signal is too low, one needs to excite the chlorophyll fluorescence in a second measurement either by blue or green excitation light to obtain a good signal. In such cases one has to assess and compare the plant tissue samples via the UV-excited blue-green fluorescence and the blue (or green) light excited chlorophyll fluorescence.

4.2.2.4 Fluorescence Ratios as Assessment Criterion

For the comparative evaluation of leaves, vegetables, and nongreen agrofood, we found it very useful to form the fluorescence ratios blue/red, blue/far-red, blue/green, and the very powerful chlorophyll fluorescence ratio red/far-red. These ratios provide valuable information on the freshness of agrofood and they change with storage or due to infections with pathogens, exposure to biotic and abiotic stress, or during an ongoing chlorophyll breakdown. For such plant tissue investigation of agrofood, the spectral region of the excitation light can mostly be kept constant at 340 or 350 nm. If needed, for example, at low chlorophyll concentration in the sample, a higher wavelength for sensing the chlorophyll fluorescence can be chosen. In both cases, the formation of the fluorescence ratios blue/red, blue/far-red, blue/green, or red/far-red provides valuable information on the assessment of plant tissues. Thus, the relative intensity, for example, of the red + far-red fluorescence bands and their ratio ($F690/F735$), the ratios of blue to red fluorescence ($F440/F690$), and blue to far-red fluorescence ($F440/F735$) are characteristic parameters that can be used to distinguish different plants, tissues, fluorophore content, fruit quality, storage times, stress intensity, and stress types (Lichtenthaler et al. 1996; Buschmann and Lichtenthaler 1998; Lenk et al. 2007a). In fact, the two fluorescence ratios $F440/F690$ and $F440/F735$ have been shown to be the most sensitive, affected by changes of environmental factors, stress constraints, developmental stages, storage or fruit ripening, as well as postharvest chlorophyll breakdown. The chlorophyll fluorescence ratio red/far-red ($F690/F735$), in turn, is a very valuable indicator of the in vivo chlorophyll content (inverse curvilinear relationship) (D'Ambrosio et al. 1992; Lichtenthaler and Babani 2004; Buschmann 2007). Changes in chlorophyll levels (e.g., breakdown during ripening) can easily be monitored via this fluorescence ratio (see below). The blue/green fluorescence ratio ($F440/F520$) of plant tissues often remains constant over time, but it tends to change considerably during stress exposure and infections or aging of leaves, roots, and fruits. In such cases, green fluorescing substances in particular may accumulate and lead to a decrease of the blue/green fluorescence ratio ($F440/F520$).

Recently, the calculation of the ratio between the red chlorophyll fluorescence excited in the UV ($^{UV}F690$) and that excited in the blue ($^{B}F690$) has been suggested to measure the sun exposure of leaves (Bilger et al. 1997). Fluorescence ratio images (Lenk et al. 2007b) were demonstrated to be a valuable tool for predicting the ripeness of grapes.

4.2.3 BASIS FOR THE VARIATION OF FLUORESCENCE SIGNATURES OF PLANT TISSUES

The fluorescence signatures of the plant tissue are influenced by several parameters as outlined below. A comprehensive knowledge of these factors is important for the correct interpretation of fluorescence measurements and to avoid the measurement of artifacts. The variation of the fluorescence signatures within a plant tissue often requires the use of imaging technology to be able to receive a clear characterization of the sample. Most of the studies on blue-green fluorescence and red + far-red chlorophyll fluorescence reported in the literature so far were focused on green and variegated leaves, but in principle these results are also fully relevant to roots, fruits, and vegetable crops.

4.2.3.1 Dependence of Fluorescence on the Concentration of Chlorophyll *a* and Ferulic Acid

4.2.3.1.1 Chlorophyll a

It seems obvious that the fluorescence intensity should steadily rise with the concentration of the fluorophore. However, in the case of chlorophylls this is true only at low to medium concentrations of the fluorophore chlorophyll *a*. At higher concentrations of chlorophyll *a* in solution, the fluorescence of chlorophyll *a* decreases with increasing concentration of chlorophyll *a* as originally detected in pigment extract solutions of leaves and in chloroplast suspensions (Lichtenthaler and Rinderle 1988). This is shown in Figure 4.2.8 for isolated chlorophyll solutions (Gitelson et al. 1998). This fluorescence decrease at higher concentrations affects both the red (Fr) + far-red chlorophyll fluorescence (Ffr); however, the red band to a stronger degree than

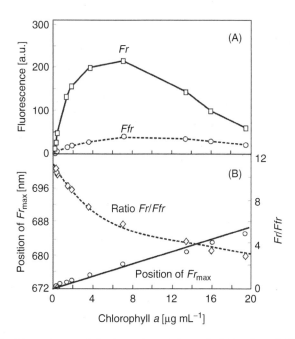

FIGURE 4.2.8 Dependence of the intensity of the red (Fr) + far-red (Ffr) chlorophyll fluorescence signals and ratios on the chlorophyll concentration. (A) Changes in the two fluorescence bands of isolated chlorophyll a, dissolved in 95% ethanol, with increasing chlorophyll a concentration of the solution. (B) Decrease of the ratio of the red + far-red chlorophyll fluorescence (Fr/Ffr) and shift of the red fluorescence maximum (Fr_{max}) to longer wavelengths (from 672 to 686 nm) with increasing chlorophyll a content of the solution. The ratio Fr/Ffr, shown here for a chlorophyll solution, is measured in intact leaves as red/far-red fluorescence ratio $F690/F740$. (Data based on Gitelson, A.A., Buschmann, C., and Lichtenthaler, H.K., *J. Plant Physiol.*, 152, 283, 1998.)

the far-red band. The reason being that the red chlorophyll fluorescence band near 672–680 nm considerably overlaps the absorption band of chlorophyll a in solution causing a substantial reabsorption of the chlorophyll band Fr. This also applies to the far-red band/shoulder near 725 nm, however, the latter is less affected by reabsorption than the red band. As a consequence, the chlorophyll fluorescence ratio Fr/Ffr continuously decreases with increasing chlorophyll a concentration of the solution from values of 10.6 to 2.8 (Figure 4.2.8). At the same time, the wavelength position of the red fluorescence emission band Fr steadily shifts from 672 nm (dilute solution) to a value of 676 nm (high chlorophyll a concentration).

Similar processes are going on in green leaf and fruit tissue where the red + far-red chlorophyll fluorescence bands are shifted to longer wavelengths near 690 and 735 nm, respectively. Again, the red chlorophyll fluorescence band $F690$ is affected to a higher degree by reabsorption through the in vivo chlorophyll absorption bands in the chlorophyll–protein complexes of chloroplasts than the long wavelength chlorophyll fluorescence band $F735$. This is due to a stronger overlapping of the red fluorescence band $F690$ with the absorption bands of in vivo chlorophyll, whereas the long wavelength fluorescence band at 735 nm is less affected as demonstrated in Figure 4.2.9. By means of correct absorption spectra measured in the same leaf, it is possible to calculate the chlorophyll fluorescence spectrum actually emitted at the site of light absorption from the chlorophyll fluorescence emission spectrum measured, as shown in Figure 4.2.9. In contrast to the chlorophyll fluorescence measured, the retrieved, actually emitted leaf chlorophyll *a* fluorescence exhibits a high maximum in the red region and a shoulder only in the far-red region near 735 nm. Thus, it is very similar to the fluorescence spectrum of a chlorophyll

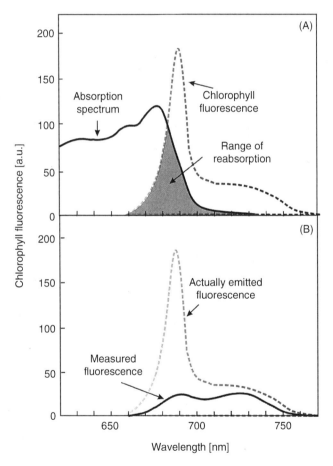

FIGURE 4.2.9 Overlapping of the chlorophyll absorption spectrum with the chlorophyll fluorescence emission spectrum in green leaves causing partial reabsorption of the chlorophyll fluorescence. (A) Red part of the absorption spectrum of a leaf, as a result of light absorption by leaf chlorophylls, and spectrum of the chlorophyll fluorescence. The overlapping part of the spectra marked in gray represents the range of reabsorption. (B) Measured chlorophyll fluorescence emission spectrum of a leaf (strongly influenced by reabsorption) and actually emitted chlorophyll fluorescence spectrum (with reabsorption eliminated by calculation, see Gitelson et al. 1998).

diluted in a solution (compare with Figure 4.2.2). On the basis of the reabsorption process within the leaf, the intensity of the fluorescence $F690$ decreases with increasing chlorophyll content of the green mesophyll cells of leaves and fruits, and consequently with it also the ratio $F690/F735$. In fact, the chlorophyll fluorescence ratio $F690/F735$ shows an inverse curvilinear relationship (Figure 4.2.10A) with the chlorophyll content of green plant tissues (D'Ambrosio et al. 1992; Lichtenthaler and Babani 2004; for a recent review see Buschmann 2007), a relationship that can be used for an in vivo chlorophyll determination without having to extract the chlorophylls. On a double logarithmic plot of the fluorescence ratio $F690/F735$ and the chlorophyll content, an inverse linear relationship is obtained (Figure 4.2.10B).

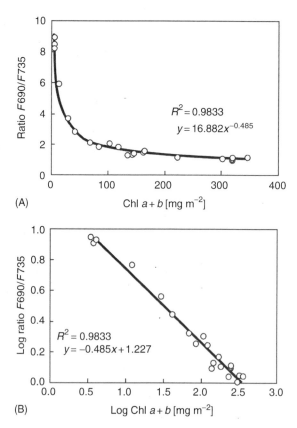

FIGURE 4.2.10 Dependence of the red/far-red chlorophyll fluorescence ratio ($F690/F735$) of wild wine (*Parthenocissus* spec.) leaves on the chlorophyll $a+b$ content per leaf area unit. The ratio was determined from the fluorescence spectra of leaves measured at room temperature. Leaves with different chlorophyll content (photometrically determined from acetone extracts) were chosen during the autumnal chlorophyll breakdown. (A) Inverse curvilinear relationship using linear scales with fit by a regression curve with a potential function, (B) double logarithmic plot with fit by a linear regression line. (Data based on Lichtenthaler, H.K., *J. Plant Physiol.*, 131, 101, 1987a; Buschmann, C., *Photosynth. Res.*, 92, 261, 2007.)

4.2.3.1.2 Ferulic Acid

In contrast to the chlorophyll *a* fluorescence, the blue-green fluorescence that predominantly derives from ferulic acid covalently bound to the cell walls (Lichtenthaler and Schweiger 1998) is little affected by quenching or reabsorption at higher concentrations of the fluorophore ferulic acid. This is obviously caused by the much lower concentration of ferulic acid as compared to chlorophyll. The higher the ferulic acid content, the higher is the blue-green fluorescence yield. Thus, the higher amounts of cell wall bound ferulic acid contained in leaves of monocotyledonous plants (e.g., wheat and maize) as compared to those of dicotyledonous plants (e.g., spinach, sugar beet) go along with a significantly higher blue-green fluorescence of the monocotyledons (Figure 4.2.11). Therefore, the higher blue-green fluorescence can be used to distinguish these two groups of flowering plants in most cases. Tobacco, a dicotyledonous plant, exhibits one of the highest

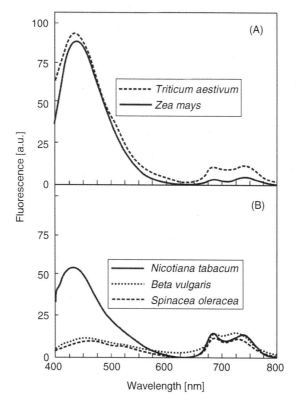

FIGURE 4.2.11 Differences in fluorescence emission spectra of leaves of monocotyledons (A) and dicotyledons (B). The emission spectra were taken at room temperature with intact leaves (excitation in the UV: 340 nm). The higher blue-green fluorescence of monocotyledonous plants as compared to dicotyledons is caused by a higher content of cell wall bound fluorophore ferulic acid (leaves of monocotyledons: 1.1–5.7 mg cell wall bound ferulic acid/g dry weight; leaves of dicotyledons: 0 to 0.8 mg cell wall bound ferulic acid per g dry weight). (Based on Lichtenthaler, H.K. and Schweiger, J., *J. Plant Physiol.*, 152, 272, 1998.)

blue-green fluorescence intensities of dicotyledonous plants, and it also possesses considerably higher ferulic acid content in its leaf cell walls than the average dicotyledonous plants.

4.2.3.2 Tissue Structure and Penetration of Excitation Light

For fluorescence emission to occur, the excitation light must be absorbed. Depending on the optical properties of the plant tissue (degree of absorption by pigments, scattering, internal reflection, and refraction) excitation light penetrates into different depths of the plant tissue (Figure 4.2.12). UV radiation may strongly be absorbed in the epidermis (this is the case for sunlight-grown plants possessing high amounts of flavonols in the vacuoles of their epidermis cells). Thus, the UV-excited fluorescence of the in vivo chlorophylls located in the mesophyll cells beneath the chlorophyll-free epidermis (except for some chlorophyll in the stomata guard cells) can strongly or fully be reduced. The absorption of UV-A radiation in the chlorophyll-free epidermis

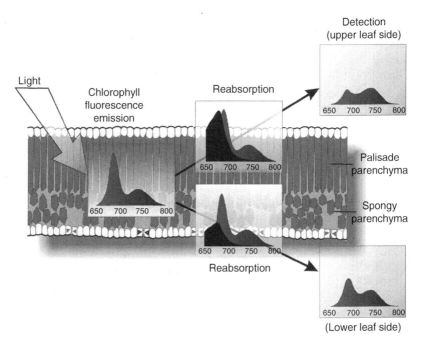

FIGURE 4.2.12 (See color insert following page 376.) Emission, partial reabsorption, and detection of the light-induced chlorophyll fluorescence of green leaves. The scheme indicates the reabsorption primarily of the red chlorophyll fluorescence band $F690$ by the in vivo absorption bands of chlorophyll inside a leaf (shown here with a bifacial leaf structure). The chlorophyll fluorescence emission spectrum (red curve) overlaps with the chlorophyll absorption spectrum (black curve) (see also Figure 4.2.9). The amount of reabsorption depends on the chlorophyll density and is higher when detecting the chlorophyll fluorescence from the upper leaf side (high chlorophyll content in the densely packed palisade parenchyma cells) as compared to the lower leaf side (less chlorophyll in the more loosely arranged cells of the spongy parenchyma layer).

of green fruits, usually grown under sunlight exposure, also explains the occasionally very low red + far-red fluorescence yield of fruits (Figure 4.2.4). The decrease of chlorophyll fluorescence emission due to an epidermal UV-shielding of the green mesophyll or other causes (rough surface structures, wax layers) leads to higher values of the fluorescence ratios blue/red ($F440/F690$) and/or blue/far-red ($F440/F735$), which are very sensitive to environmental changes and to the photon flux density of the incident light.

The effect of the UV-shielding effect of the epidermis has first been observed when measuring chlorophyll fluorescence excitation spectra of leaves and needles (Schweiger et al. 1996). The red chlorophyll fluorescence yield was very low at 350 nm, then steadily increased with longer wavelengths and was highest at blue light (400 nm) excitation. Thus, the evaluation of the chlorophyll fluorescence yield in blue light and in UV at 350 nm is a measure of the UV-shielding degree of the epidermis. The latter was higher in plants grown at high irradiance (e.g., sun exposure) as compared to plants kept at low light conditions (e.g., shade, see Lenk and Buschmann 2006), but also higher in outdoor plants as compared to greenhouse plants (Schweiger et al. 1996). The comparison of chlorophyll fluorescence yield at excitation with blue light and UV radiation is applied as a measure of the UV transmittance of the epidermis, and in some cases it can be used as a stress indicator (Bilger et al. 1997). The broad blue light absorption bands of carotenoids, together with the blue absorption of the chlorophylls, cause a diminished penetration of blue excitation light into the deeper layers of green plant tissue. Therefore, the chlorophyll fluorescence signal upon blue light excitation is emitted more closely to the sample surface than the fluorescence excited with red light. Red light qualities are absorbed only by the red chlorophyll bands, which are rather narrow compared with the broad chlorophyll absorption bands in the blue range. When exciting with green light that has a lower chance of being absorbed due to the low absorption bands of chlorophyll in this wavelength region, the chlorophyll fluorescence comes from deeper inside the green leaf or fruit sample. Since the absorption of green light by the chlorophylls is low, it is absorbed by the leaf or fruit mesophyll to a fairly strong degree (1) because it penetrates into deeper cell layers of the leaf where it excites the chlorophylls, and (2) also due to the lengthening of the light path by multiple reflecting and scattering of the green photons in the green leaf or fruit tissue cells before they are finally absorbed (detour effect). In contrast to this, there also exists a sieve effect of light. This effect is caused by the nonhomogeneous distribution of chlorophyll in the chloroplasts leading to a lower absorption, especially in spectral bands with high absorption, i.e., in case of chlorophyll in the blue and red region (McClendon and Fukshansky 1990).

The optical properties of plant tissues lead to a strong variation of absorption, transmittance, and reflectance of light quanta. All these, including the light-induced fluorescence, depend on the radiation's angle of incidence on the sample. The wavelength-dependent variation of the light path in the leaf tissue (Combes et al. 2007) also influences the intensity of the fluorescence signal, since the absorption properties determine the fluorescence excitation. This is of particular importance not only for round-shaped fruits but also for curled leaves. For instance, to acquire an image of an apple independent of the angle of incident light, one has to reduce the image size to an area where the apple is more or less flat, i.e., a circle with an area of about 300 mm^2 (Lötze et al. 2006). On the other hand, when calculating the fluorescence ratios, the large variations in the absolute fluorescence yield as a result of differences between the distance of individual sample parts (e.g., of an apple, a bell pepper, or a wavy vegetable leaf surface) as well as of different angles of the exciting light become much smaller. Hence, the fluorescence ratios facilitate the interpretation of the fluorescence signatures within one sample. In other words, one should not judge two different samples and draw conclusions based on the differences of the absolute fluorescence signals between control and sample, on a possible infection by pathogens, on tissue senescence, or on stress and ripening effects, etc. One can draw conclusions on stress, damage, infection, or aging of a second sample as compared to a first control sample only when the fluorescence differences clearly show up in the fluorescence ratio images. Such ratio differences (e.g., in the very sensitive fluorescence ratios blue/red and blue/far red) can either (1) be distributed more homogenously over the sample (e.g., storage and senescence effects affecting the whole sample) or (2) consist of various isolated small local differences in several spots of the sample (e.g., local infection of pathogens or punctures of insects) or smaller cell groups on the leaf rim (e.g., caused by water stress). It is also essential to take into consideration that the detection of small local spots with fluorescence ratio changes as compared to a control sample only shows up in a pixel to pixel division, for example, of the blue and red fluorescence yields of the individual sample pixels.

Highly reflecting surfaces and an often rather high wax cover of leaves and particular fruits can strongly increase the reflectance of the excitation light and thus considerably reduce the fluorescence yield, whereby the chlorophyll fluorescence is generally more affected than the blue-green fluorescence. These may cause problems with fluorescence detection since the large reflection of the exciting photons reduces the amount of the radiation that passes through the epidermis into the subepidermal green tissues. As a consequence, the excitation of the tissue fluorophores (e.g., chlorophyll a) is reduced and with it the intensity of the emitted fluorescence. In such cases, higher values are found for the fluorescence ratios blue/red and blue/far-red as compared with plant tissues with a matted surface.

4.2.3.3 Tissue Structure: Penetration of Fluorescence from Deeper Cell Layers

For fluorescence detection, the fluorescence emitted inside the plant tissue must penetrate to the sample surface. The optical characteristics of green plant tissue for the incident radiation (see discussion above) are also valid for the fluorescence emitted. Blue-green fluorescence from the cell walls deep inside the tissue is readily absorbed on its way to the sample surface by the blue absorption bands of the carotenoids and chlorophylls in the chloroplasts of the green mesophyll cells. Therefore, the blue fluorescence sensed of green leaf tissue, vegetable tissue, and fruit tissue is nearly exclusively emitted from the cell walls of the epidermis side facing the detector. Since the epidermis cells lack chlorophyll and carotenoids (except for the few guard cells of the stomata, see Figure 4.2.1B), they are no barrier for the emitted blue-green fluorescence.

Fluorescence

The red fluorescence with a maximum near 690 nm that can be measured at the upper adaxial leaf side after excitation by UV-A and blue light has been emitted by the chloroplasts of the green mesophyll cells below the epidermis. On its way to the leaf surface, it can strongly be reabsorbed by the chlorophylls of the mesophyll cells (Figures 4.2.9 and 4.2.12). This reabsorption effect is most pronounced for the red chlorophyll fluorescence $F690$. The far-red chlorophyll fluorescence $F735$ is much less affected by this reabsorption because the absorption spectrum of the in vivo chlorophylls declines in the far-red spectral region. Thus, the higher the chlorophyll content of a plant tissue sample, the higher the reabsorption of the red chlorophyll fluorescence and the lower its relative amounts as compared to the far-red chlorophyll fluorescence. As a consequence, the fluorescence ratio red/far-red, $F690/F735$ (also expressed as ratio $F690/F740$) decreases with increasing chlorophyll content of the plant tissue. Owing to the differential penetration depth of blue, green, and orange excitation light, the fluorescence ratio red/far red is highest at blue light excitation since in this case the chlorophyll fluorescence emanates from the upper part of the mesophyll cells close to the sample's epidermis surface and thus exhibits only a low degree of reabsorption. Applying orange-red and green excitation light, the values of the fluorescence ratio $F690/F735$ are significantly lower because the chlorophyll fluorescence is emitted also from the chloroplasts of green cells in deeper tissue layers farther away from the tissue surface. This aspect is also documented in the different shape of the fluorescence emission spectra as indicated above (Figure 4.2.7).

4.2.3.4 Photosynthetic Activity

With the absorption of energy by leaves and other green plant tissues, the photosynthetic activity is induced. Under normal physiological conditions, a high percentage of the absorbed light quanta is used for photosynthetic quantum conversion and only a very small percentage for fluorescence emission. As a general rule, the intensity of chlorophyll fluorescence is inversely related to the photosynthetic activity of the plant tissue. There is ample literature about this relation starting from the work of Kautsky in the 1930s (as reviewed by Lichtenthaler 1992) up to more recent overviews (Lichtenthaler and Babani 2004; Lichtenthaler et al. 2005a) as well as the comprehensive reviews by various authors in two books on chlorophyll fluorescence (Lichtenthaler 1988; Papageorgiou and Govindjee 2004). When photosynthesis is high, the chlorophyll fluorescence yield is low; however, when the photosynthetic processes are blocked, for example, by the herbicide diuron, the chlorophyll fluorescence yield is high (as demonstrated also by fluorescence imaging; Lichtenthaler et al. 1997). Photosynthesis comprises the sum of processes from the energy transfer between the absorbing molecules (chlorophylls, carotenoids), the two light reactions and associated electron transport processes, and the CO_2 fixation. There are many parameters of the chlorophyll fluorescence that are used to assess and investigate the photosynthetic activity of leaves: photochemical quenching (q_P), nonphotochemical quenching (q_N or NPQ), maximum quantum yield of photosystem II (F_v/F_m), actual or effective quantum yield ($\Delta F/F_m'$), and the chlorophyll fluorescence decrease ratio (R_{Fd}) (see, e.g., reviews Buschmann 1995, 1999). The parameters q_P and q_N possess inverse information: at high photosynthetic rates, the photochemical quenching of absorbed light energy is high and the nonphotochemical quenching is low. The values of the chlorophyll ratio R_{Fd} are directly correlated to the photosynthetic CO_2 fixation rates, thus their determination by fluorescence imaging has been used to determine the different CO_2 fixation rates of sun and shade leaves of various plants (Lichtenthaler et al. 2007a,b). The different chlorophyll fluorescence parameters are well-defined ratios. They are measured according to a generally accepted fixed protocol using low-intensity pulsed excitation light, stronger additional continuous actinic light, and high-intensity saturating light flashes (for a recent review see Lichtenthaler et al. 2005a). Since these parameters are strongly attached to the processes of energy transfer in the antenna systems, to the charge separation in the reaction centers, to the photosynthetic electron transport and oxygen evolution as well as to CO_2 fixation, they can be taken to monitor and investigate these basic photosynthetic processes, which are most essential for plant life. Via fluorescence imaging of these parameters it is possible to detect gradients in photosynthetic activity across the leaf area and in particular small areas with lower or higher photosynthetic activity (patchiness of activity) as well as differences in photosynthetic activity between sun and shade leaves (e.g., Lichtenthaler et al. 2007a,b).

4.2.4 Examples for the Application of Multicolor Fluorescence Imaging in Studies of Plant Tissues

The fluorescence images of fluorescence bands are most often presented in false color with the fluorescence intensity given in colors ranging from blue (low intensity), via green and yellow (medium intensity) toward red (high intensity). This is the best way to recognize the gradients and patchiness of the fluorescence yield of a sample and local disturbances, which are early indicators of stress constraints or ongoing changes due to storage or ripening. In contrast, in black-and-white presentations (gray tones) such differences and gradients in fluorescence signatures are difficult to recognize. Multicolor fluorescence images of plant tissue samples are consecutively acquired in a sequence of measurements in the four major fluorescence bands of plants, i.e., the fluorescence maxima at 440 (blue), 520 (green), 690 (red), and 740 nm (far red). After several years of comprehensive investigations on the different plant fluorescence signatures, we have selected these four wavelength regions as the most suitable for the fluorescence investigation of green leaves and other plant samples and have been applying them since the first imaging of plant fluorescence during 1994–1996 using the Strasbourg/Karlsruhe laser-induced fluorescence imaging system, the first functional multicolor fluorescence imaging system (Lang et al. 1994b, 1996; Lichtenthaler et al. 1995, 1996). In addition, by forming the ratios of the fluorescence yield sensed at the plants' four fluorescence maxima, one obtains images of the distribution of the fluorescence ratios. These are ratio images of the four fluorescence ratios blue/red, blue/far-red, red/far-red, and blue/green. The ratios are calculated from the fluorescence images for each individual picture element (pixel) by a pixel-to-pixel division, and the ratio value of each ratio image pixel is presented in false colors from blue via green and yellow to red (highest value). The fluorescence ratio images provide excellent information on the physiological and quality assessment of green and nongreen agrofood. The fluorescence ratios blue/red and blue/far-red turned out to be the most sensitive toward changes in growth and environmental conditions and due to biotic or abiotic stress constraints. The chlorophyll fluorescence ratio red/far-red is an indicator of the in vivo chlorophyll content of green leaf and vegetable tissue (see review Buschmann 2007), and the fluorescence ratio blue/green is an indicator of changes or the appearance of additional blue or green fluorescing substances, for example, under stress events. Several examples of fluorescence imaging and ratio imaging of plant tissue and agrofood are given below (Figures 4.2.12 through 4.2.17). The selected examples are typical for and representative of the investigated plants, as has been shown by comparative fluorescence imaging analysis of several other leaves or fruits of the same plant. Concerning the interpretation of multicolor fluorescence images, one has to consider that the imaging of fluorescence signals may be disturbed for some fruit and leaf samples: (1) uneven or rounded surfaces of fruits or leaves (differences in excitation and fluorescence emission), (2) strongly reflecting surfaces (e.g., waxes or hairs), (3) calculating ratios between fluorescence signals with strong differences in intensity, and (4) surface contamination with dust (often fluorescing in the blue) or soil (complete loss of fluorescence). To characterize rapid changes of growth, development, and changes by stress events it is advantageous to monitor fresh plant tissue with high metabolism. For a better characterization of the plant samples, it is advisable in the case of green leaf and fruit issue to spectrophotometrically determine the chlorophyll and carotenoid levels of the sample as a complementary measure to multicolor fluorescence imaging. Both photosynthetic pigments can be determined in the same pigment extract solution (e.g., 100% acetone) by measuring at the absorption maxima of chlorophylls and carotenoids using the predetermined standard equations (Lichtenthaler 1987b; Lichtenthaler and Buschmann 2001).

4.2.4.1 Fluorescence Imaging and Fluorescence Ratio Imaging of Plant Tissue and Fluorescence Agrofood

4.2.4.1.1 Image of a Tobacco Leaf
We have started with the description of the fluorescence images of a green, photosynthetically active tobacco leaf. The fluorescence yield at the four fluorescence bands is shown in Figure 4.2.13. From the image it is obvious that the four fluorescence bands are not homogeneously distributed across the whole leaf area. The highest blue ($F440$) and green ($F520$) fluorescence come from the leaf veins where the chlorophyll content is lower than in other leaf parts. The false color presentation also clearly demonstrates that the blue fluorescence is higher than the green fluorescence. The red ($F690$) and far red ($F740$) chlorophyll fluorescence primarily emanate from the vein-free leaf regions that also have a higher chlorophyll density. Moreover, one can easily recognize that the red fluorescence is higher than the far-red fluorescence, and both are higher than the blue and green fluorescence. When evaluating fluorescence images and ratio images it has to be taken into consideration that the scales are different for the individual fluorescence bands and ratios in most cases. The corresponding fluorescence ratio images of the same tobacco leaf are found in Lichtenthaler et al. (1996).

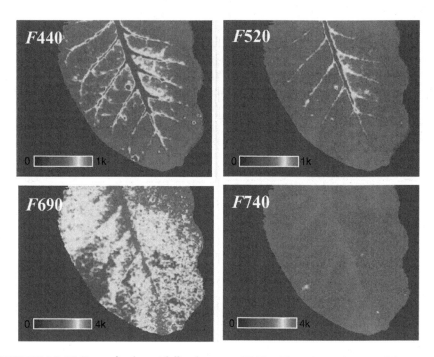

FIGURE 4.2.13 (See color insert following page 376.) Fluorescence images of the upper leaf side of a green tobacco leaf. Shown here in false colors are the intensity of the blue ($F440$) and green ($F520$) fluorescence and the red ($F690$) + far-red ($F740$) chlorophyll fluorescence. In the images, the fluorescence yield increases from blue (no fluorescence) via green and yellow to red as the highest fluorescence. The highest blue and green fluorescence emanate from the leaf veins, whereas the highest chlorophyll fluorescence comes from the vein-free leaf regions. Note that the scales for blue and green fluorescence are different from those of the red + far-red chlorophyll fluorescence. k in the scales means kilo (=1000) counts. (Changed from Lang, M., Lichtenthaler, H.K., Sowinska, M., Summ, P., and Heisel, F., *Botanica Acta* 107, 230, 1994a; Lichtenthaler, H.K., Lang, M., Sowinska, M., Heisel, F., and Miehé, J.A., *J. Plant Physiol.*, 148, 599, 1996.)

4.2.4.1.2 Imaging of Grapes
The false color presentation of the fluorescence images indicates that the blue fluorescence of the greenish desert grapes is much higher than the green or the two red + far-red chlorophyll fluorescence bands (Figure 4.2.14). The red chlorophyll fluorescence $F690$ is significantly higher than the far-red $F740$ fluorescence, an observation to be expected because the chlorophyll density of grapes is very low, and, as a consequence, the usual reabsorption of the $F690$ fluorescence is very low as well. When viewing the blue and red fluorescence images, one recognizes that the fluorescence yield of a single grape is not evenly distributed across the whole irradiated grape surface. In fact, the middle parts of the rounded grape surface that are closest and more perpendicular to the fluorescence detector, in general, exhibit a higher fluorescence yield as compared to the border areas of the grape that are farther

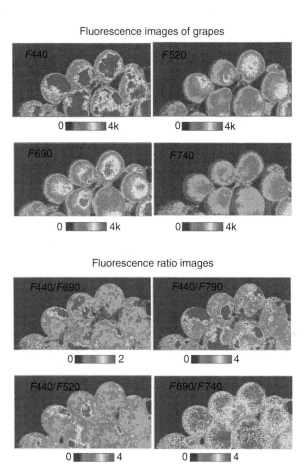

FIGURE 4.2.14 (See color insert following page 376.) Fluorescence imaging of grapes (greenish variety) obtained from a local market. Upper part: images of blue (*F*440), green (*F*520), red (*F*690), and far-red (*F*740) fluorescence. Lower part: fluorescence ratio images blue/red, blue/far-red, blue/green, and red/far-red. The patchiness in the fluorescence ratio images of individual grapes is shown by spot-like increases in the blue fluorescence and in the fluorescence ratios blue/green, blue/red, and blue/far-red and indicates differential maturity and beginning degradation of the individual grapes. The fluorescence intensities in the images are shown in false colors; the fluorescence yield increases from blue (no fluorescence) via green and yellow to red as the highest fluorescence; k in the scales means kilo (=1000) counts. The scales in the fluorescence ratio images indicate the absolute ratio value. Note that the scales for the individual fluorescence signatures are different.

away from the detector. This is a typical phenomenon for all round-shaped, nodular, and spherical fruits and other agrofood samples. In the case of the grapes shown here, one has also to take into consideration that individual grapes and grape parts were at a slightly differing maturity degree and some local regions even at a beginning degradation (see also Lenk et al. 2007b). This effect can best be observed in the fluorescence ratio images blue/red, blue/far-red, and blue/green where the ratio value distribution of single grapes exhibits a completely different behavior as compared to the absolute fluorescence intensities. In the case of the chlorophyll fluorescence ratio $F690/F740$, the ratio values are relatively high—values of up to 4 indicating low chlorophyll content. In addition, they are more evenly distributed across the different grapes indicating that the low chlorophyll content of the grapes and grape regions was more or less in the same range.

4.2.4.1.3 Imaging of Green Bell Peppers

The fully green, well-developed, and rather fresh bell peppers exhibited a high blue fluorescence $F440$, a lower green and far-red fluorescence, and a fairly low red chlorophyll fluorescence $F690$ (Figure 4.2.15, upper part). The fluorescence yield was relatively homogeneous across the middle parts of the fruit (red region of the false color image) that were directly hit by the excitation light. However, toward the borderlines, the fluorescence intensity became successively lower forming "ring-type" quasi-homogeneous zones around the central fruit part. These zones are seen in the false colors of the intensity distribution, first as a yellow zone, then as a green zone, and finally as a light blue zone, especially on the left lateral fruit side. These zones are partly caused by the impact and the radiation's angle of excitation on the sample and to a major part by the fact that the fluorescence emitted can be sensed to a higher degree from the fruit surface directly facing the detector system, but only to a much lower degree from the lateral fruit side. Although the absolute fluorescence yields exhibit these ring-type zone structures, the values of the fluorescence ratios are more homogenous over the bell pepper surface (Figure 4.2.15, lower part). The highest ratio values are found for the ratio blue/red. In contrast to the light-green dessert grapes (Figure 4.2.14) with low chlorophyll content, the chlorophyll fluorescence ratio $F690/F740$ of the fully green bell pepper exhibits much lower values, most of them ranging from 0.5 to 0.7 (maximal values up to 1), which is typical of green plant tissues with a normal chlorophyll content.

4.2.4.1.4 Images of Orange-Red Bell Peppers

Fully grown orange to red bell peppers, colored by high amounts of secondary carotenoids and practically free of chlorophylls, exhibited a significantly lower blue and green fluorescence, up to 2000 counts and 500 counts, respectively (Figure 4.2.16, upper part), as compared to the green bell peppers described above. Imaging these fruits in the chlorophyll fluorescence bands $F690$ and $F740$, we obtained very low chlorophyll fluorescence signals (predominantly in the low range of 70–110 counts). In such cases, one should immediately doubt whether genuine chlorophyll fluorescence was measured because such low-intensity signals are hardly above the background noise. The fact is that the genuine blue-green fluorescence drops down rather fast from the blue maximum to much lower values toward longer wavelengths but still possesses some very low parts in the red + far-red region. They can easily be mistaken for chlorophyll fluorescence (see also Section 4.2.4.3). At low or no chlorophyll fluorescence, the fluorescence ratio images blue/red and blue/far-red show rather high values, in the case of the orange-red bell pepper of up to 32,000 (Figure 4.2.16, lower part). At such high ratio values, one should always be cautious and double-check (e.g., fluorescence emission spectra excited by blue light), if

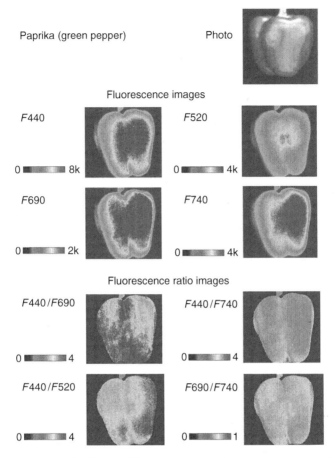

FIGURE 4.2.15 (See color insert following page 376.) Fluorescence imaging of green bell peppers (*Capsicum annuum* L.) from a local market. Upper part: images of blue ($F440$), green ($F520$), red ($F690$), and far-red ($F740$) fluorescence. Lower part: fluorescence ratio images blue/red, blue/far-red, blue/green, and red/far-red. The fluorescence intensities are shown in the images in false colors; the fluorescence yield increases from blue (no fluorescence) via green and yellow to red as the highest fluorescence; k in the scales means kilo (=1000) counts. The scales in the fluorescence ratio images indicate the absolute ratio value. Note that the scales for the individual fluorescence signatures are different.

chlorophyll and a chlorophyll fluorescence signal were really present in that fruit or plant tissue. Although the orange- and red-colored bell peppers are green in the young developmental stage, the pure yellow, orange, or red peppers have generally lost their chlorophyll by degradation at their fully matured stage.

4.2.4.1.5 Imaging of Green Bell Peppers after Storage
When green bell peppers have been stored for some time and are no longer fresh, the water content goes down and the fruits may show decay symptoms (brownish parts) usually combined with a chlorophyll breakdown. The fluorescence images and ratio

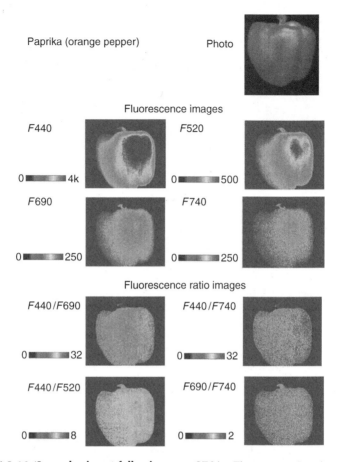

FIGURE 4.2.16 (See color insert following page 376.) Fluorescence imaging of orange-red bell peppers (*Capsicum annuum* L.) from a local market. Upper part: images of blue ($F440$), green ($F520$), red ($F690$), and far-red ($F740$) fluorescence. Lower part: fluorescence ratio images blue/red, blue/far-red, blue/green, and red/far-red. The fluorescence intensities are shown in false colors in the images; the fluorescence yield increases from blue (no fluorescence) via green and yellow to red as the highest fluorescence; k in the scales means kilo (=1000) counts. The scales in the fluorescence ratio images indicate the absolute ratio value. Note that the scales for the individual fluorescence signatures are different.

images of such an example are shown in Figure 4.2.17. The photo of that fruit demonstrates the position of the brown region. The blue and green fluorescence bands are fairly evenly distributed across the excited fruit surface except for a certain ring-type zoning toward the borderlines as can also be found in fresh bell peppers (Figure 4.2.15). The red + far-red fluorescence images show a high intensity except for the upper middle part of the fruit, which is already seen as a brown region exhibiting a lower chlorophyll content. In the fluorescence ratio images blue/red and blue/far red (Figure 4.2.17, lower part), this upper central brown fruit region is documented by much higher values (near 4000, red parts in the false color image)

Fluorescence 303

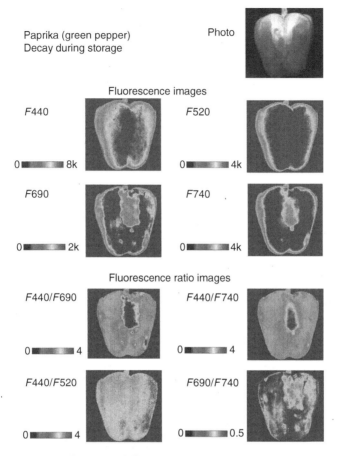

FIGURE 4.2.17 (See color insert following page 376.) Fluorescence imaging of green bell peppers (*Capsicum annuum* L.) from a local market with decay symptoms that showed up after storage at room temperature. Upper part: images of blue ($F440$), green ($F520$), red ($F690$), and far-red ($F740$) fluorescence. Lower part: fluorescence ratio images blue/red, blue/far-red, blue/green, and red/far-red. The fluorescence images and ratio images change during decay (associated with partial chlorophyll breakdown). The fluorescence intensities are shown in false colors in the images; the fluorescence yield increases from blue (no fluorescence) via green and yellow to red as the highest fluorescence; k in the scales means kilo (=1000) counts. The scales in the fluorescence ratio images indicate the absolute ratio value. Note that the scales for the individual fluorescence signatures are different.

as compared to the other fruit parts (mostly a homogeneous green false color). In the fluorescence ratio $F690/F740$, this brown fruit part is visible here and displayed by lower values; this is surprising because at a partial chlorophyll breakdown, one would have expected slightly higher values for the ratio red/far-red, an observation that would need further investigation. The fluorescence ratio blue/green, in turn, possesses a rather even distribution and does not show a specific change in that brown fruit part of the bell pepper. This once again documents that the blue and green fluorescence primarily emanate from the epidermis layer and are hardly influenced by changes in the chlorophyll content of the subepidermal green fruit cells.

4.2.4.1.6 Imaging of Apples at Harvest and after Storage
With the multicolor fluorescence imaging, one can follow the changes occurring in apples during storage and ripening. Freshly harvested light-green apples with some red, anthocyanin-containing parts (sun-exposed parts) were investigated via fluorescence imaging on the light-green side. Directly after harvest, the blue and green fluorescence images gave good signals, the intensities were, however, lower than those of the red + far-red chlorophyll fluorescence (Figure 4.2.18, upper part). During storage at 4°C for up to 6 months, the blue and green fluorescence slightly increased and became more homogeneous across the apple surface. The red chlorophyll fluorescence increased in part and then remained constant. In contrast, the far-red fluorescence band $F740$ continuously decreased indicating a loss of chlorophyll. The fact that the red $F690$ fluorescence band did not decline is a result of the red $F690$ band increasing at first due to a decrease of the reabsorption by the chlorophyll absorption bands during a large part of the breakdown phases of chlorophylls. The $F690$ band only declines in the final stage of chlorophyll degradation, which had not yet been reached during the 6 month storage time. The fluorescence ratio images showed a somewhat differing behavior. The blue/green fluorescence $F440/F520$ increased up to 3 months of storage and then slightly declined up to 6 months. The chlorophyll fluorescence ratio $F690/F740$ strongly increased after 2 months of storage and then increased very little. This demonstrates that during storage, a chlorophyll breakdown took place in the first 2 months and that thereafter the remaining chlorophyll level was fairly constant. The ratio blue/red increased significantly up to the storage time of 3 months and the blue/far red ratio even up to 6 months. Please note that the fluorescence ratio scales after 2 months had to be changed to allow a correct presentation of the increasing fluorescence ratios.

4.2.4.2 Developmental Stage

With the developmental stage of the plant tissue (aging of leaves, ripening of fruits), the amount of fluorophores and the photosynthetic activity may change. In many fruits ripening is accompanied by a loss of chlorophyll. First, the chlorophyll fluorescence increases with decreasing chlorophyll content, since the degree of reabsorption of the chlorophyll fluorescence emitted decreases with decreasing chlorophyll levels; then, at a rather low chlorophyll content of the plant tissue, the chlorophyll fluorescence declines together with the chlorophylls. These relationships had well been investigated before using the conventional fluorescence measurements (e.g., Lichtenthaler 1987a; Lichtenthaler and Rinderle 1988, Lichtenthaler and Babani 2004). Such decreases in the chlorophyll fluorescence yield have also been observed via fluorescence imaging in the ripening of lemons (Nedbal et al. 2000) and grapes (Lenk et al. 2007b) as well as for storing apples as shown in Figure 4.2.18 and also by Ciscato et al. (2001). Prediction of fruit firmness, skin chroma, and flesh hue

Fluorescence

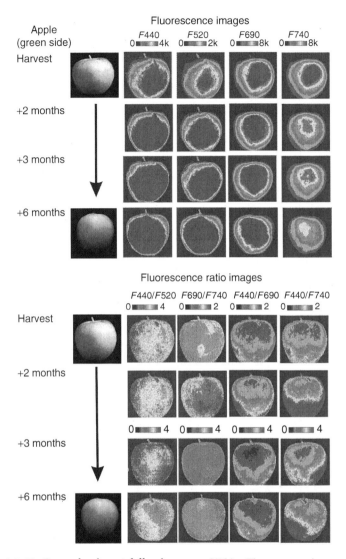

FIGURE 4.2.18 (See color insert following page 376.) Fluorescence images of apples (*Malus x domestica* 'Braeburn') measured at room temperature after different times of storage at 4°C. Upper part: Shown are fluorescence images of blue (*F*440), green (*F*520), red (*F*690), and far-red (*F*740) fluorescence. Lower part: fluorescence ratio images blue/green (*F*440/*F*520), blue/red (*F*440/*F*690), blue/far-red (*F*440/*F*740), and red/far-red (*F*690/*F*740). The fluorescence intensities are shown in false colors in the images; the fluorescence yield increases from blue (no fluorescence) via green and yellow to red as the highest fluorescence; k in the scales means kilo ($= 1000$) counts. The scales in the fluorescence ratio images indicate the absolute ratio value. Note that the scales for the individual fluorescence signatures are different.

of apples has been made via fluorescence images (Noh and Lu 2007). During the development of the different plant tissues, the tissue structure may undergo strong changes, for example, due to cell density, size of intercellular air spaces, and cellular shrinking or expansion, including changes in the water content. These changes determine the optical properties of the plant tissue and influence the penetration of the excitation light and of the fluorescence emitted. During a longer storage period of fruits, a decay may become visually detectable in fruit colors (see, e.g., Figure 4.2.17, upper right part) as a result of chlorophyll breakdown that also changes the fluorescence characteristics.

4.2.4.3 Strain and Stress

Plants undergo a wide variety of stress or strain effects and the typical responses of plant tissues have been summarized in the unifying stress concept of plants (Lichtenthaler 1996). These effects may be abiotic, caused, for example, by climate, irradiance, air, or chemicals, or biotic, caused, for example, by viruses, bacteria, fungi, or animals. As outlined above, multicolor fluorescence measurements can help in the detection of strain and stress provided that the effect changes (1) the content of the fluorophore chlorophylls and blue fluorescing substances, (2) the tissue structure, or (3) the photosynthetic activity. There is a general trend in plant tissues to respond to stress constraints and to the attack by fungal pathogens, insects, or other small animals with a formation of constitutive defense substances; often this is an accumulation of particular soluble phenolic compounds, of local lignin accumulation often accompanied by an accumulation of soluble ferulic acid or chlorogenic acid (Langsdorf 1994). Such stress responses can be associated with an increase in blue-green fluorescence. In addition, local variation of the fluorescence signatures (e.g., by biting insects) may lead to a characteristic pattern of signals that can be used for classifying stress events and types. In many cases, it is not easy or even impossible to distinguish between different stress types because they sometimes lead to the same or to similar effects with respect to the blue-green fluorescence and the chlorophyll fluorescence. Yet, by forming the different fluorescence ratios and following their development due to biotic or abiotic stress constraints, one can sort out several possible stressors and confine them to the few most probable ones. There exist reviews on biotic and abiotic stress effects that were detected using the conventional method of measuring chlorophyll fluorescence of leaf points (e.g., Lichtenthaler and Rinderle 1988; Krause and Weis 1991; Govindjee 1995; Lichtenthaler et al. 1998 and literature cited therein). In the meantime, some reviews have been presented (Buschmann and Lichtenthaler 1998; Daley 1995; Ning et al. 1995; Lichtenthaler et al. 1996; Lichtenthaler and Miehé 1997; Nedbal and Whitmarsh 2004; Oxborough 2004) on monitoring stress effects in plants via chlorophyll fluorescence imaging. Examples of studies of biotic and abiotic stress effects on fruits using chlorophyll fluorescence imaging, and in two cases (Sowinska et al. 1998; Ariana et al. 2006) also blue-green fluorescence imaging, are listed in Table 4.2.3. However, only imaging at all four fluorescence bands of green plant samples allows forming images of the four fluorescence ratios, which is currently the most reliable basis for the assessment of plant agrofood.

The changes of the fluorescence ratios $F440/F740$, $F690/F740$, and $F440/F520$ in different leaf tissues and under stress constraints (Table 4.2.4) can be explained by

TABLE 4.2.3
Changes of Fluorescence Signatures of Fruits upon Biotic and Abiotic Stress as Demonstrated by Fluorescence Imaging of Fruits

Changes of Fluorescence Signatures	Fruits	References
Increase of chlorophyll fluorescence		
Chilling stress	Mature green tomato	Abbott 1999
Fungus Infection	Lemon	Nedbal et al. 2000
Skin damage	Lemon	Nedbal et al. 2000
Increase of blue-green fluorescence		
Nitrogen deficiency	Apples	Sowinska et al. 1998
Decrease of chlorophyll fluorescence		
Insect injuries or blossom scar	Mature green tomato	Abbott 1999
Bitter pit, scald, black rot	Apples	Ariana et al. 2006; Lötze et al. 2006
Water stress	Apples	Ciscato et al. 2001
Decrease of blue-green fluorescence		
Bitter pit, scald, black rot	Apples	Ariana et al. 2006

specific changes of one or several of the parameters mentioned above. Since the changes in the fluorescence ratios blue/red and blue/far-red show the same trend in most cases, in Table 4.2.4 we have indicated only the fluorescence ratio $F440/F740$.

TABLE 4.2.4
Changes of the Three Fluorescence Ratios Blue/Far-Red ($F440/F740$), Red/Far-Red ($F690/F740$), and Blue/Green ($F440/F520$) when Comparing Different Leaf Tissues and Stressed Plants with Control Plants

No.	Treatment	$F440/F740$	$F690/F740$	$F440/F520$
1.	Lower leaf side versus upper leaf side	+++	+	0
2.	Variegated leaf versus green leaf	+++	+++	0
3.	UV A-treatment	− − −	0	−
4.	Sun exposure	+++	+	− − −
5.	Diuron treatment	− − −	+	0
6.	Heat treatment	− − −	0	−
7.	Nitrogen deficiency	+++	+	0
8.	Photoinhibition	+++	− − −	0
9.	Water deficiency	+++	0	0
10.	Mite attack	+++	0	+

Source: From Buschmann, C. and Lichtenthaler, H.K., *J. Plant Physiol.*, 152, 297, 1998.

Note: In the case of the comparisons and treatments No. 1 to No. 7 the cells across the whole leaf area respond homogeneously in the same way. In the case of treatments No. 8 to No. 10 only particular cell groups, either small leaf zones or local spots, respond in the indicated way as compared to the controls. + + +, strong increase; +, slight increase; 0, no change; − slight decrease; − − −, strong decrease.

It depends on the type of stress, treatment, etc., whether the three fluorescence ratios increase or decrease, and to what degree as compared to a control. The largest variations were found in the fluorescence ratio $F440/F740$ and also in the ratio $F440/F690$. Such changes can be a result of the differences in blue fluorescence yield ($F440$), either by a decrease or an increase of the blue fluorophores. In addition, increases in $F440/F740$ are often caused by a breakdown of chlorophylls whereby the blue fluorophore remains unchanged, as well as by an increase in UV-A absorbing flavonols in the epidermis. Since the latter functions as UV barrier, the flavonols reduce the amount of excitation radiation that passes through the epidermis causing a considerable reduction of the chlorophyll fluorescence $F740$. With the ratio $F690/F740$, changes in the chlorophyll concentration of two different plant tissue samples facing the detector can be sensed. Thus, in leaves of C_3-plants, where the fluorescence is sensed from the lower leaf side (i.e., the spongy parenchyma with large aerial interspaces and hence low chlorophyll density) the ratio red/far red is higher as compared to the ratio value sensed from the upper leaf part with its densely packed palisade parenchyma cells (Figure 4.2.12). This is also the case when a variegated leaf with less chlorophyll is compared to a fully green leaf. In addition, this increase in $F690/F740$ also occurs after diuron treatment. That herbicide blocks the photosynthetic quantum conversion, and the overall chlorophyll fluorescence yield strongly increases, whereby the $F690$ band rises significantly higher than the far-red $F740$ band. The fluorescence ratio blue/red, in turn, remains mostly unchanged. However, at full sun exposure, UV-A, or heat treatment the values decrease as compared to controls indicating that some green fluorescing substances must have been formed and accumulated in the leaf under these conditions.

Most of the treatments and comparisons listed in Table 4.2.4 (points 1 to 7) induce a homogeneous response applying to all leaf parts and cells of the sample as compared to the controls. In the case of water deficiency and mite attack the response (points 9 and 10), however, only affects a larger number of local leaf parts or leaf spots. In the case of water stress, these are located, for example, in cell groups on the leaf rim (with increased chlorophyll fluorescence) or in several small zones across the leaf area yielding a certain patchiness. In the case of mite attack, the response is restricted to very small local spots with either higher or lower fluorescence yield, and these spots are distributed irregularly across the whole leaf area. Moreover, the process of photoinhibition (point 8) can either rather homogeneously affect all green cells across the whole leaf area or concern several more local leaf zones, which are clearly distinguished by fluorescence imaging. The great advantage of the multicolor fluorescence imaging method is that it allows to see, via changes in the fluorescence ratios, small and also early local stress-induced responses before real damage of the leaves and plant occurs. With the conventional fluorescence techniques measuring successively only the absolute chlorophyll fluorescence signals on several leaf parts, such more locally restricted spot-like responses of leaves had been overlooked.

4.2.4.4 Preventive Measures in Fluorescence Imaging

Although laser-induced or flashlamp-induced fluorescence imaging of both, blue-green fluorescence and chlorophyll fluorescence, is presently the best method for the fluorescence assessment of leaves, vegetable crops, and fruits, one has to take certain precautions to guarantee a correct interpretation of the imaging results. A few examples are given here:

1. When comparing two agrofood samples it is best not to judge two samples exclusively based on their differences in the absolute fluorescence yield, for example, if one sample might be stressed, infected by pathogens or aged, and of lower quality than a control sample. The differences in absolute fluorescence yields can have many causes (see above), and variations in fluorescence yields even show in completely healthy and fresh agrofood samples. Establishing the fluorescence ratios blue/red, blue/far-red, red/far-red, and blue/green is, however, on the safe side in the assessment, since the fluorescence ratios vary to a much lower degree than the absolute fluorescence signals. At sample sites with low fluorescence yield, such as the side regions of apples or bell peppers, the blue fluorescence is lower by about the same degree as the red or far-red chlorophyll fluorescence. As a consequence, the fluorescence ratios, for example, blue/red and blue/far red, tend to show the same values on the sample parts with high- and low-fluorescence yields.
2. When determining and interpreting multicolor fluorescence images of plant tissues (leaves, fruits, roots, inflorescences), it should be taken into consideration that the fluorescence may change with storage time. Therefore, please note in advance if the sample is fresh and still physiologically active or how much time has passed since the harvest (storage time). Also, the shape of the sample (e.g., flat leaf, rounded fruit or root) and the sample surface (plane, rough, curly, matt, polished, or with wax cover) need to be taken into consideration because the yield and distribution of the fluorescence intensities across the sample surface depend on these factors as pointed out in the imaging examples in Section 4.2.4.1. Also, one has to account for the plant group and family of a sample. Some contain high amounts of cell wall bound ferulic acid, for example, most monocotyledonous plants, such as grasses and cereals. In contrast, many dicotyledonous plants exhibit rather low amounts of it, although exceptions do exist, such as tobacco and beet root (*Beta vulgaris* L.) with a higher ferulic acid level. This knowledge is essential and helps to select and set the imaging system to the correct sensitivity range (gain) of the fluorescence detector. One always needs to adjust the imaging system for the correct detection of either high or low yields of blue-green fluorescence or red + far-red chlorophyll fluorescence. In addition, one has to find out if the sample has high or low amounts of soluble blue-fluorescing substances, such as ferulic acid, chlorogenic acid, or certain flavonols that can accumulate due to biotic or abiotic stress constraints. Moreover, one should take into consideration that the UV-excited chlorophyll fluorescence signal may be low if higher amounts of UV-A absorbing substances are in the epidermis layer, such as in fully sun-exposed green plant tissues (leaves, fruits) as too little excitation radiation reaches the chloroplasts in the green subepidermal cells.
3. When imaging chlorophyll-poor or practically chlorophyll-free fruits, roots, or inflorescences (e.g., cauliflower), one will obtain besides the images of the blue and green fluorescence, also images of the red + far-red

chlorophyll fluorescence of very low intensity, which may range from only 1% to 10% of the much higher blue fluorescence intensity. Such low levels of chlorophyll fluorescence may be artificial. They may be a result of the noise of the background signals and of the fact that the blue-green fluorescence, which quickly drops down from its high peak in the blue region toward longer wavelengths, still shows very low signals in the red + far-red region. In such cases, it is highly recommended to prove via a UV-A excited fluorescence emission spectrum of the plant tissue whether chlorophyll fluorescence is really present. Chlorophyll fluorescence should be indicated by a clear peak in the red (F690) and a shoulder in the far red (F735). If the situation is not yet clear, one has to take either a blue light (e.g., 450 nm) excited chlorophyll fluorescence emission spectrum or blue light excited chlorophyll fluorescence images, since blue light results in a much higher chlorophyll fluorescence yield than UV-A radiation. If chlorophyll is present in the plant tissue that can visually not be detected, the intensities of the red + far-red chlorophyll fluorescence images should considerably increase by blue light excitation as compared to UV-A excitation. Only when chlorophyll is still present in the tissues investigated, it makes sense to establish the fluorescence ratios blue/red and blue/far red that are early and efficient indicators of ongoing changes in the composition and freshness or aging of agrofood. Another alternative is the direct extraction of chlorophylls from the tissue and its photometric determination (Lichtenthaler 1987b).

4. When establishing the fluorescence ratios there are two possibilities. One can establish the ratios by (1) pixel-to-pixel division (e.g., of blue and red fluorescence) of the more than 10,000 fluorescence signals of a sample or (2) division of the sum of the blue fluorescence yield of all pixels by the sum of the red fluorescence yield of all pixels integrated over the sample area. A pixel-to-pixel division generally is the recommended method and is an essential diagnosis characteristic for the detection of small local disturbances. In the fluorescence ratio images, it is absolutely required to show those small restricted local sites with disturbances in the fluorescence ratios that are either early indicators for biotic attacks (e.g., mites), infections by pathogens, or of such small local areas, for example, on the leaf rim with water deficit suggesting early stress long before visual damage can be detected. Establishing the fluorescence ratio only from the fluorescence information of the sum of all pixels when comparing two plant tissue samples, the differing fluorescence ratios of the few local sites with disturbances in the fluorescence yield and ratios are going to be excluded. When both calculation methods are applied and compared, a great difference of the two values indicates local irritations in the sample.

4.2.5 Fluorescence Imaging Systems for Plant Tissues

A fluorescence imaging system (FIS) to acquire multicolor fluorescence images of plants, plant tissues, and fruits requires an appropriate light source for fluorescence

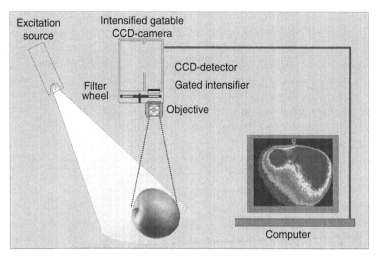

FIGURE 4.2.19 (See color insert following page 376.) Scheme of a fluorescence imaging setup for the investigation of plant leaves and fruits. The sample (here a light-green apple) irradiated by the excitation source emits fluorescence that is taken by a camera (here an intensified gatable CCD-camera). The camera is equipped with a lens, an intensifier (here a gated intensifier with a microchannel plate synchronized to the excitation pulses), and a detector (here a CCD). The camera takes black-and-white images, i.e., the fluorescence intensities are measured in counts irrespective of the wavelength range. For choosing one of the four plant tissue fluorescence bands, for example, in the blue, green, red, and far-red wavelength region, appropriate color filters of a filter wheel are inserted in front of the intensifier. The fluorescence image is transferred to a computer that displays the fluorescence intensity in false colors with an intensity scale from dark blue (zero fluorescence intensity) via green and yellow to red (highest fluorescence intensity).

excitation (e.g., a pulsed laser or a pulsed flashlight lamp), a highly sensitive camera for fluorescence detection arranged in suitable geometry (Figure 4.2.19), and an appropriate computer program to coordinate the individual steps of multicolor fluorescence imaging at the four fluorescence bands. Several fluorescence imaging systems, either laser FIS or flashlamp FIS, have been developed. Laser FIS devices have been described for leaves and fruits (Lang et al. 1994b; Lichtenthaler et al. 1996; Sowinska et al. 1998; Kim et al. 2003). Pulsed xenon flashlamp FIS devices, which allow a flexible choice of the excitation wavelengths by selecting appropriate filters, have been applied (Buschmann and Lichtenthaler, 1998; Langsdorf et al. 2000) and the former developed by an expanded form including reflectance measurements (Lenk et al. 2007a). There are also several FIS-approaches for imaging the chlorophyll fluorescence with different systems for optics, electronics, and software (see, e.g., Buschmann and Lichtenthaler 1998; Lichtenthaler and Babani 2000; Kim et al. 2001; Chen et al. 2002; Nedbal and Whitmarsh 2004; Oxborough 2004; Lenk et al. 2007a; Lichtenthaler et al. 2007a,b). Since most of these FIS imaging devices are restricted to imaging only the red chlorophyll fluorescence of leaves or leaf regions, they are primarily suited for the investigation of the photosynthetic function, such as the determination of photochemical and nonphotochemical quenching of chlorophyll fluorescence and other fluorescence parameters during the light-induced induction kinetics (cf. Lichtenthaler et al. 2005a). For early stress detection in plants and the quality assessment of agrofood, however, imaging of the red chlorophyll fluorescence band exclusively is not suitable and not sufficient for the assessment, since it provides only limited information and only on green leaves and fruits but not on nongreen agrofood. In fact, only multicolor fluorescence imaging, i.e., in the blue, green, red, and far-red emissions bands of plant tissues, allows creating several fluorescence ratio images (Buschmann and Lichtenthaler 1998) and therefore provides ample possibilities for an assessment of agrofood. Selecting the appropriate filters and defining the suitable fluorescence bands, an adequate excitation wavelength and a multifunctional laser or a pulsed flashlamp as an excitation source, such multicolor fluorescence imaging systems with broad application in agriculture, early stress detection, and quality assessment of agrofood have been established by us in the early 1990s. The first multicolor fluorescence imaging system, the Karlsruhe/Strasbourg FIS device, built in cooperation with French physicists, contained a pulsed laser as excitation source (Lang et al. 1994a, 1996; Lang 1995; Lichtenthaler et al. 1995, 1996; Heisel et al. 1996). This instrument was later further developed, replacing the expensive laser by a less expensive modulated pulsed xenon flashlamp excitation device (Buschmann et al. 2000; Langsdorf et al. 2000; Lichtenthaler and Babani 2000). This enables UV-A (355 nm) excitation and, with appropriate filters, also the excitation with different wavelengths of visible light in contrast to lasers defined to only one excitation wavelength. For simultaneous excitation of blue-green fluorescence and the red + far-red chlorophyll fluorescence, the excitation wavelength of 355 nm was chosen as a compromise. It still supplies a reasonable blue-green fluorescence and in most cases provides a sufficiently strong chlorophyll fluorescence signal. The Karlsruhe flashlamp fluorescence imaging system (flashlamp FIS) allows imaging smaller and larger leaves (spinach, chard, lettuce, cabbage), whole fruits (apples, bell peppers, grapefruits, strawberries as well as cucumbers, avocados, egg plant, zucchinis, etc.), potatoes, tubers, roots (carrots), or inflorescences (cauliflower, broccoli). It senses images in close distance and in the correct scale and thus possesses broad application possibilities in the quality assessment of all kinds of agrofood. This self-constructed imaging instrument was recently further developed by inclusion of the possibility to sense, together with the four plant fluorescence bands, also the reflectance of the plant sample (Lenk et al. 2007a).

The traditional fluorescence microscopy (Rost 1995; Gilroy 1997) introduced already at the beginning of the twentieth century by Max Haitinger (1868–1946) has meanwhile found a wide field of applications, but at a much smaller scale, i.e., the microscopic scale. Today, fluorescence microscopes are nearly exclusively connected to camera systems with computer-aided image analysis and also allow, in combination with the appropriate filters, the imaging of the different plant fluorescence bands. Microscopic fluorescence investigation of all four fluorescence bands and the formation of the four fluorescence ratio images of intact leaves or other plant tissue have not yet been performed as far as we know. Fluorescence microscopy of biological tissues and cells is now often applied in context with labeling fluorescent dyes that are used for the localization of molecules (e.g., antibodies, fluorescent proteins, fluorescent tags in DNA sequencing). However, these particular cytology techniques are not suited for the quality assessment of leaves, vegetables, fruits, and other agrofood.

At present, only few fluorescence imaging systems are commercially available that take images of larger samples (e.g., leaves, fruits, or small plants) or higher numbers of small plants in one measurement. Most of these FIS instruments only detect images of the red chlorophyll fluorescence excited by pulses of red light-emitting diodes (LEDs) (Nedbal and Whitmarsh 2004). Pulse mode excitation with synchronized camera detection, as performed in the Karlsruhe multicolor flashlamp fluorescence imaging system, allows distinguishing the low fluorescence from a high background of reflected excitation light and ambient light. This design strongly increases the dynamic range of the fluorescence. Such research prototype systems are more flexible in the wavelength range of both excitation and fluorescence detection and thus are able to provide detailed spectral information by one or several discrete broad band-pass filters as in the multispectral system of Lenk et al. (2007a), or by narrow spectral resolution in the order of 10 nm (hyperspectral systems), which senses in the green (530 nm) and red fluorescence bands (685 nm) (Kim et al. 2001). The selection of the optimum wavelength range of excitation and fluorescence detection can help in the interpretation of an image. For this purpose, the determination of an excitation–emission matrix plot adapted to the specific need of a sample and the properties to be studied is of great advantage (Nakauchi 2005). The camera taking the fluorescence images contains a charge-coupled device (CCD) detector, which is more sensitive than the detectors used in conventional video cameras. In many cases, an intensifier (microchannel plate) is used that amplifies the fluorescence signal taking into account the time window in which the pulses excite the fluorescence. The signal can be optimized by increasing the integration time of the detector or the number of accumulated images. High differences in signal intensity within one sample (a peduncle of an apple with high chlorophyll fluorescence) may overextend the dynamic range of the instrumentation and thus cause problems in finding the proper amplification for the whole image area. Careful calibration of the instrumentation has to be carried out, for example, correcting for heterogeneity of the excitation light and spectral sensitivity of the camera.

There are laboratory imaging systems that are able to sense thermography images (Chaerle et al. 2006) and reflectance images (Ariana et al. 2006) in parallel to chlorophyll fluorescence images. The only system that yields the four multicolor fluorescence images and, in addition, reflectance images is that of Lenk et al. (2007a), which is based on the Karlsruhe flashlamp imaging system (flashlamp FIS). The latter is also using automated sample changes (e.g., by an indexing table). Sample changing by robotized cameras for a chlorophyll fluorescence imaging system has been described as well (Chaerle et al. 2006).

A multicolor fluorescence imaging system for blue-green and for field measurements from short distance of several meters up to ~50 m with laser excitation has been described and successfully applied (Edner et al. 1995; Sowinska et al. 1999). Moreover, systems for measuring sunlight-excited chlorophyll fluorescence in the short distance (but no imaging) have been reviewed by Moya and Cerovic

(2004). An airborne remote sensing chlorophyll fluorescence imaging system for aircrafts and possibly satellites with sunlight excitation is on its way (Smorenburg et al. 2002). Sunlight-excited fluorescence is detected in the Fraunhofer lines, i.e., very narrow spectral lines in which the sunlight is extinguished by gases of the atmosphere (cf. McFarlane et al. 1980; Lichtenthaler et al. 1992; Smorenburg et al. 2002).

Multicolor or red chlorophyll fluorescence measurements on the macroscopic scale have up to now mostly been carried out as point measurements, resulting in one signal integrated over a small area of a green leaf or fruit. In contrast, imaging systems can help recognize and localize the heterogeneity and patchiness of the signals across the sample. The imaging technique can also be used to measure several samples at a time, thus increasing the statistical accuracy of the results. Moreover, fluorescence imaging systems with an additional analysis of component features (shape, size, density), pattern recognition, and contrast image analysis possibilities (Codrea et al. 2004; Leinonen and Jones 2004; Ariana et al. 2006; Matouš et al. 2006) can also be developed and applied as a tool for classifying and sorting samples. In fact, automated fluorescence image processing guarantees rapid and equally correct monitoring of fruits and vegetables, which is of crucial importance for modern agriculture and horticulture.

4.2.6 CONCLUSION

Multicolor fluorescence imaging of plant samples is an excellent method and much superior to the point measurements applied so far. Although imaging of chlorophyll fluorescence and photosynthetic chlorophyll fluorescence parameters can provide some basic and mainly photosynthetic information on green leaves and fruits, a full assessment of green agrofood is only possible via an inclusion of the blue and green fluorescence and in the case of chlorophyll fluorescence not only by imaging the red band $F690$ but also the far-red band $F740$. Multicolor fluorescence imaging permits to form images of the fluorescence ratios blue/red, blue/far-red, red/far-red, and blue/green providing ample possibilities for the follow-up of changes of agrofood in chemical composition during the development, storage, ripening, and under stress constraints. This fluorescence ratio imaging technique is presently the most suited method and due to its high statistical significance (high pixel numbers) a most reliable basis for the assessment of all forms of green and nongreen plant agrofood. Multicolor fluorescence imaging, including essential spectral information of the samples, can be regarded as an artificial eye that recognizes valuable fluorescence information that remains hidden to our eyes.

REFERENCES

Abbott, J.A. 1999. Quality measurement of fruits and vegetables. *Postharvest Biology and Technology* 15:207–225.

Ariana, D., D.E. Guyer, and B. Shrestha. 2006. Integrating multispectral reflectance and fluorescence imaging for detection on apples. *Computers and Electronics in Agriculture* 50:148–161.

Barber, J., S. Malkin, and A. Telfer. 1989. The origin of chlorophyll fluorescence in vivo and its quenching by the photosystem II reaction centre. *Philosophical Transactions of the Royal Society of London* B323:227–239.

Bilger, W., M. Veit, L. Schreiber, and U. Schreiber. 1997. Measurement of leaf epidermal transmittance of UV radiation by chlorophyll fluorescence. *Physiologia Plantarum* 101:754–763.

Buschmann, C. 1995. Variation of the quenching of chlorophyll fluorescence under different intensities of the actinic light in wildtype plants of tobacco and in an aurea mutant deficient of light-harvesting-complex. *Journal of Plant Physiology* 145:245–252.

Buschmann, C. 1999. Photochemical and non-photochemical quenching coefficients of the chlorophyll fluorescence: Comparison of variation and limits. *Photosynthetica* 37:217–224.

Buschmann, C. 2007. Variability and application of the chlorophyll fluorescence emission ratio red/far-red of leaves. *Photosynthesis Research* 92:261–271.

Buschmann, C. and H.K. Lichtenthaler. 1998. Principles and characteristics of multi-colour fluorescence imaging of plants. *Journal of Plant Physiology* 152:297–314.

Buschmann, C., G. Langsdorf, and H.K. Lichtenthaler. 2000. Imaging of the blue, green and red fluorescence emission of plants. *Photosynthetica* 38:483–491.

Chaerle, L., M. Pineda, R. Romero-Aranda, D. Van der Straeten, and M. Barón. 2006. Robotized thermal and chlorophyll fluorescence imaging of pepper mild mottle virus infection in *Nicotiana benthamiana*. *Plant and Cell Physiology* 47:1323–1336.

Chappelle, E.W., F.M. Wood, Y.E. McMurtrey, and W.W. Newcomb. 1984. Laser induced fluorescence of green plants 1: A technique for remote detection of plant stress and species differentiation. *Applied Optics* 23:134–138.

Chen, Y.R., K. Chao, and M.S. Kim. 2002. Machine vision technology for agriculture applications. *Computers and Electronics in Agriculture* 36:173–191.

Ciscato, M., M. Sowinska, M. van de Ven et al. 2001. Fluorescence imaging as a diagnostic tool to detect physiological disorders during storage of apples. *Acta Horticulturae* 553:507–512.

Codrea, M.C., O.S. Nevalainen, E. Tyystjarvi, M. van de Ven, and R. Valcke. 2004. Classifying apples by the means of fluorescence imaging. *International Journal of Pattern Recognition and Artificial Intelligence* 28:157–174.

Combes, D., L. Bousqut, S. Jacquemoud, H. Sinoquet, C. Varlet-Grancher, and I. Moya. 2007. A new spectrogoniophotometer to measure leaf spectral and directional optical properties. *Remote Sensing of Environment* 109:107–117.

Daley, F. 1995. Chlorophyll fluorescence analysis and imaging in plant stress and disease. *Canadian Journal of Plant Pathology* 17:167–173.

D'Ambrosio, N., K. Szabó, and H.K. Lichtenthaler. 1992. Increase of the chlorophyll fluorescence ratio F690/F735 during the autumnal chlorophyll breakdown. *Radiation and Environmental Biophysics* 31:51–62.

Edner, H., J. Johannsen, S. Svanberg et al. 1995. Remote multi-colour fluorescence imaging of selected broad-leaf plants. *EARSel Advances in Remote Sensing* 3:2–14.

Gilroy, S. 1997. Fluorescence microscopy of living plant cells. *Annual Reviews of Plant Physiology* 48:165–190.

Gitelson, A.A., C. Buschmann, and H.K. Lichtenthaler. 1998. Leaf chlorophyll fluorescence corrected for re-absorption by means of absorption and reflectance measurements. *Journal of Plant Physiology* 152:283–296.

Goodwin, R.H. 1953. Fluorescent substances in plants. *Annual Reviews of Plant Physiology* 4:283–304.

Govindjee. 1995. Sixty three years since Kautsky: Chlorophyll a fluorescence. *Australian Journal of Plant Physiology* 22:131–160.

Harris, P.J. and R.D. Hartley. 1976. Detection of bound ferulic acid in cell walls of the Gramineae by ultraviolet fluorescence microscopy. *Nature* 259:508–510.

Heisel, F., M. Sowinska, J.A. Miehé, M. Lang, and H.K. Lichtenthaler. 1996. Detection of nutrient deficiencies of maize by laser-induced fluorescence imaging. *Journal of Plant Physiology* 148:622–631.

Kim, M.S., Y.R. Chen, and P.M. Mehl. 2001. Hyperspectral reflectance and fluorescence imaging system for food quality and safety. *Transactions of the ASAE* 44:721–729.

Kim, M.S., M.L. Alan, and C. Yud-Ren, 2003. Multispectral laser-induced fluorescence imaging system for large biological samples. *Applied Optics* 42:3927–3934.

Krause, G.H. and E. Weis. 1991. Chlorophyll fluorescence and photosynthesis: The basics. *Annual Reviews of Plant Physiology and Plant Molecular Biology* 42:313–349.

Lang, M. 1995. Studies on the blue-green and chlorophyll fluorescence of plants and their application for fluorescence imaging of leaves. *Karlsruhe Contributions to Plant Physiology* 19:1–124.

Lang, M., F. Stober, and H.K. Lichtenthaler. 1991. Fluorescence emission spectra of plant leaves and plant constituents. *Radiation and Environmental Biophysics* 30:333–347.

Lang, M., H.K. Lichtenthaler, M. Sowinska, P. Summ, and F. Heisel. 1994a. Blue, green and red fluorescence signatures and images of tobacco leaves. *Botanica Acta* 107:230–236.

Lang, M., H.K. Lichtenthaler, M. Sowinska et al. 1994b. Sensing of plants using the laser-induced fluorescence imaging system. In *6th International Symposium on Physical Measures and Signatures in Remote Sensing*, Val d'Isère, pp. 945–952. CNES, Toulouse.

Lang, M., H.K. Lichtenthaler, M. Sowinska, F. Heisel, J.A. Miehé, and F. Tomasini. 1996. Fluorescence imaging of water and temperature stress in plant leaves. *Journal of Plant Physiology* 148:613–621.

Langsdorf, G. 1994. Bildung von Abwehrstoffen in Nutzpflanzen. In *Giftpflanzen- Pflanzengifte*, eds. L. Roth, M. Daunderer, and K. Kormann, pp. 980–1013. Ecomed, Landsberg, ISBN: 3-609-61810-4.

Langsdorf, G., C. Buschmann, F. Babani et al. 2000. Multicolour fluorescence imaging of sugar beet leaves with different N-status by flash lamp UV-excitation. *Photosynthetica* 38:539–551.

Leinonen, I. and H.G. Jones. 2004. Combining thermal and visible imagery for estimating canopy temperature and identifying plant stress. *Journal of Experimental Botany* 55:1423–1431.

Lenk, S. and C. Buschmann. 2006. Distribution of UV-shielding of the epidermis of sun and shade leaves of the beech (*Fagus sylvatica* L.) as monitored by multi-colour fluorescence imaging. *Journal of Plant Physiology* 163:1273–1283.

Lenk, S., L. Chaerle, E. Pfündel et al. 2007a. Multi-colour fluorescence and reflectance imaging at the leaf level and its application possibilities. *Journal of Experimental Botany* 58:807–814.

Lenk, S., C. Buschmann, and E. Pfündel. 2007b. In vivo assessing flavonols in white grape berries (*Vitis vinifera* L. cv. Pinot Blanc) of different degrees of ripeness using chlorophyll fluorescence imaging. *Functional Plant Biology* 34:1092–1104.

Lichtenthaler, H.K. 1987a. Chlorophyll fluorescence signatures of leaves during the autumnal chlorophyll breakdown. *Journal of Plant Physiology* 131:101–110.

Lichtenthaler, H.K. 1987b. Chlorophylls and carotenoids, the pigments of photosynthetic biomembranes. *Methods in Enzymol*ogy 148:350–382.

Lichtenthaler, H.K. 1988. *Applications of Chlorophyll Fluorescence*. Kluwer Academic Publishers, Dordrecht, ISBN 90-274-3787-7.

Lichtenthaler, H.K. 1992. The Kautsky effect: 60 years of chlorophyll fluorescence induction kinetics. *Photosynthetica* 27:45–55.

Lichtenthaler, H.K. 1996. Vegetation stress: An introduction to the stress concept in plants. *Journal of Plant Physiology* 148:4–14.

Lichtenthaler, H.K. and U. Rinderle. 1988. The role of chlorophyll fluorescence in the detection of stress conditions in plants. *CRC Critical Reviews in Analytical Chemistry* 19:S29–S85.

Lichtenthaler, H.K. and J.A. Miehé. 1997. Fluorescence imaging as a diagnostic tool for plant stress. *Trends in Plant Sciences* 2:316–320.

Lichtenthaler, H.K. and J. Schweiger. 1998. Cell wall bound ferulic acid, the major substance of the blue-green fluorescence emission of plants. *Journal of Plant Physiology* 152:272–282.

Lichtenthaler, H.K. and F. Babani. 2000. Detection of photosynthetic activity and water stress by imaging the red chlorophyll fluorescence. *Plant Physiology and Biochemistry* 38:889–895.

Lichtenthaler, H.K. and C. Buschmann. 2001. Chlorophylls and carotenoids—Measurement and characterisation by UV—VIS. *Current Protocols in Food Analytical Chemistry (CPFA)*, (Supplement 1), F4.3.1–F 4.3.8. John Wiley, New York.

Lichtenthaler, H.K. and F. Babani. 2004. Light adaption and senescence of the photosynthetic apparatus: Changes in pigment composition, chlorophyll fluorescence parameters and photosynthetic activity during light adaptation and senescence of leaves. In *Chlorophyll Fluorescence: A Signature of Photosynthesis,* eds. G.C. Papageorgiou and Govindjee, pp. 713–736. Springer, Dordrecht, ISBN: 1-4020-3217-X.

Lichtenthaler, H.K., P. Karunen, and K.H. Grumbach. 1977. Determination of prenylquinones in green photosynthetically active moss and liver moss tissues. *Physiologia Plantarum* 40:105–110.

Lichtenthaler, H.K., F. Stober, and M. Lang. 1992. The nature of the different laser-induced fluorescence signatures of plants. *EARSeL Advances in Remote Sensing* 1:20–32.

Lichtenthaler, H.K., M. Lang, F. Stober, M. Sowinska, F. Heisel, and J.A. Miehé. 1995. Detection of photosynthetic parameters and vegetation stress via a new high resolution fluorescence imaging system. In *Proceedings of the International Colloquium Photosynthesis and Remote Sensing,* ed. G. Guyot, pp. 103–112. August 28–30, 1995, Montpellier, EARSeL, Paris, ISBN 2-90885-16-6.

Lichtenthaler, H.K., M. Lang, M. Sowinska, F. Heisel, and J.A. Miehé. 1996. Detection of vegetation stress via a new high resolution fluorescence imaging system. *Journal of Plant Physiology* 148:599–612.

Lichtenthaler, H.K., M. Lang, M. Sowinska, P. Summ, F. Heisel, and J.A. Miehé. 1997. Uptake of the herbicide diuron (DCMU) as visualized by the fluorescence imaging technique. *Botanica Acta* 110:158–163.

Lichtenthaler, H.K., O. Wenzel, C. Buschmann, and A. Gitelson. 1998. Plant stress detection by reflectance and fluorescence. *Annals of New York Academy Sciences* 851:271–285.

Lichtenthaler, H.K., C. Buschmann, and M. Knapp. 2005a. How to correctly determine the different chlorophyll fluorescence parameters and the chlorophyll fluorescence decrease ratio R_{Fd} of leaves with the PAM-fluorometer. *Photosynthetica* 43:379–393.

Lichtenthaler, H.K., G. Langsdorf, S. Lenk, and C. Buschmann. 2005b. Chlorophyll fluorescence imaging of photosynthetic activity with the flash-lamp fluorescence imaging system. *Photosynthetica* 43:355–369.

Lichtenthaler, H.K., F. Babani, and G. Langsdorf. 2007a. Chlorophyll fluorescence imaging of photosynthetic activity in sun and shade leaves of trees. *Photosynthesis Research* 93:235–241.

Lichtenthaler, H.K., A. Ac, M.V. Marek, J. Kalina, and O. Urban. 2007b. Differences in pigment composition, photosynthetic rates and chlorophyll fluorescence images of sun and shade leaves of four tree species. *Plant Physiology and Biochemistry* 45:577–588.

Lötze, E., C. Huybrechts, A. Sadie, K.I. Theron, and R. Valcke. 2006. Fluorescence imaging as a non-destructive method for pre-harvest detection of bitter pit in apple fruit (*Malus domestica* Borkh.). *Postharvest Biology and Technology* 40:287–294.

Matouš, K., Z. Benediktyová, S. Berger, T. Roitsch, and L. Nedbal. 2006. Case study of combinatorial imaging: What protocol and what chlorophyll fluorescence image to use when visualizing infection of *Arabidopsis thaliana* by *Pseudomonas syringae*? *Photosynthesis Research* 90:243–253.

McClendon, J.H. and L. Fukshansky. 1990. On the interpretation of absorption spectra of leaves—I. Non-absorbed ray of the sieve effect and the mean optical pathlength in the remainder of the leaf. *Photochemistry and Photobiology* 51:211–216.

McFarlane, J.C., R.D. Watson, A.F. Theisen et al. 1980. Plant stress detection by remote measurement of fluorescence. *Applied Optics* 19:3287–3289.

Morales, F., Z.G. Cerovic, and I. Moya. 1996. Time-resolved blue-green fluorescence of sugar beet (*Beta vulgaris* L.) leaves. Spectroscopic evidence for the presence of ferulic acid as the main fluorophore of the epidermis. *Biochimica et Biophysica Acta* 1273:251–262.

Morales, F., A. Cartelat, A. Álvarez-Fernández, I. Moya, and Z.G. Cerovic. 2005. Time-resolved spectral studies of blue-green fluorescence of artichoke (*Cynara cardunculus* L. var. Scolymus) leaves: Identification of chlorogenic acid as one of the major fluorophores and age-mediated changes. *Journal of Agricultural and Food Chemistry* 53:9668–9678.

Moya, I. and Z.G. Cerovic. 2004. Remote sensing of chlorophyll fluorescence: Instrumentation and analysis. In *Chlorophyll a Fluorescence—A Signature of Photosynthesis*, eds. G.C. Papageorgiou and Govindjee, pp. 429–445. Springer, Dordrecht, ISBN: 1-4020-3217-X.

Nakauchi, S. 2005. Spectral imaging technique for visualizing the invisible information. In *Image Analysis*, eds. H. Kalviainen, J. Parkkinen, and A. Kaarna, pp. 55–64. Springer, New York, ISBN: 3540263209.

Nedbal, L. and J. Whitmarsh. 2004. Chlorophyll fluorescence imaging of leaves and fruits. In *Chlorophyll a Fluorescence—A Signature of Photosynthesis*, eds. G.C. Papageorgiou and Govindjee, pp. 389–407. Springer, Dordercht, ISBN: 1-4020-3217-X.

Nedbal, L., J. Soukupová, J. Whitmarsh, and M. Trtílek. 2000. Postharvest imaging of chlorophyll fluorescence from lemons can be used to predict fruit quality. *Photosynthetica* 38:571–579.

Ning, L., G.E. Edwards, G.A. Strobel, L.S. Daley, and J.B. Callis. 1995. Imaging fluorometer to detect pathological and physiological change in plants. *Applied Spectroscopy* 49:1381–1389.

Noh, H.K. and R. Lu. 2007. Hyperspectral laser-induced fluorescence imaging for assessing apple fruit quality. *Postharvest Biology and Technology* 43:193–201.

Oxborough, K. 2004. Using chlorophyll a fluorescence imaging to monitor photosynthetic performances. In *Chlorophyll a Fluorescence—A Signature of Photosynthesis*, eds. G.C. Papageorgiou and Govindjee, pp. 409–428. Springer, Dordercht, ISBN: 1-4020-3217-X.

Papageorgiou, G.C. and Govindjee. 2004. *Chlorophyll a Fluorescence—A Signature of Photosynthesis*. Springer, Dordercht, ISBN: 1-4020-3217-X.

Rost, F.W.D. 1995. *Fluorescence Microscopy*. Vol. 2. Cambridge University Press, Cambridge, ISBN 0-521-41088-6.

Schweiger, J., M. Lang, and H.K. Lichtenthaler. 1996. Differences in fluorescence excitation spectra of leaves between stressed and non-stressed plants. *Journal of Plant Physiology* 148:536–547.

Smorenburg, K., G. Bazalgette Courrèges-Lacoste, M. Berger et al. 2002. Remote sensing of solar induced fluorescence of vegetation. *Proceedings of SPIE* 4542:178–190.

Sowinska, M., T. Deckers, C. Eckert, F. Heisel, R. Valcke, and J.A. Miehé. 1998. Evaluation of nitrogen fertilization effect on apple-tree leaves and fruit by fluorescence imaging. *Proceedings of SPIE* 3382:100–110.
Sowinska, M., B. Cunin, A. Deruyver et al. 1999. Near-field measurements of vegetation by laser-induced fluorescence imaging. *Proceeding of SPIE* 3868:120–131.
Stober, F. 1993. Investigations on the laser-induced blue, green and red fluorescence of plants by means of an optical multichannel analyzer (OMAIII). *Karlsruhe Contributions to Plant Physiology* 25:1–201.
Stober, F. and H.K. Lichtenthaler. 1993a. Studies on the constancy of the blue and green fluorescence yield during the chlorophyll fluorescence induction kinetics (Kautsky effect). *Radiation and Environmental Biophysics* 32:357–365.
Stober, F. and H.K. Lichtenthaler. 1993b. Studies on the localization and spectral characteristics of the fluorescence emission of differently pigmented wheat leaves. *Botanica Acta* 106:365–370.
Stober, F., M. Lang, and H.K. Lichtenthaler. 1994. Studies on the blue, green and red fluorescence signatures of green, etiolated and white leaves. *Remote Sensing of Environment* 47:65–71.
Truscott, T.G. 1990. The photophysics and photochemistry of carotenoids. *Journal of Photochemistry and Photobiology B: Biology* 6:359–371.
Valeur, B. 2002. *Molecular Fluorescence—Principles and Applications*. Wiley-VCH, Weinheim, ISBN 3-527-29919-X.

4.3 MONITORING RAW MATERIAL BY LASER-INDUCED FLUORESCENCE SPECTROSCOPY IN THE PRODUCTION

YASUNORI SAITO

4.3.1 BASICS OF LASER-INDUCED FLUORESCENCE SPECTROSCOPY

In this section, the basics of laser-induced fluorescence spectroscopy (LIFS) and its application to in-vivo monitoring of agricultural products are described. Some native compounds in living plants reemit photons in response to laser irradiation (called laser-induced fluorescence (LIF)). The shape and intensity of a LIF spectrum depend on the plant species and the amount of fluorophores. On the other hand, quenching phenomena and reabsorbance of the photons emitted by fluorophores influence the apparent excitation and emission intensities. In the following paragraph, examples of the LIF spectrum of several compound standards are provided to point out the potential of LIF spectroscopy in analyzing the nutritional status of plant organs. Whether there are possibilities for measuring nutritionally valuable compounds for human beings in agricultural crops is critically discussed. The focus will be on blue-green (BG)-LIF.

The report by Measures et al. at the University of Toronto (Measures et al. 1973) about sugar maple trees appears to be the first on monitoring remotely BG-LIF of plants. However, the BG fluorescence of plant tissues had already been seen before in the fluorescence microscopy (Harris and Hartley 1976). Subsequently, a series of reports about LIF of green plants published by Chappelle and his group at the NASA Goddard Space Flight Center (Chappelle et al. 1984a, 1984b, 1985), further advanced by the group of Lichtenthaler (Lichtenthaler et al. 1990, 1991a,b,

1992; Lang et al. 1992; Stober and Lichtenthaler 1993a,b), triggered the widespread application of LIFS for monitoring plants. It should be added that the fluorescence described in this section is autofluorescence; materials stained with fluorophores such as green fluorescence protein are not considered. The time behavior of LIF, including the induction kinetics of the red chlorophyll fluorescence (Kautsky effect) and the fluorescence lifetimes, is also not treated here. In contrast to chlorophyll fluorescence, the BG fluorescence and BG-LIF signatures are stable during excitation and do not show any induction kinetic (Stober and Lichtenthaler 1993c).

4.3.1.1 Fluorescence Mechanism

Fluorescence is one kind of energy transition that occurs inside molecules. Figure 4.3.1 shows a schematic diagram (known as a Jablonski diagram) of the energy transitions of a single molecule. Since transitions between energy levels are governed by the Franck–Condon principle, namely that an electronic transition preserves the internuclear distance of the vibrational motion of the nuclei, the Jablonski diagram that does not show the electronic potential curve as a function of internuclear distance is adequate for understanding the transitions.

The fluorescence process starts by a molecule-absorbing energy; in LIFS, the source of energy is a photon from a laser light source. A molecule in the ground state (S_0) is accelerated by a photon to excited states ($S_{1,2,3,\ldots,n}$), where the level of the excited state (E[eV]) is determined by the energy that the photon has ($E = h\nu = \lambda c/\nu$, where h = Planck's constant 6.626×10^{-34} [Js], ν = frequency [Hz = 1/s], c = speed of light 2.998×10^{10} [cm/s], λ = wavelength of the photon [m]). The excited states are very unstable so that the molecule returns to the ground state through a variety of interactions, such as collisions with other molecules, energy transfer to other excited molecules, and thermal relaxation. The molecule then descends to the first singlet state (S_1). From the excited state, a fluorescence transition may occur to one of the many energy states, which are vibrational and rotational levels, within the ground singlet state (S_0), while reemitting photons (fluorescence). Nonfluorescence transitions also occur between S_1 and triplet state (T_1). The latter is described by the term phosphorescence. Here, the emission lasts over a period of several milliseconds. By contrast, the time period for fluorescence is generally shorter than a nanosecond. As can be seen in Figure 4.3.1, the energy of fluorescence transitions is lower than that of absorption; hence, fluorescence occurs at longer wavelengths than absorption (Stokes effect). Absorption spectrum (fluorescence excitation spectrum) and reemission spectrum (fluorescence emission spectrum) generally appear as a mirror image to the other.

Since fluorescence occurs via excited states that are unstable, fluorescence itself is readily affected by the environment in which the molecule is located. Environmental factors change the fluorescence quantum yield (i.e., the emission intensity), the spectral shape, and the fluorescence lifetime measured on leaves. The factors include viscosity, pH, temperature (Lang et al. 1996), leaf hairs (Lang and Schindler 1994), light conditions, growth stress, and nutrient supply (Heisel et al. 1996;

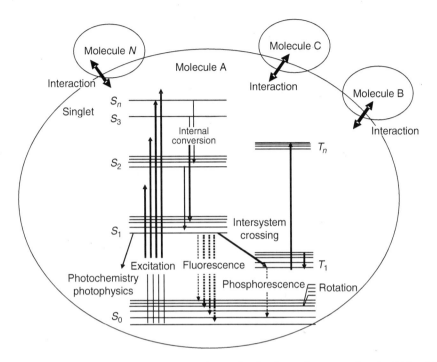

FIGURE 4.3.1 Fluorescence processes of a molecule shown by energy levels (Jablonski diagram) with interactions between different levels: S, singlet state; T_1, triplet state; S_0, ground state; $S_{1,2,3,\ldots,n}$, excited states.

Cerovic et al. 1999). On the other hand, plants have to vary their physiological status in response to environmental impacts. In many cases, they achieve this by changing the species and contents of native compounds that partly belong to the group of fluorophores.

For example, LIF spectra obtained from a tomato leaf during growth period were shown by Saito (2007). Shoot leaves were located on the top of the seedling, mature leaves in the middle, older leaves in the lower sections, and withered leaves at the bottom. Tomato leaves exhibited both BG fluorescence that is in the range 400–650 nm and red + far-red (RFR) chlorophyll fluorescence that is in the range 650–750 nm. The intensity of RFR chlorophyll fluorescence is dependent on the chlorophyll levels of leaves (Stober and Lichtenthaler 1993b). It was greater in mature tomato leaves, while the intensity of BG fluorescence became large in leaves during senescence parallel to the chlorophyll breakdown. The source of the RFR fluorescence is the pigment chlorophyll a, which is related to photosynthesis and plant productivity. Potential sources of the BG fluorescence are mainly ferulic acid, further cinnamic acid derivatives, and also other compounds of the phenylpropanoid family presenting the vacuoles of plant tissues (Chappelle et al. 1991; Lang et al. 1991; Morales et al. 1996; Cerovic et al. 1999). However, the major substance appears to be ferulic acid, covalently bound to all plant cell walls, as has been demonstrated by hydrolyzing experiments (Harris and Hartley 1976; Lichtenthaler and Schweiger 1998). Monitoring changes in the fluorescence emission spectrum of plant tissues can provide data on leaf development and plant responses related to environmental stress.

4.3.1.2 Instrumentation for Fluorescence Experiments

A basic construction of a fluorescence monitoring system (Figure 4.3.2) consists of a light source, a detector, and a processing instrument.

Because of recent developments, laser technology is more widely used as the excitation source. The monochromaticity of laser increases the efficiency of fluorescence emission and simplifies the analysis of fluorescence data since the excited state can be determined to a precision that lies within the laser's spectral width. In the near future, ultraviolet (UV) light-emitting diodes, laser diodes, and laser-diode pumped tunable all-solid-state lasers (see Section 3.1) will be available in miniaturized form. In particular, AlGaN (Adivarahan et al. 2004) and AlN (Taniyasu et al. 2006) deep UV light-emitting diodes are already very attractive light sources for LIFS since they are compact, robust, have low power consumption, long lifetime, and are easy to use.

Spectrophotometers equipped with a charge-coupled device (CCD) line sensor with a large number of pixels have become very popular and are especially suitable

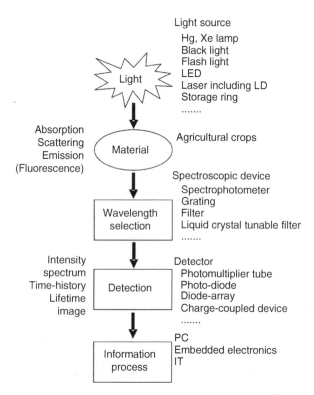

FIGURE 4.3.2 Flow diagram of the set-up for fluorescence analyses.

Fluorescence

for outdoor applications or online monitoring of production lines in factories. Alternatively, if the fluorescence wavelength to be monitored is known, an optical filter for the specific wavelength can be employed. This is a practical solution for fluorescence imaging that provides information on spatial variations in plant physiological activities. Several multicolor fluorescence imaging system (FIS), either laser FIS or flashlamp FIS, have been developed. Laser FIS devices have been described by a few working groups (Lang et al. 1994; Lichtenthaler et al. 1996; Sowinska et al. 1999; Kim et al. 2001, 2003; Saito et al. 2005a). Pulsed xenon-flashlamp FIS devices, which allow a flexible choice of the excitation wavelengths by selecting appropriate filters, have been applied (Buschmann and Lichtenthaler 1998). Fluorescence readings can be extended by additional reflectance measurements (Kuczynski 2003; Lenk et al. 2007). However, a correction of the BG fluorescence emission regarding shielding and reabsorption effects in the complex matrix by means of the reflectance spectrum did not improve calibrations on single phenolic compounds in apples and strawberries (Wulf et al. 2005, 2007).

Multiple-wavelength fluorescence imaging in the approaches mentioned above can be performed using several filters attached to a rotating wheel. A liquid-crystal tunable filter is one alterative for such applications (Saito et al. 2005a). Wavelength of the filter can be tuned by varying the voltage applied to the filter; it thus enables imaging having desired fluorescence wavelength without any mechanical movement.

Photomultiplier tubes (PMTs) and CCD are generally used as detectors in LIFS systems. PMTs have high sensitivities and low noise making them ideal for detecting the low-intensity autofluorescence of living plants, since generally the maximum autofluorescence quantum efficiency of living plants is in the range of 2%–3%. CCDs are effective for imaging applications, but it is currently still difficult to use them to detect low-intensity fluorescence like plant LIF. Image-intensified CCD (ICCD) detectors that can amplify photoelectrons are powerful imaging devices for the low-intensity fluorescence applications. CCD detectors can integrate a signal over a period of time to detect low-intensity signal. By integrating over a long time period, a large increase in the total fluorescence intensity can be obtained, but the signal-to-noise ratio does not increase by the same amount since the noise is also integrated. Cooled ICCD equipped with a Peltier thermoelectric cooling device is effective for such applications. Gated ICCDs and multichannel plate PMTs that can operate in a gate mode are quite effective for synchronized detection with a short-pulse (<10 ns) laser. Such an arrangement can reduce the amount of ambient light such as sunlight in field that appears as background in the spectra. Furthermore, using a streak scope can enable three-dimensional (i.e., time, wavelength, and intensity) monitoring.

4.3.2 BLUE-GREEN LIF SPECTRA OF NUTRITIONAL VALUABLE COMPOUNDS

Native compounds such as polyphenols and vitamins (riboflavin, tocopherol) found in agricultural products are strong fluorophores. A Q-switched pulsed laser (wavelength: 355 nm, pulse energy: 0.3 mJ, pulse length: 7 ns, repetition rate: 10 Hz), a spectrometer with a linear CCD sensor (range: 200–800 nm, resolution: 5 nm, 1024 pixels), and a PC were used for the readings of standard solutions (Figures 4.3.3

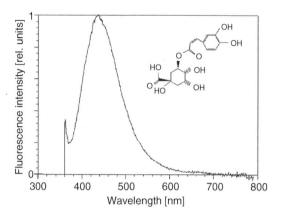

FIGURE 4.3.3 LIF spectrum of chlorogenic acid in methanol, 1.00×10^{-5} mol/L.

through 4.3.6). The solvent spectrum was subtracted in each case. The spectrum of each fluorophore was normalized using its most intense peak. Other data processing methods such as smoothing was not applied to the spectra. It should be added that some spectra are quite noisy because of the low LIF intensity owing to low quantum yield of the pigments.

In plants, the provitamin β-carotene with the chemical formula $C_{40}H_{56}$ (molecular weight: 536.89) is found together with other carotenoids (xanthophylls) in many fruits and vegetables. The carotenoids, orange photosynthetic pigments, transfer the absorbed energy to chlorophyll in the chloroplasts of green cells and thus do not show fluorescence, although reported in some studies. For β-carotene, a BG fluorescence had been assumed by several authors, but as chromatographically purified

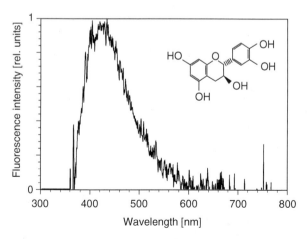

FIGURE 4.3.4 LIF spectrum of D-catechin in methanol, 1.00×10^{-5} mol/L.

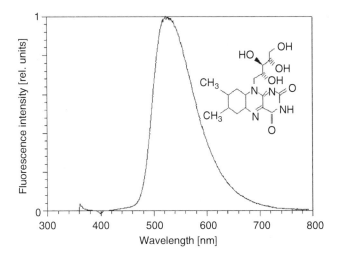

FIGURE 4.3.5 LIF spectrum of riboflavin in distilled water, 1.06×10^{-6} mol/L.

compound it does not show any BG fluorescence (Lang et al. 1991; Buschmann et al. 2008). On the other hand, carotenoids are strong absorbers of fluorescence intensities influencing the apparent fluorescence signal.

Out of the group of polyphenols, for instance, chlorogenic acid (Figure 4.3.3) with chemical formula $C_{16}H_{18}O_9$ (molecular weight: 354.31) as found in leaves and fruits is one of the major soluble phenolic compounds in the plant vacuoles; it is also used in pharmaceuticals, food, and food additives. Another example is the D-catechin (Figure 4.3.4) with chemical formula $C_{15}H_{14}O_6$ (molecular weight: 290.27) that is found in numerous plant species and derived products, including fruits, vegetables, and tea leaves. These compounds exhibit BG fluorescence and can modify, depending

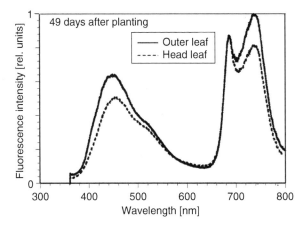

FIGURE 4.3.6 LIF spectra of lettuce 'Shinano summer' lettuce after 49 days after planting, outer leaf (solid line) and head leaf (dotted line).

on their concentration, the major BG fluorescence signal that comes from the cell-wall bound ferulic acid.

An example for vitamin fluorescence is riboflavin (Figure 4.3.5) with chemical formula $C_{17}H_{20}N_4O_6$ (molecular weight: 176.13) that can be found in leafy green vegetables, legumes, milk, cheese, and other foods. Its trivial name is vitamin B_2 and it is used in a variety of food such as cereals, pasta, drinks mainly for coloring owing to its yellow or orange-yellow color.

The BG-LIF spectra of ferulic acid, nicotinamide adenine dinucleotide phosphate, flavin adenine dinucleotide, and other native plant compounds with additional subscriptions were reported earlier (Bargmann et al. 1987; Lang et al. 1991; Jorgensen et al. 1992; Lichtenthaler and Schweiger 1998; Cerovic et al. 1999).

4.3.3 Application in Agricultural Product Monitoring

Frequently in the cultivation of agricultural products, judgments are made on the basis of an individual's experiences. Such subjective judgments are formed based on plant physical parameters such as size, shape, weight, and color, and they often differ between individuals. To monitor the plant growth process precisely and to apply rapid and appropriate treatment if required, more objective criteria that do not depend on individual experience are required. Such a technique is essential for realizing a stable supply of agricultural products that are guaranteed to be of a certain quality for all food markets and industries. Such a technique will be a key technology for achieving precision agriculture in the near future.

LIFS is a potential candidate for this approach. In this section, examples of BG-LIF spectra of some agricultural products and plant leaves are provided and we discuss the feasibility of monitoring agricultural products using LIF data. Lettuce, napa cabbage, and pear fruit are investigated. LIF monitoring system is the same as shown in Figures 4.3.9 and 4.3.13, while the optical geometry is a front-face design.

4.3.3.1 Examples of BG-LIF Spectra of Agricultural Products

4.3.3.1.1 Vegetables

In 'Shinano summer' lettuce (*Lactuca sativa* L.), the BG-LIF and RFR-LIF spectra for autumn cropping were recorded. The intensity of the fluorescence from outer leaves is greater and varied more than that from head leaves (Figure 4.3.6).

Napa cabbage (*Brassica rapa* L.) LIF spectra of a normal core and a rotten core show greater intensities in the green wavelength range of the LIF spectrum of the rotten core in comparison to that of the normal core (Figure 4.3.7). The integrated peak area between 450 and 600 nm for the rotten core was more than twice that for the normal core.

4.3.3.1.2 Fruit

In pear (*Pyrus communis* L.), differences were found in the LIF spectra from the surface of a young pear and a ripe pear (Figure 4.3.8). The intensity of the blue region (400–500 nm) of the LIF spectrum in the ripe pear surface was increased in comparison to the pear picked at harvest time in a still unripe stage.

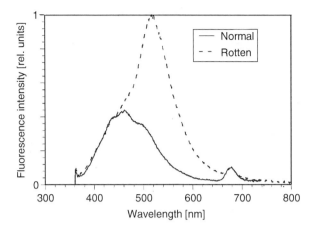

FIGURE 4.3.7 LIF spectra of normal (solid line) and rotten (dotted line) cores of napa cabbage.

4.3.3.2 What Information Can Be Extracted from LIF Spectra?

4.3.3.2.1 Growth Process

Outside monitoring of the growth of 'Shinano summer' variety lettuces until harvest season was performed for four successive years (Ishizawa et al. 2002; Saito 2007). The lettuces were cultivated outside under natural conditions and were treated with agricultural agents and insecticides.

The LIF monitoring system used in laboratory was improved for the use in field (outside). The mobile LIF monitoring system is shown in Figure 4.3.9 (Saito 2006a). Size of the system was reduced so that it could be transported by one person on a small cart. Laser was delivered through a transmitting optical fiber to a lettuce leaf

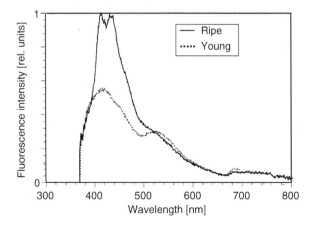

FIGURE 4.3.8 LIF spectra of a young pear surface on September 11 (dotted line) and a ripe pear surface on October 5 (solid line).

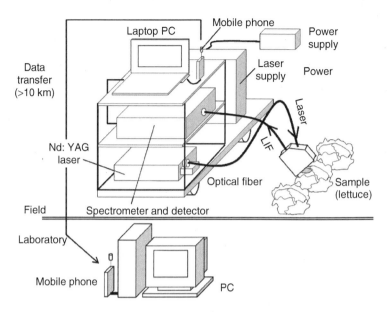

FIGURE 4.3.9 Mobile LIF system as used for monitoring in the lettuce field (compare Figure 4.2.13). The data obtained in the field were transferred by a mobile phone to our laboratory in Shinshu University, which was 10 km from the system.

and the LIF from the leaf was collected and sent to a spectrometer via a receiving fiber. Using two separate fibers improved manipulation of light handling, the configuration was convenient especially in field. The inlet of the receiving fiber was a 10 mm diameter circle to match the emission pattern of the LIF, and the outlet shape was rectangular to match the entrance slit of the spectrometer. The outlet of the transmitting fiber and the inlet of the receiving fiber were combined in a single holder that was gently pressed onto the lettuce leaves. A soft sponge was attached to this holder to block out ambient sunlight.

The fluorescence of the outer leaves was investigated since they had a larger LIF intensity than the head leaves (see Figure 4.3.6). The LIF variation is more accurately described in terms of fluorescence ratios than absolute values, and the blue-LIF intensity to red-LIF intensity ratio appears to be the most sensitive parameter for LIF. Daily variation appeared in the ratio of the intensity of the outer lettuce leaves at 460 nm ($F460$) in blue-LIF to that at 685 nm ($F685$) in red-LIF (Figure 4.3.10).

The variation pattern in the relative intensity can be explained as follows:

1. The relative intensity decreased during the first 4 weeks after planting; there was a rapid decrease in the first 2 weeks and then a more gradual decrease over the following 2 weeks. During the 4 weeks, the intensity of the 685 nm fluorescence increased relative to that of the 460 nm fluorescence. It is suggested that the chlorophyll contents (origin of the RFR fluorescence) increased in the first 2 weeks, while in the following 2 weeks the lettuce became mature and the chlorophyll accumulation rate was smaller than the

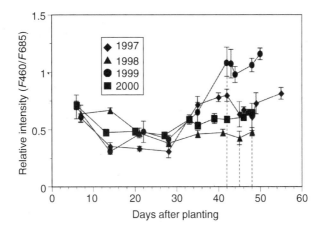

FIGURE 4.3.10 Daily variation in relative LIF intensity of outer leaves of lettuce at 460 nm ($F460$) to that at 685 nm ($F685$), the harvest time for each year is indicated by a dotted line.

former weeks. The lettuce went into the next stage. It should be added that head formation started at around these 4 weeks.

2. Then the gradient of the relative BG fluorescence intensity turned to positive and the relative intensity of apparent fluorescence returned to almost the same value as the initial one at the time of planting. The basic growing process probably ended and the accumulation of soluble secondary metabolic products in the vacuole with blue or green fluorescence dominated and could possibly increase the overall BG fluorescence yield. A slight breakdown of chlorophyll is also a possible cause for the observed change in the relative proportions of the BG to the RFR fluorescence.
3. The relative fluorescence intensity peaked at about 41 days after planting and it had a minimum at around 45 days. The harvest times as judged by a horticultural specialist nearly coincided with this period. Even though the reason of this decrease for a short period is unclear, this may be related to a change in the lifecycle of a lettuce, for example, from the vegetative period to the reproductive period.
4. After passing through the minimum, a subsequent increase appeared, which is assumed to be due to withering. The same pattern in variation was observed every year during the 4 year monitoring period, although small differences are apparent. This LIF variation data will help modeling a standard growth pattern of lettuce, which is useful information for cultivation.

Blue-green fluorescence primarily derives from the BG fluorophore ferulic acid covalently bound to all plant cell walls as has been directly shown by a detailed chemical analysis of hydrolyzed cell walls (Harris and Hartley 1976; Lichtenthaler and Schweiger 1998) and by an indirect method using a temperature gradient (Morales et al. 1996). However, soluble phenols in the vacuoles can modify this primary fluorescence emission of plant tissue, for example, chlorogenic acid that affects the shape of the BG-LIF spectrum.

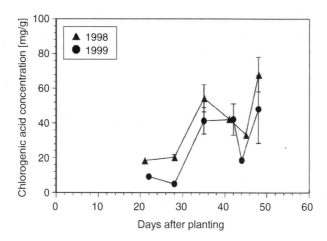

FIGURE 4.3.11 Variation of chlorogenic acid concentration in 1998 and 1999 measured with an HPLC.

The concentration of this polyphenol was measured by high-performance liquid chromatography (HPLC) (Figure 4.3.11) pointing out the variation of the chlorogenic acid concentration in 1998 and 1999 (Ishizawa et al. 2002; Saito 2005b). A comparison of the spectral area of the chlorogenic acid LIF spectrum was analyzed and the concentration is given in Figure 4.3.12 (Ishizawa et al. 2002).

4.3.3.2.2 Disease and Contaminants

In all food industries, it is highly desirable to be able to detect rotten or spoiled areas inside vegetables. The results shown in Figure 4.3.7 provide a basis for such detection. In case of napa cabbage, it is known that cabbages can easily become

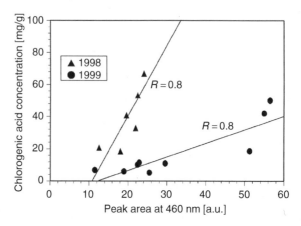

FIGURE 4.3.12 Comparison of the analyzed area of LIF spectrum centered at around 450 nm and concentration of chlorogenic acid measured using an HPLC.

diseased if the soil contains an excessive amount of nutrients or the soil is in a bad condition. LIF spectral data can be utilized subsidiary for land development resulting in the production of high-quality foods.

The LIF spectra of apple seedlings pointed out the possibility of using LIFS to detect fungal infection (Ludeker et al. 1996). Application of this technique to gardening, horticulture, and forest management is anticipated.

LIF images of apples artificially contaminated with dilute animal feces have been obtained (Kim et al. 2003). An uncontaminated apple surface emitted fluorescence at 450 and 550 nm, but the areas treated with feces, which could not be visually distinguished from uncontaminated areas, did not fluoresce in the BG region. Detecting contamination based on the absorption of LIF by the contamination gives potential applications in food-safety inspection. Tissue browning of apples and banana was detected by means of BG fluorescence with excitation at 337 nm (Zude 2004).

4.3.4 FUTURE PROSPECTS OF LASER-INDUCED FLUORESCENCE SPECTROSCOPY IN FIELD MEASUREMENTS

Food materials are often processed in vessels, containers, or rooms that are subject to extreme conditions such as high/low temperatures and pressures, and sterilized atmospheres. These often require high voltages and currents, and devices/machines that are liable to explode. The characteristics of optical analytical methods are amenable to satisfying this requirement. The coherency of laser, which is one of its most important characteristics, allows photon energy to be delivered effectively to the location where the material is to be processed and monitored. Optical fiber delivering is the most practical one that realizes in a noncontact way.

Combining laser monitoring with laser processing is another attractive application. Since laser beams can be easily focused down to very small areas that are several tens micrometers in diameter, laser vaporization can eliminate undesired areas of the raw materials that are very small but that might cause a capital problem. The beam area is so small that processing can be applied without damaging other normal areas. UV-laser irradiation is also expected to impart a bactericidal effect. Production of new nutrients by optically driving chemical reactions within a specific area is a challenging but attractive application. Variations in the LIF spectrum caused by such processing can be monitored and data thus obtained can be fed back to control the process. This will be especially useful in the supplement and drug industries. For these purposes, multiple-wavelength LIF imaging (Kim et al. 2003; Saito et al. 2005a) should be investigated further. This technique can offer spatial information, for example, visualization of differences between processed areas and nonprocessed areas.

Another application of LIFS that is of considerable importance is use in the field. The high directivity of laser beams gives the monitoring system capability of remote monitoring at distances of several 100 m or greater. Vineyards, orange groves, wheat fields, etc. generally extend over such distances. It is quite difficult for both to perform chemical and physical monitoring using sensors over such wide areas successively. Satellite monitoring that uses reflection of natural sunlight is becoming

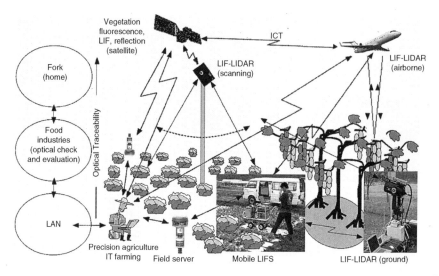

FIGURE 4.3.13 Concept of optical farming based on LIFS and LIF-LIDAR over a global field. For schematic view on monitoring system see Figure 4.3.9.

popular. Reflection data reveal the physical properties of plants and vegetation. However, from the point of managing plant cultivation, data must contain physiological information.

One possible way to achieve this is to combine LIFS with a laser remote-sensing technique known as light detection and ranging (LIDAR), imaging LIF-LIDAR (Hans et al. 1994; Edner et al. 1995; Johansson et al. 1996; Saito et al. 1997, 2002; Sowinska et al. 1999) appears particularly suitable for this application. This is illustrated in Figure 4.3.13, in which optical farming as a combined style of LIFS and Information Communications Technology (ICT) is proposed. A ground-based scanning LIDAR system (Saito et al. 2005b) continually collects plant physiological information. Mobile LIF-LIDAR systems transported using a vehicle (Cecchi et al. 1994; Hans et al. 1994; Sowinska et al. 1999), an airplane (Hoge et al. 1983), or a ship (Alberotanza et al. 1995) are also utilized. Satellite-based LIF-LIDAR has already been proposed (Guzzi 2007) and feasibility study on vegetation fluorescence (not laser-induced) monitoring from space has been going on. Plant in-vivo information monitored by them can be collected using an advanced ICT system such as a field server (Saito et al. 2006) and is delivered through the Internet or a satellite communication system. Network of optical traceability based on optical check and evaluation can be expanded through the entire distribution field, namely field to fork.

4.3.5 Conclusion

Monitoring of food-related materials based on LIFS has been summarized along with current and future applications. Discussion on LIFS monitoring technique and practical usage of LIF spectral data have confirmed the great potential and practicality of using LIFS. The ability of a nondestructive, real-time, and remote monitoring has been demonstrated in field.

Research into the fluorescence phenomena in laboratory has a long history but practical applications of BG-LIFS had to wait until the development of UV lasers. This complains the fact that experimental demonstrations are not adequate at the present time. Creating a fruitful database on BG-LIF spectra of a large variety of products in various physiological statuses is left as emergent assignment. In the last decade, much progress in the assessment of plant tissues and agro-food has been made via multiple fluorescence imaging.

Assembling of all optical phenomena including LIF lifetime (time-resolved LIFS), Raman spectroscopy, absorption, scattering, and reflection is the next step for optical monitoring. Fusion with ICT will improve optical monitoring to optical sensing, which is a sublimated structure of optical monitoring accompanying good judgment and decision, information flow, and distribution. Instantaneous and global data handling with the optical sensing system has the potential to address the stability and balance of food supply on a global scale.

ACKNOWLEDGMENTS

The author would like to express his thanks to Dr. Akiko Takeuchi of Nagaoka National College of Technology. A large portion of the data shown in this section was obtained while she was a student in our laboratory. Dr. Fumitoshi Kobayashi of the Faculty of Engineering, Shinshu University supported the development of the system. Professor Naoto Inoue of the Faculty of Agriculture and Professor Hiroaki Ishizawa of the Faculty of Textile Science and Technology, Shinshu University have been working together as partners of a part of the joint research program on this. Kazuhiro Komatsu of the Nagano Vegetable and Ornamental Crops Experimental Station supplied agricultural products. Professor Takaharu Kameoka of Mie University, and Dr. Masayuki Hirafuji and Dr. Seishi Ninomiya of the National Agricultural Research Center supported the research budget and proposed new ideas based on his agricultural expertise. Professors Akio Nomura and Takuya Kawahara of Shinshu University participated in the discussion.

A portion of this material is based on the research funded by a Grant-in-Aid for Scientific Research (B), 17360191 (2005–2008) supported by the Ministry of Education, Science, Sports, and Culture and Technology; Synergetic Information Systems with Distributed Database and Models Project (2001–2005) supported by the Ministry of Agriculture, Forestry and Fisheries of Japan; and Gakucho-Sairyo-Keihi Research Project (2006–2007) supported by Shinshu University.

REFERENCES

Adivarahan, V., W.H. Sun, A. Chitnis, M. Shatalov, S. Wu, H.P. Marusa, and M. Asif Khan. 2004. 250 nm AlGaN light-emitting diodes. *Applied Physics Letters* 85: 2175–2177.

Alberotanza, L., P.L. Cova, C. Ramasco, S. Vianello, M. Bazzani, G. Cecchi, L. Pantani, V. Raimondi, P. Ragnarson, S. Svanverg, and E. Wallinder. 1995. Yellow substance and chlorophyll measurements in the Venice lagoon using laser-induced fluorescence. *EARSeL Advances Remote Sensing* 3: 102–111.

Bargmann, W.R., M.D. Barkley, R.W. Hemingway, and W.L. Mattice. 1987. Heterogeneous fluorescence decay of (4 6)-ans (4 8)-linked dimmers of (+)-catechin and (−)-epicatechin as result of rotational isomerism. *American Chemical Society* 109: 6614–6619.

Buschmann, C. and H.K. Lichtenthaler. 1998. Principles and characteristics of multi-colour fluorescence imaging of plants. *Journal of Plant Physiology* 152: 297–314.

Buschmann, C., G. Langsdorf, and H.K. Lichtenthaler. 2008. The blue, green, red and far-red fluorescence signatures of plants, their multicolour fluorescence imaging and applications in agrofood assessment. In *Optical Methods for Monitoring Fresh and Processed Agricultural Crops*, ed. M. Zude. CRC Press (Taylor & Francis Group), Boca Raton, FL.

Cecchi, G., P. Mazzinghi, L. Pantani, R. Valentini, D. Tirelli, and P.D. Angelis. 1994. Remote sensing of chlorophyll *a* fluorescence of vegetation canopies. 1: Near and far field measurement techniques. *Remote Sensing of Environment* 47: 18–28.

Cerovic, Z.G., G. Samson, F. Morales, N. Tremblay, and I. Moya 1999. Ultraviolet-induced fluorescence for plant monitoring: Present state and prospects. *Agronomie* 19: 543–578.

Chappelle, E.W., F.M. Wood, J.E. McMurtrey, and W.W. Newcomb. 1984a. Laser-induced fluorescence of green plants. 1: A technique for the remote detection of plant stress and species differentiation. *Applied Optics* 23: 134–138.

Chappelle, E.W., J.E. McMurtrey, F.M. Wood, and W.W. Newcomb. 1984b. Laser-induced fluorescence of green plants. 2: LIF caused by nutrient deficiencies in corn. *Applied Optics* 23: 139–142.

Chappelle, E.W., F.M. Wood, J.E. McMurtrey, and W.W. Newcomb. 1985. Laser-induced fluorescence of green plants. 3: LIF spectral signature of five major plant types. *Applied Optics* 24: 74–80.

Chappelle, E.W., J.E. McMurtrey III, and M.S. Kim. 1991. Identification of the pigment responsible for the blue fluorescence band in the laser induced fluorescence (LIF) spectra of green plants, and the potential use of this band in remotely estimating rates of photosynthesis. *Remote Sensing of Environment* 36: 213–218.

Edner, H., J. Johansson, P. Ragnarson, S. Svanberg, and E. Wallinder. 1995. Remote multicolour fluorescence imaging of selected broad-leaf plants. *EARSel Advances in Remote Sensing* 3: 2–14.

Guzzi, R. 2007. Advanced lidar technologies for active remote sensing of vegetation fluorescence from space. In *3rd International Workshop on Remote Sensing of Vegetation Fluorescence*, Florence, Italy.

Hans, E., J. Johansson, S. Svanberg, and E. Wallinder. 1994. Fluorescence lidar multicolor imaging of vegetation. *Applied Optics* 33: 2471–2479.

Harris, P.J. and R.D. Hartley. 1976. Detection of bound ferulic acid in cell walls of the Gramineae by ultraviolet fluorescence microscopy. *Nature* 259: 508–510.

Heisel, F., M. Sowinska, J.A. Miehe, M. Lang, and H.K. Lichtenthaler. 1996. Detection of nutrient deficiencies of maize by laser induced fluorescence imaging. *Journal of Plant Physiology* 148: 622–631.

Hoge, F.E., R.N. Swift, and J.K. Yungel. 1983. Feasibility of airborne detection of laser-induced fluorescence emissions from green terrestrial plants. *Applied Optics* 22: 2991–3000.

Ishizawa, H., Y. Saito, T. Amemiya, and K. Komatsu. 2002. Non-destructive monitoring of agricultural products (lettuce) based on Laser-induced fluorescence. *Journal of the Japanese Society of Agricultural Machinery (Nogyokikai Gakkaishi)* 64: 89–94.

Johansson, J., M. Anderson, H. Edner, J. Mattsson, and S. Svanberg. 1996. Remote fluorescence measurements of vegetation spectrally resolved and by multi-colour fluorescence imaging. *Journal of Plant Physiology* 148: 632–637.

Jorgensen, K., H. Stapelfeldt, and L.H. Skibsted. 1992. Fluorescence of carotenoids. Effect of oxygenation and cis/trans isomerization. *Chemical Physics Letters* 190: 514–519.

Kim, M.S., Y.R. Chen, and P.M. Mehl. 2001. Hyperspectral reflectance and fluorescence imaging system for food quality and safety. *Transactions of the American Society of Agricultural Engineers* 44: 721–729.

Kim, M.S., M.L. Alan, and C. Yud-Ren. 2003. Multispectral laser-induced fluorescence imaging system for large biological samples. *Applied Optics* 42: 3927–3934.

Kuczynski, A.P. 2003. Exploring reflectance spectra of apple slices and their relation to active phenolic compounds. *Polish Journal of Food Nutritional Science* 53: 151–158.

Lang, M. and C. Schindler. 1994. The effect of leaf-hairs on blue and red fluorescence emission and on zeaxanthin cycle performance of *Senecio medley* L. *Journal of Plant Physiology* 144: 680–685.

Lang, M., F. Stober, and H.K. Lichtenthaler. 1991. Fluorescence emission spectra of plant leaves and plant constituents. *Radiation and Environmental Biophysics* 30: 333–347.

Lang, M., P. Siffel, Z. Braunova, and H.K. Lichtenthaler. 1992. Investigations of the blue-green fluorescence emission of plant leaves. *Botanica Acta* 105: 435–440.

Lang, M., H.K. Lichtenthaler, M. Sowinska, P. Summ, and F. Heisel. 1994. Blue, green and red fluorescence signatures and images of tobacco leaves. *Botanica Acta* 107: 230–226.

Lang, M., H.K. Lichtenthaler, M. Sowinska, F. Heisel, and J.A. Miehe. 1996. Fluorescence imaging of water and temperature stress in plant leaves. *Journal of Plant Physiology* 148: 613–621.

Lenk, S., L. Chaerle, E.E. Pfündel G. Langsdorf, D. Hagenbeek, H.K. Lichtenthaler, D.V.D. Straeten, and C. Buschmann. 2007. Multi-colour fluorescence and reflectance imaging at the leaf level and its application possibilities. *Journal of Experimental Botany* 58: 807–814.

Lichtenthaler, H.K. and J. Schweiger. 1998. Cell wall bound ferulic acid, the major substance of the blue-green fluorescence emission of plants. *Journal of Plant Physiology* 152: 272–282.

Lichtenthaler, H.K., F. Stober, C. Buschmann, U. Rinderle, and R. Hák. 1990. Laser-induced chlorophyll fluorescence and blue fluorescence of plants. In *International Geoscience and Remote Sensing Symposium, IGARSS '90*, D.C. Washington, Vol. III. University of Maryland, College Park, MD, pp. 1913–1918.

Lichtenthaler, H.K., M. Lang, and F. Stober. 1991a. Laser-induced blue fluorescence and red chlorophyll fluorescence signatures of differently pigmented leaves. In *5th International Colloquium on Physical Measurements and Signatures in Remote Sensing, Courchevel*. ESA Publications Division, Noordwijk, The Netherlands, pp. 727–730.

Lichtenthaler, H.K., M. Lang, and F. Stober. 1991b. Nature and variation of blue fluorescence spectra of terrestrial plants. In *International Geoscience and Remote Sensing Symposium, IGARSS '91, Espoo*, Vol. IV. Helsinky University of Technology, Espoo, Finland, pp. 2283–2286.

Lichtenthaler, H.K., F. Stober, and M. Lang. 1992. The nature of the different laser-induced fluorescence signatures of plants. *EARSeL Advances in Remote Sensing* 1: 20–32.

Lichtenthaler, H.K., M. Lang, M. Sowinska, F. Heisel, and J.A. Miehé. 1996. Detection of vegetation stress via a new high resolution fluorescence imaging system. *Journal of Plant Physiology* 148: 599–612.

Ludeker, W., H.-G. Dahn, and K.P. Gunther. 1996. Detection of fungal infection of plants by laser-induced fluorescence: An attempt to use remote sending. *Journal of Plant Physiology* 148: 579–585.

Measures, R.M., W. Houston, and M. Bristow. 1973. Development and field tests of a laser fluorosensor for environmental monitoring. *Canadian Aeronautics and Space Journal* 19: 501–506.

Morales, F., Z.G. Cerovic, and I. Moya. 1996. Time-resolved blue-green fluorescence of sugar beet (*Beta vulgaris* L.) leaves. Spectroscopic evidence for the presence of ferulic acid as the main fluorophore of the epidermis. *Biochimica et Biophysica Acta* 1273: 251–262.

Saito, Y. 2005. Laser-induced fluorescence as an index for monitoring plant activity and productivity related to photosynthesis, Chapter 11. In *Recent Progress of Bio/Chemiluminescence and Fluorescence Analysis in Photosynthesis*, eds. N. Wada and M. Mimuro. Research Signpost, Kerala, India, pp. 235–251.

Saito, Y. 2006. Plant health status monitoring using laser technique. *Bionics* March: 70–71.

Saito, Y. 2007. Laser-induced fluorescence spectroscopy/technique as a tool for field monitoring of physiological status of living plants. *Proceeding of SPIE* 6604: 66041W-1–66041W-12.

Saito, Y., K. Hatake, E. Nomura, T.D. Kawahara, A. Nomura, N. Sugimoto, and T. Itabe. 1997. Range-resolved image detection of laser-induced fluorescence of natural trees for vegetation distribution monitoring. *Japanese Journal of Applied Physics* 36: 7024–7027.

Saito, Y., K. Kurihara, H. Takahashi, F. Kobayashi, T. Kawahara, A. Nomura, and S. Takeda. 2002. Remote estimation of the chlorophyll concentration of living trees using laser-induced fluorescence imaging lidar. *Optical Review* 9: 37–39.

Saito, Y., T. Matsubara, T. Koga, F. Kobayashi, T.D. Kawahara, and A. Nomura. 2005a. Laser-induced fluorescence imaging of plants using a liquid crystal tunable filter and charge coupled device imaging camera. *Review of Scientific Instruments* 76: 106103-1–106103-3.

Saito, Y., H. Kurata, Y. Hara, F. Kobayashi, T. Kawahara, and A. Nomura. 2005b. Three dimensional monitoring of stack plume dynamics by a scanning Mie lidar system as a plume watchdog station. *Optical Review* 12: 328–333.

Saito, Y., T. Suzuki, K. Kobayashi, K. Sato, M. Hirafuji, T. Fukatsu, R. Yashiro, S. Takeuchi, K. Yuasa, S. Watanabe, F. Kobayashi, T. Kawahara, and T. Kameoka. 2006. Field server monitoring system for construction of IT farming and agri-tourism—Trial report from Obuse-town, Nagano, Japan. In *Proceedings of SICE-ICASE International Joint Conference*. Busan, Korea, pp. 4848–4851.

Sowinska, M., B. Cunin, F. Heisel, and J.A. Miehe. 1999. New UV-A laser-induced fluorescence imaging system for near-field remote sensing of vegetation: Characteristics and performances. *Proceedings of SPIE* 3707: 91–102.

Stober, F. and H.K. Lichtenthaler. 1993a. Characterisation of the laser-induced blue, green and red fluorescence signatures of leaves of wheat and soybean leaves grown under different irradiance. *Physiologia Plantarum* 88: 696–704.

Stober, F. and H.K. Lichtenthaler. 1993b. Studies on the localization and spectral characteristics of the fluorescence emission of differently pigmented wheat leaves. *Botanica Acta* 106: 365–370.

Stober, F. and H.K. Lichtenthaler. 1993c. Studies on the constancy of the blue and green fluorescence yield during the chlorophyll fluorescence induction kinetics (Kautsky effect). *Radiation and Environmental Biophysics* 32: 357–365.

Taniyasu, Y., M. Kasu, and T. Makimot. 2006. An aluminium nitride light-emitting diode with a wavelength of 210 nanometers. *Nature* 444: 325–328.

Wulf, J.S., M. Geyer, B. Nicolaï, and M. Zude. 2005. Non-destructive assessment of pigments in apple fruit and carrot by laser-induced fluorescence spectroscopy (LIFS) measured at different time-gate positions. *Acta Horticulturae (ISHS)* 682: 1387–1394.

Wulf, J.S., S. Rühmann, I. Regos, I. Puhl, D. Treutter, M. Zude. 2008. Analyses of phenolic compounds in strawberry fruits (Fragaria x ananassa) by chromatographic and non-destructively applied laser-induced fluorescence spectroscopy. *Journal of Agricultural and Food Chemistry* 56(9): 2875–2882.

Zude, M. 2004. Detection of fruit tissue browning using laser-induced fluorescence spectroscopy. *Acta Horticulturae* 628: 85–91.

4.4 FRONT-FACE FLUORESCENCE ANALYSIS TO MONITOR FOOD PROCESSING AND NEOFORMED CONTAMINATION

JAD RIZKALLAF, LYES LAKHAL, AND INÈS BIRLOUEZ-ARAGON

4.4.1 INTRODUCTION

Fluorescence spectroscopy is a sensitive analytical technique particularly suitable for monitoring the impact of processing on food quality parameters. Many heat- or oxidation-sensitive nutrients such as vitamin E, B6, and B2 are fluorescent. Consequently, their fluorescence is expected to correlate with process conditions or storage (Mas et al. 2004; Cheikhousman et al. 2005; Christensen et al. 2005; Riccio et al. 2006). Peptide tryptophan fluorescence is also a good indicator for heat-induced protein denaturation that occurs during spray drying or pasteurization of milk because of modification of protein percentage of soluble proteins and binding conditions inducing a red-shifted emission of the molecule (Birlouez-Aragon et al. 2002). Therefore, monitoring the intensity and shape of emission fluorescence allows characterization of such physicochemical changes, which are also often associated to nutritional damage of food during processing (Birlouez-Aragon and Zude 2004). In addition, some neoformed molecules formed during Maillard reaction between natural food components, namely sugars or lipids with proteins or amino acids (Leclère and Birlouez-Aragon 2001; Matiacevich and Buera 2006), are also fluorescent. This is the case for heterocyclic amines, which are formed during meat and fish grilling at high temperatures (Skog et al. 2000), aromatic polycyclic hydrocarbons (Sharma et al. 2007) produced by partial carbonization of food during barbecuing, and more generally advanced Maillard products, also called AGEs (for advanced glycated end products) formed in the advanced stages of the reaction (not published). Many fluorescent AGEs have been identified, such as pentosidine, vesperlysine, and argpyrimidine, all formed by reaction between lysine or arginine and a carbonyl compound derived from sugars or lipids (Zamora and Hidalgo 1995; Ranjan and Beedu 2006). Consequently, food processing considerably changes the fluorescence fingerprint acquired on a food product, since it simultaneously induces a decrease and red shift of some naturally occurring fluorescence intensities, and an increase of fluorescence attributed to neoformed molecules (Birlouez-Aragon and Zude 2004).

Conventionally, fluorescence spectra or excitation–emission matrix (EEM) arrays are acquired on diluted samples with an optical density (OD) lower than 0.1. This implies that extraction of aqueous part or organic fraction of the food product is performed before signal acquisition. For example, the FAST method in milk was developed to monitor the formation of fluorescent Maillard products in soluble protein fraction at pH 4.6 to quantify the sterilization impact on milk and evaluate the process quality (Birlouez-Aragon et al. 1998). In vegetable oils, changes in vitamin E or polyphenol fluorescence intensity upon storage (Cheikhousman et al. 2005) or heating (Mas et al. 2004) can be monitored via dilution in adequate solvent (e.g., propanol). However, food dilution often leads to information loss given that only a small fraction of the complete and complex food product is considered and the initial physicochemical structure is degraded. This problem can be circumvented using front-face fluorescence acquisition geometry, which allows direct acquisition on intact samples. This technique is thus faster, solvent pollution free, and the resulting signal contains exhaustive physicochemical information on the food matrix. Nevertheless, fluorescence spectral data acquired through this nondestructive technique must be interpreted with great caution because of the non-Beer–Lambert absorption behavior of foods (Qin and Lu 2007). The fluorescence emitted at the surface of intact (turbid) food sample is deeply distorted because of absorption and scattering phenomena at both the excitation and emission wavelengths as well as quenching (Lakowicz 1999; Bradley and Thorniley 2006). Consequently, the fluorescence signals are difficult to interpret, and no direct quantification is allowed.

Information contained in front-face fluorescence fingerprints can be exploited using two different approaches. The logistic approach attempts to recover the specific reemission spectrum of the fluorophore under question from the distorted spectrum acquired on intact sample, for instance, by means of the transfer function (Ramanujam 2000). However, this approach requires the knowledge of the optical properties of the sample (see Section 4.4.2).

In the statistical chemometric approach, variance information of front-face fluorescence signals is directly extracted and used for discrimination or calibration purposes via multivariate or multiway decomposition and regression tools. Chemometric tools allow identification of emission wavelength discriminative of the process steps and parameters, and construction of prediction models of some quality parameters using an indirect correlations approach.

Multiway parallel factor (Parafac) analysis decomposition models are particularly relevant for extraction of main fluorescence profiles contained in bidimensional front-face fluorescence landscape such as EEM. The resulting intensities of the individual fluorescence profiles (Parafac scores) can be further used to describe the time–temperature impact or to predict neoformed contaminants (NFC). In the European ICARE project (COLL-CT-2005-516415), the chemometric approach was applied to develop a prototype equipped with selected light sources and algorithms allowing prediction of process-induced NFCs. Highly satisfactory prediction models (with cross-validation errors lower than 10%) of acrylamide content in cookies, potato crisps, or roasted malt, or carboxymethyllysine (CML) content in cookies and infant formulas were achieved. The following two sections detail the two approaches with illustrations and examples taken from applications on processed food products.

4.4.2 Physical Approach of Process-Induced Food Physicochemical Changes

4.4.2.1 Description of the Matrix-Induced Distortions

To understand the effects of the matrix on fluorescence measurements, Figure 4.4.1 compares three EEMs from turbid medium containing the same fluorophores in the presence of chromophores absorbing in different regions of ultraviolet and visible spectra. In fact, the variation of media-optical properties at each spectrum wavelength induces a distortion of fluorescence excitation and emission spectra,

Fluorescence

thus impeding clear identification and quantification of the fluorophores responsible for the observed fluorescence excitation (Figure 4.4.2). Consequently, changes in the fluorescence intensity because of variations in fluorophore concentration cannot easily be distinguished from those arising from variations in absorption and scattering.

It is worth noting, however, that sample scattering and absorption properties themselves may yield significant information. For example, changes in absorption can reflect evolution in chromophores concentration, and modification of scattering properties is indicative of changes in media structure.

Clearly, quantitative and unambiguous information about sample fluorophore concentration can only be determined if correction techniques are applied to compensate for scattering and absorption distortion effects. Currently available correction techniques can be classified into three categories: empirical techniques, measurement-based techniques, and theory-based techniques.

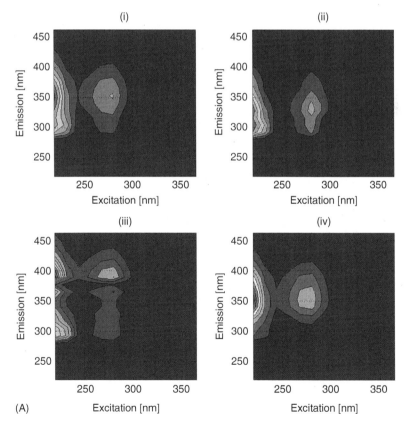

FIGURE 4.4.1 (A) EEM contour plots of an amino acid mixture (tryptophan, tyrosine, and phenylalanine) simulated in different medium: (i) dilute and in three different dyes solutions (ii) auramine, (iii) lucifer yellow, and (iv) crystal violet.

(continued)

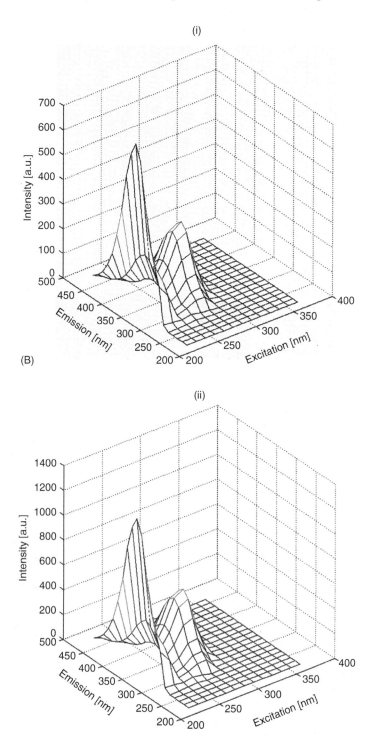

FIGURE 4.4.1 (continued) (B) EEM three-dimensional plots of an amino acids mixture of (tryptophan, tyrosine, and phenylalanine) simulated in (i) dilute media and (ii) pure scattering media. Note that a pure scattering media induce an increase of fluorescence intensity and preserves the spectral line shapes.

Fluorescence

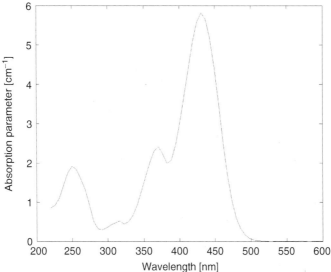

FIGURE 4.4.2 Absorption spectra $\mu_a(\lambda)$ used for Monte Carlo calculation of matrix transfer function. Note: Values of scattering parameter $\mu_s = 30$ cm^{-1} and anisotropy parameter $g = 0.8$.

Empirical techniques generally involve a combination of measurements related to the intrinsic fluorescence but independent of the matrix optical properties. The theoretical basis for this method was put forward by Jobsis et al. (1971), who postulated that blood vessels would almost totally absorb both excitation and emission light. Changes in fluorescence were therefore expected to be in direct proportion to changes in reflectance. This relationship was confirmed experimentally over a limited range of blood pressure levels and haemocrits. However, the constant of proportionality was found to vary considerably between different tissue preparations and optical systems. Experimental work coupled with Monte Carlo simulation confirmed that the confocal technique was effective in reducing distortion. However, the small region explored by this technique represents the main limit of this methodology. Small regions may not be representative of the entire tissue because of local variations in fluorophore concentrations and tissue-optical properties, for example, because of the presence of blood vessels. It may be necessary, therefore, to carry out measurements in a number of different areas to gain a more comprehensive assessment of tissue state.

Measurement-based techniques essentially use selective recording of the least attenuated fluorescence bands. A number of techniques have been proposed to selectively record fluorescence photons that have only traveled a relatively short distance through the matrix. The intensity will be less affected by distortion since the photons will have experienced relatively few interactions.

Theory-based techniques generally need the calculation of the transfer function relating pure intrinsic fluorescence to apparent distorted fluorescence measured.

4.4.2.2 Assessment of the Transfer Function

In general, a monodimensional model describes the attenuation of the excitation beam passing through the turbid sample and the attenuation of the fluorescence as it travels from the point of generation to the surface of the sample. These effects can be described mathematically by considering a turbid sample as a dilute solution with a small enough thickness to be compared to the photon mean free pathlength. Dividing a turbid sample of thickness L_{tur} into a number of thin layers of thickness dz along the optical axis allows calculating the contribution to the fluorescence, dP_{tur}, collected from the front surface of a thin layer of the sample, with a thickness dz, located at a depth z as follows (Richards-Kortum 1995; Gardner et al. 1996):

$$dP_{total}^{tur}(\lambda_{ex},\lambda_{em},z) = P_0(\lambda_{ex}) \left[\sum_{f=1}^{F} \ln(10)\varepsilon_f(\lambda_{ex}) C_f \Phi_f(\lambda_{em})(\lambda_{ex},\lambda_{em}) \right]$$

$$\times H_{in}[\mu_a(\lambda_{ex}), \mu_s(\lambda_{ex}), g(\lambda_{ex}), z, L_{tur}]$$

$$\times H_{out}[\mu_a(\lambda_{em}), \mu_s(\lambda_{em}), g(\lambda_{em}), z, L_{tur}] \left(\frac{\Omega}{2\pi} \right) \quad (4.4.1)$$

where
  H_{in} refers to the distribution of the excitation radiation within the tissue
  H_{out} is the fluorescence escaping function

This is illustrated in Figure 4.4.3 for a plane-wave illumination.

The total fluorescence emitted from the front surface of a sample of thickness L_{tur} is given by integrating the function over the turbid sample thickness:

$$P_{total}^{tur}(\lambda_{ex},\lambda_{em}) = \int_0^{L_{tur}} dP_{total}^{tur}(\lambda_{ex},\lambda_{em},z)$$

$$= P_0(\lambda_{ex}) \left(\frac{\Omega}{2\pi} \right) \int_0^{L_{tur}} dz \left[\sum_{f=1}^{F} \ln(10)\varepsilon_f(\lambda_{ex}) C_f \Phi_f(\lambda_{em}) \right]$$

$$\times H_{in}[\mu_a(\lambda_{ex}), \mu_s(\lambda_{ex}), g(\lambda_{ex}), z, L_{tur}]$$

$$\times H_{out}[\mu_a(\lambda_{em}), \mu_s(\lambda_{em}), g(\lambda_{em}), z, L_{tur}] \quad (4.4.2)$$

If fluorophores are distributed homogeneously within the sample, the summation can be taken outside the integral yielding to the following equation:

$$P_{total}^{tur}(\lambda_{ex},\lambda_{em}) = P_0(\lambda_{ex}) \left(\frac{\Omega}{2\pi} \right) \left[\sum_{f=1}^{F} \ln(10)\varepsilon_f(\lambda_{ex}) C_f \Phi_f(\lambda_{em}) \right]$$

$$\times \int_0^{L_{tur}} dz H_{in}[\mu_a(\lambda_{ex}), \mu_s(\lambda_{ex}), g(\lambda_{ex}), z, L_{tur}] H_{out}[\mu_a(\lambda_{em}), \mu_s(\lambda_{em}), g(\lambda_{em}), z, L_{tur}]$$

$$(4.4.3)$$

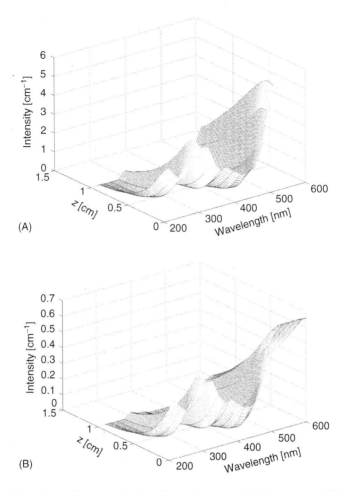

FIGURE 4.4.3 Monte Carlo calculation of one-dimensional (A) fluence rate $H_{in}(z)$ and (B) fluorescence escape probability $H_{out}(z)$ corresponding to the optical properties of Figure 4.4.2.

Considering the EEM of the turbid sample EEM_{tur}, we can define the ijth element of EEM_{tur} as the fraction of the fluorescence emitted from the sample front surface, normalized to the incident irradiance and sample thickness:

$$\begin{aligned} EEM_{tur}(\lambda_{ex}, \lambda_{em}) &= \frac{P_{tur}^{Total}(\lambda_{ex}, \lambda_{em})}{P_0(\lambda_{ex})L_{Tur}} \\ &= \left(\frac{\Omega}{2\pi}\right)\left(\frac{1}{L_{tur}}\right)\left[\sum_{f=1}^{F} \ln(10)\varepsilon_f(\lambda_{ex})C_f\Phi_f(\lambda_{em})\right] \\ &\quad \times \int_0^{L_{tur}} dz H_{in}[\mu_a(\lambda_{ex}), \mu_s(\lambda_{ex}), g(\lambda_{ex}), z, L_{tur}] \\ &\quad \times H_{out}[\mu_a(\lambda_{em}), \mu_s(\lambda_{em}), g(\lambda_{em}), z, L_{tur}] \end{aligned} \qquad (4.4.4)$$

For a dilute solution with isotropically emitting fluorophores, H_{in} is 1 and H_{out} is 1/2, simplifying the equation as follows:

$$\text{EEM}_{\text{Dil}}(\lambda_{\text{ex}}, \lambda_{\text{em}}) = \left(\frac{\Omega}{4\pi}\right)\left[\sum_{f=1}^{F} \ln(10)\varepsilon_f(\lambda_{\text{ex}})C_f\Phi_f(\lambda_{\text{em}})\right] \quad (4.4.5)$$

Combining Equations 4.4.4 and 4.4.5 allows relating EEM_{tur} to EEM_{Dil} and calculating the transfer function (TF):

$$\text{EEM}_{\text{tur}}(\lambda_{\text{ex}}, \lambda_{\text{em}}) = \text{EEM}_{\text{Dil}}(\lambda_{\text{ex}}, \lambda_{\text{em}}) \frac{2\text{TF}[\mu_a(\lambda_{\text{ex}}, \lambda_{\text{em}}), \mu_s(\lambda_{\text{ex}}, \lambda_{\text{em}}), g(\lambda_{\text{ex}}, \lambda_{\text{em}})]}{L_{\text{tur}}}$$

(4.4.6)

$$\text{TF}[\mu_a(\lambda_{\text{ex}}, \lambda_{\text{em}}), \mu_s(\lambda_{\text{ex}}, \lambda_{\text{em}}), g(\lambda_{\text{ex}}, \lambda_{\text{em}})]$$
$$= \int_0^{L_{\text{tur}}} dz H_{\text{in}}[\mu_a(\lambda_{\text{ex}}), \mu_s(\lambda_{\text{ex}}), g(\lambda_{\text{ex}}), z, L_{\text{tur}}] \times H_{\text{out}}[\mu_a(\lambda_{\text{em}}), \mu_s(\lambda_{\text{em}}), g(\lambda_{\text{em}}), z, L_{\text{tur}}]$$

(4.4.7)

Thus, the turbid EEM is related to the EEM of the equivalent dilute solution by a wavelength dependent transfer function TF.

This fundamental result provides the key for extracting the true fluorescence signal, which is independent of media optics, EEM_{Dil}, from the measured fluorescence of a turbid sample. It is important to note that the transfer function only depends on the average optical properties of the sample and can be evaluated from measurements of sample reflectance and transmission without requiring knowledge of the details of the sample composition. Equation 4.4.6 is independent on the light propagation in the sample. In the next section, the transfer function will be evaluated using the Monte Carlo simulation, a powerful stochastic method to quantify photon migration and scattering in turbid media.

4.4.2.3 Extraction of Pure Fluorescence Related to Native and Neoformed Compounds

The extraction of the pure fluorescence emitted by a sample is based on the calculation of the transfer function by Monte Carlo simulation.

Monte Carlo refers to a well-known statistical simulation technique, first proposed by Metropolis and Ulam (1949) to simulate physical processes using a stochastic model. In Monte Carlo method applied to radiative transport, photons path is recorded taking into account their scattering and absorption probabilities in a specific medium.

The Monte Carlo method is attractive because it is easily implemented and sufficiently flexible allowing complex medium to be modeled. Theoretically, Monte Carlo solutions can be obtained for any accuracy level. However, the accuracy level is proportional to $1/\sqrt{N}$ where N is the number of photons propagated.

Fluorescence

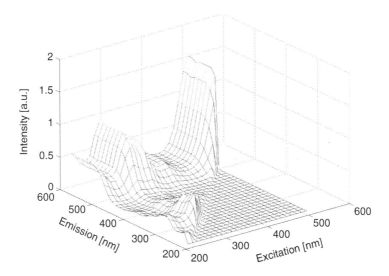

FIGURE 4.4.4 Transfer function calculated by numerical integration of Equation 4.4.7.

The method begins by launching a photon downward into the matrix at the origin. The photon is initially directed directly downward into the matrix. Once launched, the photon moves through a distance ΔL where it may be scattered, absorbed, undisturbed while propagated, internally reflected, or transmitted out of the matrix. The photon progresses in the matrix until it either escapes from or is absorbed by the matrix. If the photon escapes from the matrix, the reflection or transmission of the photon is recorded. If the photon is absorbed, the position of the absorption is recorded. Once this first complete step is achieved, a new photon is launched at the origin. This process is repeated until the desired number of photons has been propagated. The recorded reflection, transmission, and absorption profile will approach true values (for a media with specified optical properties) as the number of photons propagated approaches infinity.

The physical process described in Figure 4.4.4 is simulated using the procedure described above. The optical parameters of the medium at excitation and emission wavelengths are considered to be $\mu_a(\lambda_{ex})$; $\mu_s(\lambda_{ex})$ and $g(\lambda_{ex})$; $\mu_a(\lambda_{em})$, $\mu_s(\lambda_{em})$, and $g(\lambda_{em})$. Absorption probability $H_{in}(\lambda_{em}, z)$ and emission probability $H_{out}(\lambda_{ex}, z)$ are then calculated (Figure 4.4.5). The transfer function is obtained by numerical integration of Equation 4.4.7.

4.4.3 CHEMOMETRIC ANALYSIS OF FRONT-FACE FLUORESCENCE SIGNAL: ASSESSMENT OF NEOFORMED CONTAMINATION IN HEAT-TREATED OILS AND STARCH-BASED PRODUCTS

4.4.3.1 Introduction

The following section describes the methods applied and some results obtained in the framework of the European collective research ICARE project. Two work packages

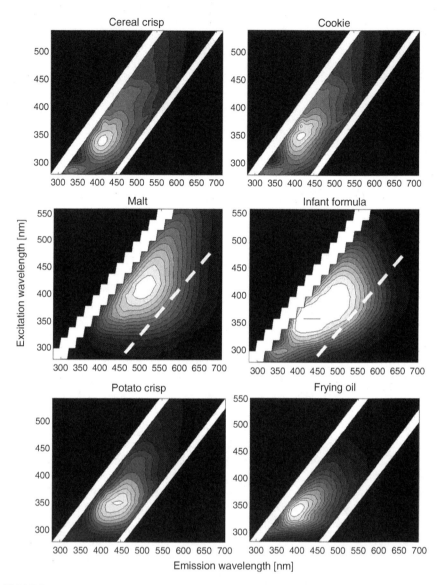

FIGURE 4.4.5 Examples of EEMs* acquired on commercial samples from each food product. *Note:* In the EEMs acquired in conventional mode using fixed excitations (malt and infant formula), dashed white lines were added to assist in comparison with EEMs acquired in synchronous mode. White ribbons represent missing values, while black areas outside the EEMs are zero values.

are dedicated to the development of a rapid analytical method and prototype allowing SMEs (small and medium enterprises) to control the heat treatment applied during food processing and monitor NFC level on production lines. Front-face fluorescence spectroscopy has a good potential in achieving NFC prediction and fits the specification of an industrial plant-adapted analytical method, i.e., the technique is rapid, cost efficient, pollution free, and can be adjusted for online or in-situ use (Wold et al. 2002; Christensen 2005). Front-face fluorescence signal reflects the sample's fluorescence properties (species and concentration of fluorophores) shaped relatively to the sample's optical properties, notably through quenching and scattering phenomena resulting from the physicochemical structure of the food matrix (Lakowicz 1999; Christensen 2005). Therefore, any technological process inducing modifications on either the fluorescence or optical properties of the food sample, such as heat treatment, will inevitably lead to changes in the native front-face fluorescence signal. Resulting, the acquired signal can be calibrated on quality parameters such as NFC content via an indirect statistical correlation approach. To achieve this objective, chemometric analysis of front-face fluorescence landscape obtained from bidimensional signal acquisition as EEM and calibration over conventionally (chromatographically) measured NFC content was applied. Through multiway and multivariate chemometric tools, systematic front-face fluorescence information was extracted and used to discriminate formulation and technological process, monitor thermal treatment, and calibrate the chromatographically measured NFC content of various food products.

4.4.3.2 Methodology

4.4.3.2.1 Sample Description

Analyzed samples were divided into two types. The first type encompassed commercial samples obtained from various European markets, while the second type included experimentally designed samples and samples acquired at consecutive steps of industrial process lines (Table 4.4.1). Experimentally designed samples were based on controlled factors known to influence NFC levels such as heat treatment severity and variation in ingredients (modification of NFC precursors). Preprocessing, analysis of spectral data, and prediction of NFC were performed on each data set (product) separately.

4.4.3.2.2 Front-Face Fluorescence Acquisition

Front-face fluorescence acquisition was directly performed on powders, grinded samples, and liquids placed in acryl cuvettes. Two-dimensional front-face fluorescence EEMs were obtained either through acquisition of emission spectra at consecutive fixed excitations or after shifting the two-dimensional synchronous spectra acquired at consecutive delta ($\Delta = \lambda_{em} - \lambda_{ex}$). In general, front-face fluorescence signal was recorded between excitation wavelengths 280–550 nm and emission wavelengths 280–650 nm. The two-dimensional spectra were acquired in duplicates or triplicates on different sides of the cuvettes to appraise the homogeneity of the samples. Examples of front-face fluorescence EEMs acquired on commercial samples and evolution of EEMs with increasing heat treatment are given in Figures 4.4.5 and 4.4.6.

4.4.3.2.3 Spectra Pretreatment

Scatter signal should be reduced or removed if possible from EEM matrices before chemometric modeling, as this information is chemically irrelevant and interferes

TABLE 4.4.1
Samples Analyzed and Type of Spectral Analysis Applied

Product	N Samples	Origin (Country)	Acquisition Fluorometer	Recorded as	N Replicates	Measured NFC
Sampled from European market						
Cereal crisp	24	France	Varian	2DSS	3	HMF
Cookies	62	Spain	Varian	2DSS	2	—
Malt	22	France	Fluorolog	EEMs	3	Acr, Furan, HMF, Furfural
Infant formula	6	Czechoslovakia	Fluorolog	EEMs	3	—
Potato crisp	10	Spain	Varian	2DSS	2	Acr
Frying oils	74	France	Varian	2DSS	3	polar content
Experimentally designed						
Cereal crisp	68	Italy	Varian	2DSS	3	HMF
Cookies	148	France	Varian	2DSS	3	CML–HMF
Cookies sampled from Industrial processes	41	France	Varian	2DSS	3	CML–HMF
Potato crisp	117	Spain	Varian	2DSS	3	Acr

Note: 2DSS, two-dimensional synchronous spectra acquired using Varian Cary-4000 front-face angle around 35°; EEMs, excitation emission matrices acquired using Horiba Jobin Yvon Fluorolog-3 front-face angle 25°; Acr, acrylamide; CML, carboxymethyllysine; HMF, hydroxymethylfurfural; N replicates is the number of two-dimensional signal acquired on different sides of the cuvettes for each sample.

with the fluorescence signal and consequent models. To deal with scatter, ribbons of zeroes and missing values were used to replace scatter and spectral data not conforming to fluorescence (Thygesen et al. 2004; Tomasi and Bro 2005). In the ribbon strategy we applied, EEM values not conforming to fluorescence were replaced by zeroes. EEM regions displaying an overlap between fluorescence and scatter signal were replaced by missing values to avoid possible spectral artifacts created by inserting zeroes next to nonzero fluorescence intensities.

Two additional pretreatments were applied on unfolded matrices according to the excitation mode (concatenated emissions). Scaling of data was performed using the norm of each unfolded emission spectra. In multiplicative scatter correction (MSC) (Dhanoa et al. 1994), each unfolded emission spectrum is corrected using the intercept and the slope of the regression between the respective unfolded emission spectra and the mean unfolded emission spectra. Pretreated spectral data were subsequently folded back and arranged into three-way arrays before chemometric modeling.

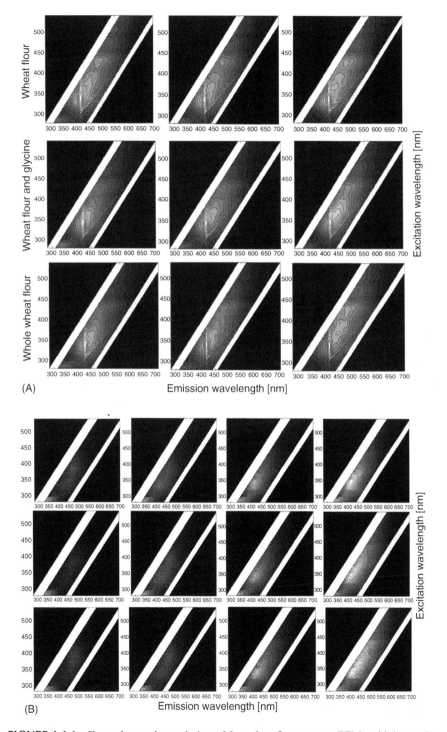

FIGURE 4.4.6 Examples on the evolution of front-face fluorescence EEMs with increasing heat treatment*: (A) Cereal crisp with different flour formulation (baking at 180°C). (B) Experimentally baked cookies.

(*continued*)

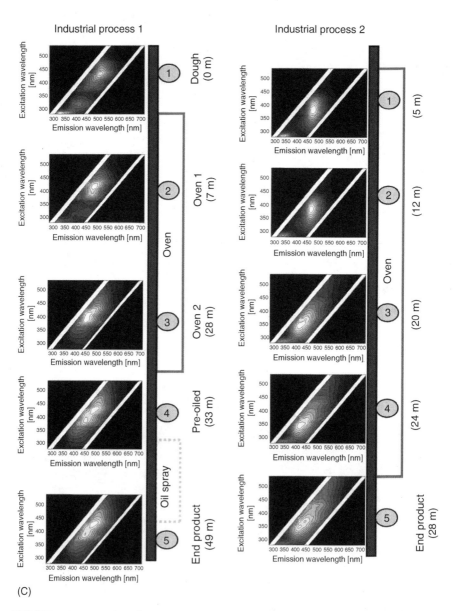

FIGURE 4.4.6 (continued) (C) Evolution of front-face fluorescence EEMs of cookies during baking in two different industrial processes.

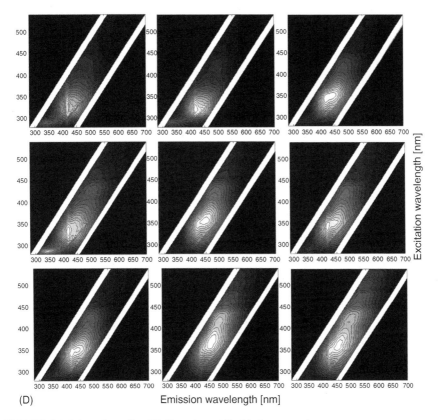

FIGURE 4.4.6 (continued)   (D) Experimentally fried potato crisp.

4.4.3.2.4 Multiway Decomposition

Front-face fluorescence three-way arrays of each data set were decomposed into sets of vectors corresponding to excitation and emission profiles (spectral loadings) varying only in intensity (scores) between samples by means of multiway Parafac models (Harshman 1970; Bro 1997, 1998). The parameters of the models were estimated by alternating least square fit algorithm with a nonnegativity constraint imposed on all parameters. The models were refitted several times starting with random values to ensure stability and uniqueness of solutions. Conventional diagnosis tools such as residual analysis, deviation of trilinearity using the core consistency diagnostic (Corcondia), and the interpretability (physical appropriateness) of the Parafac spectral loadings were used as indicators of the number of Parafac components to retain and to evaluate the quality of the models. In some data sets, the uniqueness of the models was further validated through decompositions performed on several segments of the data set (split analysis).

In addition to three-way arrays, higher order arrays were also created in experimentally designed samples through arrangement of the EEMs relative to the controlled factors. Parafac decomposition of these arrays allowed identification and quantification linear variation in the profiles (loadings) of EEMs with respect to the levels of controlled factors.

4.4.3.2.5 Regression of Parafac Scores over Neoformed Compounds Content

Generalized linear models (GLMs), $f(\mu_y) = b_0 + b_1 x_1 + b_2 x_2 + \cdots + b_n x_n + e$ (McCullagh and Nelder 1989), were used to build models prediction of the chromatographically measured NFC content using the front-face fluorescence scores of Parafac models. In the equation, the link function $f(\mu_y)$ was defined as log function of the expected chromatographically measured NFC content. Regression coefficients, b_i, were estimated by means of Fisher scoring algorithm (iteratively reweighed least squares) for maximum likelihood estimation. Depending on the number of samples, the root mean square errors of a leave one or two sample (and the respective replicates) out cross validations were used as estimates of GLM models prediction error. Parafac scores and NFC content of cross-validation samples were estimated using the Parafac loadings and NFC regression coefficients predetermined from calibration samples. Leverages and Pearson residuals were used to detect possible outliers and influential samples in the regression models.

Processing of spectral data was performed using MatLab (Version 7, The MathWorks Inc., United States). Multiway model and GLM were fitted using the Nway toolbox for MatLab (Andersson and Bro 2000) available online and MatLab's Statistics toolbox, respectively.

4.4.3.3 Results

4.4.3.3.1 Multiway Parafac Decomposition

Synchronous spectra are acquired by varying the emission and excitation monochromators, and simultaneously keeping a positive wavelength difference (Δ) between the two monochromators ($\lambda_{em} = \lambda_{ex} + \Delta$). The advantage of synchronous scanning compared to the standard fixed excitation/emission method is that all recorded intensities conform physically to fluorescence phenomena ($\lambda_{em} > \lambda_{ex}$). Two-dimensional synchronous spectra are obtained by measuring the excitation spectra at consecutive delta values. The intensities in the resulting matrix are then shifted to fixed emission wavelength positions to reconstruct the conventional EEM. Reconstructing the EEM before Parafac modeling is mandatory because data in the two-dimensional synchronous matrix are not bilinear of low rank and therefore cannot be approximated by the outer product of two vectors as stipulated by Parafac models. The benefit of using Parafac model compared to other decomposition tools such as Tucker models is the simplicity and uniqueness of the Parafac decomposition parameters. The component profiles presented in Figure 4.4.7 are the outer product of the spectral loadings extracted from Parafac decomposition of front-face fluorescence three-way arrays arranged for each data sets after unfold scaling pretreatment. Parafac spectral profiles were named in agreement with pure fluorophores known from literature to be present in analyzed samples and having fluorescence profiles in right angle measurements similar to those obtained in the Parafac decomposition. The Parafac spectral profiles seem to correspond to native tryptophan (Trp with maximum intensities $\lambda_{ex}/\lambda_{em} \approx 290/350$ nm), riboflavin (Rf with maximum intensities $\lambda_{ex}/\lambda_{em} \approx 350/550$ nm),

Fluorescence

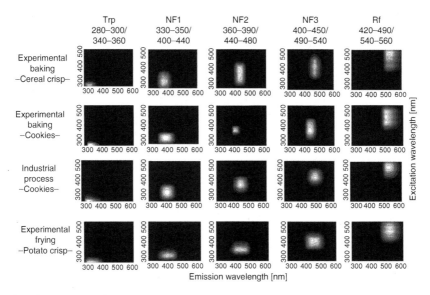

FIGURE 4.4.7 Parafac main component landscapes of experimental samples (explain over 95% of data variance).

and several fluorescence profiles generally associated to compounds deriving from lipid oxidation and Maillard reaction occurring in the food matrix submitted to heat process (Bro 1999; Christensen 2005). These contaminations were denoted NF for neoformed fluorescence since they were not present in unprocessed raw ingredients: NF1 (maximum $\lambda_{ex}/\lambda_{em} \approx 330$–$350/400$–$440$ nm), NF2 (maximum $\lambda_{ex}/\lambda_{em} \approx 360$–$390/440$–$480$ nm), and NF3 (maximum $\lambda_{ex}/\lambda_{em} \approx 400$–$450/490$–$540$ nm).

Of course, the nomenclature is only used for convenience and to facilitate comparison between different data sets, the extracted profiles do not represent pure fluorophore landscapes since the shape and intensity of the loadings depend on the nature and concentration of the fluorophores present as well as the sample's optical properties. Therefore, the parameters of the models are set dependent and the solution is only valid for samples of the same type with similar fluorescence and optical properties. The number of significant factors was chosen using conventional validation tools such as residual analysis, Corcondia values, and estimating the measuring uncertainty by means of validation. The interpretability, physical appropriateness, and accordance of extracted profiles with prior knowledge also played a central role in the assessment of models validity.

Parafac decomposition of front-face fluorescence arrays resulted in factors explaining 90%–98% of the array's variance with Corcondia values ranging between 30% and 98%, depending on the data set being analyzed. Generally, higher deviation from trilinearity and reduced captured variance were observed in commercial samples compared to experimentally designed samples. This is explained by higher variability in fluorescent and optical properties owing to more pronounced differences in ingredient formulations and technological processes applied in commercial samples. This resulted in an increase of spectral shifts, overlaps, and presence of some unique profiles in commercial samples leading to the reduction in explained variance and deviations of Corcondia values.

As mentioned earlier, the Parafac profile scores are not proportional to the true fluorophore concentrations as these scores are highly influenced by the scattering and quenching properties of the samples. Nevertheless, the Parafac scores represent a measure of variability in the front-face fluorescence signal expressed in terms of parallel profiles. Given that such variability reflects to a great deal in the differences in process and formulation, the scores can be used in calibration over NFC, which accumulate in the course of heat treatment and depend on the presence of NFC precursors.

In almost all analyzed data sets, the structure of the scores improved by means of reduced distance between replicates and colinearity, and a better evolution with process, when scaling and MSC pretreatments were applied. Scaling and MSC also resulted in faster convergence of the Parafac models and reasonable Corcondia values, when the trilinear models were fitted with a relatively high number of components. The shape of the extracted spectral parameters did not significantly change upon pretreatment. MSC usually outperformed scaling in terms of improved score structure and Corcondia in Parafac models fitted on experimentally designed samples, while the contrary applied in commercial samples. This observation can be rationalized using the same arguments explaining the decrease in Corcondia values in commercial samples. Explicitly, the presence of pronounced spectral shifts, overlaps, and dissimilar profiles in commercial samples owing to different formulations significantly biased the mean spectrum of the set-dependent MSC pretreatment and subsequent results.

Global decomposition results show that front-face fluorescence profiles remain to a large extent bilinear of low rank in data sets having limited signal variability such as in samples of the same nature and experimentally designed samples. Hence, front-face fluorescence arrays resultant of such data sets can be well approximated using Parafac models. Parafac models allowed extraction of parallel profiles of each data set and identification of common profiles present in various products. The respective Parafac scores can now be used in calibration over NFC content measured using conventional chemical methods.

4.4.3.3.2 Regression of Parafac Scores Over Neoformed Compounds Content

It is generally known that heat treatment generates NFC in food. Heat treatment also modifies the fluorescence and optical properties of the samples, and accordingly, the acquired front-face fluorescence signal. Using these indirect correlations because of the presence of a common factor (heat treatment), the chromatographically measured NFC content in samples can be calibrated via systematic information present in front-face fluorescence signal.

GLM models were fitted on front-face fluorescence Parafac scores via log-link function of the likelihood estimate of NFC content measured using conventional chemical separation (chromatographic) and quantification methods. The reason GLM was used instead of multiple linear regression (MLR) is that linear regression models resulted in tailed residuals, i.e., samples with extreme NFC content had higher residues compared to the other samples, indicating that a nonlinear fit is needed.

Depending on the data set, three to five components were used in the prediction (the choice was guided by cross-validation results). Leverages and other cross-validation segment results were used to identify possible influential and outlier samples. Better NFC regression models were generally obtained on experimentally designed and industrially processed samples as compared to the models fitted on commercial products. This is mostly due to the high variability in both the fluorescence (nature and concentration of fluorophores) and optical properties of samples prepared using different industrial processes and formulations. The other reason making NFC calibration difficult in samples acquired from the market is the absence of the reference EEMs acquired on raw samples. Nevertheless, acceptable prediction models were achieved in commercial samples, when only limited nonparallel (Tucker) variation was observed in the front-face fluorescence signal or when the optical properties (color and texture) of the samples were well correlated with the measured NFC level. In some case, the problem associated to the high variability of acquired signals was circumvented using separate regression models for groups split according to their formula (Figure 4.4.8) after Parafac decomposition of front-face fluorescence data.

The effect of scaling and MSC pretreatments of arrays on regression models was similar to the one observed on Parafac models. Other modeling strategies capable of handling nonlinearity such as neural networks or other nonlinear models could also be used. When moderate nonlinear trend was observed between the Parafac scores and NFC content, orthogonal signal correction before n-way partial least squares regression (OSC-NPLS) also achieved good results. However, simple models such as GLM based on the robust and unique Parafac scores were always favored since they better comply with the parsimony principle.

The results of NFC calibration models applied on commercial samples along with the results obtained on experimentally designed and industrially processed samples (Table 4.4.2) allowed identification of the most relevant parallel profiles for NFC prediction. This enabled the selection of some excitation and emission wavelength pairs that will be afterward used in a simplified fluorometer prototype intended for industrial applications.

The overall results show that through multiway decomposition of preprocessed front-face fluorescence data coupled with multivariate regression tools, industrial process control and robust NFC calibration models can be achieved.

4.4.4 Conclusion

It has been shown that the signal acquired from conventional destructive fluorescence gives only partial information indicative of the fluorescence sources present in the sample and transparent extract. On the other hand, front-face fluorescence provides information on fluorescence properties, as well as on the optical properties of the matrix of intact samples. Given that any technological process inducing modifications on either the fluorescence or optical properties of the food sample inevitably leads to changes of the apparent front-face fluorescence signal, the non-destructively acquired signal is capable of exploration of the nutritional and health value of processed foods.

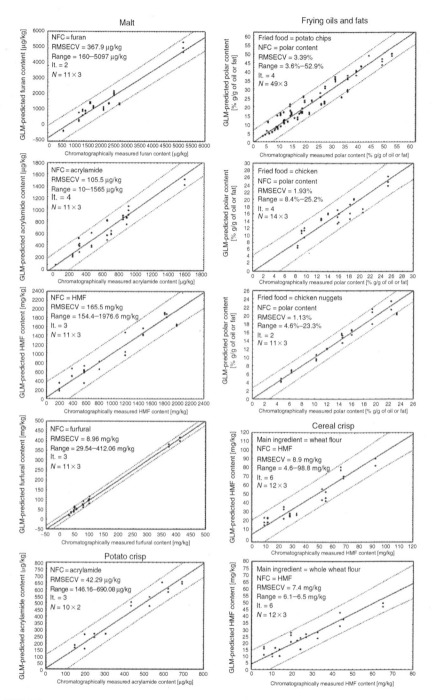

FIGURE 4.4.8 GLM prediction plots (y-axis) of chromatographically measured NFC content in commercial samples (x-axis) using front-face fluorescence Parafac scores. *Note:* Dashed lines in the plots represent 95% prediction interval. RMSECV is the root mean square error of cross validation. Range is the minimum and maximum chromatographically measured NFC values. "N" is the number of samples included in the regressions × number of EEM replicates acquired on different sides of the cuvette and "It." is the number of Fisher scoring iterations.

TABLE 4.4.2
GLM Regression Models of Chromatographically Measured NFC Content in Experimentally (Laboratory Scale) and Industrially Processed Samples Based on Four Front-Face Fluorescence Parafac Scores

Product	Subgroup	NFC	RMSECV	Range
Bread crisp	Wheat flour	HMF (mg/kg)	15.18	3.1–281.3
	Wheat and glycine flour	HMF (mg/kg)	8.52	2.8–87.4
	Whole wheat flour	HMF (mg/kg)	1.37	0.3–33.2
Cookies	Experimentally processed	CML (mg/kg)	2.8	0–35.9
	Experimentally processed	HMF (mg/kg)	1.18	0–15
	Industrially processed	CML (mg/kg)	5.65	5–67
	Industrially processed	Acr (μg/kg)	14.75	2–316
Potato crisp	Bintje cultivar	Acr (μg/kg)	848.87	2.5–8183
	Hermes cultivar	Acr (μg/kg)	243.24	7.2–2507.3

Note: RMSECV, root mean square error of cross validation; Acr, acrylamide; CML, carboxymethyllysine; HMF, hydroxymethylfurfural. Neoformed contaminants were analyzed using harmonized conventional chromatographical methods, which were evaluated in a ring test organized in the scope of the ICARE project and COST927 analytical activity.

Applying chemometric tools on this complex signal allows extraction of systematic physicochemical information related to the deep structural and chemical changes occurring in the food sample during processing.

In conclusion, the approach offered by developments in chemometric and physical models of light–matter interaction, associated to front-face fluorescence signal acquisition, allows a better understanding on the food processing impact on quality. Sensors based on autofluorescence or reflectance analysis are under development for implementation on industrial production lines. Better quality control is expected to be achieved, leading to an improved nutritional and safety quality of processed foods.

REFERENCES

Andersson, C.A. and R. Bro. 2000. The N-way toolbox for MATLAB. *Chemometrics and Intelligent Laboratory Systems* 52: 1–4.

Birlouez-Aragon, I. and M. Zude. 2004. Fluorescence fingerprints as a rapid predictor of the nutritional quality of processed and stored food. *Czech Journal of Food Sciences* 22: 68–71.

Birlouez-Aragon, I., M. Nicolas, A. Metais, N. Marchond, J. Grenier, D. Calvo. 1998. A rapid fluorimetric method to estimate the heat treatment of liquid milk. The FAST method. *International Dairy Journal* 8: 771–777.

Birlouez-Aragon, I., P. Sabat, and N. Gouti. 2002. A new method for discriminating milk heat treatment. *International Dairy Journal* 12: 59–67.

Bradley, R.S. and M.S. Thorniley. 2006. A review of attenuation. Correction techniques for tissue fluorescence. *Journal of the Royal Society Interface* 3: 1–13.

Bro, R. 1997. PARAFAC tutorial and applications. *Chemometrics and Intelligent Laboratory Systems* 38: 149–171.

Bro, R. 1998. Multi-way analysis in the food industry. Theory, algorithms and applications. PhD Thesis, University of Amsterdam, Amsterdam, The Netherlands.

Bro, R. 1999. Exploratory study of sugar production using fluorescence spectroscopy and multi-way analysis. *Chemometrics and Intelligent Laboratory Systems* 46: 133–147.

Cheikhousman, R., M. Zude, D. Jouan-Rimbaud-Bouveresse, D.N. Rutledge, and I. Birlouez-Aragon. 2005. Fluorescence spectroscopy for monitoring extra virgin olive oil deterioration upon heating. *Analytical and Bioanalytical Chemistry* 382: 1438–1443.

Christensen, J. 2005. Autofluorescence of intact food—An exploratory multi-way study. PhD Thesis, The Royal Veterinary and Agricultural University, Denmark.

Christensen, J., E. Miquel Becker, and C.S. Frederiksen. 2005. Fluorescence spectroscopy and PARAFAC in the analysis of yogurt. *Chemometrics and Intelligent Laboratory Systems* 75: 201–208.

Dhanoa, M.S., S.J. Lister, R. Sanderson, and R.J. Barnes. 1994. The link between multiplicative scatter correction (MSC) and standard normal variate (SNV) transformations of NIR spectra. *Journal of Near Infrared Spectroscopy* 2: 43–47.

Gardner, C.M., S.L. Jacques, and A.J. Welch. 1996. Fluorescence spectroscopy of tissue: Recovery of intrinsic fluorescence from measured fluorescence. *Applied Optics* 35: 1780–1792.

Harshman, R.A. 1970. Foundations of the PARAFAC procedure: Models and conditions for an "explanatory" multimodal factor analysis. *UCLA Working Papers in Phonetic* 16: 1–84.

Jobsis, F.F., M. O'Connor, A. Vitale, and H. Vreman. 1971. Intracellular redox changes in functioning cerebral cortex. I. Metabolic effects of epileptiform activity. *Journal of Neurophysiology* 24: 735–749.

Lakowicz, J.R. 1999. *Principles of Fluorescence Spectroscopy*, 2nd edn. Kluwer Academic/Plenum Publishers, New York.

Leclère, J. and I. Birlouez-Aragon. 2001. The fluorescence of advanced Maillard products is a good indicator of lysine damage during Maillard reaction. *Journal of Agricultural and Food Chemistry* 49: 4682–4687.

Mas, P.-A., D. Bouveresse, and I. Birlouez-Aragon. 2004. Fluorescence spectroscopy for monitoring rapeseed oil quality upon heating. *Czech Journal of Food Sciences* 22: 127–129.

Matiacevich, S.B. and M.P. Buera. 2006. A critical evaluation of fluorescence as a potential marker for the Maillard reaction. *Food Chemistry* 95: 423–430.

McCullagh, P. and J.A. Nelder. 1989. An outline of generalized linear models. In *Generalized Linear Models*, 2nd edn. Chapman & Hall Publishers, New York, pp. 21–44.

Metropolis, N. and S. Ulam. 1949. The Monte Carlo method. *Journal of the American Statistical Association* 44: 335–341.

Qin, J. and R. Lu. 2007. Measurement of the absorption and scattering properties of turbid liquid foods using hyperspectral imaging. *Applied Spectroscopy* 61: 388–396.

Ramanujam, N. 2000. Fluorescence spectroscopy in vivo. In *Encyclopedia of Analytical Chemistry*, ed. R.A. Meyers. John Wiley & Sons Ltd., Chichester, United Kingdom, pp. 20–56.

Ranjan, M. and S.R. Beedu. 2006. Spectroscopic and biochemical correlations during the human lens aging. *BMC Ophthalmology* 6: 10–15.

Riccio, F., C. Mennella, and V. Fogliano. 2006. Effect of cooking on the concentration of Vitamins B in fortified meat products. *Journal of Pharmaceutical and Biomedical Analysis* 41: 1592–1595.

Richards-Kortum, R. 1995. Fluorescence spectroscopy of turbid media. In *Optical-Thermal Response of Laser-Irradiated Tissue*, eds. A.J. Welch and M.J.C. van Gemert. Plenum Press, New York, pp. 667–707.

Sharma, H., V.K. Jain, and Z.H. Khan. 2007. Identification of polycyclic aromatic hydrocarbons (PAHs) in suspended particulate matter by synchronous fluorescence spectroscopic technique. *Spectrochimica Acta Part A—Molecular and Biomolecular Spectroscopy* 68: 43–49.

Skog, K., A. Solyakov, and M. Jägerstad. 2000. Effects of heating conditions and additives on the formation of heterocyclic amines with reference to amino-carbolines in a meat juice model system. *Food Chemistry* 68: 299–308.
Tomasi, G. and R. Bro. 2005. PARAFAC and missing values. *Chemometrics and Intelligent Laboratory Systems* 75: 163–180.
Thygesen, L.G., A. Rinnan, S. Barsberg, and J.K.S. Møller. 2004. Stabilizing the PARAFAC decomposition of fluorescence spectra by inserting zeros outside the data area. *Chemometrics and Intelligent Laboratory Systems* 71: 97–106.
Wold, J.P., K. Jørgensen, and F. Lundby. 2002. Nondestructive measurement of light induced oxidation products by fluorescence spectroscopy and imaging. *Journal of Dairy Science* 85: 1693–1704.
Zamora, R. and F.J. Hidalgo. 1995. Linoleic acid oxidation in the presence of amino compounds produces pyrroles by carbonyl amine reactions. *Biochimica et Biophysica Acta—Lipids and Lipid Metabolism* 1258: 319–327.

4.5 INTEGRATED SYSTEM DESIGN

TAKAHARU KAMEOKA AND ATSUSHI HASHIMOTO

Quality control and evaluation of the agricultural products are important to produce consistently high-quality products for marketing, while carefully using resources to keep an environmentally friendly cultivation management. Two types of integrated system designs by using optical sensors are introduced in this section. One is a field monitoring server and the other is a tasting robot. The former has been studied for sustainable and high-quality production of agricultural products by applying sensing and information techniques. The latter has been developed for improvement of our dietary life.

4.5.1 APPROACHES IN MULTIBAND SPECTROSCOPY

4.5.1.1 X-Ray Fluorescence and Infrared Spectroscopy

Valuable information on plant nutrition needs to address not only the nutrient contents but also their balance in the plant organs over the entire period of plant and harvest product development. A quantitative monitoring should be carried out by a simple-to-use, nondestructive, simultaneous, and rapid method. In addition, the measured data need to be retrieved for future simulation using various modeling software. While most often, the visible wavelengths range is used in fluorescence spectroscopy and the visible and near-infrared wavelength range in reflectance spectroscopy, in the future of nondestructive sensing it is also an option to measure with an excitation in the ultrashort and mid-infrared wavelength range. Application of x-ray fluorescence (XRF) in combination with mid-infrared (MIR) spectroscopy shows a high potential for process integration in the entire supply chain. Incidentally, due to recent compound developments of portable spectrometer devices, both spectroscopic methods provide substantial innovation as quantitative tools in the field to analyze the plant vigor and product quality.

This subsection describes a development process of a simultaneous and quantitative method for evaluating the elements necessary for high plant vigor using the XRF and MIR spectroscopies (Hashimoto et al. 2005a). The XRF and MIR information on the inorganic nutrients such as K, Ca, P, S, and N in different modes were studied in tomato leaf for receiving on-site data of the nutritional status of the plant.

Generally, the XRF spectrum was significantly affected by the geometrical structure near the irradiated points of the sample. Therefore, pressing the leaf sample with a ring would moderate the problems of the deflection near the irradiated point. To quantitatively analyze the XRF leaf spectra, the measured intensities were standardized using the peak intensity of Rh$K\alpha$ (Figure 4.5.1), since Rh$K\alpha$ should be constant for all sample without the Rh element (Hashimoto et al. 2004a). The differences of the spectral features between the peaks of Rh$K\alpha$ and Rh$K\beta$ for all samples were negligible both qualitatively and quantitatively. Additionally, the peaks relating to P$K\alpha$, K$K\alpha$, and Ca$K\alpha$ were observed, respectively. The intensity of each peak became higher with increase in the chemical element content.

A high linear correlation was found between the peak areas for the standardized XRF spectra and the concentrations of the nutrient element mixtures (Figure 4.5.2) (Hashimoto et al. 2006).

On the other hand, Figure 4.5.3 shows the MIR spectra of the fresh tomato leaf and the prepared leaf samples impregnated with proteinic and nitrate solutions (tissue phantom). The leaf spectrum represented very similar spectral features in comparison to the fresh tomato leaf. Weak peaks are observed at approximately 1350 cm^{-1}, where the absorption band of the NO stretching mode of the nitric functional groups exists. Furthermore, absorption peaks of the amino functional groups at 1650 and

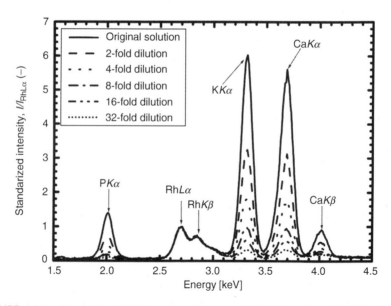

FIGURE 4.5.1 Standardized XRF spectra of impregnated leaf samples with different concentrations of nutrient elements.

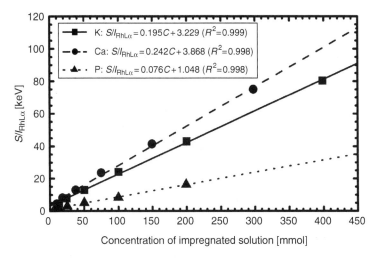

FIGURE 4.5.2 Calibration curves based on peak areas of standardized XRF spectra of impregnated leaf samples (tissue phantoms) for $PK\alpha$, $KK\alpha$, and $CaK\alpha$ and concentrations of nutrient element mixtures impregnated into them.

1550 cm^{-1} could not be easily recognized, as the very high peak due to the OH vending of water is observed in the same band pass.

The MIR spectra of the impregnated leaf sample after the spectral subtraction of the leaf sample impregnated by means of water provided the MIR spectroscopic information of the components under question. At 1350 cm^{-1}, the order of the absorbance related to the nitric acid concentration of the impregnated leaf was measurable (Figure 4.5.4). Resulting, the nitrate nitrogen content in the leaf sample could be determined by the MIR spectroscopic method. Furthermore, the simultaneous

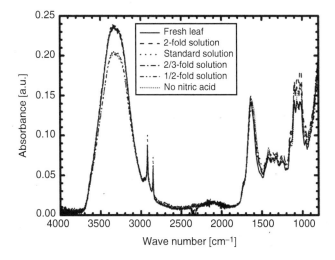

FIGURE 4.5.3 MIR spectra of fresh tomato leaf and the impregnated leaf samples.

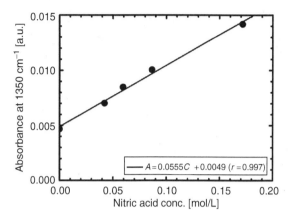

FIGURE 4.5.4 Relationship between absorbance at 1350 cm^{-1} of MIR spectra of leaves and nitric acid concentration of nitrogen mixture solution impregnated into leaf samples.

determination of the nitrate and proteinic nitrogen would be possible by using the ratio of the nitrate content to the proteinic nitrogen content.

These examples show the high potential to obtain nutrient information on the chemical elements (such as P, K, and Ca) and the nitrogen applying both spectroscopic methods. Additionally, the XRF spectral information could desirably be associated with the other information relating to the plant vigor (Oka et al. 2004), especially with the image information obtained in the visible wavelength range by a digital camera (Hashimoto et al. 2001) in an easy and inexpensive way.

4.5.1.2 Infrared and Terahertz Spectroscopy

With the recent advances in terahertz (THz) technology, the spectroscopic analysis of metabolites has been attracting much interest (Nishizawa et al. 2003). In MIR region, many spectroscopic studies on metabolites have been reported (Nakanishi et al. 2003; Kanou et al. 2005), and the theoretical and experimental approaches have been performed. This subsection describes spectroscopic characteristics of mono/disaccharides in the MIR and THz frequency regions (Hashimoto et al. 2007a), because they play important roles in biological system due to various functions in the metabolism. Monosaccharides are the basic unit and structural elements of the other saccharides, and the disaccharides are the basic unit with one glycosidic linkage (Nelson and Cox 2000).

Within the THz region, absorbance of monosaccharides is measurable (Figure 4.5.5). The spectra of the crystallite samples in polyethylene pellets were collected using a Fourier transform infrared (FT-IR) spectrometer at 300 K. Glucose, mannose, and galactose have a pyranose ring but with different configurations of the hydroxyl groups. Fructose has a furanose ring that is different from the others. The spectra show their fingerprint features, which could spectroscopically identify the specified materials, as well as the MIR spectra (Vasko et al. 1971).

Figure 4.5.6 shows the glucose spectra obtained by an FT-IR/ATR (attenuated total reflection; Harrick 1967) method and by an FT-IR/PAS (photoacoustic

Fluorescence

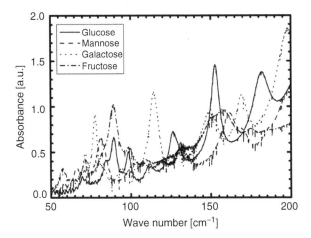

FIGURE 4.5.5 THz spectra of crystallite monosaccharides in polyethylene pellets.

spectroscopy) measurement of the crystallite sample, respectively. From the ATR spectra, we could extract the chemical component information influenced by the geometrical structure of sample. On the other hand, the influences of the geometrical structure on the PAS spectrum are relatively negligible. The peaks that characterize glucose in the aqueous solution were observed at the same wave numbers as those of crystallite glucose, but the spectral features were significantly different from each other. Thus, the spectral differences displayed would be affected by the interaction between glucose and water. Generally, metabolites could choose the most suitable molecular situation for its physical, chemical, and biological environments in vivo. For the application of the THz wave to the biological and agricultural fields, the

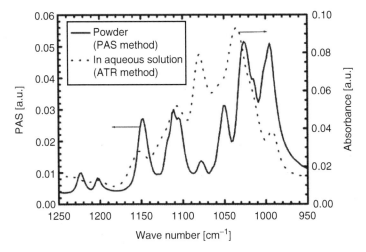

FIGURE 4.5.6 ATR spectra of glucose in aqueous solution and PAS spectra of crystallite glucose.

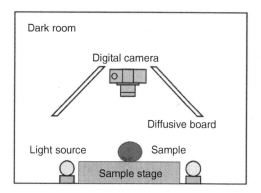

FIGURE 4.5.7 Scheme of color image acquisition system.

experimental grasp of the sugar–water interaction in the THz frequency vibrational modes would be required as well as the assignments of the absorption peaks.

4.5.1.3 Color Calibration for Image Analysis

The color of agricultural products is determined by the pigment components and the geometrical structure and is an important quality factor. Most often, color parameters in the $L^*a^*b^*$ color system, the RGB color coordinate system, or the hue-angle Equation 4.5.1 are used for the color evaluation of agricultural products (Chapter 2). Figure 4.5.7 shows the standard color acquisition system at the laboratory for the field acquisition. The time courses of the hue values of the post-ripening tomato fruit color and the color calibration effects were addressed with a simple setup (Figure 4.5.10). For the images acquired using the single-lens reflex camera, the calibrated values consistently agreed with those taken in the standard color acquisition system, and the color dispersion among the image was negligible. The lines drawn in Figure 4.5.8 were successfully fitted by the following Boltzmann equation that can express the surface color change of the tomato fruit from green to red based on the pigment synthesis and degradation.

$$\text{Hue}(t) = \frac{\text{Hue}_{\text{ini}} - \text{Hue}_{\text{fin}}}{1 + \exp\left(\frac{t-t_0}{w}\right)} + \text{Hue}_{\text{fin}} \qquad (4.5.1)$$

where
hue and t are the hue value and time, respectively
t_0 and w are the inflection point of the time course and the time constant ticking its curve, respectively
subscripts ini and fin indicate the value at the initial and final stages, respectively

The coefficient of determination (R^2) were all very high, but were different in each acquisition condition; 0.996 for the ordinary single-lens reflex camera by the raw format recording and 0.982 for the ordinary webcam by the JPEG format recording.

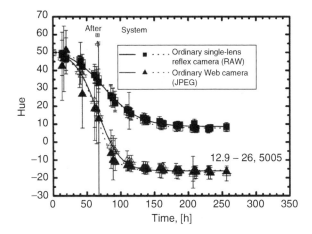

FIGURE 4.5.8 Time course of average and dispersion of hue value of tomato surface color during post-ripening process.

In addition, though the color was excellently calibrated using the developed method, the absolute values and the fitting results of the hue changes varied tremendously. Moreover, for the ordinary single-lens reflex camera by the raw format recording, the time course before the color calibration only roughly coincided with that after calibration.

4.5.2 Field Server Application

In this subsection, the recent challenge on the real-time monitoring and sensing technology in the agricultural field is described because real-time high-resolution spatial and temporal data is needed in the bioprocess monitoring.

4.5.2.1 Concept of Field Server for Plant Monitoring

A field server (FS) (Fukatsu and Hirafuji 2005; Fukatsu and Hirafuji 2006) is composed of a rugged case, multiple sensors, field server engine (FSE) for data acquisition and Web server, and network cameras. The air–temperature sensor and the humidity sensor are in airflow as Asmann psycrometer. Simultaneously, the airflow cools the circuit boards in the case. Figure 4.5.9 shows the concept of FS.

An FSE is the core of FS that controls other devices such as the network camera. The current FSE (version 3.1) is equipped with two A/D converters (24 bit resolution: 8 channel, 10 bit resolution: 8 channel), direct digital synthesizer (DDS), two power photo-MOS relays, and an RS-232C interface. All functions of devices on the FSE are controlled as Web services. For example, we can change the IP address of FSE, the password, and input voltage range of the A/D converter on the Web page. External devices such as power supplies, network cameras, and an LED lighting system are controlled. Furthermore, 24 channels for A/D inputs of FS enable us to connect to commercially available additional sensors.

The FS consists presently of three Web servers, which control an FSE, a wireless LAN access-point card, and a network camera. It is a fundamental concept of field

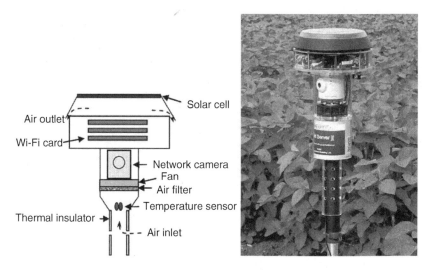

FIGURE 4.5.9 Components of the field server.

servers that all devices should be Web servers, and measured information is collected by an external agent system. Measured data are accessible through the Web page of the main Web server. An agent system running on an external PC-cluster collects the data through a VPN connection, and then the agent system converts the data to an XML file and stores them on a data storage server. Figure 4.5.10 shows the network architecture of an FS sensor-network. Stored data are opened on an FS Web site (Field Server Home, Tsukuba, Ibaraki, Japan, http://model.job.affrc.go.jp/FieldServer/default.htm). Users can browse the data by using data viewer (Figure 4.5.11) and image viewer. The latter works like webcams that are found nowadays showing the field area under requests.

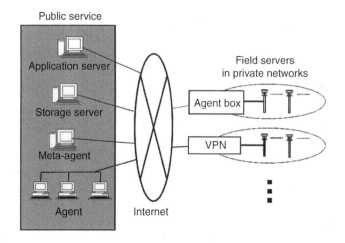

FIGURE 4.5.10 Network architecture of a field server sensor-network.

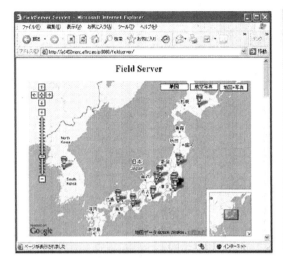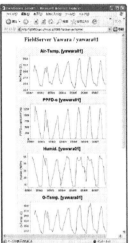

FIGURE 4.5.11   Data viewer.

4.5.2.2   Sensor Network

Sensor networks might be a breakthrough to accelerate the progress of field science by capturing high-resolution spatial data. The needs for FSs have gradually increased, and several projects were organized to improve the FS and related sensing technologies. So far, pilot lots of FSs have been deployed in Japan, The United States, Thailand, China, Denmark, Syria, Korea, and Fiji. Observed data, which is collected on a data storage server of Ministry of Agriculture, Forestry and Fisheries Network in Japan, are combined by MetBroker, which is a middleware of network service for data grid. The data can be a kind of scientific common data resource.

If new FSs are installed, the measured data are also available without any revises in source codes. Current measured data by FSs at fields are shown by browsers such as Internet Explorer via the Internet and FS's wireless-LAN. The data, which is a Web page (html), is collected by Fieldserver-Agent automatically, and stored in public database servers such as a PC-cluster in a computer center. The agent is accessing FSs every 1–10 min, and storing the data as distributed XML database. The data can be used easily through MetBroker. MetBroker combines databases of conventional weather databases (>22,000 weather databases) and data of field monitoring servers.

By combining FS, MetBroker, Fieldserver-Agent, and Fieldserver-Gateway, wireless smart sensor networks could be created. This application shows distribution of data provided by MetBroker on a map.

In the FS, the role of a built-in camera has been getting more important. In the crop, monitoring experiments were carried out on tomatoes in a greenhouse (Hashimoto et al. 2007b). The images were taken by three kinds of digital cameras through the maturing process for 14 days and were sent to the data server via VPN with the field environmental information collected by the field server. The standard RGB

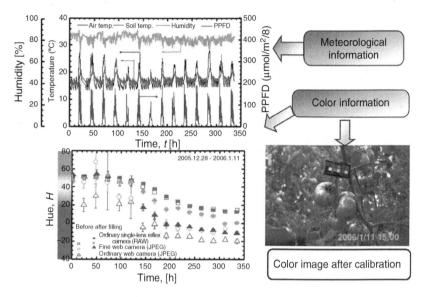

FIGURE 4.5.12 (See color insert following page 376.) Information on remote-monitored tomatoes in greenhouse.

color chart on the black matte was set behind the cluster of the tomato fruits. The method introduced in Section 4.5.1.3 is applied to the built-in camera of FS for the continuous remote monitoring of the surface color changes of agricultural products with the atmosphere condition data during cultivation. Figure 4.5.12 shows the weather data, the time course of average and dispersion of hue value of tomato surface color during maturing process, and the calibrated surface color image.

4.5.2.3 Optical Farming

An integrated field monitoring system (Kameoka et al. 2002) has been studied for acquiring and analyzing growth information of crops in cultivation process. The farming by using this system was named optical farming. Usually, such data are acquired by means of many different devices under the various conditions for the huge number of cultivars. Hence, the criteria for the database construction and the analysis of the information should be necessary. In this system, BIX (Bio information eXchange), which is the XML standard for the data exchange between sensed data of the information arising on the spot and database, aims the standardization of the format on all information arising on the spot by opening the format to the public (Figure 4.5.13).

The FS plays the most important role for measuring environmental conditions with the inexpensive soil moisture sensor. In the cultivation management, soil moisture management is a key for producing high-quality fruits. Excellent farmers control irrigation system properly with monitoring sugar contents of fruits and their experiences. Soil moisture sensor (ECH2O, Decagon Devices Inc., Pullman, WA, United States) connected to the FS can measure the variation of soil moisture conditions under the ground in the farm (Ito et al. 2004) and saves the data in the databases.

Fluorescence

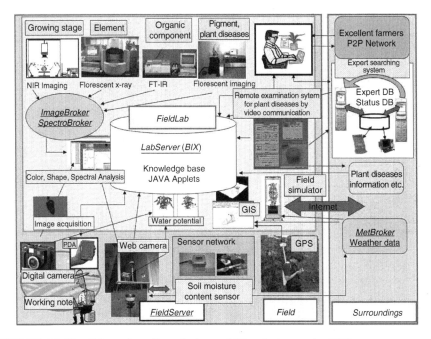

FIGURE 4.5.13 Schematic outline of advanced IT agriculture using BIX system.

The BIX Image Broker System (BIX-IBS) aiming the exchange of a variety of digital image data for agricultural products plays another important role (Hashimoto et al. 2007b). This system was developed as hybrid type P2P system. It distributes and manages the meta data and the image data. We are also able to handle data storage that managed FS image by using BIX-IBS. Users easily search and get images of FS by using a user-interface of BIX-IBS. On database of BIX-IBS, users can register file name, image name, shooting date and time, place, comment, etc. as meta data of images, and users can search images by multi-items of meta date. As mentioned in Section 4.5.1, a series of multi-band optical sensing of the vigor of the crops in the agricultural fields is made independently in cooperation with FS. The successive accumulation of the above optical sensing data will be helpful in the future in acquiring the precise information of the plant vigor and help the farmers select the optimum cultivation way.

4.5.3 Tasting Robot

This subsection describes an artificial sense of taste and the world's first "tasting robot" (Shimazu et al. 2005; Shimazu et al. 2007), which was developed by the collaboration work by NEC System Technologies, Ltd. and Mie University under the support by the New Energy and Industrial Technology Development Organization (NEDO) for the robot project: prototype robot exhibition at EXPO 2005 Aichi, Japan.

Lendl and coworkers proposed the "optical-tongue" concept based on the spectroscopic information of foods (Edelmann and Lendl 2002). We have developed an optical tongue with an IR sensor. It is attached to a personal robot, PaPeRo (Fujita 2002). The optical tongue is the integration of IR spectroscopic technologies with pattern recognition technologies.

4.5.3.1 Concept of Tasting Robot

4.5.3.1.1 Tasting

The robot could be expected to get at least three major potentialities when it had a sense of taste. The first potentiality is to qualitatively and quantitatively analyze the major components in a food sample such as sugar and fat. Because a human cannot analytically distinguish certain compounds, the robot with this ability can give useful advice to humans in various situations.

The second potentiality is to express the similar taste feelings of a food sample to a human's ones. Humans have various vocabularies to express their feelings of tastes; "I am sure you like this wine and it matches well with Sushi." A robot with this ability can become a critique of food and a human can enjoy discussion on food or drink with the robot.

The last potentiality is to identify the names of food products. Such application usually requires an analytical sensory panel built by human beings with special skills and training. If a robot has the ability, humans can have their own sommelier and get useful and interesting advice. In addition, if the name of a food product is spectroscopically identified, its main components and average contents are then estimated by referring the literatures.

4.5.3.1.2 Approaches to Tasting

There are various approaches to provide the information of the sense of taste by analyzing the food and its available components. One is the use of transducers of the sensor composed of lipids immobilized with polyvinyl chloride. The sensor has a concept of global selectivity, which is the ability to classify enormous kinds of chemical substances into basic taste groups such as saltiness, sourness, bitterness, umami, and sweetness (Toko 2001). The approach is effective to realize the second and third abilities defined in Section 4.5.3.1.1. The size of the sensor is now similar to a desktop PC. As a more specific approach, the spectroscopic fingerprint information of foods can be used: an optical-tongue technique (Kameoka et al. 1998; Nakanishi et al. 2003; Hashimoto et al. 2004b; Shimazu et al. 2007). Handheld sensors are already commercialized and advertised to work properly, and it is now possible to attach a sensor to a personal robot. Although, the complex matrix provides some challenges due to varying influences on the apparent signal.

4.5.3.1.3 Architecture of the Tasting Robot

The architecture of tasting robot consists of three modules (natural language dialog module, taste analysis module with an IR sensor and its controller, and advice generation module) as shown in Figure 4.5.14. A commercial robot software development tool, RoboStudio (version 1.0, NEC, Japan), served in the programming steps. It provides the natural language dialog module. By using RoboStudio, a human designer defines a set of scenarios that describe a sequence of conversation dialogs between a human and a robot.

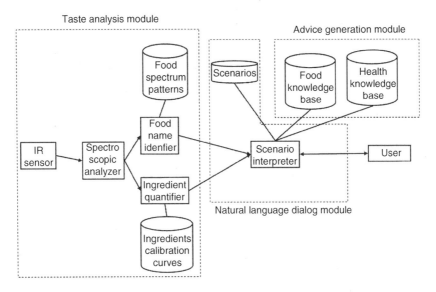

FIGURE 4.5.14 Architecture of the tasting robot.

4.5.3.1.4 Appearance of the Tasting Robot
We assumed that the tasting robot will be placed at a dining kitchen or on a dining table to taste the foods and drinks. So its size must be similar to a thermos bottle. PaPeRo is also a small partner robot. It is about 0.5 m tall with a speech recognition and generation functions. Because it satisfies our requirements, we decided to use PaPeRo by attaching an arm with an IR sensor as shown in Figure 4.5.15.

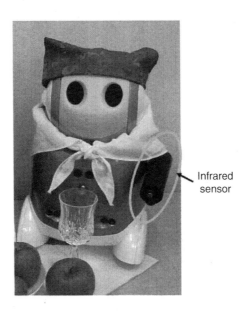

FIGURE 4.5.15 Appearance of the tasting robot.

4.5.3.2 Spectroscopic Data of Foods

It is known that each food has different IR spectral characteristics; the spectroscopic fingerprints. For example, the differences of the spectral pattern among the varieties and of the intensities of the peaks characterizing each variety are observed (see Section 3.3 for more information on NIRS). We could then grasp the quantitative spectroscopic fingerprint information of foods. Therefore, there are two independent methods for analyzing a food spectrum. Both of these are stored in the taste analysis module shown in Figure 4.5.16. The spectroscopic quantification was performed by studying the concentration dependencies of the spectra of the main components such as sugar, acid, and fat in a sample food and by making the calibration curves. The quantification method was acceptable for simple foodstuff, but it was difficult to complete robust calibration curves for the chemically and structurally complex foods.

On the other hand, the pattern recognition techniques were introduced to identify food names. The IR spectroscopic fingerprint information of varieties of food samples was accumulated and was stored with their food names in a pattern database. If a food sample is given, its spectrum pattern is compared with the stored spectrum patterns and the distances calculated. A food name with a closest distance is assumed to be identical to the food sample. The basic algorithm of the food identification analysis consists of two steps. Step 1 selects a set of wave numbers that are effective in calculating the distances between a spectrum of a sample food and a stored one. Because a spectrum pattern covers a wide range of wave numbers, the calculating distances at all points over the wave numbers is very time consuming and

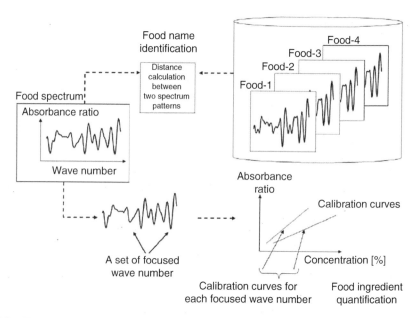

FIGURE 4.5.16 Scheme of analytical process of spectral information.

ineffective. Step 2 uses canonical correlation analysis where its input variables are the absorbance ratio at selected wave numbers of a spectrum, and its output variable is a food name.

4.5.3.3 Sommelier Robot in the Future

One of the most difficult foods we experimented was wine. The spectral differences among the tested wines were strikingly smaller than those among the other types of foods. For the robot to be able to achieve the ability to differentiate wine, we made several improvements as follows:

1. Improvements were made in the resolution using an ATR accessory.
2. Discrimination of samples is conducted by focusing on the automatic extracted point.
3. Robot is now equipped with the functionality of asking questions to determine a customer's wine preferences like a human sommelier.

In this subsection, the description is focused on the spectroscopic analysis on extraction of the wine fingerprints for the identification (Hashimoto et al. 2005b). The spectral patterns of the wines show the qualitative and quantitative agreements with those of the ethanol aqueous solution, especially in the wave number ranges of 3100–2800 cm^{-1} and of 1500–900 cm^{-1}. The influence of ethanol, which is the major component, in the wine on the spectral information is too dominant to identify a brand of wine based on the absorbance one. The spectral extraction of the wine components except ethanol and water was then examined.

By studying the spectral characteristics of the correlation coefficients between the second derivatives of the ethanol in the aqueous solution and the concentrations and the vibrational modes of the peaks characterizing them, the multiple linear regression analysis model for the ethanol content determination using the second derivative values at 1981, 1419, and 1085 cm^{-1} was constructed. The ethanol contents estimated for the alcoholic drinks consistently agreed with those obtained by a high-performance liquid chromatography method. Then the ethanol spectrum in the wine, which was calculated based on the estimated ethanol content in consideration of the interactions between the ethanol and water, was subtracted from the wine spectrum. Moreover, by subtracting the water spectrum from the above one, the spectral information of the wine components other than ethanol and water could be experimentally and simulatively extracted. The absorption bands characterizing the wine components other than ethanol and water are clearly observed and the fingerprint information is able to be confirmed. By the above improvements, the robot could identify several tens of the wine brands.

The above is an important step in the development of the quality and taste of wines based on the IR spectroscopic fingerprint and the robot development was selected in *TIME* as one of the best inventions 2006 (TIME. http://www.time.com/time/2006/techguide/bestinventions/inventions/meals5.html). However, to develop feasible applications in various fields, there is a part where the improvement becomes necessary in the future. We will continue to develop in these fields.

REFERENCES

Edelmann, A. and B. Lendl. 2002. Toward the optical tongue: Flow-through sensing of tannin-protein interactions based on FTIR spectroscopy. *Journal of the American Chemical Society* 124:14741–14747.

Fujita,Y. 2002. Personal robot PaPeRo. *Journal of Robotics Mechatronics* 14:60–63.

Fukatsu, T. and M. Hirafuji. 2005. Field monitoring using sensor-nodes with a web server. *Journal of Robotics Mechatronics* 17:164–172.

Fukatsu, T. and M. Hirafuji. 2006. An agent system for operating web-based sensor nodes via the internet. *Journal of Robotics Mechatronics* 18:186–194.

Harrick, N.J. 1967. *Internal Reflection Spectroscopy*. Harrick Scientific Corporation, New York.

Hashimoto, A., H. Kondou, Y. Motonaga, H. Kitamura, K. Nakanishi, and T. Kameoka. 2001. Evaluation of tree vigor by digital camera based on fruit color and leaf shape. In: *Proceedings of 1st World Congress of Computers in Agriculture and Natural Resources*. pp. 70–77.

Hashimoto, A., T. Niwa, M. Rahman et al. 2004a. X-ray fluorescent spectroscopic analysis of tomato leaf. In: *Proceeding of AFITA/WCCA2004 Joint Congress on IT in Agriculture*. pp. 874–879.

Hashimoto, A., H. Mori, M. Kanou, A. Yamanaka, and T. Kameoka. 2004b. Mid-infrared spectroscopic analysis on brewed coffee characteristics. In: *Proceedings of 10th Asian Pacific Confederation of Chemical Engineering Conference* (CD-ROM version). 3P-01-068.

Hashimoto, A., T. Niwa, T. Yamamura et al. 2005a. X-ray fluorescent and mid-infrared spectroscopic measurement of leaf model. In: *Proceedings of EFITA/WCCA 2005 Joint Conference*. pp. 252–259.

Hashimoto, A., M. Kanou, A. Yamanaka, T. Kameoka, K. Kobayashi, and H. Shimazu. 2005b. Mid-infrared spectroscopic analysis for characterizing wines. In: *Abstracts of 3rd International Conference of Vibrational Spectroscopy*, P11.14.

Hashimoto, A., T. Niwa, T. Yamamura et al. 2006. X-ray fluorescent and mid-infrared spectroscopic analysis of tomato leaves. In: *Proceedings of SICE-ICASE International Joint Conference 2006*. pp. 3559–3562.

Hashimoto, A., M. Kanou, K. Nakanishi et al. 2007a. Mid-infrared and THz spectroscopic analysis of mono- and disaccharides. In: *Abstracts of 4th International Conference on Advanced Vibrational Spectroscopy*.

Hashimoto, A., R. Ito, K. Nakanishi et al. 2007b. An integrated field monitoring system for sustainable and high-quality production of agricultural products based on BIX concept with field server. In: *Proceedings of the 2207 International Symposium on Applications and the Internet* (SAINT 2007), CD-ROM, ISBN 0-7695-2757-4.

Ito, R., M. Harada, A. Michida, et al. 2004. Soil moisture monitoring using near-infrared sensing technique and the internet in a coffee plantation field. In: *Proceedings of AFITA/WCCA 2004*. pp. 463–469.

Kameoka, T., Okuda, T., A. Hashimoto, A. Noro, Y. Shiinoki, and K. Ito. 1998. FT-IR/ATR analysis of sugars in aqueous solution using ATR method. *Nippon Shokuhin Kagaku Kogaku Kaishi*. 45:192–198 (in Japanese).

Kameoka, T., M. Harada, A. Hashimoto et al. 2002. Accurate sensing of bioinformation by optical method with multiband spectra and its structured data handling. In: *Proceedings of 6th International Symposium on Fruit, Nut, and Vegetable Production Engineering*, eds. M. Zude, B. Herold, and M. Geyer, pp. 549–554. Leibniz-Institut für Agrartechnik Potsdam-Bornim, ISBN 3-00-008305-7.

Kanou, M., K. Nakanishi, A. Hashimoto, and T. Kameoka. 2005. Influences of monosaccharides and its glycosidic linkage on infrared spectral characteristics of disaccharides in aqueous solutions. *Applied Spectroscopy* 59:885–892.

Nakanishi, K., A. Hashimoto, M. Kanou, T. Pan, and T. Kameoka. 2003a. Mid-infrared spectroscopic measurement of ionic dissociative materials in metabolic pathway. *Applied Spectroscopy* 57:1510–1516.

Nelson, D.L. and M.M. Cox. 2000. *Lehninger Principles of Biochemistry*, 3rd ed., Worth Publishers. pp. 239–324.

Nishizawa, J., K. Suto, T. Sasaki, T. Tanabe, and T. Kimura. 2003. Spectral measurement of terahertz vibrations of biomolecules using a GaP terahertz-wave generator with automatic scanning control. *Journal of Physics D: Applied Phyics* 36:2958–2961.

Oka, M., T. Kameoka, A. Hashimoto et al. 2004. Application of some sensing techniques for evaluating a condition of a coffee tree. In: *Proceedings of AFITA/WCCA2004 Joint Congress on IT in Agriculture*, pp. 470–476.

Shimazu, H., K. Kobayashi, A. Hashimoto, and T. Kameoka. 2005. Tasting robot: A personal robot with an optical-tongue. In: *Proceedings of 36th International Symposium on Robotics.* CD-ROM.

Shimazu, H., K. Kobayashi, A. Hashimoto, and T. Kameoka. 2007. Tasting robot with an optical tongue: Real time examining and advice giving on food and drink. In: *Proceedings of 12th International Conference on Human–Computer Interaction*, pp. 950–957.

Toko, K. 2001. *Kansei Bio Sensor, Asakura Shoten* (in Japanese), Tokyo, ISBN:4-254-20109-5.

Vasko, P.D., J. Blackwell, and J.L. Koenig. 1971. Infrared and Raman spectroscopy of carbohydrates: Identification of O—H and C—H-Related vibrational modes for D-glucose, maltose, cellobiose, and dextran by deuterium-substitution methods. *Carbohydrate Research* 19:297–310.

5 Spectroscopic Methods for Texture and Structure Analyses

Pilar Barreiro Elorza, Natalia Hernández Sánchez, Renfu Lu, Jesús Ruiz-Cabello Osuna, and David G. Stevenson

CONTENTS

5.1 Brief Overview on Approaches for Nondestructive Sensing of Food Texture 378
    5.1.1 Introduction on Food Texture 378
    5.1.2 Fruit and Vegetable Cell Walls in Relation to Texture 380
    5.1.3 Spectroscopic Techniques to Measure Texture of Starch-Based Cereal Foods 381
    5.1.4 Spectroscopy for Fruit Texture and Structure Determination 382
    5.1.5 Vegetable Texture Measured by Spectroscopy 384
    5.1.6 Texture of Dairy Products Measured by Spectroscopy 385
    5.1.7 Spectroscopy for Measuring Texture of Nondairy Processed Foods 386
    5.1.8 Summary 386
References 387
5.2 Spectroscopic Technique for Measuring the Texture of Horticultural Products: Spatially Resolved Approach 391
    5.2.1 Introduction 391
    5.2.2 Light Propagation in Scattering-Dominant Biological Materials ... 393
        5.2.2.1 Scattering and Absorption 393
        5.2.2.2 Diffusion Theory Model 395
        5.2.2.3 Steady-State Solutions 396
    5.2.3 Hyperspectral Imaging Technique for Measuring the Optical Properties of Horticultural Products 398
        5.2.3.1 Principle and Instrumentation 399
        5.2.3.2 Procedures of Determining the Optical Properties 402
    5.2.4 Applications 405
        5.2.4.1 Optical Properties of Fruits and Vegetables 405
        5.2.4.2 Evaluation of Apple Fruit Firmness 409

		5.2.4.3	Estimation of Light Penetration Depths in Fruit...............	409

                5.2.4.3 Estimation of Light Penetration Depths in Fruit............... 409
                5.2.4.4 Monte Carlo Simulation of Light Propagation
                         in Apple Fruit... 411
        5.2.5   Light-Scattering Technique Feasible for Assessing Fruit
                Firmness in Practice.. 413
                5.2.5.1 Wavelengths Selection.. 414
                5.2.5.2 Instrumentation... 414
                5.2.5.3 Mathematical Description of Light-Scattering Profiles..... 416
                5.2.5.4 Fruit Firmness Assessment..................................... 418
        5.2.6   Conclusions and Needs for Future Research................................. 419
Acknowledgment.. 420
References... 421
5.3   NMR for Internal Quality Evaluation
      in Horticultural Products... 423
        5.3.1   Overview on Applications in Fruits and Vegetables..................... 424
        5.3.2   Basics of NMR Relaxometry and NMR Spectroscopy.................. 426
                5.3.2.1 Magnetic Moment of Nucleus and Its Excitation............ 426
                5.3.2.2 Relaxation of Nucleus after Excitation........................... 429
                5.3.2.3 Signal Detection during Relaxation................................. 432
                5.3.2.4 NMR Relaxometry and NMR Spectroscopy..................... 432
        5.3.3   MRI Fundamentals... 435
                5.3.3.1 Image Acquisition and Reconstruction........................... 436
                5.3.3.2 Effect of Movement on Image Quality........................... 438
                5.3.3.3 Sequence Parameters and Their Effects
                         on MR Image Quality... 440
                5.3.3.4 Fast and Ultrafast MRI Sequences.................................. 441
        5.3.4   Detailed View of Applications in Fruits.. 445
                5.3.4.1 Maturity in Avocados.. 445
                5.3.4.2 Pit in Cherries and Olives.. 447
                5.3.4.3 Internal Browning in Apples.. 448
                5.3.4.4 Mealiness in Apples and Wooliness in Peaches............... 452
                5.3.4.5 Internal Breakdown in Pears... 453
                5.3.4.6 Freeze Injury in Citrus.. 458
                5.3.4.7 Seed Identification in Citrus.. 460
        5.3.5   Concluding Remarks... 462
References... 464

5.1 BRIEF OVERVIEW ON APPROACHES FOR NONDESTRUCTIVE SENSING OF FOOD TEXTURE

DAVID G. STEVENSON

5.1.1 INTRODUCTION ON FOOD TEXTURE

Food texture is an often underestimated quality trait that determines consumer satisfaction and the likelihood for the food product. One bad incidence of consumers experiencing texture of foods that fail to meet their expectations can curtail future purchases of that food. Each type of food has specific textural attributes considered ideal, which frequently vary for consumers residing in different geographical locations. While the sensations experienced during mastication of food define the texture properties of food, visual cues and mechanical properties during handling provide the mind with information about the perceived texture before the food ever enters the mouth. Visually, the shape, size, color, and incidence of cell structure or air spaces preempt our expectation of food texture. Touching or cutting the food instantly provides information on the likely firmness, plus depending on response to this mechanical force, brittleness, adhesiveness, cohesiveness, elasticity, and springiness of food may be gauged. Slicing into food may also provide sounds providing further cues about the food texture such as crunchiness. Once food enters the mouth, initial touch attempts to deform food and provide some information about firmness. Slight shear from tongue movements provides stimuli reflecting the springiness, viscosity, and adhesiveness. The first few chews provide information on hardness and brittleness, and subsequent chews mix saliva with the food, forming a cohesive bolus that experiences higher shear allowing advanced detection of brittleness, moistness, crispness, graininess, smoothness, creaminess, and adhesiveness.

The concepts of food texture are one of the easiest fields of science to comprehend, but may be one of the toughest to obtain meaningful results. Since food texture is determined by our sensory perception, it is obvious to have panelists judging the textural attributes to optimize measurements. However, it is challenging to obtain accurate texture measurements from a sensory panel, because every individual perceives texture differently and it is difficult to exclude bias. Frequent problems encountered include the following: (1) training of sensory panelists can be time consuming and expensive; (2) insufficient number of participants to enable accurate measurements of the true perceived textural properties of the total population; (3) to get sufficient panel size, participants are often recruited from within institution conducting sensory panel and therefore have bias because of having too much prior knowledge about the sensory panel objectives; (4) food products such as fruits and vegetables tend to be seasonal and therefore difficult to study storage effects as panelists know when produce is typically harvested; and (5) often difficult to provide panelists with control samples during storage studies to ensure panelists are consistent with their rating over time.

In an effort to save time and potentially obtain textural analysis from a greater number of food samples, instrumentation that attempts to simulate the sensory perception of panelists was developed. Currently, the primary instruments used to measure food texture are the Instron universal testing machine (TAXT2 texture analyzer) and Zwicki (1120, Zwick Materialpruefung Co. Ltd.). These instruments allow for food to undergo both compression or tensile stress, with potential to vary the shape and size of the probe contacting food, speed probe travels on contact with food, the percentage compression or strain, and the number of times the probe contacts the food. The meaning of each food textural term is subjective, but to assist development of instrumental texture analysis, standardized definitions were established (Szcześniak 1963) and a procedure involving double compression of the food was developed (Freidman et al. 1963). Correlation between instrumental analysis of food texture and sensory panel perception of texture has been well established (Szcześniak 1987), but instrumentation still has the disadvantage of being a destructive test.

Ideally, distributors of fresh produce and food manufacturers want a rapid test that can nondestructively evaluate the textural quality of food so that non-acceptable food products can be removed from market, avoiding food producers from having a tarnished reputation. The need for rapid, nondestructive evaluation of food texture has resulted in the development of spectroscopic methods such as spatially resolved spectroscopic methods (see Section 5.2) and nuclear magnetic resonance (NMR) spectroscopy and its associated magnetic resonance imaging (MRI) (see Section 5.3).

5.1.2 Fruit and Vegetable Cell Walls in Relation to Texture

Plant material consists of epidermis, vascular bundles, parenchyma, and thick-walled supporting collenchyma and sclerenchyma cells. Parenchyma tissue is mainly used for food consumption because of its acceptable texture. Size and shape of cells, cytoplasm to vacuole ratio, intercellular space volume, cell-wall thickness, osmotic pressure, and solutes present all influence texture of fruit and vegetable parenchyma (Ilker and Szcześniak 1990). Adjacent cells are cemented together by the middle lamella, which when stronger than the cell walls, release liquid contents when compressed. Pectin provides mechanical strength for the cell walls and adhesion between cells. Solid-state NMR has been demonstrated to effectively measure polysaccharide mobility in plant cell walls (Jarvis and Apperley 1990).

Degradation of pectin resulting in softening of fruit and vegetable can be detected by decreases in infrared band and magnetic resonance signals specific to pectic substances. Pectin has characteristic infrared spectra with bands at 1101, 1026, and 957 cm^{-1} from uronic acids and absorption band at 1066 cm^{-1} from sugars. Antisymmetrical stretching of glycosidic bonds is observed at 1154 cm^{-1}. The main ^{13}C-NMR signals from tomato pericarp were attributed to galacturonic acid in pectin, its methyl-ester substituent as well as the arabinose and galactose side chains, and rhamnose and acetyl esters. It was demonstrated from ^{13}C-NMR spectra that different tomato varieties were enriched with loosely interacting pectic galactan or methyl-ester side chains, which are important in cell-wall rigidity and adhesion in determining texture. Uronic acids and protein peaks from near-infrared (NIR) spectra of tomato fruit were highly correlated with mealiness (negatively) and juiciness (positively). NMR spectra detected differences in pectic arabinan side chains, which have been linked to cell adhesion defects and perceived mealiness.

In a study of cell walls of lemons and maize, percentage crystallinity and crystal size of cellulose microfibrils influence NMR and wide-angle x-ray scattering (WAXS), with higher crystallinity and larger crystal size having improved textural properties (Rondeau-Mouro et al. 2003). Smaller angles of reflection using WAXS indicate regions of food that are shifting from crystalline to amorphous. While plant cells have complex organization, there is some repeatability in arrangement of molecules within cells. This highly ordered structure has made whole fruit and vegetable tissues attractive food products to study the potential of spectroscopic techniques to nondestructively elucidate their textural attributes. Presently, less attention has been given to study implementation of spatially resolved spectroscopic techniques to determine texture of processed foods because many molecules are modified during processing, water is redistributed, and majority of molecules are intertwined resulting in a considerably greater heterogeneous material to study that has greater problems with absorption and scattering of signals.

5.1.3 Spectroscopic Techniques to Measure Texture of Starch-Based Cereal Foods

Many studies have utilized NMR, MRI, and NIR to study water holding capacity in starch-based foods, primarily of cereal origin. Starch is composed of essentially linear amylose and highly branched amylopectin. MRI can experience problems with cereal products owing to reduced transverse relaxation times. Proton NMR spectroscopy found that hardness of cooked rice was strongly correlated with spin–spin relaxation constants of protons (Ruan et al. 1997). Increasing starch concentration results in decreasing T_2 values in gels, reflecting the shorter distance water molecules have on average to diffuse before interacting with starch, which consequently enhances proton decay because of decrease in water mobility, proton exchange, and cross-relaxation mechanisms (Hills et al. 1990; Yakubu et al. 1993).

In other studies involving rice, NIR spectra obtained in optical geometry for transmittance readings could be successfully correlated to apparent amylose content, an important influence of cooked rice texture. White rice provided stronger correlations than brown rice probably because of bran in the latter (Villareal et al. 1994). Rice grains that were immature, slender, damaged, or dark had higher variance in the spectral intensities in the NIR as well as visible wavelength range.

Near-infrared reflectance spectroscopy (NIRS) data compared with textural attributes of cooked white rice, rated by sensory panelists, were able to correlate adhesiveness, hardness, cohesiveness of mass (three chews), and toothpack, while cohesiveness of mass (eight chews), roughness of mass, and toothpull were not well correlated (Meullenet et al. 2002). NIRS of 62 Chinese rice flour samples successfully predicted the paste viscosity parameters of setback and breakdown, but textural parameters of chewiness, hardness, and gumminess were less related (Bao et al. 2001). Texture attributes of 77 different short-, medium-, and long-grain rice cultivars have been correlated with NIRS data. Hardness, initial starchy coating, cohesiveness of mass, slickness, and stickiness as rated by sensory panelists were predicted by NIR spectral data. Important wavelengths contributing to the prediction models were amylose, protein, and lipid contents. Several wavelengths contributing to models for stickiness contributed to model for amylose content, whereas several wavelengths contributing to slickness model contributed to the protein model. Several wavelengths contributing to models of amylose, protein, and lipid correlations contributed to hardness model. Wavelengths at regions 1654–1666 nm and 1966–1996 nm contributed to initial starchy coating, cohesiveness of mass, and stickiness.

Strong positive correlations between cooked rice cohesiveness or adhesiveness and NIR measurements are because of observed absorbance of O—H and C—H owing to moisture (958 nm) and starch (878 and 979 nm), respectively. Negative correlations between NIR spectra and hardness, springiness, gumminess, and chewiness are owing to absorbance of N—H of proteins (1018 nm) and C—H of oil (1212 nm).

Multiphase behavior in water relaxation studied by NMR has been observed in other starchy food systems such as flour dough (Leung et al. 1979) and bread (Chen et al. 1997; Engelsen et al. 2001). Amylose retrogradation contributes to increased firmness of foods and occurs so rapidly during cooling that often NMR analysis cannot be accomplished (Choi and Kerr 2003).

NMR has also been used to determine the texture of pasta. Stronger NMR signal intensity in the center of cooked lasagna sheets was shown to correspond with softer texture and water migration influx, while lower NMR signal intensity at exterior reflected regions where water migrated from and had stiff texture (González et al. 2000). Spin–spin relaxation time of water protons from MRI was used to detect moisture distribution in cooked spaghetti, and detected pasta with soft or brittle texture owing to moisture homogeneity (Irie et al. 2004). Transverse relaxation time T_2 images and maps of white salted noodles demonstrated differences in water status among noodles (Lai and Hwang 2004). Increased cooking time resulted in decreased T_2 values and softer noodles.

NIR spectra of wheat bread were highly correlated to firmness measured by texture analyzer (TA) (Xie et al. 2003). Less variation in data was obtained from NIR spectra among different bread batches compared with TA; therefore, NIRS was more precise at measuring both physical and chemical changes in bread staling, with extra advantage of being nondestructive. Physical changes are due to alterations in scattering properties as crystallinity develops during bread aging. NIRS also measures chemical modification during bread staling such as water loss and starch structural changes, and this is the most likely reason NIRS has superior detection results of bread staling. Moisture loss and changes in starch crystallinity will affect firmness of bread.

Carbon-13 cross-polarization magic-angle-spinning nuclear magnetic resonance (^{13}C-CP-MAS-NMR) spectra of bread crumb and crust found narrowing of anomeric carbon atom (C-1) peak, which also experiences displacement to a larger chemical shift, and loss of triplet characteristics could detect differences between rusk and crispy textures. The observed peak is most likely due to amylose–lipid complexes, but amorphous amylopectin and amylose could also contribute (Primo-Martín et al. 2007).

5.1.4 Spectroscopy for Fruit Texture and Structure Determination

A high proportion of the studies utilizing spectroscopic techniques for the determination of texture has focused on fruit, especially apples. Watercore, distinctive to pipfruit, results in reduced firmness because of the fruit while still on the tree experiences saturated fluid in intercellular airspaces adjacent to vascular strands. Multi-slice MRI of fruit was shown to successfully locate regions that had higher signal intensity, reflecting where watercore was prevalent (Wang et al. 1988). Longer spin–spin T_2 relaxation times and marginally shorter spin–lattice T_1 relaxation times have been reported for watercore apples relative to unaffected apples (Clark and Richardson 1999). Quantification of spatial variation of gray tones from MRI has been used to detect apples that varied in hardness, elasticity, and incidence of bruises (Létal et al. 2003).

Mealy apples have unpleasant texture in which cells have become separated without rupturing and have weak resistance to compression from biting. Cells are pushed to the side rather than damaged and texture appears as soft and dry. Time-resolved laser remittance spectroscopy (TRS) has been used to detect mealiness of apples by utilizing visible and NIR lasers (compare Section 1.3.2). Prediction of mealy apples was found to be more accurate using absorption and scattering TRS coefficients than the textural parameters of soft or dry (Valéro et al. 2004). Using the time-resolved approach also the effective pathlength can be calculated. In a sensor fusion approach of time-resolved and continuous wave spectroscopy the Lambert–Beer law was used directly for analysing fruit and vegetables quality (Zude et al. 2008). Nondestructively recorded VIS-NIR spectra correlated well with roughness, crunchiness, and mealiness of apples (Mehinagic et al. 2003). Apple cultivars with different texture were discriminated by detection of two large peaks in NIR range at 1440 and 1940 nm, which most likely correspond to water. Roughness of apples was highly correlated to a coefficient at wavelengths 1886 and 2050 nm, which most likely corresponds to starch and protein contents, respectively. Crunchiness of apples was correlated positively, while mealiness had a negative relationship with spectroscopic data in the 680–710 nm range and at 980 nm, which corresponds to chlorophyll absorbance bands as well as starch and water contents, respectively.

Apple cultivar differences were observed for correlation between compressive properties and laser scatter parameters (Cho and Han 1999). Bio-yield and rupture forces correlated well with laser light migration in the tissue. Red laser consistently performed better than green for determination of apple firmness, which may be explained by laser beam absorption ability penetrating below apple skin. Prediction of firmness and, after wavelength selection (Qing et al. 2007a), simultaneously, of firmness and soluble solid content (Qing et al. 2007b) is possible. Laser light optical power is not important for determining apple firmness, as higher power lasers did not reduce erroneous results. Hyperspectral scattering readings provide useful spectral and spatial images of apple fruit that enables successful prediction of firmness (Lu 2007) (see Section 5.2).

Internal quality defects of tropical fruit crops (Jagannathan et al. 1994) as well as pipfruit, stone fruit, and citrus (Chen et al. 1989) have been detected utilizing MRI. Rupturing of cells by freezing, which can occur in interior but hidden by external layers, results in fruit softening and detection has been demonstrated in blueberries, kiwifruit, zucchini, and apples by measuring the reduction in magnetic susceptibility because of changes in T_2 relaxation from loss of cellular integrity (Duce et al. 1992; Gamble 1994; McCarthy et al. 1995; Kerr et al. 1997). Chilling injury, a tissue breakdown disorder that occurs at above freezing temperatures in some subtropical fruits, was not detected during cold storage for persimmons using MRI, but was observed when left at room temperature (Clark and Forbes 1994).

Peach woolliness is an undesirable texture trait similar to apple mealiness with soft flesh that has absence of crispness and juiciness. Peach woolliness has been studied using MRI with less skewed MRI spectra T_2 maps found for fruits with woolliness. NIRS in conjunction with nondestructive impact tests was able to also identify peaches exhibiting woolliness (Ortíz et al. 2001). Controlled atmospheric storage can result or prevent undesirable fruit textural changes. MRI has been demonstrated to be effective in detecting tissue breakdown in cores of pears (Wang and Wang 1989) stored at low oxygen, and illustrate prevention of woolly breakdown in nectarines stored at low oxygen and elevated carbon dioxide (Sônego et al. 1995).

Water-soaking is a physiological disorder of melons characterized by a glassy texture that can be detected in fruits using NMR imaging. Use of Fourier transform infrared (FTIR) diffuse reflectance spectroscopy (Fu et al. 2007) and time-domain diffuse reflectance spectroscopy (see Section 1.3.2) can both effectively predict firmness of kiwifruit noninvasively.

5.1.5 Vegetable Texture Measured by Spectroscopy

Likewise for fruit texture, spectroscopy techniques have been investigated for their potential to measure vegetable texture, especially potatoes. MRI has been shown to differentiate raw potatoes that varied in some textural attributes when cooked. Three-quarters of variation in hardness and 50%–54% of adhesiveness and moistness of cooked potatoes were predicted by MRI of raw potatoes, but mealy and grainy textures were poorly correlated with MR images. Prediction of mealiness was better from vertical MR images rather than from entire or horizontal tuber regions (Thybo et al. 2000, 2004). Relaxation curves obtained from low-field ^{1}H-NMR showed good correlation with texture of cooked potatoes because NMR detected water content that was inversely related to starch content. Starch content was in fact the predominant determinant of cooked potato texture. Adhesiveness and springiness of cooked potatoes were better predicted by NMR than compositional analysis (Thygesen et al. 2001), but later studies modeling NMR relaxation data of cooked potatoes found good correlations with hardness, cohesiveness, adhesiveness, mealiness, graininess, and moistness (Povlsen et al. 2003). NIR spectra predicted well the moistness, waxiness, firmness, and mealiness of steamed potatoes adjudged by sensory panelists (Boeriu et al. 1998). Although O—H, N—H, and C—H stretching overtones attributed to polysaccharides and proteins can be traced in spectrum, water contributed the greatest to NIR spectra of potatoes. Another study found NIR spectra of raw potatoes were related to textural attributes (van Dijk et al. 2002). NIR spectra were able to measure dry matter and starch content of potatoes, both important contributors to cooked texture. Correlations were established between NIR spectral data and moistness, mealiness, crumbliness, waxiness, graininess, ability to mash, and firmness of cooked potatoes.

NIR spectra from cooked carrots were influenced by rate of water absorption and absorption of carotenoids in visible light region (De Belie et al. 2003; Zude et al. 2007). The phloem region of carrots provided better NIR spectra for determining texture compared with xylem tissues, possibly because phloem has larger volume, and higher sugar and carotenoid content. NMR relaxation and MRI can measure the abundance and spatial distribution of free and bound water. NIR spectra correlated well with cooked carrot hardness, crispness, and juiciness. Water is one of the strongest absorbers of infrared, with three broad peaks in the NIR spectra observed for fruits and vegetables with high water content. The peaks are at 970, 1450, and 1940 nm corresponding to the stretch of O—H molecular bonds of the second and first overtones and O—H deformation, respectively. For FTIR, water prominent bands are at 3360 cm^{-1} for H—O stretching, 2130 cm^{-1} for water association, and 1640 cm^{-1} for H—O—H bending vibration (Šáfář et al. 1994). A peak in visible region (450 nm) is due to absorption of carotenoids that decreased during cooking. The magnitude of reflectance change signifies differences in texture and related optical scattering properties of cooked carrots.

5.1.6 TEXTURE OF DAIRY PRODUCTS MEASURED BY SPECTROSCOPY

Unlike fruit, vegetable, and meat texture, many dairy products have been processed and consist of a heterogeneous medium, creating additional challenges for relating spectroscopic techniques to texture. Cheese texture, in particular, can be difficult to measure using traditional methods. Two-dimensional spin warp MRI was shown to effectively detect water and liquid-lipid portions of cheddar, brie, and Danish blue cheeses during ripening by changes in relaxation time (Duce et al. 1995). Cavities in cheese, such as *fromage frais*, could be detected due to their lack of response using three-dimensional missing pulse steady-state free precession NMR and Dixon chemical shift resolved imaging sequences.

Another method to nondestructively measure cheese texture is to use spectra from tryptophan fluorescence that is able to predict surface state, dry to watery ratio, texture length, and pastiness of soft cheeses by detecting the organization of protein networks (Dufour et al. 2001). Different textures (firmness, disintegration, pastiness, graininess, springiness, crumbliness, and oval holes) of soft cheeses owing to differences in molecular organization have also been identified using surface fluorescence spectra. Semi-hard cheese texture during ripening was measured using fluorescence and infrared spectra that detected differences in protein network (Mazerolles et al. 2001). NIR spectra have been used to determine springiness, pastiness, coherence, and hardness of semi-hard cheeses (Sørensen and Jepsen 1998). Fluorescence spectra of Salers cheese were also found to have good correlation with firmness, adhesiveness, and springiness (Lebecque et al. 2001). NIR spectra of 20 Emmental cheeses correlated well with firmness and adhesiveness (Karoui et al. 2003). Crumbly texture of cheddar cheeses has been correlated well with NIR spectra (Downey et al. 2005). Models developed based on NIR spectra from processed cheeses were successful in predicting chewy, melting, creamy, fragmentable, firmness, rubbery, greasy, and mouthcoating textural attributes adjudged by sensory panelists. NMR can noninvasively determine microstructure of Grana Padano cheese (De Angelis Curtis et al. 2000).

Viscosity of milk can be measured using MRI. Signal-to-noise ratio (SNR) of chocolate milk is decreased and velocity profile of flow is blurred compared with strawberry milk because chocolate milk contains solid cocoa particles, whereas magnetic resonance signal originates from aqueous protons (McCarthy et al. 2006).

Mid-infrared reflectance (MIR) spectrum represents absorption of all chemical bonds having infrared activity between 4000 and 400 cm^{-1}. Phospholipid-phase transition to a firmer gel texture can be detected by MIR, as increasing temperature results in shift of bands associated with C—H and carbonyl stretching (Casal and Mantsch 1984). FTIR in conjunction with attenuated reflectance circumvents sampling problems allowing investigation of textural changes resulting from aggregation and gelation of β-lactoglobulin (Dufour et al. 1998). Fluorescence spectroscopy can provide information on development of milk coagulation because of fluorescence of vitamin A located in the fat globule–protein interactions (Dufour and Riaublanc 1997).

5.1.7 Spectroscopy for Measuring Texture of Nondairy Processed Foods

Spectroscopy techniques have been less explored for measuring texture of processed foods because of interweaving of molecules creating a less homogeneous medium. Cell wall softening measured by spatially resolved spectroscopy, which has been discussed previously, occurs after food is heated, frozen and thawed, brined, or air-dried. Pectin solubilization occurs in acidic and alkaline conditions, and demethoxylation occurs during heating. Carboxyl groups can form salt linkages resulting in increased firmness.

French fries are desirable when texture is crisp and moist rather than soggy. Time-domain NMR is a rapid and accurate nondestructive technique that can measure the nuclear spin–spin T_2 relaxation times enabling detection of water and oil mobility and location within French fries and development of optimum frying conditions that produce desirable texture (Hickey et al. 2006).

Three-dimensional MR images of chocolate confectioneries can completely resolve the internal and external structure, thereby providing potential tool to assess texture of this heterogeneous food (Miquel et al. 1998).

Static light scattering can be utilized to determine droplet size in oil in water emulsions with starch or gums added as thickeners. Droplet size influences texture, and polysaccharides added stabilize emulsion against creaming by enhancing continuous phase viscosity owing to gel network formation (Quintana et al. 2002). Decrease in average fat globule size reduces creaming velocity and increases emulsion stability (Desrumaux and Marcand 2002). Light scattering detects changes in oil globule size owing to Brownian motion and probability of collision and coalescence both increase during processing that alters texture.

Small-angle x-ray scattering (SAXS) can detect textural changes because of agglomeration of protein and carrageenan. Lower intensity indicates small differences in electronic intensity between scattered particles and their surrounding (Mleko et al. 1997). T_2 parameter of starch-based heated sauces measured by NMR relaxometry was found to be a good indicator of viscosity (Thebaudin et al. 1998).

5.1.8 Summary

Measuring food texture accurately is very challenging and conventionally has been achieved by either sensory panels or instrumentation. Food industry's desire for rapid nondestructive evaluation of food products has led to the emergence of noninvasive texture analysis utilizing spatially resolved spectroscopy such as NIR, FTIR, NMR, and MRI. Significant advancements have already been made in establishing relationships between these nondestructive tests and texture determined by sensory panelists and instruments, thereby leading the way for a future in which spatially resolved spectroscopic techniques are standard quality-control procedures installed on conveyor belts of fresh produce and processed foods to ensure desired texture is delivered to consumers without defects.

REFERENCES

Bao, J.S., Y.Z. Cai, and H. Corke. 2001. Prediction of rice starch quality parameters by near-infrared reflectance spectroscopy. *Journal of Food Science* 66: 936–939.

Boeriu, C.G., D. Yüksel, R. van der Vuurst De Vries, T. Stolle-Smits, and C. van Dijk. 1998. Correlation between near infrared spectra and texture profiling of steam cooked potatoes. *Journal of Near Infrared Spectroscopy* 6: A291–A297.

Casal, H.L. and H.H. Mantsch. 1984. Polymorphic phase behavior of phospholipid membranes studied by infrared spectroscopy. *Biochimica et Biophysica Acta* 779: 382–401.

Chen, P., M.J. McCarthy, and R. Kauten. 1989. NMR for internal quality evaluation of fruits and vegetables. *Transactions of the American Society of Agricultural Engineers* 32: 1747–1753.

Chen, P.L., Z. Long, R. Ruan, and T.P. Labuza. 1997. Nuclear magnetic resonance studies of water mobility in bread during storage. *Lebensmittel-Wissenschaft und -Technologie* 30: 178–183.

Cho, Y.-J. and Y.J. Han. 1999. Nondestructive characterization of apple firmness by quantitation of laser scatter. *Journal of Texture Studies* 30: 625–638.

Choi, S.G. and W.L. Kerr. 2003. Water mobility and textural properties of native and hydroxypropylated wheat starch gels. *Carbohydrate Polymers* 51: 1–8.

Clark, C.J. and S.K. Forbes. 1994. Nuclear magnetic resonance imaging of the development of chilling injury in Fuyu persimmon (*Diospyros kaki*). *New Zealand Journal of Crop and Horticultural Science* 22: 209–215.

Clark, C.J. and C.A. Richardson. 1999. Observation of watercore dissipation in 'Braeburn' apple by magnetic resonance imaging. *New Zealand Journal of Crop and Horticultural Science* 27: 47–52.

De Angelis Curtis, S., R. Curini, M. Delfini, E. Brosio, F. D'Ascenzo, and B. Bocca. 2000. Amino acid profile in ripening of Grana Padano cheese: An NMR study. *Food Chemistry* 71: 495–502.

De Belie, N., D.K. Pedersen, M. Martens, R. Bro, L. Munck, and J. De Baerdemaker. 2003. The use of visible and near-infrared reflectance measurements to assess sensory changes in carrot texture and sweetness during heat treatment. *Biosystems Engineering* 85: 213–225.

Desrumaux, A. and J. Marcand. 2002. Formation of sunflower oil emulsions stabilized by whey proteins with high-pressure homogenization (up to 350 mPa): Effect of pressure on emulsion characteristics. *International Journal of Food Science and Technology* 37: 263–269.

Downey, G., E. Sheehan, C. Delahunty, D. O'Callaghan, T. Guinée, and V. Howard. 2005. Prediction of maturity and sensory attributes of Cheddar cheese using near-infrared spectroscopy. *International Dairy Journal* 15: 701–709.

Duce, S.L., T.A. Carpenter, and L.D. Hall. 1992. Nuclear magnetic resonance imaging of fresh and frozen courgettes. *Journal of Food Engineering* 16: 165–172.

Duce, S.L., M.H.G. Amin, M.A. Horsfield, M. Tyszka, and L.D. Hall. 1995. Nuclear magnetic resonance imaging of dairy products in two and three dimensions. *International Dairy Journal* 5: 311–319.

Dufour, É. and A. Riaublanc. 1997. Potentiality of spectroscopic methods for the characterisation of dairy products. I: Front-face fluorescence study of raw, heated and homogenised milks. *Lait* 77: 671–681.

Dufour, É., P. Robert, D. Renard, and G. Llamas. 1998. Investigation of β-lactoglobulin gelation in water/ethanol solutions. *International Dairy Journal* 8: 87–93.

Dufour, É., M.F. Devaux, P. Fortier, and S. Herbert. 2001. Delineation of the structure of soft cheeses at the molecular level by fluorescence spectroscopy—Relationship with texture. *International Dairy Journal* 11: 465–473.

Engelsen, S.B., M.K. Jensen, H.T. Pedersen, L. Nørgaard, and L. Munck. 2001. NMR-baking and multivariate prediction of instrumental texture parameters in bread. *Journal of Cereal Science* 33: 59–69.

Freidman, H., J. Whitney, and A. Szcześniak. 1963. The texturometer: A new instrument for objective texture measurement. *Journal of Food Science* 28: 390–396.

Fu, X., Y. Ying, H. Lu, H. Xu, and H. Yu. 2007. FT-NIR diffuse reflectance spectroscopy for kiwifruit firmness detection. *Sensing and Instrumentation for Food Quality and Safety* 1: 29–35.

Gamble, G.R. 1994. Non-invasive determination of freezing effects in blueberry fruit tissue by magnetic resonance imaging. *Journal of Food Science* 59: 573–610.

González, J.J., K.L. McCarthy, and M.J. McCarthy. 2000. Textural and structural changes in lasagna after cooking. *Journal of Texture Studies* 31: 93–108.

Hickey, H., B. MacMillan, B. Newling, M. Ramesh, P. van Eijck, and B. Balcom. 2006. Magnetic resonance relaxation measurements to determine oil and water content in fried foods. *Food Research International* 39: 612–618.

Hills, B.P., S.F. Takács, and P.S. Belton. 1990. A new interpretation of proton NMR relaxation time measurements of water in food. *Food Chemistry* 37: 95–111.

Ilker, R. and A.S. Szcześniak. 1990. Structural and chemical bases for texture of plant foodstuffs. *Journal of Texture Studies* 21: 1–36.

Irie, K., A.K. Horigane, S. Naito, H. Motoi, and M. Yoshida. 2004. Moisture distribution and texture of various types of cooked spaghetti. *Cereal Chemistry* 81: 350–355.

Jagannathan, N.R., R. Jayasundar, V. Govindaraju, and P. Raghunathan. 1994. Applications of high resolution magnetic resonance imaging (MRI) and spectroscopy (MRS) techniques to plant materials. *Indian Academy of Sciences Chemical Science* 106: 1595–1604.

Jarvis, M.C. and D.C. Apperley. 1990. Direct observation of cell wall structure in living plant tissue by solid-state ^{13}C NMR spectroscopy. *Plant Physiology* 92: 61–65.

Karoui, R., G. Mazerolles, and É. Dufour. 2003. Spectroscopic techniques coupled with chemometric tools for structure and texture determinations in dairy products. *International Dairy Journal* 13: 607–620.

Kerr, W.L., C.J. Clark, M.J. McCarthy, and J. De Ropp. 1997. Freezing effects in fruit tissue of kiwifruit observed by magnetic resonance imaging. *Scientia Horticulturae* 69: 169–179.

Lai, H.-M. and S.-C. Hwang. 2004. Water status of cooked white salted noodles evaluated by MRI. *Food Research International* 37: 957–966.

Lebecque, A., A. Laguet, M.F. Devaux, and É. Dufour. 2001. Delineation of the texture of Salers cheese by sensory analysis and physical methods. *Lait* 81: 609–623.

Létal, J., D. Jirák, L. Šudelová, and M. Hájek. 2003. MRI "texture" analysis of MR images of apples during ripening and storage. *Lebensmittel-Wissenschaft und-Technologie* 36: 719–727.

Leung, H.K., J.A. Magnuson, and B.L. Bruinsma. 1979. Pulsed nuclear magnetic resonance study of water mobility in flour doughs. *Journal of Food Science* 44: 1408–1411.

Lu, R. 2007. Nondestructive measurement of firmness and soluble solids content for apple fruit using hyperspectral scattering images. *Sensing and Instrumentation for Food Quality and Safety* 1: 19–27.

Mazerolles, G., M.F. Devaux, G. Duboz, M.H. Duployer, M. Riou, and É. Dufour. 2001. Infrared and fluorescence spectroscopy for monitoring protein structure and interaction changes cheese ripening. *Lait* 81: 609–623.

McCarthy, M.J., B. Zion, P. Chen, S. Ablett, A.H. Darke, and P.J. Lillford. 1995. Diamagnetic susceptibility change in apple tissue after bruising. *Journal of the Science of Food Agriculture* 67: 13–20.

McCarthy, M.J., Y.J. Choi, A.G. Goloshevsky, J.S. De Ropp, S.D. Collins, and J.H. Walton. 2006. Measurement of fluid viscosity using microfabricated radio frequency coils. *Journal of Texture Studies* 37: 607–619.

Mehinagic, E., G. Royer, D. Bertrand, R. Symoneaux, F. Laurens, and F. Jourjon. 2003. Relationship between sensory analysis, penetrometry and visible-NIR spectroscopy of apples belonging to different cultivars. *Food Quality and Preference* 14: 473–484.

Meullenet, J.-F., A. Mauromoustakos, T.B. Horner, and B.P. Marks. 2002. Prediction of texture of cooked white rice by near-infrared reflectance analysis of whole-grain milled samples. *Cereal Chemistry* 79: 52–57.

Miquel, M.E., S.D. Evans, and L.D. Hall. 1998. Three dimensional imaging of chocolate confectionery by magnetic resonance methods. *Lebensmittel-Wissenschaft und-Technologie* 31: 339–343.

Mleko, S., E.C.Y. Li-Chan, and S. Pikus. 1997. Interactions of κ-carrageenan with whey proteins in gels formed at different pH. *Food Research International* 30: 427–433.

Ortíz, C., P. Barreiro, E. Correa, F. Riquelme, and M. Ruiz-Altisent. 2001. Non-destructive identification of woolly peaches using impact response and near-infrared spectroscopy. *Journal of Agricultural Engineering Research* 78: 281–289.

Povlsen, V.T., Å. Rinnan, F. van den Berg, H.J. Andersen, and A.K. Thybo. 2003. Direct decomposition of NMR relaxation profiles and prediction of sensory attributes of potato samples. *Lebensmittel-Wissenschaft und-Technologie* 36: 423–432.

Primo-Martín, C., N.H. van Nieuwenhuijzen, R.J. Hamer, and T. van Vleit. 2007. Crystallinity changes in wheat starch during the bread-making process: Starch crystallinity in the bread crust. *Journal of Cereal Science* 45: 219–226.

Qing, Z.S., B.P. Ji, and M. Zude. 2007a. Wavelengths selection for predicting physicochemical apple fruit properties based on near infrared spectroscopy. *Journal of Food Quality* 30: 511–526.

Qing, Z.S., B.P. Ji, and M. Zude. 2007b. Predicting soluble solids content and firmness in apple fruit by means of laser light backscattering image analysis. *Journal of Food Engineering* 82: 58–67.

Quintana, J.M., A.N. Califano, N.E. Zaritzky, P. Partal, and J.M. Franco. 2002. Linear and nonlinear viscoelastic behavior of oil-in-water emulsions stabilized with polysaccharides. *Journal of Texture Studies* 33: 215–236.

Rondeau-Mouro, C., B. Bouchet, B. Pontoire, P. Robert, J. Mazoyer, and A. Buléon. 2003. Structural features and potential texturising properties of lemon and maize cellulose microfibrils. *Carbohydrate Polymers* 53: 241–252.

Ruan, R.R., C. Zou, C. Wadhawan, B. Martínez, P.L. Chen, and P. Addis. 1997. Studies of hardness and water mobility of cooked wild rice using nuclear magnetic resonance. *Journal of Food Processing and Preservation* 21: 91–104.

Šáfář, M., P.R. Bertrand, M.F. Devaux, and C. Génot. 1994. Characterization of edible oils, butters and margarines by Fourier transform infrared spectroscopy with attenuated total reflectance. *Journal of the American Oil Chemists' Society* 71: 371–377.

Sônego, L., R. Ben-Arie, J. Raynal, and J.C. Pech. 1995. Biochemical and physical evaluation of textural characteristics of nectarines exhibiting woolly breakdown: NMR imaging, X-ray computed tomography and pectin composition. *Postharvest Biology and Technology* 5: 187–198.

Sørensen, L.K. and R. Jepsen. 1998. Assessment of sensory properties of cheese by near-infrared spectroscopy. *International Dairy Journal* 8: 863–871.
Szczesniak, A. 1963. Classification of textural characteristics. *Journal of Food Science* 28: 385–389.
Szczesniak, A. 1987. Correlating sensory with instrumental texture measurements: An overview of recent developments. *Journal of Texture Studies* 18: 1–15.
Thebaudin, J.-Y., A.-C. Lefèbvre, and A. Davenel. 1998. Determination of the cooking rate of starch in industrial sauces: Comparison of nuclear magnetic resonance relaxometry and rheological methods. *Sciences Des Aliments* 18: 283–291.
Thybo, A.K., I.E. Bechmann, M. Martens, and S.B. Engelsen. 2000. Prediction of sensory texture of cooked potatoes using uniaxial compression, near infrared spectroscopy and low field ^{1}H NMR spectroscopy. *Lebensmittel-Wissenschaft und-Technologie* 33: 103–111.
Thybo, A.K., P.M. Szczypiński, A.H. Karlsson, S. Dønstrup, H.S. Stødkilde-Jørgensen, and H.J. Andersen. 2004. Prediction of sensory texture quality attributes of cooked potatoes by NMR-imaging (MRI) of raw potatoes in combination with different image analysis methods. *Journal of Food Engineering* 61: 91–100.
Thygesen, L.G., A.K. Thybo, and S.B. Engelsen. 2001. Prediction of sensory texture quality of boiled potatoes from low-field ^{1}H NMR of raw potatoes. The role of chemical constituents. *Lebensmittel-Wissenschaft und-Technologie* 34: 469–477.
Valéro, C., M. Ruiz-Altisent, R. Cubeddu, et al. 2004. Detection of internal quality in kiwi with time-domain diffuse reflectance spectroscopy. *Applied Engineering and Agriculture* 20: 223–230.
van Dijk, C., M. Fischer, J. Holm, J.-G. Beekhuizen, T. Stolle-Smits, and C. Boeriu. 2002. Texture of cooked potatoes (*Solanum tuberosum*). 1. Relationships between dry matter content, sensory-perceived texture, and near-infrared spectroscopy. *Journal of Agricultural and Food Chemistry* 50: 5082–5088.
Villareal, C.P., N.M. De La Cruz, and B.O. Juliano. 1994. Rice amylose analysis by near-infrared transmittance spectroscopy. *Cereal Chemistry* 71: 292–296.
Wang, S.Y., P.C. Wang, and M. Faust. 1988. Non-destructive detection of watercore in apple with nuclear magnetic resonance imaging. *Scientia Horticulturae* 35: 227–234.
Wang, C.Y. and P.C. Wang. 1989. Nondestructive detection of core breakdown in 'Bartlett' pears with nuclear magnetic resonance imaging. *HortScience* 24: 106–109.
Xie, F., F.E. Dowell, and X.S. Sun. 2003. Comparison of near-infrared reflectance spectroscopy and texture analyzer for measuring wheat bread changes in storage. *Cereal Chemistry* 80: 25–29.
Yakubu, P.I., E.M. Ozu, I.C. Bäianu, and P.H. Orr. 1993. Hydration of potato starch in aqueous suspension determined from nuclear magnetic studies by ^{17}O, ^{2}H, and ^{1}H NMR: Relaxation mechanisms and quantitative analysis. *Journal of Agricultural and Food Chemistry* 41: 162–167.
Zude, M., I. Birlouez, J. Paschold, and D.N. Rutledge. 2007. Nondestructive spectral-optical sensing of carrot quality during storage. *Postharvest Biology and Technology* 45: 30–37.
Zude, M., L. Spinelli, and A. Torricelli, 2008. Approach for nondestructive pigment analysis in model liquids and carrots by means of time-of-flight and multiwavelength remittance readings. *Analytica Chimica Acta* 623: 204–212.

5.2 SPECTROSCOPIC TECHNIQUE FOR MEASURING THE TEXTURE OF HORTICULTURAL PRODUCTS: SPATIALLY RESOLVED APPROACH

RENFU LU

5.2.1 INTRODUCTION

Texture is an important component in evaluating the overall quality of fresh horticultural products. People with different cultural, ethnic, and age backgrounds would have different expectations for the texture of individual horticultural products. Since texture is the human's perception of the physical and mechanical properties of a food by the feeling of touch, it cannot be accurately described using single well-defined engineering parameters. Despite the difficulty in defining texture, there are certain common textural properties that are considered important by consumers. For example, it is generally agreed that firmness is one of the most important textural properties of horticultural products. The importance of firmness is underlined by the fact that it is a key parameter in the grading standards for many horticultural products. Like many terminologies used in describing food texture, firmness is an elusive term; it has different meanings to different people. Food sensory scientists would define fruit firmness as the degree of force perceived by a person in crushing a piece of fruit using his or her molar teeth. Horticulturists, on the other hand, may consider the force required to penetrate a cylindrical probe into a fruit sample for a specific depth to be a measure of firmness. Engineers may use elastic modulus or failure strength to indicate the firmness of horticultural samples. Hence, it is not surprising that many different methods have been used to measure the firmness of horticultural products. The Magness–Taylor (MT) firmness tester is the most widely used in the horticultural industry and research laboratories.

MT firmness testing is performed using a cylindrical steel probe with a curved end surface and a specific diameter to penetrate the flesh tissue for a specific depth, and the maximum force recorded during the penetration is used to indicate the firmness of the product item. MT firmness measurements involve a complex form of mechanical load, including compression, shearing, and tension; they essentially reflect the composite mechanical failure strength of the test samples, which is difficult to quantify using engineering mechanics theory. MT firmness measurements are susceptible to operational error and may not yield consistent results (Harker et al. 1996). Despite these drawbacks, the MT firmness tester and its variants are simple and easy to use, and correlate well with the human perception of firmness (Bourne 2002). Hence, the MT firmness tester has been long used as a standard method for measuring, monitoring, and inspecting the quality of horticultural products during harvest, postharvest handling and storage, and marketing. A new nondestructive technique under development is often evaluated against the destructive MT firmness measurement. Researchers have attempted to develop and use more objective, easier-to-define firmness measurement methods to replace the existing MT testing method, but these alternative methods have not been widely adopted (Abbott et al. 1997).

Researchers have explored and developed many nondestructive firmness measurement techniques for horticultural products; most of them have not been used by the industry either because they are not practical or because they cannot provide accurate, reliable measurement of firmness, especially when they are evaluated against the destructive MT firmness measurement (Abbott et al. 1997). Early research was mainly focused on nondestructive mechanical methods such as quasi-static force/deformation, vibration, impact, and sonic resonance (Abbott et al. 1997; Lu and Abbott 2004). Among these, impact and sonic techniques look particularly promising for nondestructive evaluation of the texture of horticultural products because they would cause little or no damage to product items and because they are rapid and suitable for online sorting and grading. Impact and sonic techniques have been used recently in some modern commercial packinghouses for sorting and grading fruit. However, firmness measurements by the impact and sonic techniques tend to be influenced by fruit size and shape or geometry. The techniques seem to work better for certain type of fruits, but they may not work well for firm products such as apple (Shmulevich and Howarth 2003). Moreover, since impact and sonic techniques essentially measure elastic parameters, they may not correlate with MT firmness well. Hence, research on the development of more effective nondestructive firmness measurement techniques is continuing to draw considerable interest from researchers around the world.

With the rapid developments in computer and optical technologies, especially the detecting devices over the past decade, computer imaging and spectroscopy are now being widely used for quality inspection of horticultural and food products. NIR spectroscopy is one such prominent technology that has recently gained increasing applications in monitoring, grading, and measuring the quality of fresh horticultural products. NIR spectroscopy measures an aggregate amount of light diffusely reflected or transmitted from a product item over a spectral range between 780 and 2500 nm (more broadly, it may also include the visible range between 400 and 780 nm). According to the Beer–Lambert law (Williams and Norris 2001), the absorption of light at a specific wavelength is proportional to the concentration of a chemical component in the sample. Hence NIR spectra are shaped by the absorption of chemical components in the product item at specific wavelengths or bands. Since different chemical components often absorb light at different wavelengths or bands, they could be measured simultaneously from NIR spectra.

Numerous studies have been reported of using NIR spectroscopy for assessing the texture of food products. NIR spectroscopy, e.g., has been used for measuring the hardness of wheat kernels, the texture of bread, and the tenderness of meat (see Section 5.1 for an overview of NIR and other optical technologies for measuring the texture of foods). Studies have also been reported of using NIR spectroscopy to measure the firmness of fresh horticultural products (Lammertyn et al. 1998; McGlone and Kawano 1998; Lu et al. 2000; Lu 2001). These studies have showed that NIR technology is still not viable for evaluation of the firmness of fresh horticultural products. Such finding is not surprising in view of the complexity of firmness measurement and the underlining principle of NIR measurement. Firmness reflects the composite mechanical failure strength for a product item, and is determined by its structural and physical properties, which are in turn related to the physiological activities and maturity level. When light enters a turbid material, scattering and absorption occur. Absorption is primarily related to the chemical properties of the material, whereas scattering is influenced by the density and structural characteristics. Scattering is dominant in turbid biological materials over the visible and short-wave NIR region of 500–1300 nm. Hence measurement and separation of the scattering and absorption properties could provide more useful information on the chemical composition and structural characteristics of a horticultural product. Since conventional NIR spectroscopy cannot measure and separate absorption and scattering, its ability for assessing firmness is restricted.

Considerable research has been reported on the measurement of the optical properties of human tissues as a diagnostic tool for medical applications (Tuchin 2000; Vo-Dinh 2003). Several measurement techniques have been developed, which include spatially resolved, time-resolved, and frequency-domain techniques (Tuchin 2000). But only limited research has been reported on the measurement of the optical properties (absorption and scattering) of food and agricultural products (Birth 1978; Birth et al. 1978; Cubeddu et al. 2001; Qin and Lu 2006; Xia et al. 2007). This is because the techniques for measuring the optical properties are still not well developed, and are too expensive and not sufficiently reliable and robust for food and agricultural products. Hence, much research is needed in the development of a practical, cost-effective technique for measuring the optical properties of food and agricultural products, quantification of the light–plant tissue interactions, and establishment of the relationship between the measured optical properties and quality attributes or properties of these products.

This section reviews the latest developments and applications of spatially resolved spectroscopic techniques for measuring the optical properties and texture or firmness of horticultural products. The section first provides a brief overview of the theory of light transfer in turbid biological materials and the principle of spatially resolved techniques for measuring the optical properties. It then presents a hyperspectral imaging technique along with the data analysis methods and procedures for determining the optical properties of horticultural products and for quantifying fruit firmness and light propagation in apples. Thereafter, the section reviews the latest developments of a light-scattering technique and mathematical methods for characterizing the spectral scattering profiles of horticultural products to assess fruit firmness. The section ends with some concluding remarks on future research needs for these emerging technologies.

5.2.2 Light Propagation in Scattering-Dominant Biological Materials

5.2.2.1 Scattering and Absorption

Most biological materials including food and agricultural products are optically opaque in the visible and NIR region between 500 and 1300 nm because photons traveling within them will undergo multiple scattering. Scattering is a physical phenomenon that causes change in the traveling directions of photons in the turbid material without loss of their electromagnetic energy, i.e., photons remain the same wavelength or the same level of energy after scattering. Biological materials are heterogeneous at the microstructural level, with individual components being different in their optical properties. Scattering will occur in the form of refraction or the redirection of a light wave that is propagating from one material or component to a second material with a different refractive index. In a scattering-dominant material, photons will undergo multiple scattering before being absorbed or reemerging from the material. Scattering depends on the density and structural characteristics of the biological material, and hence it could be used for ascertaining its physical characteristics and properties.

Absorption, on the other hand, is a process involving the extraction of energy from photons by the molecules and atoms of the turbid material, which is subsequently converted into other forms of energy (such as heat and photochemical energy) or is released in the form of lower energy (longer-wavelength) light such as fluorescence. According to the quantum theory, atoms and molecules only absorb photons in specific transition, and the absorbed energy is used to increase their internal energy states. The regions of the spectrum where this energy absorption occurs are known as absorption bands; these bands are specific to particular molecular species. Absorption of photons by molecules and atoms may take place in three different patterns: (1) electronic, (2) vibrational, and (3) rotational. Electronic transitions occur in both molecules and atoms, whereas vibrational and rotational transitions occur in molecules. NIR spectroscopy is primarily based on the fundamental vibrational transition mechanism.

Consider a simple case, in which a collimating beam is incident upon a layer of homogeneous medium of thickness (L) [cm], the differential change of light intensity (dI) of the collimating light beam traversing an infinitesimal distance (dz) through the medium may be described as follows:

$$dI = -(\mu_a + \mu_s)I\,dz = -\mu_t I\,dz \qquad (5.2.1)$$

where
  I [W cm^{-2}] is the intensity of transmitted light
  μ_a [cm^{-1}] is the absorption coefficient
  μ_s [cm^{-1}] is the scattering coefficient
  μ_t [cm^{-1}] is the extinction coefficient (or total attenuation coefficient)

Integrating over the layer's thickness L gives the well-known Beer–Lambert law:

$$I = I_0 \exp(-\mu_t L) \qquad (5.2.2)$$

which may also be expressed as

$$I = I_0 \exp(-\varepsilon c_i L) \qquad (5.2.3)$$

where
   I_0 [W cm^{-2}] is the incident intensity
   ε is molar absorptivity
   c_i is molar concentration of the absorber

Transmission, T, is a commonly used quantity, which is defined as the ratio of transmitted intensity I to incident density I_0,

$$T = \frac{I}{I_0} \qquad (5.2.4)$$

The absorbance (A) or optical density (OD) of an attenuating medium is given by

$$A = \mathrm{OD} = \log_{10}\left(\frac{I_0}{I}\right) = \log_{10}\left(\frac{1}{T}\right) = \varepsilon c_i L \qquad (5.2.5)$$

Equation 5.2.5 is the basis of modern NIR spectroscopy, which states that the concentration of an absorbing constituent is linearly proportional to the absorbance or the logarithm of transmission. Strictly speaking, Equation 5.2.5 is only valid for one-dimensional cases in which the incident light beam is collimated. Most NIR applications are implemented in one of the three sensing modes: diffuse reflectance, interactance, and transmission (Abbott et al. 1997; Schaare and Fraser 2000). When reflectance (or transmission or interactance) measurements are performed on an intact product item in which scattering is dominant, Equation 5.2.5 can only provide an estimate of total light attenuation in the product item, and it cannot be used to determine μ_a and μ_s separately (see also Sections 1.3.1 and 1.3.2).

5.2.2.2 Diffusion Theory Model

Light interaction with a turbid medium is rather complicated. A rigorous treatment of the light propagation in a scattering medium will require invoking the radiation transfer equation (also known as the Boltzmann equation), which can be derived from the balance of energy flowing into and out the scattering medium (Tuchin 2000). The radiation transfer equation, which is expressed in a general integro-differential form, is difficult to solve analytically.

For many biological materials, scattering is dominant over the visible and short-wave NIR region (approximately 500–1300 nm). Under the assumption of scattering dominance ($\mu_s \gg \mu_a$), the transport of photons in the turbid material may be approximated to be a diffusion process. The radiation transfer equation can then be expressed by an approximate diffusion theory model, which is given as

$$\frac{1}{c}\frac{d\Phi(\vec{r},t)}{dt} = D\nabla^2\Phi(\vec{r},t) - \mu_a\Phi(\vec{r},t) + S(\vec{r},t) \qquad (5.2.6)$$

where

∇^2 is the Laplacian operator

c [m s^{-1}] is the velocity of light in the medium

$\Phi(\vec{r},t) = \int_{4\pi} I(\vec{r},\vec{\Omega},t)\,d\Omega$ is the photon density (also called the fluence rate [photons cm$^{-2}\cdot$s^{-1}])

$D = [3(\mu_a + \mu'_s)]^{-1}$ is the diffusion coefficient

$S(\vec{r}, t)$ is the source in the scattering medium that gives the number of photons emitted at position $\vec{r}$ and time t per unit volume per unit time. μ'_s is called the reduced scattering coefficient, which is related to the scattering coefficient (also called the anisotropic scattering coefficient), by the following relationship:

$$\mu'_s = (1-g)\mu_s \qquad (5.2.7)$$

where g is the anisotropy factor that indicates the number of scattering events required before the initial photon propagation direction is completely randomized.

The value of g ranges from -1 to 1; $g=0$ represents isotropic scattering, $g=1$ represents total forward scattering, while $g=-1$ represents total backward scattering. For most biological materials, the value of g is between 0.60 and 0.99.

The rate of increase of photons (the left-hand side of Equation 5.2.6) within a sample volume is equal to the number of photons scattered into the volume element from its surroundings (the first term of the right-hand side of Equation 5.2.6) minus the number of photons absorbed within the volume element (the second term of the right-hand side of Equation 5.2.6), plus the number of photons emitted from sources within the volume element at position $\vec{r}$ and t. With the knowledge of μ_a and μ'_s coupled with appropriate boundary conditions, we may solve Equation 5.2.6 either analytically or numerically to quantify light propagation or photon distributions in a turbid biological material.

Analytical solutions of the approximate diffusion theory model have been obtained for simple geometries of turbid media, such as slabs and semi-infinite media, for different forms of light sources (i.e., steady-state or continuous-wave [cw] light, pulsed light, and intensity-modulated light). The next subsection presents a solution for a steady-state light illumination case that is of great practical interest and is directly related to the spatially resolved hyperspectral imaging technique described in Section 5.2.3 for measuring the optical properties of food and agricultural products.

5.2.2.3 Steady-State Solutions

Consider a semi-infinite homogeneous turbid medium as shown in Figure 5.2.1, in which scattering is dominant ($\mu'_s \gg \mu_a$). A steady-state (or cw) point light is impinged vertically on the surface of the medium. Assume that no photon source exists in the medium, $S(\vec{r}, t) = 0$. Then the approximate diffusion theory model (Equation 5.2.6) is reduced to

$$DV^2\Phi(\vec{r},t) = \mu_a \Phi(\vec{r},t) \tag{5.2.8}$$

A solution to Equation 5.2.8 for the problem described has been derived by Farrell et al. (1992). The diffuse reflectance at the surface of the semi-infinite medium, which can be derived from the photon fluence rate (Φ), is expressed as a function of distance from the light source:

$$R_f(r;\mu_a,\mu'_s) = \frac{a'}{4\pi}\left[\frac{1}{\mu'_t}\left(\mu_{eff} + \frac{1}{r_1}\right)\frac{\exp(-\mu_{eff}r_1)}{r_1^2}\right.$$
$$\left. + \left(\frac{1}{\mu'_t} + \frac{4A}{3\mu'_t}\right)\left(\mu_{eff} + \frac{1}{r_2}\right)\frac{\exp(-\mu_{eff}r_2)}{r_2^2}\right] \tag{5.2.9}$$

where
  r is the distance from the incident point
  a' is the transport albedo, $a' = \mu'_s/(\mu_a + \mu'_s)$
  μ_{eff} is the effective attenuation coefficient, $\mu_{eff} = [3\mu_a(\mu_a + \mu'_s)]^{1/2}$
  μ'_t is the total attenuation coefficient, $\mu'_t = \mu_a + \mu'_s$

The variables r_1 and r_2 are given by the following two equations:

$$r_1 = \left[\left(\frac{1}{\mu'_t}\right)^2 + r^2\right]^{\frac{1}{2}} \tag{5.2.10}$$

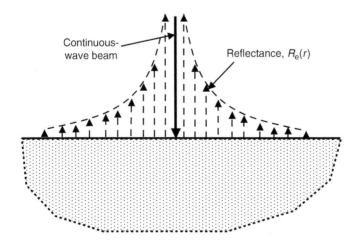

FIGURE 5.2.1 Principle of the steady-state spatially resolved technique for measuring the optical properties of biological materials.

$$r_2 = \left[\left(\frac{1}{\mu'_t} + \frac{4F}{3\mu'_t}\right)^2 + r^2\right]^{\frac{1}{2}} \tag{5.2.11}$$

where F is an internal reflection coefficient determined by the mismatch of relative refractive index (n_r) at the interface, and it may be calculated by the following empirical equation (Groenhuis et al. 1983):

$$F = \frac{1 + r_d}{1 - r_d} \tag{5.2.12}$$

in which

$$r_d \approx -1.44 n_r^{-2} + 0.710 n_r^{-1} + 0.668 + 0.0636 n_r \tag{5.2.13}$$

$$n_r = \frac{n_s}{n_{air}} \tag{5.2.14}$$

The specific refractive index for turbid biological materials (n_s) varies slightly with wavelength; however, in many practical applications, we may treat n_s constant (Nichols et al. 1997; Dam et al. 1998). Thus, once μ_a and μ'_s are known, the shape of the spatial reflectance profile is uniquely determined by Farrell's diffusion theory model (Equation 5.2.9). Conversely, if the reflectance profile resulting from a point light source over the surface of the turbid medium is measured, μ_a and μ'_s may be determined by applying an inverse algorithm to Equation 5.2.9. Farrell model provides an excellent description of the spatial diffuse reflectance profiles for turbid biological materials in which scattering is dominant (Farrell et al. 1992; Nichols et al. 1997; Dam et al. 1998; Gobin et al. 1999). It provides the mathematical basis for the spatial resolved technique that is described in Section 5.2.3.

5.2.3 Hyperspectral Imaging Technique for Measuring the Optical Properties of Horticultural Products

Hyperspectral imaging is a relatively new technology that has emerged as a powerful method for quality evaluation and safety inspection of food and agricultural products in the past decade. It possesses the features of imaging and spectroscopy, and thus enables us to obtain both spectral and spatial information from an object. As a result, the technique is particularly useful for detecting quality attributes and chemical components in a product item that are spatially variable and hence may be difficult to ascertain using either imaging or spectroscopic techniques. Hyperspectral imaging has been used for detecting quality of fruit (such as bruises and sugar distributions in the fruit) and diseases or wholesomeness of poultry and meat products (Lu and Chen 1998; Martinsen and Schaare 1998; Lu 2003; Park et al. 2002). In this section, we present a novel application of hyperspectral imaging technique for measuring the optical properties of horticultural products and for assessing fruit texture or firmness.

5.2.3.1 Principle and Instrumentation

The measurement of the optical properties of horticultural products by the hyperspectral imaging technique is based on the spatially resolved (SR) spectroscopic principle, which is schematically shown in Figure 5.2.1. As a small, collimated broadband beam is incident upon the surface of a semi-infinite turbid medium, the light is scattered into different directions, and some light is absorbed. A fraction of the light will backscatter and exit from the surface, generating diffuse reflectance at the surface of the medium. By capturing the reflectance profiles from the sample surface using an optical device for individual wavelengths, we may determine the optical properties (μ_a and μ_s') using an inverse algorithm for Farrell's diffusion theory model (Equation 5.2.9) to fit the individual scattering profiles. Clearly, to measure the optical properties over a range of wavelengths using a spatially resolved method, one needs an optical system that is capable of acquiring both spectral and spatial information from the sample. Hyperspectral imaging is thus ideally suited to accomplish this task.

The hyperspectral imaging technique is commonly implemented in one of the two sensing modes: (1) the line-scanning or push-broom mode and (2) the band-pass filter-based mode. A line-scanning hyperspectral imaging system scans the object one line at a time. Each pixel from the scanning line is represented as a spectrum on the area-array charge-coupled device (CCD) detector. A 3-D hyperspectral image cube is created by sequentially scanning the entire surface of the sample. A filtered-based hyperspectral imaging system is commonly equipped with a liquid crystal tunable filter (LCTF) or an acousto-optic tunable filter (AOTF) to acquire 2-D spatial images for individual wavelengths or bands in sequence. Each imaging mode has its merits and shortcomings. In the line-scanning mode, there should be relative movement between the sample and the imaging system to obtain 3-D hyperspectral images for the entire object. This mode is advantageous for online implementation because the moving object will present itself to the hyperspectral imaging system for the sequential scanning of its entire surface. The filter-based hyperspectral imaging system, on the other hand, does not require the relative movement between the sample and the imaging system. However, since 3-D hyperspectral images are created via sequential acquisitions of spectral images for each wavelength, this sensing mode is not appropriate for real-time, online applications. Moreover, spectral image calibrations for the filter-based mode can be more complicated than those for the line-scanning mode. This section describes a line scanning-based hyperspectral imaging system for acquiring spatially resolved spectral scattering profiles from intact fruit and vegetable samples, and appropriate methods and algorithms for extracting the optical parameters from the scattering profiles.

A schematic of a hyperspectral imaging system for measuring the optical properties of horticultural products is shown in Figure 5.2.2. The system consists of three main units: a hyperspectral imaging unit, a light source unit, and a sample handling unit. The hyperspectral imaging unit has a high-performance CCD camera with a control device, an imaging spectrograph that line scans the sample and disperses the light from the scanning line into different wavelengths while preserving

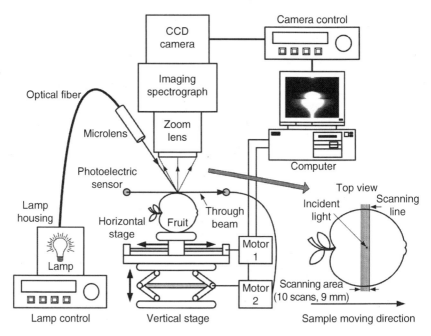

FIGURE 5.2.2 Schematic of a hyperspectral imaging system for acquiring spatially resolved backscattering images from a horticultural sample. (From Qin, J., Ph.D. dissertation, Michigan State University, East Lansing, MI, 2007.)

its original spatial information, and a zoom lens. To acquire high-quality hyperspectral images, it is important that the CCD camera have a large dynamic range (a 16-bit format is preferred) with good SNRs and high photon efficiencies over the wavelength range of interest. The hyperspectral imaging unit needs to be calibrated both spectrally and spatially. If spectral and spatial distortions (i.e., spectral and spatial lines on the image are not straight and/or parallel) occur to the hyperspectral imaging unit, it will require full-scale spectral and spatial calibrations pixel by pixel. The procedure of performing full-scale spatial and spectral calibrations can be quite complicated and time consuming (Lawrence et al. 2003). If spectral and spatial distortions are small (i.e., no more than one pixel) or negligible, a simpler procedure for spectral and spatial calibrations should be sufficient (Lu and Chen 1998).

The light source unit consists of a lamp housing installed with a broadband lamp operating in cw mode, a computer-controlled device to provide a highly stable direct current (DC) output to the lamp, and an optic fiber assembly composed of an optic fiber and a collimating lens. A quartz tungsten halogen lamp is presently a preferred choice because it has smooth spectral responses over the visible and NIR region. A DC-regulated light is required to maintain a stable light output. In addition, a light intensity feedback control device is recommended so that the output level of the light can be monitored and adjusted in real time for high reproducibility (Peng and Lu 2006a). The optic fiber assembly delivers a light beam to the sample. The light beam should be small (not more than 1.5 mm in diameter) so that the assumption made in deriving Farrell's diffusion theory model (Equation 5.2.9) is not violated. Finally, the sample-handling unit moves the sample to the desired position for the imaging unit to take images at a speed that is synchronized with the imaging acquisition speed.

After the light beam enters the sample, a diffuse reflectance image will be generated at the surface of the sample as a result of light scattering and propagation. If the sample is homogeneous and sufficiently large in size, the diffuse reflectance image will be symmetric to the beam incident point. In this case, we may take advantage of the symmetry feature by only taking single line scans, instead of scanning the entire scattering surface of the sample. This arrangement will only create one particular 2-D scattering image for each sample, and thus can save considerable time in the image acquisition.

To accurately measure the optical parameters, it is necessary to perform instrument response calibrations to correct any nonuniform responses that may occur in the hyperspectral imaging system. Nonuniform instrument responses refer to the phenomenon in which the light intensities recorded on each row of CCD pixels (corresponding to a specific wavelength) are not equal when a scanning image is taken from a uniform flat surface that is evenly illuminated. They are caused by the imperfection or specific settings of the optical components (e.g., uneven responses of the CCD pixels, a minor variation of the slit of the imaging spectrograph, and the varying viewing angle of the lens) or the light source fluctuation. In most hyperspectral imaging applications, we can correct this nonuniform response problem by collecting a scanning image from a reference panel with the uniform spectral responses over the spectral region of interest under the same lighting condition, and then calculating the relative reflectance between the sample and the reference on the pixel-by-pixel basis. However, this correction approach is inappropriate for the hyperspectral imaging system shown in Figure 5.2.2, because of its special lighting arrangement and the more stringent requirements for nonuniform response corrections. Nonuniform responses are instrument specific and may change as the settings for the imaging system change. Figure 5.2.3 shows nonuniform response curves for three wavelengths obtained using the hyperspectral imaging system shown in Figure 5.2.2. The imaging system exhibited a considerable degree of nonuniform response; the nonuniformity increases to as high as about 40% at a distance of 20 mm from the beam incident point. The nonuniform instrument response corrections should be performed, using a standardized laboratory setup, for all spatial distances (rows of pixels) that are used in the acquisition of scattering images for all wavelengths covered by the imaging system (Qin 2007).

After the nonuniform instrument response corrections have been completed, the imaging system should be tested against the reference samples with known optical properties to calibrate the system on reproducible values. Depending on application needs, either liquid or solid reference samples (phantoms) may be used. Liquid phantoms are commonly made of absorbing solution (such as black ink or other dyes) and scattering material (such as Intralipid, an emulsion made of fat that would

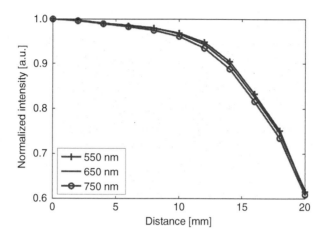

FIGURE 5.2.3 Normalized nonuniform instrument response as a function of distance from the light beam center for the hyperspectral imaging system at three selected wavelengths. (From Qin, J. and Lu, R., *Appl. Spectrosc.*, 61, 388, 2007a.)

act as pure scattering materials). Such liquid reference samples are easy to produce, but they may not resemble the geometry of actual samples to be measured. Solid phantoms can be prepared to resemble actual samples of complex geometry/structure, and they can be made of either transparent medium (such as polymers, silicone, or gelatin) or inherently scattering materials like wax. Ideally, phantoms should be prepared so that they would resemble actual samples to be measured in shape, size, and other physical characteristics. At least a few of these reference samples should be prepared to cover a typical range of optical properties that would be expected from the actual samples to be measured (Tuchin 2000).

Once the hyperspectral imaging system is validated with phantoms, it can then be used for measuring the optical properties of real-world samples. To have better SNRs and to minimize the effect of local tissue property variation on the scattering image acquisition, multiple scattering images should be considered. The hyperspectral imaging unit shown in Figure 5.2.2 would acquire up to 10 scattering images from each fruit sample and these images are then averaged to obtain one average scattering image.

5.2.3.2 Procedures of Determining the Optical Properties

Figure 5.2.4 shows a hyperspectral scattering image acquired from a 'Golden Delicious' apple, which is displayed in both 2-D (Figure 5.2.4a) and 3-D (Figure 5.2.4b) formats. The useful spectral range was between 500 and 1000 nm and the spatial distance for each side of the scattering profile was 10 mm. A line taken from the image in the horizontal direction represents a spatial scattering profile for a specific wavelength (Figure 5.2.4c), whereas a vertical line taken from the image represents a spectrum for a pixel from the scanning line at a specific distance

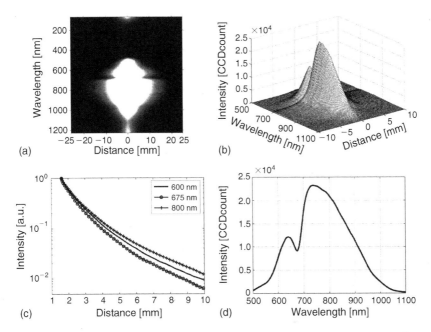

FIGURE 5.2.4 Original hyperspectral scattering image for an apple fruit in (a) 2-D format (b) 3-D format, (c) the spatial scattering profiles at three wavelengths, and (d) a spectral profile taken at the distance of 1.6 mm to the beam incident point. (From Qin, J., Ph.D. dissertation, Michigan State University, East Lansing, MI, 2007.)

from the light incident center (Figure 5.2.4d). Hence the hyperspectral image in Figure 5.2.4a may be viewed as either composed of hundreds of spectra, each representing a pixel of a different distance from the scanning line, or composed of hundreds of spatial scattering profiles for different wavelengths. The optical parameters will be determined from each spatial scattering profile (Figure 5.2.4c).

The following stepwise protocol was used to determine the optical properties of fruit samples (Figure 5.2.5):

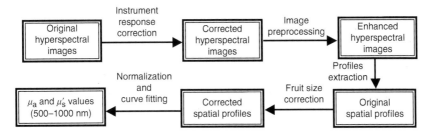

FIGURE 5.2.5 Procedure for determining the optical properties from the hyperspectral scattering images of horticultural samples.

1. Correcting the original hyperspectral image for the nonuniform instrument responses for each wavelength.
2. Preprocessing the corrected image (e.g., the averaging of several rows of pixels) to enhance image quality.
3. Extracting spatial scattering profiles from the enhanced image and reducing the two-sided spatial scattering profiles to one-sided scattering profiles (by utilizing the symmetry feature).
4. Correcting the scattering profiles, as needed, for the effect of fruit size (see the detailed description of this step in Section 5.2.4.1).
5. Performing normalizations for the scattering profiles (see further discussion below) and determining the values of μ_a and μ_s' for each wavelength via a nonlinear curve-fitting algorithm to fit each normalized scattering profile with Equation 5.2.9.

The spatial scattering profiles extracted in step 3 do not represent the actual absolute reflectance intensities from the sample, and they are not on the same scale as those described by Farrell's diffusion theory model (Equation 5.2.9). Hence it is necessary to normalize both experimental scattering profiles (designated as R_e) and Farrell model (designated as R_f) with respect to a specific scattering distance. The normalized experimental and Farrell's scattering profiles are given by the following equations:

$$R_{e,n}(r) = \frac{R_e(r)}{R_e(r_{normal})} \quad (5.2.15)$$

$$R_{f,n}(r) = \frac{R_f(r)}{R_f(r_{normal})} \quad (5.2.16)$$

where

$R_{e,n}$ and $R_{f,n}$ are the normalized experimental and Farrell model profiles, respectively

r_{normal} is the distance chosen for normalization (it is convenient to use the point closest to the light incident center)

The normalization process avoids the need of measuring absolute reflectance profiles, which are more difficult to measure experimentally.

In performing the nonlinear curve-fitting procedure in step 5, a three-step procedure is recommended. The procedure includes (1) treating both μ_a and μ_s' as unknown variables and obtaining their values through the nonlinear curve-fitting algorithm; (2) fitting the μ_s' values obtained in the first step with the following wavelength-dependent function:

$$\mu_s' = a\lambda^{-b} \quad (5.2.17)$$

where a and b are parameters for the power series model, and they are related to the density and the average size of the scattering particles, respectively (Mourant et al.

1997); and (3) inserting the results of μ'_s from step 2 into the normalized Farrell model, and repeating the same fitting procedure in step 1 with μ_a as the only unknown. This three-step curve-fitting procedure would reduce the fitting noise for both μ_a and μ'_s, thus improving the accuracies over the one-step procedure (Qin and Lu 2007a).

5.2.4 Applications

In this section, several application examples of the spatially resolved hyperspectral imaging technique are given, which include the measurement of the optical properties of selected fruits and vegetables, evaluation of the firmness of apple fruit, and quantitative analysis of light propagation regarding the penetration depths and light distributions in apple fruit.

5.2.4.1 Optical Properties of Fruits and Vegetables

Measurement of the optical properties of horticultural products can help us better understand the light–tissue interactions and develop more effective optical sensing techniques for assessing product quality. Horticultural products vary in size and shape, and their surface usually is not flat. Hence, there will be limitations in applying the diffusion theory model to horticultural products, and certain assumptions and simplifications are needed. In applying Farrell's diffusion theory model (Equation 5.2.9) to horticultural products, we assume that the samples are homogeneous in structure and sufficiently large in size compared with the light-scattering distances so that they may be considered as a semi-infinite medium. However, actual horticultural products are inhomogeneous, and they have a protective surface layer (skin) whose structural characteristics or properties are quite different from those of the interior tissue. Hence, in using the Farrell model to determine μ_a and μ'_s, we implicitly assume that the optical properties measured are the combination of those of the skin and the flesh. It remains to be further investigated to what extent each component contributes to the measured optical properties or if a more complex diffusion model for multilayered media should be employed. Despite these limitations, the homogeneity assumption should provide a reasonable start in measuring the optical properties of horticultural products.

Although it seems reasonable to treat fruits and vegetables (apple, peach, pear, etc.) as a semi-infinite medium for the purpose of using Farrell's diffusion model, the actual effect of the curved sample surface on the measurement of diffuse reflectance may not be neglected. Figure 5.2.6 shows how a curved surface would cause distortions on the measurement of diffuse reflectance profiles from a sample of circular shape. Because of the finite distance between the imaging device and the sample, the actual acceptance angle for each point at the curved surface varies with the distance from the central axis. As a result, the measured diffuse reflectance would underestimate the actual reflectance, which becomes more significant as the distance r increases or if the size of samples becomes smaller. Hence, the intensities of the measured reflectance should be corrected to take into account the effect of individual

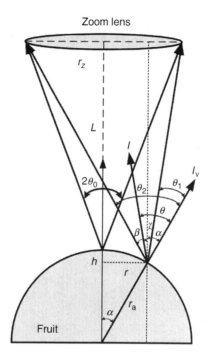

FIGURE 5.2.6 Effect of fruit size and scattering distance on the measurement of diffuse reflectance profiles (I, the reflectance component at an angle of θ from the normal direction; I_v, reflectance component in the normal direction; L, distance between the sample and imaging lens; r_z, lens radius; r, horizontal distance from the original axis; r_a, fruit radius; θ_0, one-half acceptance angle at the original point $r = 0$; $\theta_1 = \alpha - \gamma$; $\theta_2 = \alpha + \beta$; $\alpha = a\sin(r/r_a)$; $\beta = a\tan[(r_z + r)/(L+h)]$; $\gamma = a\tan[(r_z - r)/(L+h)]$; $h = r_a - [r_a^2 - r^2]^{1/2}$).

samples' size. The shape of many fruits and vegetables may be considered circular or spherical. Under this simplification, we can calculate the actual diffuse reflectance from the measured diffuse reflectance using the geometric relationship shown in Figure 5.2.6. Kienle and coworkers (1996) showed that the angular diffuse reflectance intensity for a scattering-dominant material obeys the Lambertian cosine law (Kortüm 1969). This implies that the reflectance component I with an angle θ with respect to the surface normal can be calculated as $I = I_v \cos\theta$. For a spatial point with a distance of r from the light incident center, the reflectance measured by the imaging system is equal to the integration of the reflectance I over the acceptance angle $(\theta_2 - \theta_1)$ of the zoom lens (Figure 5.2.6), and it can be calculated by the following equation (Lu and Peng 2007):

$$R_e(r) = \int_{\theta_1}^{\theta_2} I_v dS \cos^2\theta \, d\theta = I_v \, dS \left[\left(\frac{\theta_2}{2} + \frac{\sin 2\theta_2}{4} \right) - \left(\frac{\theta_1}{2} + \frac{\sin 2\theta_2}{4} \right) \right] \quad (5.2.18)$$

where

dS represents a small line element that is perpendicular to the normal direction
$\theta_1 = \alpha - \gamma$
$\theta_2 = \alpha + \beta$
$\alpha = a\sin(r/r_a)$
$\beta = a\tan[(r_z + r)/(L_s + h)]$
$\gamma = a\tan[(r_z - r)/(L_s + h)]$
$h = r_a - (r_a^2 - r^2)^{1/2}$ (see Figure 5.2.6 for the meaning of individual symbols)

Equation 5.2.18 shows that once the setting of the imaging system (i.e., the distance between the sample and the lens and the lens size) is chosen, the measured reflectance R_e would be dependent on the scattering distance and the curvature of the fruit. The acceptance angle for the measured reflectance at scattering distance $r = 0$ is equal to $2\theta_0$, where $\theta_0 = a\tan(r_z/L)$. If the scattering distance is greater than zero, the acceptance angle will change, which will in turn affect the measured reflectance. Ideally, the imaging system should collect the reflectance covering the same acceptance angle of $2\theta_0$ at each scattering distance. This cannot be achieved with the current imaging system. However, we may calculate the correct or actual reflectance at any scattering distance based on the knowledge of the normal reflectance component I_v, which is given by the following equation:

$$R(r) = \int_{-\theta_0}^{\theta_0} I_v \, dS \cos^2 \theta \, d\theta = I_v \, dS \left(\theta_0 + \frac{\sin 2\theta_0}{2} \right) = E(r) R_e(r) \qquad (5.2.19)$$

where $E(r)$ is the correction factor given by

$$E(r) = \frac{\theta_0 + \dfrac{\sin 2\theta_0}{2}}{\left(\dfrac{\theta_2}{2} + \dfrac{\sin 2\theta_2}{4}\right) - \left(\dfrac{\theta_1}{2} + \dfrac{\sin 2\theta_1}{4}\right)} \qquad (5.2.20)$$

Equation 5.2.19 was used to correct the measured reflectance profiles for all test samples before the procedure of determining the μ_a and μ_s' values was started.

Figure 5.2.7 shows the spectra of μ_a and μ_s' from one sample for each of the nine fruits and vegetables over the spectral region of 500–1000 nm. It should be noted that the spectra of μ_a and μ_s' presented in Figure 5.2.7 do not necessarily represent typical value ranges for these products. Great variations in the optical properties exist among samples of the same type of products or even the same variety as shown later for 'Golden Delicious' apples. Nevertheless, these spectra reflect the general trends and major features for these horticultural products. The absorption spectra for the three varieties of apple ('Golden Delicious', 'Fuji', and 'Red Delicious'), the peach, the pear, and the kiwifruit all peaked at 675 and 970 nm, which were attributed to the absorption by chlorophyll and water in the fruit tissue, respectively. The kiwifruit had the highest chlorophyll absorption, which could have been expected from its greenish flesh. The white flesh peach had the lowest value of the absorption coefficient among the five fruit samples (Figure 5.2.7a), which could be

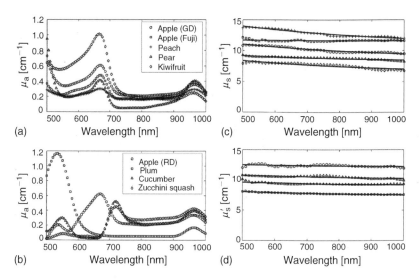

FIGURE 5.2.7 Optical properties measured for selected fruit and vegetable samples: (a) and (b) spectra of the absorption coefficient; and (c) and (d) spectra of the reduced scattering coefficient (GD, 'Golden Delicious'; RD, 'Red Delicious').

due to its low amount of chlorophyll content, as suggested by its white flesh. For the spectral range less than 550 nm, absorption due to carotenoids in the fruit tissue, which has one absorption peak at 480 nm (Merzlyak et al. 2003), became more evident for 'Golden Delicious' and 'Fuji' apples, the peach, the pear, and the kiwifruit. For wavelengths between 730 and 900 nm, the absorption values for all fruit and vegetable samples were relatively small and stable.

No discernable chlorophyll absorption peak at 675 nm was observed from the μ_a spectra for the plum, cucumber, and zucchini squash samples (Figure 5.2.7b). A strong absorption peak was found at 535 nm for the plum due to absorption by anthocyanins in the fruit tissue (Merzlyak et al. 2003), and at the same spectral position, a smaller absorption peak was also observed for the 'Red Delicious' apple for the same reason. The cucumber and the zucchini squash showed similar patterns for their absorption spectra, which, however, were conspicuously different from those of other fruit samples. The μ_a spectra of these two vegetable samples peaked at 550 and 720 nm, which were probably attributed to the absorption by turmeric and other pigments in the vegetable tissue.

The reduced scattering spectra of the fruit and vegetable samples did not show particular spectral features other than the trend of steadily decreasing values with the increase of wavelength (Figure 5.2.7c and d). The peach sample had the highest μ'_s values over the entire spectral region from 500 to 1000 nm, while the μ'_s values were lower for the kiwifruit and the plum. Other samples had intermediate values for the reduced scattering coefficient.

Figure 5.2.8 shows the mean and ±1.0 standard deviation (SD) spectra for the absorption coefficient from 600 'Golden Delicious' apples, which were measured after they had been kept in controlled atmosphere (CA) environment (at 0°C with 2% O_2 and 3% CO_2) for about 5 months (unpublished data). There were great

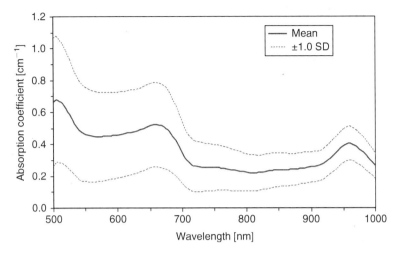

FIGURE 5.2.8 Mean and ±1.0 SD spectra for the absorption coefficient obtained from 600 'Golden Delicious' apples.

variations in the values of μ_a among the samples. The variation of μ_a values among the apples was especially evident in the visible spectral region. Similarly, large variations in the μ_s' spectra were also observed for these apples over the wavelengths of 500–1000 nm.

5.2.4.2 Evaluation of Apple Fruit Firmness

Since firmness is related to the structural characteristics of a fruit, one can expect that absorption and scattering parameters would be useful for evaluating fruit firmness. Spectra of the 600 'Golden Delicious' apples (Figure 5.2.8) were used to predict the MT fruit firmness. The firmness of each fruit was measured using the MT firmness tester right after the scattering images were acquired. Firmness prediction models relating μ_a or μ_s' to the MT firmness were developed using partial least squares regression coupled with cross validation for the 400 calibration samples. In addition, the combined data of μ_a and μ_s' were used for predicting fruit firmness using a stepwise multi-linear regression method. Model validation results for the 200 validation samples are shown in Figure 5.2.9. Both μ_a and μ_s' were correlated to fruit firmness, with $r = 0.83$ and 0.70, respectively. Moreover, when μ_a and μ_s' are combined, better predictions of fruit firmness were achieved with $r = 0.86$. These results indicate that both absorption and scattering properties are related to fruit firmness and better firmness predictions can be achieved with the combined data of μ_a and μ_s' than using single optical parameters. Hence, measurement of the absorption and scattering properties can provide a new means for evaluating internal quality of fruit and other food products.

5.2.4.3 Estimation of Light Penetration Depths in Fruit

Knowledge of light distribution and penetration depth in a particular type of fruit can provide a guide in designing an effective optical sensing configuration for quality evaluation. For example, this knowledge could be used to determine the optimal

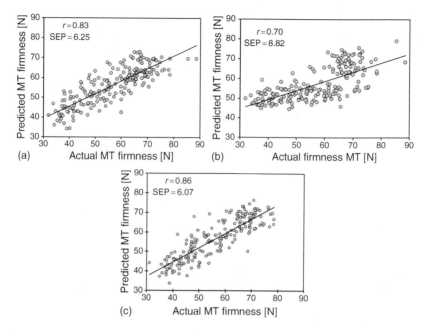

FIGURE 5.2.9 Prediction of the measured MT fruit firmness for 'Golden Delicious' using the (a) absorption, (b) reduced scattering coefficients, and (c) their combined data.

distance between the light incident area and the detecting area, so that the light signals acquired would have reached the target tissue of the fruit to be measured.

Direct measurement of light penetration depths in a fruit can be difficult and may not be practical without destroying the sample. Researchers investigated light penetration depths in several types of fruits including mandarin and apple (Lammertyn et al. 2000; Fraser et al. 2002). The light penetration depths reported in the literature differ in their values due to different instrument setups and different definitions used. With the knowledge of optical parameters, one can easily estimate light penetration depths without conducting expensive experiments. For instance, based on diffusion theory, the light penetration depth (δ) may be estimated using the following equation (Wilson and Jacques 1990):

$$\delta = \frac{1}{\mu_{\text{eff}}} = \frac{1}{\sqrt{3\mu_a(\mu_a + \mu_s')}} \quad (5.2.21)$$

Light penetration depth is defined as the distance required for the light intensity level to reduce to a factor of $1/e$, or ~37% (Equation 5.2.21). Other researchers defined the light penetration depth to be at 1% relative transmission intensity in the fruit tissue (Fraser et al. 2002). Since μ_a and μ_s' are wavelength- and fruit-dependent, light penetration depths will change with wavelength and vary from fruit to fruit. Figure 5.2.10 shows the mean light penetration depths for 'Golden Delicious' apples and ±1.0 SD using the 1% transmission attenuation (~4.6 × δ), based on the data

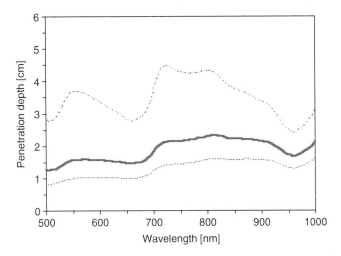

FIGURE 5.2.10 Light penetration depths for 600 'Golden Delicious' apples estimated from the mean values of the absorption and reduced scattering coefficients and their ±1.0 SD using Equation 5.2.21 (or $4.6 \times \delta$ at 1% transmittance).

for the absorption coefficient presented in Figure 5.2.8 and the corresponding reduced scattering coefficient (not presented). The maximum light penetration depths were in the NIR region between 700 and 900 nm. For a given wavelength, there was a large variation in the penetration depth among the apple samples. For example, the estimated penetration depth at 800 nm using ±1.0 SD of the mean μ_a and μ_s' ranged from 1.6 to 4.3 cm among the apples. The light penetration depth was greatly affected by the absorption of light by pigments and water in the fruit.

5.2.4.4 Monte Carlo Simulation of Light Propagation in Apple Fruit

While analytical solutions to the diffusion equation may be obtained for turbid media of simple geometries and specific types of light source (such as the one described in Figure 5.2.1), many problems of practical interest cannot be solved analytically. For general light propagation problems, we have to resort to numerical methods and the Monte Carlo (MC) method is the most widely used. Monte Carlo simulation is a stochastic numerical approach that employs random numbers in solving the problems of light transfer in turbid media. In applying an MC approach to photon transport, single photons are traced through each step and the distribution of light in the medium is built based upon the trajectories of individual photons. The parameters of each step are calculated using functions whose arguments are random numbers. As the number of photons increases to infinity, the MC prediction for the light transfer approaches an exact solution of the diffusion equation. Several MC simulation programs are already publicly available for studying light transfer in turbid media (e.g., MCML and CONV [Wang et al. 1995a,b] and tMCImg [Boas et al. 2002]).

MC simulations were performed for quantifying diffuse reflectance and photon absorption inside an apple fruit illuminated by a cw point light source as described in

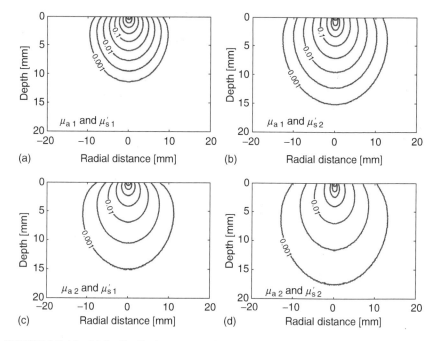

FIGURE 5.2.11 Light distribution patterns in apple fruit estimated by Monte Carlo simulations. The absorption (μ_a) and reduced scattering (μ_s') coefficients used for the simulations were $\mu_{a,1} = 0.39$ cm^{-1}, $\mu_{a,2} = 0.04$ cm^{-1}; $\mu_{s,1}' = 21.45$ cm^{-1}, $\mu_{s,2}' = 8.63$ cm^{-1}. (From Qin, J. and Lu, R., ASABE Paper No. 073058, American Society of Agricultural and Biological Engineers, St. Joseph, MI, 2007b.)

Figure 5.2.2 (Qin and Lu 2007b). Four pairs of μ_a and μ_s' values were chosen for MC simulations to demonstrate how the change of μ_a and μ_s' values would change the pattern of light distributions in the fruit.

Figure 5.2.11 shows light absorption power densities as contour maps versus sample depth and radial distance inside a 'Golden Delicious' apple with each of the four pairs of μ_a and μ_s' values. The pattern of absorption was greatly shaped by the combinations of μ_a and μ_s' values. The fruit tissue with a large μ_a value absorbed light energy rapidly in a shorter distance, making it more difficult for the light to propagate into the deeper and broader areas (Figure 5.2.11a and b). When the μ_a value is constant, the tissue with a larger μ_s' value would prevent light from penetrating the deeper layers of the fruit tissue (Figure 5.2.11a versus b and Figure 5.2.11c versus d). In summary, a fruit with small values of μ_a and μ_s' would have smaller attenuation or a larger light penetration depth, whereas a fruit with larger values of μ_a and μ_s' would have greater attenuation or a smaller light penetration depth. Light penetration depth is determined by the combination of μ_a and μ_s' values for the fruit.

Monte Carlo simulation can help to estimate the effective detecting distance for a specific sensing configuration, i.e., when the specific light source and the detector are already selected. MC simulation results are also useful in determining an optimal sensing configuration if the sample conditions and detection distance are already known (Qin and Lu 2007b).

5.2.5 LIGHT-SCATTERING TECHNIQUE FEASIBLE FOR ASSESSING FRUIT FIRMNESS IN PRACTICE

The spatially resolved hyperspectral imaging technique described in Section 5.2.3 allows us to measure and separate the absorption and scattering parameters. The optical property data are useful for quantitative analysis of light propagation and distribution in the sample and estimation of light penetration depths. Moreover, the optical property data can be used for assessing quality attributes such as firmness and soluble solids content. However, this fundamental approach requires sophisticated algorithms that may be susceptible to fitting error in extracting the values of μ_a and μ'_s for each wavelength. The technique is not suitable for rapid assessment of the texture of individual product items, which is required in product sorting and grading. This section presents a practical light-scattering technique for acquiring spectral scattering images from individual fruit and the mathematical models to quantify spectral scattering characteristics for prediction of fruit firmness.

The concept of utilizing light-scattering characteristics to assess product quality was first proposed by Birth and colleagues in the 1970s (Birth 1978, 1982; Birth et al. 1978). A transmission technique with a laser at 632 nm as a light source was used to measure the scattering and absorption parameters of the Kubelka–Munk model. Birth and coworkers (1978) reported that the scattering coefficient could be used for evaluating pork quality. However, the method and technique proposed were inconvenient and rather primitive by today's standard. Little progress was made on light-scattering research in the 1980s. Interest in the technique was renewed in the 1990s, and several studies were reported on the measurement of fruit quality/maturity by analyzing the reflectance images or profiles from an intact fruit generated by a visible or NIR laser. Tu and coworkers (2000) performed a study of using a He–Ne laser at 670 nm to generate scattering images at the surface of tomatoes to assess their maturity levels. They reported that the total number of pixels recorded by the red band CCD above a specific threshold value was related to the maturity levels of the tomatoes. McGlone and associates (1997) studied the intensities of scattered light from kiwifruit, which was generated by a diode laser at 864 nm, in relation to fruit firmness. They reported that the intensities of the scattered light from the fruit increased with decreasing firmness, especially at large distances. However, since firmness is such a complex phenomenon, light scattering at single wavelengths cannot provide sufficient information about the structural characteristics of the fruit, and thus is unable to provide accurate measurement of fruit firmness.

Lu (2004) introduced the concept of multispectral scattering for assessing the quality attributes, i.e., firmness and soluble solids content, of apples. The concept was based on the premise that scattering at multiple wavelengths would provide more information about the structural characteristics of a fruit sample and thus could lead to better assessment of fruit firmness. A broadband light source was used to generate scattering images at the surface of apple fruit. Scattering images were acquired at five wavelengths using a high-performance CCD camera coupled to a mechanical filter wheel installed with five band-pass filters. One-dimensional scattering profiles were obtained from the 2-D scattering images based on their symmetry to the light incident point. Individual scattering profiles were input into a neural network for predicting the firmness and soluble solids content of apples. Relatively good corrections were obtained for both firmness and soluble solids content. In subsequent studies, Peng and Lu (2005, 2006b,c, 2007) proposed improved hardware designs and mathematical models for measuring and analyzing spectral scattering images and obtained good predictions of fruit firmness.

To effectively predict fruit firmness and other quality attributes with the multi-spectral scattering technique, it is important to select appropriate wavelengths and use proper mathematical models, which are described in the following subsections.

5.2.5.1 Wavelengths Selection

Selection of appropriate wavelengths is the first step in implementing the multispectral scattering technique for assessing fruit firmness and other quality attributes such as soluble solids content. Ideally, the number of wavelengths selected should be small, e.g., not more than four to five, and they can provide maximum information about the structural and chemical properties of a particular type or variety of fruit. Considerable works in NIR spectroscopy have been reported on measurement of the firmness and solid content of fruits, and they can provide some guide in the preliminary selection of wavelengths for the multispectral scattering method. Because of the complexity of light–tissue interactions and large variations in properties and characteristics for different varieties, different sets of wavelengths have been reported. For example, Lu (2004) selected five wavelengths (680, 880, 905, 940, and 1060 nm) in his study on apples. The wavelength of 680 nm was related to the absorption of light by chlorophyll in the fruit. Chlorophyll content is an important indicator of fruit maturity and thus has an important implication on fruit firmness. Light at the wavelengths of 880 and 905 nm has better penetration into the fruit tissue and hence would be helpful for assessment of the properties of fruit flesh. These wavelengths are also useful for soluble solids measurement. The wavelengths of 940 and 1060 nm are useful for firmness and soluble solids prediction. Qing et al. (2007) also used a set of five wavelengths (680, 780, 880, 940, and 980 nm) in their study of assessing apple fruit quality.

A better, but more expensive and time-consuming, approach to wavelengths selection is to conduct a complete wavelength search by acquiring spectral scattering images at individual wavelengths over a spectral range of interest. Peng and Lu (2006b,c) acquired 36 scattering images between 650 and 1000 nm for 'Golden Delicious' and 'Red Delicious' apples using a monochromatic CCD camera coupled with an LCTF. By applying stepwise multi-linear regression analysis coupled with cross validation, they found a set of optimal wavelengths. The top five wavelengths for firmness prediction were in the spectral range between 690 and 990 nm.

5.2.5.2 Instrumentation

After the wavelengths are determined, one needs to choose an instrumentation setup for acquiring spectral scattering images. The instrumentation setup for scattering measurement will depend on application needs and research goals. A typical spectral scattering system consists of an imaging device with a computer, a band-pass filter, and a light source. Both color and monochromatic CCD cameras have been used.

Since light attenuates rapidly in the fruit tissue, the intensities of reflectance images will decrease drastically in a short distance from the beam incident point. Hence, it is important that the CCD imager have a large dynamic range. The imager should also have high photon efficiencies especially in the short-wave NIR range (700–1000 nm), which is needed for scattering measurement. Monochromatic CCD imagers are preferred because they have larger dynamic ranges, e.g., 12 or 16 bits versus 8 bits for the color CCD detectors.

Selection of a light source is critical for the measurement of light scattering. Two types of light sources are currently in use: monochromatic or laser light and broadband light. To better quantify the scattering characteristics, the light beam should be small in size (e.g., 2 mm or smaller). Laser is an excellent monochromatic light source (Lu and Peng 2007; Qing et al. 2007); it can deliver more light power per unit area for a given wavelength than a broadband light source, which is needed for fast acquisition of scattering images. However, to acquire scattering images at multiple wavelengths, one will need multiple lasers, which can be expensive and inconvenient. A broadband light source can provide all the needed wavelengths (Lu 2004; Peng and Lu 2006b). The main drawback of the broadband light source is its low output power per unit area because considerable light would be lost when the light from the lamp is coupled to a single optic fiber cable.

Figure 5.2.12 shows a multispectral imaging system equipped with an LCTF for acquiring spectral scattering images between 650 and 1000 nm. The system consists of a multispectral imager and a broadband light source that is coupled to an optic fiber with a collimating lens to generate a focused light beam (~1.5 mm in size). The LCTF is electronically tunable for rapid selection of any specific wavelength. Since individual spectral scattering images are acquired one at a time, the system shown in

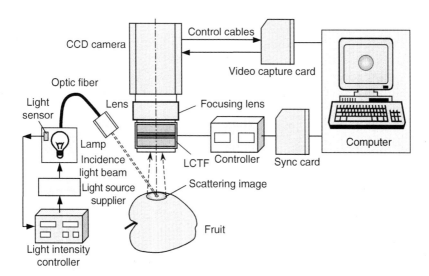

FIGURE 5.2.12 Schematic of a multispectral imaging system installed with an LCTF for acquiring spectral scattering images from apple fruit. (From Peng, Y. and Lu, R., *Postharvest Biol. Technol.*, 41, 266, 2006d.)

Figure 5.2.12 is not suitable for real-time or online sorting and grading applications. Lu and Peng (2007) developed a laboratory laser-based multispectral imaging prototype for real-time assessment of apple firmness. A common aperture CCD camera was coupled with a multispectral imaging spectrograph for simultaneous acquisition of spectral scattering images at four selected wavelengths. A custom-built multi-laser unit was used for the prototype, which had four lasers (680, 880, 905, and 940 nm) that were coupled into a single optic fiber with a micro-lens to generate one single beam of 1 mm size. The output power of the four lasers could be adjusted individually to achieve desirable scattering images at the four wavelengths. The exposure time for the scattering image acquisition could be as short as 10 ms, which would meet the requirement for online sorting and grading applications.

5.2.5.3 Mathematical Description of Light-Scattering Profiles

After scattering images have been acquired, it is critical that appropriate mathematical methods and procedures be used to describe the scattering characteristics. Figure 5.2.13 shows a typical scattering image acquired from an apple fruit and the method/procedure of processing and analyzing the scattering image. Preprocessing of the scattering images may be needed to remove noisy signals or pixels and

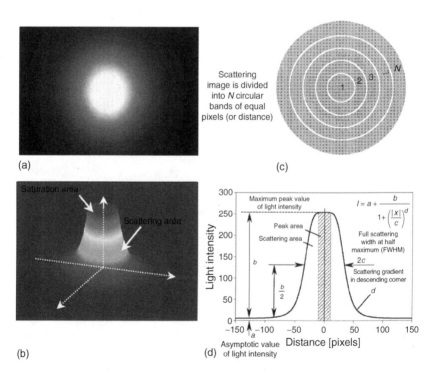

FIGURE 5.2.13 Spectral scattering images of an apple fruit in (a) 2-D display format and (b) 3-D display format; (c) the radial averaging of scattering image pixels; and (d) the resultant 1-D scattering profile fitted by the modified Lorentzian distribution function. (From Peng, Y. and Lu, R., *Trans. ASABE*, 49, 259, 2006b.)

Spectroscopic Methods for Texture and Structure Analyses 417

improve image quality. For instance, many apples have small, isolated spots on their skin, whose pigments are distinctly different from the rest of the fruit. The presence of these spots at the fruit surface could cause unusually low (dark) or high (bright) reflectance on the 2-D scattering image, which in turn would affect the resultant 1-D scattering profiles. Peng and Lu (2006d) suggested a filtering method to remove these abnormal pixels from the scattering images to improve the image quality.

Scattering images acquired from a fruit are generally symmetric to the beam incident center. Hence after the preprocessing of the scattering images, we can utilize the symmetry feature to reduce each 2-D scattering image to a 1-D scattering profile. This is achieved by first dividing the scattering images into a number of concentric rings of equal distance and then performing the radial averaging of all pixels within each circular ring (Figure 5.2.13c). The resultant 1-D scattering profile (Figure 5.2.13d) consists of two sections: a saturation section and an unsaturated, scattering section. Both saturated and unsaturated sections contain information about the scattering characteristics of a fruit and hence are useful for firmness assessment. Before further analysis of the scattering profiles, it is recommended that each scattering profile be corrected for the effect of fruit size using Equation 5.2.19 or using a simpler equation proposed by Peng and Lu (2006d).

Different mathematical functions may be used to describe the scattering profiles of fruit. The modified Lorentzian distribution function in Equation 5.2.22 and the modified Gompertz function in Equation 5.2.23 are two functions that were found to be appropriate for characterizing the scattering profiles of fruit (Peng and Lu 2006a,b,d, 2007).

$$R_L(r) = a + \frac{b}{1 + \left|\frac{r}{c}\right|^d} \quad (5.2.22)$$

and

$$R_G(r) = \alpha + \beta\left[1 - e^{-\exp(\varepsilon - \delta r)}\right] \quad (5.2.23)$$

where
    a, b, c, and d are Lorentzian parameters
    α, β, ε, and δ are Gompertz function parameters

Lorentzian functions are commonly used for describing light or signal distribution patterns in the optical and electric research fields, while Gompertz functions have been used for describing animal and organism growth (Peng and Lu 2007). Each parameter in the above equations represents certain characteristics of the scattering profile. In the modified Lorentzian form of Equation 5.2.22, the parameter a represents an asymptotic value, b is the peak value, c is related to the scattering width, and d reflects the slope of the scattering profile. Similarly, each parameter in Equation 5.2.23 is associated with a specific characteristic of the scattering profile. Examination and comparison of these parameters and their contributions to shaping the scattering profiles and firmness prediction is presented in Peng and Lu (2007).

Simpler function forms or variants with three or two parameters can be derived from the four-parameter Lorentzian distribution function and the four-parameter Gompertz function.

After a mathematical function is selected, we can use a nonlinear curve-fitting algorithm to fit the function to each scattering profile, from which values of individual function parameters are obtained. Comparison of different functions showed that the four-parameter Lorentzian function (Equation 5.2.22) and the four-parameter Gompertz function (Equation 5.2.23) could better describe the scattering profiles than their simpler function variants with two or three parameters and other functions such as Gaussian function. The modified Gompertz function (Equation 5.2.23) was slightly better in the prediction of fruit firmness than the modified Lorentzian function, but the former had slower convergence rates in the curve-fitting process (Peng and Lu 2007).

Other methods of analyzing the scattering profiles have also been proposed. One method was to use the entire scattering profiles as inputs to a neural network system for predicting fruit firmness (Lu 2004). Instead of performing radial averaging, the histograms of pixels for the scattering images can be calculated above a given threshold value and used to describe the scattering characteristics (Qing et al. 2007). This pixel–histogram calculation approach is simpler and faster compared with the mathematical modeling approach described above.

5.2.5.4 Fruit Firmness Assessment

Scattering parameters or scattering image features are useful for predicting fruit firmness and other quality attributes. Depending on the methods used to process and analyze the scattering images/profiles, different model prediction approaches may be used. A simple multilinear regression model can be established relating the scattering function parameters for each wavelength to the firmness of apples. Figure 5.2.14 shows results on using Lorentzian parameters (Equation 5.2.22) at seven wavelengths for predicting the MT firmness of 'Golden Delicious' and 'Red Delicious' apples for the calibration and validation sets, respectively. The apples were harvested in 2004, and they had been kept either in refrigerated air or CA environment for several months before the testing. Scattering images were acquired at eight wavelengths for 'Golden Delicious' (650, 680, 700, 740, 820, 880, 910, and 990 nm) and at seven wavelengths for 'Red Delicious' (680, 700, 740, 800, 820, 910, and 990 nm). These wavelengths were considered optimal for each variety based on the results from a previous study (Peng and Lu 2006c). The scattering images were obtained using a low resolution, 8-bit monochromatic CCD camera equipped with an LCTF device (Figure 5.2.12). The filtering method discussed earlier was used to remove low- and high-noise pixels and the scattering profiles were corrected for the fruit size effect using a simplified correction equation (Peng and Lu 2006d). The results presented in Figure 5.2.14 are quite remarkable considering the fact that a low resolution (512×512 pixel), 8-bit CCD camera was used in the study. Peng and Lu (2006d) compared the firmness prediction results from 'Golden Delicious' and 'Red Delicious' apples from multispectral scattering with those from NIR measurements

Spectroscopic Methods for Texture and Structure Analyses

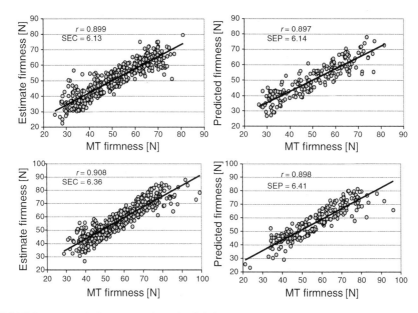

FIGURE 5.2.14 Calibration (left) and validation (right) results on prediction of the MT fruit firmness for (a) 'Golden Delicious' and (b) 'Red Delicious' apples using the four parameters of Lorentzian distribution function (Equation 5.2.22) for the scattering profiles (SEC, standard error for calibration; SEP, standard error for prediction or validation). (From Peng, Y. and Lu, R., *Postharvest Biol. Technol.*, 41, 266, 2006d.)

over the spectral region of 500–1000 nm. Even with no corrections for fruit size and noise pixels, multispectral scattering had better firmness predictions ($r = 0.82$ and 0.81 for 'Red Delicious' and 'Golden Delicious', respectively) than NIR spectroscopy ($r = 0.50$ and 0.48). Using the histogram of scattering image pixels to characterize the scattering features, good firmness prediction results for 'Elstar' and 'Pinova' apples were obtained with $r = 0.90$ (Qing et al. 2007).

5.2.6 Conclusions and Needs for Future Research

The spatially resolved hyperspectral imaging technique described in this section provides a new way of measuring the absorption and scattering properties of horticultural products. The optical parameters determined from the fruit and vegetable samples allow us to analyze light propagation and distributions in these products under different lighting and sensing configurations. They are also useful for predicting a quality attribute such as fruit firmness. Compared with time-resolved and frequency-domain techniques, the spatially resolve technique is simpler, easier to use, and more suitable for horticultural and food products. The technique is, however, based on a diffusion model that is derived for homogeneous semi-infinite turbid media, and thus it has limitations. Many horticultural products have a surface layer (skin) whose optical properties are quite different from those of the interior layers. Hence, care should be taken in interpreting the optical properties measured by the spatially resolved technique since they actually reflect both the skin and flesh. Further research should be performed to answer the question of how much fruit skin would contribute to the measured optical properties. Moreover, a diffuse theory model for two-layered or multilayered scattering media should be considered to better measure the properties of interior tissue of horticultural products. This will inevitably present more challenges than the simple diffusion model described in this section, but could lead to more accurate characterization of the optical properties of the interior layers of horticultural products.

For practical applications, light scattering at multiple wavelengths has shown great potential for assessing the texture or firmness of horticultural products. This emerging technique provides better characterization of light scattering in such horticultural products as apples than does conventional NIR spectroscopy. The light-scattering method is relatively simple and fast, and can be potentially implemented for online sorting and grading of fruit for firmness. There are, however, a few issues that should be considered in further research. First, improved instrumentation designs are needed so that the imaging system can acquire scattering images at multiple wavelengths simultaneously and rapidly, which is critical for product sorting and grading. Second, the shape of a fruit will influence the measurement of actual reflectance intensities from the sample. Although the fruit size correction equations have been developed, they are only suitable for the fruit of spherical shape. A better size/shape correction method is needed when product items to be inspected are of complicated geometry or deviate significantly from the assumed shape. Third, like those obtained with the hyperspectral imaging technique, the scattering profiles measured with the light-scattering technique are also affected by the surface layer of the sample. An appropriate method should be developed for minimizing the effect of fruit skin on the measurement of light-scattering profiles. Finally, the light-scattering technique, similar to NIR spectroscopy, relies on the establishment of appropriate calibration models to achieve accurate predictions of fruit firmness. The transferability of the calibration models from one instrument to another is an issue that has not been addressed so far. The prediction models developed in reported studies were usually validated against the samples that had the pre- and postharvest histories that were identical or similar to the calibration samples. The robustness of the calibration model for predicting samples from a new population (influence of season, location, climate, etc.) needs to be investigated and an effective method of updating the model should be considered. With proper resolution of these issues, the light-scattering technique can become an important tool for nondestructive assessment of postharvest quality of horticultural products.

ACKNOWLEDGMENT

The author wishes to thank Dr. Jianwei Qin and Dr. Yankun Peng, who were formerly with the Department of Biosystems and Agricultural Engineering at Michigan State University, East Lansing, Michigan, for their assistance in the preparation of this section.

REFERENCES

Abbott, J.A., R. Lu, B.L. Upchurch, and R.L. Stroshine. 1997. Technologies for nondestructive quality evaluation of fruits and vegetables. In: *Horticultural Reviews* 20, eds. J. Janick, John Wiley & Sons Inc., ISBN: 0-471-18906-5, pp. 1–120.
Birth, G.S. 1978. The light scattering properties of foods. *Journal of Food Science* 16:916–925.
Birth, G.S., C.E. Davis, and W.E. Townsend. 1978. The scattering coefficient as a measure of pork quality. *Journal of Animal Science* 46:639–645.
Birth, G.S. 1982. Diffuse thickness as a measure of light scattering. *Applied Spectroscopy* 36:675–682.
Boas, D.A., J.P. Culver, J.J. Stott, and A.K. Dunn. 2002. Three dimensional Monte Carlo code for photon migration through complex heterogeneous media including the adult human head. *Optics Express* 10:159–170.
Bourne, M.C. 2002. *Food Texture and Viscosity*, 2nd Edn., Academic Press, San Diego, USA. ISBN: 0-12-11062-5.
Cubeddu, R., C. D'Andrea, A. Pifferi et al. 2001. Nondestructive quantification of chemical and physical properties of fruits by time-resolved reflectance spectroscopy in the wavelength range 650–1000 nm. *Applied Optics* 40:538–543.
Dam, J.S., P.E. Andersen, T. Dalgaard, and P.E. Fabricius. 1998. Determination of tissue optical properties from diffuse reflectance profiles by multivariate calibration. *Applied Optics* 37:772–778.
Farrell, T.J., M.S. Patterson, and B. Wilson. 1992. A diffusion-theory model of spatially resolved, steady-state diffuse reflectance for the noninvasive determination of tissue optical-properties in vivo. *Medical Physics* 19:879–888.
Fraser, D.G., R.B. Jordan, R. Künnemeyer, and V.A. McGlone. 2002. Light distribution inside mandarin fruit during internal quality assessment by NIR spectroscopy. *Postharvest Biology and Technology* 27:185–196.
Gobin, L., L. Blanchot, and H. Saint-Jalmes. 1999. Integrating the digitized backscattered image to measure absorption and reduced-scattering coefficient in vivo. *Applied Optics* 38:4217–4227.
Groenhuis, R.A.J., H.A. Ferwerda, and J.J. Tenbosch. 1983. Scattering and absorption of turbid materials determined from reflection measurements–1. Theory. *Applied Optics* 22:2456–2462.
Harker, F.R., J.H. Maindonald, and P.J. Jackson. 1996. Penetrometer measurement of apple and kiwifruit firmness: Operator and instrument differences. *Journal of ASHS* 121:927–936.
Kienle, A., L. Lilge, M.S. Patterson, R. Hibst, R. Steiner, and B.C. Wilson. 1996. Spatially resolved absolute diffuse reflectance measurements for noninvasive determination of the optical scattering and absorption coefficients of biological tissue. *Applied Optics* 35:2304–2314.
Kortüm, G. 1969. *Reflectance Spectroscopy: Principles, Methods, Applications*. Springer-Verlag, LCCCN: 79-86181.
Lammertyn, J., B. Nicolaï, K. Ooms, V. De Smedt, and J. De Baerdemaeker. 1998. Non-destructive measurement of acidity, soluble solids, and firmness of Jonagold apples using NIR-spectroscopy. *Transactions of the ASAE* 41:1089–1094.
Lammertyn, J., A. Peirs, J. De Baerdemaeker, and B. Nicolaï. 2000. Light penetration properties of NIR radiation in fruit with respect to non-destructive quality assessment. *Postharvest Biology and Technology* 18:121–132.
Lawrence, K.C., B. Park, W.R. Windham, and C. Mao. 2003. Calibration of a pushbroom hyperspectral imaging system for agricultural inspection. *Transactions of the ASAE* 46:513–521.
Lu, R. and Y.R. Chen. 1998. Hyperspectral imaging for safety inspection of food and agricultural products. *SPIE Proceedings* 3544: 121–133.
Lu, R., D.E. Guyer, and R.M. Beaudry. 2000. Determination of firmness and sugar content of apples using near-infrared diffuse reflectance. *Journal of Texture Studies* 31:615–630.

Lu, R. 2001. Predicting firmness and sugar content of sweet cherries using near-infrared diffuse reflectance spectroscopy. *Transactions of the ASAE* 44:1265–1271.

Lu, R. 2003. Detection of bruises on apples using near-infrared hyperspectral imaging. *Transactions of the ASAE* 46:523–530.

Lu, R. 2004. Multispectral imaging for predicting firmness and soluble solids content of apple fruit. *Postharvest Biology and Technology* 31:147–157.

Lu, R. and J.A. Abbott. 2004. Force/deformation techniques for measuring texture. In: *Texture in Food: Volume 2: Solid Foods*, Ed. D. Kilcast, Woodhead Publishing Limited, ISBN: 1-85573-724-8, pp. 109–145.

Lu, R. and Y. Peng. 2007. Development of a multispectral imaging prototype for real-time detection of apple fruit firmness. *Optical Engineering* 46(12), 123201.

Martinsen, P. and P. Schaare. 1998. Measuring soluble solids distribution in kiwifruit using near-infrared imaging spectroscopy. *Postharvest Biology and Technology* 14:271–281.

McGlone, V.A., H. Abe, and S. Kawano. 1997. Kiwifruit firmness by near infrared light scattering. *Journal of Near Infrared Spectroscopy* 5:83–89.

McGlone, V.A. and S. Kawano. 1998. Firmness, dry-matter and soluble-solids assessment of postharvest kiwifruit by NIR spectroscopy. *Postharvest Biology and Technology* 13:131–141.

Merzlyak, M.N., A.E. Solovchenko, and A.A. Gitelson. 2003. Reflectance spectral features and non-destructive estimation of chlorophyll, carotenoid and anthocyanin content in apple fruit. *Postharvest Biology and Technology* 27:197–211.

Mourant, J.R., T. Fuselier, J. Boyer, T.M. Johnson, and I.J. Bigio. 1997. Predictions and measurement of scattering and absorption over broad wavelength ranges in tissue phantoms. *Applied Optics* 36:949–957.

Nichols, M.G., E.L. Hull, and T.H. Foster. 1997. Design and testing of a white-light, steady-state diffuse reflectance spectrometer for determination of optical properties of highly scattering systems. *Applied Optics* 36:93–104.

Park, B., K.C. Lawrence, W.R. Windham, and R.J. Buhr. 2002. Hyperspectral imaging for detecting fecal and ingesta contaminants on poultry carcasses. *Transactions of the ASAE* 45:2017–2026.

Peng, Y. and R. Lu. 2005. Modelling multispectral scattering profiles for prediction of apple fruit firmness. *Transactions of the ASAE* 48:235–242.

Peng, Y. and R. Lu. 2006a. Improving apple fruit firmness predictions by effective correction of multispectral scattering images. *Postharvest Biology and Technology* 41:266–274.

Peng, Y. and R. Lu. 2006b. An LCTF-based multispectral imaging system for estimation of apple fruit firmness: Part I. Acquisition and characterization of scattering images. *Transactions of the ASABE* 49:259–267.

Peng, Y. and R. Lu. 2006c. An LCTF-based multispectral imaging system for estimation of apple fruit firmness: Part II. Selection of optimal wavelengths and development of prediction models. *Transactions of the ASABE* 49:269–275.

Peng, Y. and R. Lu. 2006d. Improving apple fruit firmness predictions by effective correction of multispectral scattering images. *Postharvest Biology and Technology* 41:266–274.

Peng, Y. and R. Lu. 2007. Prediction of apple fruit firmness and soluble solids content using characteristics of multispectral scattering images. *Journal of Food Engineering* 82:142–152.

Qin, J. and R. Lu. 2006. Measurement of the optical properties of apples using hyperspectral diffuse reflectance imaging. ASABE Paper No. 063037, American Society of Agricultural and Biological Engineers, St. Joseph, MI.

Qin, J. 2007. Measurement of the optical properties of horticultural and food products by hyperspectral imaging. Ph.D. dissertation, Michigan State University, East Lansing, MI.

Qin, J. and R. Lu. 2007a. Measurement of the absorption and scattering properties of turbid liquid foods using hyperspectral imaging. *Applied Spectroscopy* 61:388–396.

Qin, J. and R. Lu. 2007b. Monte Carlo simulation of light propagation in apples. ASABE Paper No. 073058, American Society of Agricultural and Biological Engineers, St. Joseph, MI.

Qing, Z., B. Ji, and M. Zude. 2007. Predicting soluble solid content and firmness in apple fruit by means of laser light backscattering image analysis. *Journal of Food Engineering* 82:58–67.

Schaare, P.N. and D.G. Fraser. 2000. Comparison of reflectance, interactance, and transmittance modes of visible-near-infrared spectroscopy for measuring internal properties of kiwifruit. *Postharvest Biology and Technology* 20:175–184.

Shmulevich, I. and M.S. Howarth. 2003. Non-destructive dynamic testing of apples for firmness evaluation. *Postharvest Biology and Technology* 29:287–299.

Tu, K., P. Jancsok, B. Nicolaï, and J. De Baerdemaeker. 2000. Use of laser-scattering imaging to study tomato fruit quality in relation to acoustic and compression measurements. *International Journal of Food Science and Technology* 35:503–510.

Tuchin, V. 2000. *Tissue Optics: Light Scattering Methods and Instruments for Medical Diagnosis*. SPIE Press, Bellingham, WA, USA. ISBN: 9780819434593.

Vo-Dinh, T. 2003. *Biomedical Photonics Handbook*. CRC Press, Boca Raton, FL, USA. ISBN: 0-8493-1116-0.

Wang, L.H., S.L. Jacques, and L.Q. Zheng. 1995a. MCML—Monte-Carlo modeling of light transport in multilayered tissues. *Computer Methods and Programs in Biomedicine* 47:131–146.

Wang, L.H., S.L. Jacques, and L.Q. Zheng. 1995b. CONV—convolution for responses to a finite diameter photon beam incident on multi-layered tissues. *Computer Methods and Programs in Biomedicine* 53:141–150.

Williams, P. and K. Norris. 2001. *Near-Infrared Technology in the Agricultural and Food Industries*, 2nd Edn., AACC, St. Paul, MN, USA. ISBN: 1-891127-24-1.

Wilson, B.C. and S.L. Jacques. 1990. Optical reflectance and transmittance of tissues: principles and applications. *IEEE Journal of Quantum Electronics* 26:2186–2199.

Xia, J., E.P. Berg, J.W. Lee, and G. Yao. 2007. Characterizing beef muscles with optical scattering and absorption coefficients in VIS-NIR region. *Meat Science* 75:78–83.

5.3 NMR FOR INTERNAL QUALITY EVALUATION IN HORTICULTURAL PRODUCTS

NATALIA HERNÁNDEZ SÁNCHEZ, PILAR BARREIRO ELORZA, AND JESÚS RUIZ-CABELLO OSUNA

This section aims at summarizing the applicability of NMR in the context of internal quality assessment in fruits and vegetables. It has been structured in five sections; some of them are legible for both skilled and unskilled readers, while others may be harder in the first attempt.

It has been written in such a way as to make sections reading independent. The overview and applications section is clearly a matter of interest for a general reader, while the NMR basics and MRI fundamentals may refer to physical and mathematical concepts that will slow down the reading. Unfamiliar readers should feel free to skip those sections in the first approach.

The need for overcoming a large amount of hindrances for a competent industrial application reinforces the inclusion of such physical and mathematical concepts. It is the intention of the authors to provide a wide discussion of some technical details that may have an important role on the success for the practical use of NMR and MRI in fresh fruits and vegetables, whether this goal is achieved or not is something to be stated by the reader.

5.3.1 Overview on Applications in Fruits and Vegetables

Nuclear magnetic resonance spectroscopy (MRS) and magnetic resonance relaxometry (MRR) together with MRI have been explored since the 1980s to evaluate their applicability to the inspection of internal quality aspects in fruits and vegetables.

NMR is an especially useful monitoring technique since the signal emitted from a sample is sensitive to the density of certain nuclei, chemical structure, molecular or atomic diffusion coefficients, reaction rates, chemical exchange, and other phenomena (McCarthy 1994), which give an enormous scope for applications.

The aqueous protons (^{1}H) in soft tissues, as well as ^{13}C, ^{15}N, ^{19}F, ^{23}Na, and ^{31}P nuclei can be detected, although ^{1}H shows by far the highest applicability to fruit and vegetables (Clark et al. 1997). Correlation between quality parameters or disorders and NMR measurements has been found for many products, as reviewed by Hills and Clark (2003) and Butz et al. (2005), and as pointed out in recent publications: Létal et al. 2003; Hernández-Sánchez et al. 2004, 2006, 2007; Thybo et al. 2004; Chayaprasert and Stroshire 2005; Gambhir et al. 2005; Hernández et al. 2005; Marigheto et al. 2005; Raffo et al. 2005; Brescia et al. 2007; Goñi et al. 2007; Tu et al. 2007. Table 5.3.1 summarizes the NMR applications aimed at addressing the correlation between quality factors or disorders and NMR measurements in fruits and vegetables.

TABLE 5.3.1
Summary of Quality Attributes and Disorders in Fruits and Vegetables Studied by MRR, MRS, and MRI Techniques (Including PD Maps)

Product	Maturity/ Sugar Content	Bruises/ Voids/ Seeds	Tissue Breakdown	Heat Injury	Chill/ Freeze Injury	Infections
Fruit						
Apple	MRR/MRI	MRI	MRI/MRR PD maps			
Avocado	MRS/MRI					
Banana	MRR					
Cherimoya	MRR/MRI					
Durian	MRS/MRI		MRI			
Kiwifruit	MRR				MRI/MRR	MRI
Mandarin	MRR PD maps	MRI				
Mango	MRS/MRR			MRI		MRI
Mangosteen	MRI					MRI
Melon	MRS	MRI	MRI			
Nectarine			MRI			MRR PD maps
Orange	MRS/MRR	MRI			MRI	
Papaya				MRR		
Peach		MRI			MRR/MRI	

TABLE 5.3.1 (continued)
Summary of Quality Attributes and Disorders in Fruits and Vegetables Studied by MRR, MRS, and MRI Techniques (Including PD Maps)

Product	Maturity/ Sugar Content	Bruises/ Voids/ Seeds	Tissue Breakdown	Heat Injury	Chill/ Freeze Injury	Infections
Pear		MRI	MRR/MRI			
Persimmon					MRI	
Pineapple	MRR/MRI					
Tangerine	MRI	MRI	MRI			
Watermelon	MRS	MRI				
Berries						
Blueberry	MRI				MRI	
Grape	MRS/MRR					
Strawberry						MRI
Small, stone fruits (drupes)						
Cherry	MRR/MRS	MRI				
Olive	MRI	MRI				
Plum/prunes	MRS					
Vegetables						
Courgette/Zucchini					MRI	
Cucumber					MRI	MRI
Onion		MRI				
Potato	MRR/MRI	MRI	MRI			MRI
Tomato	MRR/MRI					

Sources: Extracted from Hills, B.P. and Clark, C.J., *Annu. Rep. NMR Spectrosc.*, 50, 75–120, 2003; and complemented by Butz Jahreszahl fehlt; Létal, J., Jirák, D., Šuderlová, L., and Hájek, M., *Lebensmittel-Wiss. Technol.*, 36, 719, 2003; Hernández-Sánchez, N., Barreiro, P., Ruiz-Altisent, M., Ruiz-Cabello, J., and Fernández-Valle, M.E., *Appl. Magn. Resonance*, 26, 431, 2004; Thybo, A.K., Szczypinski, P.M., Karlsson, A.H., Donstrup, S., Stodkilde-Jorgensen, H.S., and Andersen, H.J., *J. Food Eng.*, 61, 91, 2004; Chayaprasert, W. and Stroshire, R., *Postharvest Biol. Technol.*, 36, 291, 2005; Gambhir, P.N., Choi, Y.J., Slaughter, D.C., Thompson, J.F., and McCarthy, M.J., *J. Sci. Food Agric.*, 85, 2482, 2005; Hernández, N., Barreiro, P., Ruiz-Altisent, M., Ruiz-Cabello, J., and Fernández-Valle, M.E., *Concepts Magn. Resonance Part B: Magn. Resonance Eng.*, 26B, 81, 2005; Marigheto, N., Duarte, S., and Hills, B.P., *Appl. Magn. Resonance*, 29, 687, 2005; Raffo, A., Gianferri, R., Barbieri, R., and Brosio, E., *Food Chem.*, 89, 149, 2005; Hernández-Sánchez, N., Barreiro, P., and Ruiz-Cabello, J., *Biosystems Eng.*, 95, 529, 2006; Brescia, M.A., Pugliese, T., Hardy, E., and Sacco, A., *Food Chem.*, 105, 400, 2007; Goñi, O., Muñoz, M., Ruiz-Cabello, J., Escribano, M.I., and Merodio, C., *Postharvest Biol. Technol.*, 45, 147, 2007; Hernández-Sánchez, N., Hills, B.P., Barreiro, P., and Marigheto, N., *Postharvest Biol. Technol.*, 44, 260, 2007; Tu, S.S., Young, J.C., McCarthy, M.J., and McCarthy, K.L., *Postharvest Biol. Technol.*, 44, 157, 2007.

Note: MRR, nuclear magnetic resonance relaxometry; MRS, nuclear magnetic resonance spectroscopy; MRI, nuclear magnetic resonance imaging; PD, proton density.

Most of the studies referred to in Table 5.3.1 focused on fruits, such as apple to detect internal browning, mealiness, bruising, and watercore by means of NMR equipment operating at magnetic field strengths from 0.13 to 4.7 T, and making use of relaxometry and imaging techniques. These techniques have also been implemented in a number of studies on internal browning in pears. Maturity evaluation has been the target for a number of studies in cherimoya, durian, kiwifruit, mandarin, mango, pineapple, and tomato. In these cases, as for previous apple studies, differences in relaxation times and in signal intensities within MR images allowed the characterization of the samples.

NMR spectroscopy has been used to quantify chemical compounds such as soluble solids in a variety of species such as kiwifruit, melon, watermelon, orange, grape, and cherry; and oil in avocado. Detection of internal structures has also achieved encouraging results by means of one-dimensional (1D) images for pits in cherries or olives, and by means of two-dimensional (2D) MR images for seed detection in citrus.

Most of aforementioned works have been undertaken with commercial NMR equipment designed for medical purposes, which are not conceived to deal with the practical constrains that apply for the food industry. Such equipments operate at high-magnetic field strength of more than 2 T with high-performance hardware, which involves difficulties of setting in an industrial environment and large investments. In most of the cases, samples remained stationary during data acquisition, which, on the other hand, was performed without time-consumption restriction, being only of use for off-line examination. Practical on-line monitoring requires sample inspection under motion conditions and shortening of data acquisition time. Nevertheless, these studies comprise a revealing approach to a multisensing tool for fruit and vegetable inspection.

Only a few studies focus on the effect of sample motion, reduction of the data acquisition time, or use of low-magnetic field strength. However, the results obtained highlight the great potential of NMR techniques for internal quality monitoring and encourage further works for developing feasible on-line NMR systems.

The following two sections provide the basics of the NMR phenomenon and the NMR signal acquisition as the basis of the characterization of quality attributes, unfamiliar readers may directly skip to Section 5.3.4 where further explanation is provided for the quantification of success for most relevant applications, together with some transferability remarks with regard to industrial needs.

5.3.2 Basics of NMR Relaxometry and NMR Spectroscopy

5.3.2.1 Magnetic Moment of Nucleus and Its Excitation

NMR is based on the phenomenon that nuclei consisting of unpaired nuclear particles, that is, odd number of protons and neutrons possess an intrinsic property called spin and therefore have angular momentum. The angular momentum and the nuclear charge confer a magnetic moment on the nucleus, which can interact with an external magnetic field. Such nuclei include ^{1}H (proton), ^{2}H, ^{13}C, ^{15}N, ^{19}F, ^{23}Na, ^{31}P, etc., among which ^{1}H represents the highest quantity and biological abundance mainly as protons in water molecules, and the highest NMR signal. Proton NMR is a valuable tool for fruit quality evaluation as water influences many quality-related characteristics and metabolic processes, and therefore, we will focus on such nucleus.

The interactions of the atoms with external magnetic fields can be described by quantum mechanics, although the overall behavior is characterized in classical physics. A combination of both descriptions is normally used to illustrate the phenomena, and the latter will be what we adopt in the next paragraphs.

In the absence of an external magnetic field, the magnetic moments of the nuclei are randomly oriented having the same energy and zero net magnetization. In the presence of a static magnetic field B_0, the magnetic moments will line up with the magnetic field at different angles determined by the spin number (Figure 5.3.1). The possible orientations correspond to different energy levels being the energy differential directly related to the static magnetic field strength.

For ^1H, two orientations are possible: parallel and antiparallel. Parallel proton orientation corresponds to a slightly lower energy level in comparison to antiparallel proton orientation (Figure 5.3.1). At thermal equilibrium, a slight excess of protons resides in the parallel state giving rise to a net macroscopic magnetization (M), which produces the NMR signal. Curie's law (Equation 5.3.1) establishes the equilibrium longitudinal magnetization as function of the static magnetic field, B_0, through the static nuclear magnetic susceptibility, χ_0, which is characteristic of each nucleus and inversely dependent on the temperature.

$$M = \chi_0 B_0 \qquad (5.3.1)$$

The macroscopic magnetization increases as the magnetic moments align with the external magnetic field until the thermal equilibrium is achieved. The characteristic time constant for the equilibrium establishment is T_1, the so-called longitudinal relaxation time. The net macroscopic magnetization, M, is the sum of the contributions of all the magnetic moments of the individual protons. The component parallel

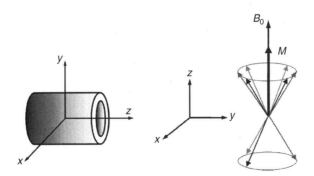

FIGURE 5.3.1 On the left, magnet axis established by convention. On the right, net magnetization (M) along the direction of the main magnetic field (B_0) as a result of the parallel and antiparallel nuclear magnetic moments orientation.

to the magnetic field direction is called longitudinal magnetization; from now onward we will refer to it as M_z.

Because of the angular momentum each magnetic moment undergoes a torque and precesses around the static magnetic field axis describing a circular orbit. The angular frequency of nuclear precession (ω_0) is known as the Larmor frequency (Equation 5.3.2) and is directly proportional to the external magnetic field strength via the gyromagnetic ratio (γ_i), which is dependent on the nucleus. The gyromagnetic ratio describes the relationship between the nucleus-specific magnetic momentum ($\vec{\mu}$) and angular momentum ($\vec{S}$), $\vec{\mu} = \gamma \vec{S}$ (Table 5.3.2).

$$\omega_0 = \gamma_i B_0 \tag{5.3.2}$$

The behavior of the magnetic momentum in the presence of a time-independent magnetic field $\vec{B} = B_0 \vec{z}$ is well known and can be described by the Schrödinger equation. The solution for this equation justifies the use of classical precession for the proton spin motion. Moreover, the discreteness of the proton's intrinsic angular momentum leads to the discreteness of the energy levels of its interaction with a magnetic field, and thus to the parallel and antiparallel states for a proton (or any spin 1/2 nuclei). This is an example of the general Zeeman effect, where nuclear magnetic moment in the presence of an external magnetic field leads to splitting in nuclear energy levels.

The phenomenon of resonance occurs when an electromagnetic pulse is applied with energy enough to induce transitions between these states. The frequency at which the nucleus will absorb such energy is precisely the Larmor frequency. In proton NMR, the necessary energy evolves in the radiofrequency (RF) range. Excitation RF pulse is applied with an oscillator coil by producing an alternating field B_1 perpendicular to B_0 with much smaller magnitude compared to it. In classical physics, the behavior of the nuclei is described such as the longitudinal magnetization precesses around the B_1 axis at the angular frequency $\omega_1 = \gamma \cdot B_1$. Here, the net magnetization moves from the z-axis toward the transversal plane, where signal measurement is carried out (Figure 5.3.2). Depending on the product of the

TABLE 5.3.2

Gyromagnetic Ratio γ of Some Nuclei of Interest in Megahertz per Tesla

Nucleus	γ_i [MHz/T]
^{1}H	42.58
^{2}H	6.54
^{13}C	10.71
^{15}N	3.08
^{19}F	40.08
^{23}Na	11.27
^{31}P	17.25

Spectroscopic Methods for Texture and Structure Analyses

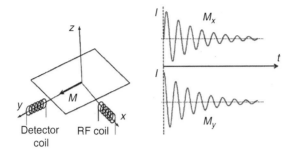

FIGURE 5.3.2 On the left, depiction of the spiral-like progress with time of the net magnetization vector (M) in x–y plane where RF pulse is applied by the RF coil along the x direction, and signal is detected by the detector coil along the y direction. On the right, evolution along the time of the x and y components of M after an RF pulse, note that M_x is zero and M_y is maximum after excitation.

amplitude and length of the applied RF pulse the so-called flip angle varies. A 90° RF pulse refers to a right flip angle so that the longitudinal net magnetization (M) tilts completely on x–y plane. According to quantum mechanics concept, when the RF pulse reaches sufficient strength or duration, an equal distribution of nuclei at the two energy states is achieved so that the longitudinal magnetization disappears. In addition to the transitions between the energy levels, immediately after the RF pulse, the precession of the nuclei is synchronized and, as a result, a magnetization perpendicular to the axis of the external magnetic field appears.

The magnetization on x–y plane is called transverse magnetization (M_{xy}) and this is the component susceptible of being detected in NMR analyses. The magnitude of the NMR signal is proportional to the number of proton nuclei in the tissue and provides a description of the latter in terms of density and composition as it reflects the distribution of air gaps, water, and metabolites.

5.3.2.2 Relaxation of Nucleus after Excitation

When the RF pulse is turned off, the magnetic moments return to the parallel lower energy level with simultaneous emission of energy at the Larmor frequency. Thus, the net magnetization swings back toward the positive z-axis until the equilibrium longitudinal magnetization is recovered. This process is called longitudinal or spin–lattice relaxation, referring to the exchange of energy in the form of thermal motions between the spins and their surroundings. To transfer the excess of energy, there is a need of receivers that allow receptivity (tuning) to the precession frequencies of the excited nucleus.

The rate of return to the initial equilibrium state is exponential, characterized by the constant T_1, longitudinal relaxation time (Equation 5.3.3). At T_1, the longitudinal magnetization has recovered 63% of its initial value M_0. The magnitude of T_1 directly depends on the static magnetic field B_0.

$$M_z(t) = M_0(1 - e^{-t/T_1}) \qquad (5.3.3)$$

Longitudinal relaxation time T_1 is different for different tissues and is a valuable source of contrast in quality-evaluation measurements. Differences arise as a result of the different environments at the microscopic and molecular levels, giving rise to a changed magnetization characteristics, which may be described by a multiple exponential decay function. When molecules in the actual environment have a movement rate or tumbling close to the Larmor frequency, the energy exchange is favored. The longitudinal relaxation becomes more efficient, that is, T_1 shortens. As differences in the thermal motion increase, efficient exchange is not allowed, which involves a subsequent T_1 lengthening. Such differences appear in relation to the molecular mobility and size. Small molecules, such as inorganic salts and water, tumble or move at much higher rates than the Larmor frequency being free to collide and so facilitating energy transfer. In contrast, larger and more rigid molecules, such as lipid, fat, protein, and complex sugars, move slower preventing collisions and so energy dissipation. The distribution of the molecular motions depends on temperature and viscosity. T_1 is expected to increase at lower temperature and for more viscous media.

As the longitudinal component M_z recovers its original value, the transverse magnetization M_{xy} decays as a result of the loss of phase coherence of the excited protons. The cause is normally associated with magnetic field inhomogeneities. Over time, the individual transverse magnetization vectors begin to cancel out and the net magnetization vector gradually reduces. The rate of the decay is also exponential with the time constant T_2, transverse relaxation time (Equation 5.3.4).

$$M_{xy}(t) = M_0 \cdot e^{-t/T_2} \qquad (5.3.4)$$

The origin of the loss of phase coherence is caused by two factors. On one hand, there is an external origin related to the imperfections in the magnetic field homogeneity of the NMR spectrometer. The second one refers to mechanisms of T_2 relaxation inherent to the sample.

It is worthy to note that since the external magnetic field is rarely completely homogeneous, an effective relaxation time shorter than T_2 governs the decaying, this is the T_2^*. The difference resides in that the dephasing induced by the external field inhomogeneities can be reversed, whereas T_2 effect is not reversible.

As for the inherent causes of loss of phase coherence, protons can be regarded as magnetic dipoles that cause intrinsic magnetic fields within the sample with intensities proportional to their magnetic susceptibility. Magnetic susceptibility variations and interactions among the nuclei themselves create local inhomogeneous magnetic fields. At the molecular level, internal and tumbling motions of the molecules and chemical exchange between proton pools characterized by different precession frequencies are sources of T_2 relaxation. Diffusion and tissue microstructure present their effects at a microscopic level.

In a viscous or solid medium, molecules are relatively fixed, which involves the presence of relatively fixed local magnetic fields that cause local inhomogeneities in the protons surroundings. This environment could lead to differences in precession frequencies that increase the rate of dephasing. In liquids, the local magnetic fields from neighboring molecules fluctuate rapidly being averaged to a small value so that dephasing is less favored and, consequently, T_2 is longer. Proton exchange occurs between water and exchangeable protons on cell metabolites, such as sugars, cell wall biopolymers, or starch granules (Hills and Clark 2003). A slow rate of proton exchange between chemical sites causes the vectors to loose phase coherence because of local magnetic field differences. Viscosity facilitates proton exchange as nuclei are held in close proximity for an extended period.

The evolution of the metabolites concentration during fruit and vegetable ripening exerts an important effect on the relaxation times. For unripe fruit, starch granules are major relaxation sinks for water protons. As fruit ripen, starch granules break down into sugars, decreasing the size and the number of granules. However, T_2 decreases since more hydroxyl groups are available to exchange with the water protons (Keener et al. 1997).

At a microscopic level, cell characteristics and tissue microstructure present a major influence on the transverse relaxation process, which is associated with the resultant effect of the water diffusion between subcellular organelles and intercellular gaps. Consequently, T_2 is a powerful source of information on cellular structure integrity. Water is normally compartmentalized into vacuole, cytoplasm, and extracellular space with increasing mobility restriction, respectively, that leads to decreasing T_2 values (Hills and Remigereau 1997). Depending on the cell morphology and size, and the membranes permeability, the diffusion of water between compartments will bring different levels of magnetization averaging and hence, loss of the contrast information. Similarly, loss of membranes integrity results in changes in water compartmentation and in overall T_2 values, which could be detected by analyzing the distribution of the transverse relaxation times and the influence on macroscopic contrast. When water diffuses through regions with different magnetic susceptibilities, the local field gradients that appear as consequence of such discontinuities produce magnetization dephasing and consequently T_2 shortens. This situation takes place especially at air–liquid interfaces and it is enhanced at high-static magnetic field strengths, since local field gradients are intensified. Table 5.3.3 summarizes the

TABLE 5.3.3
Summary of the Main Parameters Affecting the Time Constants that Characterize the Longitudinal and the Transverse Relaxation Processes of the Magnetic Moments (T_1 and T_2, Respectively)

Longitudinal Relaxation Time (T_1)	Transverse Relaxation Time (T_2)
Molecular motion	Molecular motion
Molecular size	Chemical exchange
Molecular complexity	Cell morphology and size
Viscosity	Cell compartmentation
	Membrane permeability
	Tissue microstructure
	Diffusion

parameters affecting relaxation times. For further reading on NMR relaxation in food products, the book by Hills (1998) is recommended.

5.3.2.3 Signal Detection during Relaxation

Following the RF pulse, the precessing transverse magnetization M_{xy} generates a small electromotive force that, according to Faraday's law of induction, induces a current along the receiver coils. Such coils are situated in the transverse plane (Figure 5.3.2) and separate out the signals along the x and y axes. According to their shape, there are volume and surface coils. The same coil can be used to transmit and to receive. The receiver coil observes an oscillating wave signal with a phase described by $(i\omega t)$, where ω is the nucleus resonance frequency (Figure 5.3.2). The signal amplitude decreases to zero in exponential fashion with time as $\exp(-t/T_2)$. The detected time-domain signal is called free induction decay (FID) because of its damping nature. The signal is recorded as an oscillating voltage, which is amplified and digitized by an analogue-to-digital converter, and then stored as an array of complex data, which contains the information on the magnitude and the phase of the reading.

5.3.2.4 NMR Relaxometry and NMR Spectroscopy

As aforementioned, the magnetization of hydrogen nuclei in different physical or chemical environments decays at different rates. Moreover, different chemical environments of otherwise equivalent hydrogens lead to slight differences in their precession frequencies. MRR identifies nuclei populations distinguishable on the FID because of the different decay time constants. Values can be computed for local spots as well as it is possible to spatially resolve the assignation of times. For the latter, the outcome is the so-called relaxation maps. As for MRS, the precession frequency encodes the chemical groups that give rise to the NMR signal. The outcome is an NMR spectrum where intensity is plotted versus frequency.

5.3.2.4.1 NMR Relaxometry
The determination of the longitudinal relaxation time T_1 is usually performed with the inversion-recovery sequence. The sequence consists of a 180° RF pulse that inverse the longitudinal magnetization M_z. During an inversion time interval (TI), M_z is allowed to recover. Then, it is tilted toward the transverse plane by a 90° RF pulse and the magnitude of M_{-z} is measured, which is proportional to the PD and T_1. This process is repeated for a number of different TIs. After several measurements, the constant time T_1 can be determined from Equation 5.3.5.

$$M_{T_1} = M_z(1 - 2e^{-TI/T_1}) \qquad (5.3.5)$$

T_2 determination is implemented by means of the Carr–Purcell–Meiboom–Gill (CPMG) sequence. First, a 90° RF pulse is applied to flip the longitudinal magnetization toward the transverse plane. After a time interval (τ), a train of 180° RF pulses is applied with a pulse spacing of 2τ. The 180° RF pulse refocuses the

Spectroscopic Methods for Texture and Structure Analyses

dephased transverse magnetizations as the slower precessing vectors are now in front of the faster precessing vectors, so that the latter overtake the slower ones. This fact causes what is called spin echo (SE) with the vectors reachieving coherence, losing it afterward. Maximum signal is achieved at time interval τ after each refocusing pulse application. The peak intensity is proportional to the PD and T_2. The envelope of the peak echo intensities decays along the echo train that allows calculation of T_2 (Equation 5.3.6).

$$M_{n\text{TE}} = M_z e^{-tn\text{TE}/T_2} \qquad (5.3.6)$$

The effects of the inhomogeneities in the magnetic field on the loss of phase coherence, that is, the effective T_2 (T_2^*) decay, are cancelled out.

When more than one single relaxing component (as those corresponding to vacuoles, cytoplasm, and intercellular spaces) are expected, deconvolution of the echo decay envelope is done either by assuming higher order exponential decay or by presenting the data as a continuous distribution of relaxation times and deconvolution with an inverse Laplace transform (Hills et al. 2004). See example on Figure 5.3.3. Thus, MRR also provides microstructure information on the sample.

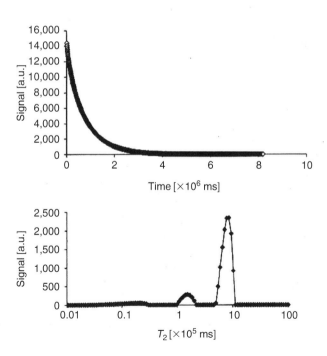

FIGURE 5.3.3 Example of a relaxation curve measured on apple tissue (top) and result of T_2 continuous distribution fitting (bottom) where two peaks are clearly resolved corresponding to different subcellular compartments, that is, water in vacuole and in cytoplasm, respectively. (Reprinted from Barreiro, P., Moya, A., Correa, E., et al., *Appl. Magn. Resonance*, 22, 387, 2002. With permission.)

Relaxometry provides basic information regarding tissue characteristics that can readily be applied for enhancing MRI contrast. Besides it may be directly employed as a mean for whole fruit quality assessment and for tissue microstructure analysis, as it will be discussed in Section 5.3.4.

5.3.2.4.2 NMR Spectroscopy

The differences in the precession frequencies in different chemical environments originate in the electron orbitals that appear around the nuclei. The circulation of electron probability in the orbitals causes a small magnetic field at the nucleus that opposes the externally applied one, this is the nuclear shielding. Thus, the effective magnetic field affecting the nucleus is generally less than the applied field by a fraction σ (shielding constant) as indicated in Equation 5.3.7.

$$B = B_0(1 - \sigma) \tag{5.3.7}$$

The electron density around each nucleus in a molecule varies according to the types of nuclei and bonds in the molecule. Therefore, protons with different chemical environments will resonate at slightly different frequencies from that defined by the applied external field B_0. This is called the chemical shift phenomenon.

The chemical shift of a nucleus (δ) is the difference between the resonance frequency of the nucleus (ν) and a reference resonance frequency (ν_{REF}) divided by ν_{REF} (Equation 5.3.8). Since the resonance frequency depends on the magnetic field strength, the relative scale removes the field dependence so that comparison between different NMR equipments becomes feasible. Chemical shift value is very small and it is generally reported in parts per million [ppm] (Table 5.3.4).

TABLE 5.3.4
Characteristic Proton Chemical Shifts for Functional Groups of Interest

Functional Groups	Structure	Chemical Shift [ppm]
Aromatic	Ar—H	6.0–8.5
Benzylic	Ar—C—H	2.2–3.0
Alcohols	H—C—OH	3.4–4.0
Ethers	H—C—OR	3.3–4.0
Esters	RCOO—C—H	3.7–4.1
Esters	H—C—COOR	2.0–2.2
Acids	H—C—COOH	2.0–2.6
Carbonyl compounds	H—C—C=O	2.0–2.7
Aldehydic	R—(H—)C=O	9.0–10.0
Hydroxylic	R—C—OH	1.0–5.5
Phenolic	Ar—OH	4.0–12.0
Enolic	C=C—OH	15.0–17.0
Carboxylic	RCOOH	10.5–12.0
Amino	RNH_2	1.0–5.0

Spectroscopic Methods for Texture and Structure Analyses

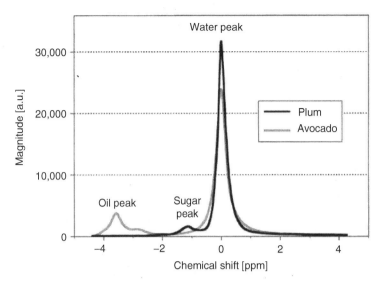

FIGURE 5.3.4 NMR spectra of an avocado and a plum showing the water, sugar, and oil peaks at the chemical shift of the corresponding protons. The horizontal axis represents the chemical shift expressed in parts per million as a relative difference in frequency from the resonance frequency of water. (Courtesy of P. Chen, unpublished data.)

$$\delta = (\nu - \nu_{REF})10^6/\nu_{REF} \quad (5.3.8)$$

The scale of the chemical shift establishes a relation between position and atom or group of atoms, which allows the identification of different compounds in the analyzed sample on the basis of frequency information (Figure 5.3.4).

Such frequency information is extracted from the time-domain signal, by applying the Fourier transformation (FT) under its abbreviated form (fast Fourier transform, FFT) to the FID. The measured time-domain signal is a superposition of individual FIDs and contains all of the frequencies coming from the excited sample, which are measured simultaneously. The FT is a mathematical tool that separates the contribution of each nucleus by its resonance frequency and identifies the corresponding intensities of the components of the FID. As we said above, the frequency-domain spectrum is a plot of intensity versus frequency (Figure 5.3.4). The areas of the resulting spectral peaks are proportional to the nuclei concentration and number of chemically equivalent nuclei of each molecule. The peak widths are inversely related to their transverse relaxation times and play a key role in peaks resolution. For most fruits, the resolved types of hydrogen nuclei are associated with water, oil, and carbohydrates.

5.3.3 MRI Fundamentals

This section describes the obtention of images by means of spatial encoding using magnetic field gradient, a smart idea that has became the basis of several Nobel prizes, and which has led to a full discipline. It requires some skill to approach the underlying physical and mathematical concepts for a deep understanding of the issue. Practical approach is also widespread by means of simple comparisons of images corresponding to characteristical samples. The former aspects are described in this section, while the latter are detailed in Section 5.3.4.

5.3.3.1 Image Acquisition and Reconstruction

For imaging, the NMR signal must carry the spatial information related to the location within the magnet. As illustrated in Equations 5.3.9 through 5.3.11, it is done by the application of a linear magnetic field gradient (G) so that the magnetic field becomes dependent on the position (s) and so does the frequency at which the spins precess (ω_s).

$$B_s = B_0 + G \cdot s \qquad (5.3.9)$$

$$\omega_s = \gamma \cdot B_s \qquad (5.3.10)$$

$$\omega_s = \gamma \cdot (B_0 + G \cdot s) = \omega_0 + \gamma \cdot G \cdot s \qquad (5.3.11)$$

When the frame of reference rotates at the angular frequency of the spectrometer, that is, $\omega_0 = \gamma B_0$, the frequency of precession at position s is given by $\gamma < G < s$. The resultant phase Φ_s is a function of the length of time during which the gradient is on.

In the conventional MRI sequences, three linear magnetic field gradients are imposed along the x, y, and z directions, respectively, to spatially encode the signal. When selecting a slice perpendicular to the main magnetic field direction, the first step consists of turning on a field gradient (G_s) along the B_0 direction, which is the z direction by convention. At the same time, a slice-selective RF pulse is applied to excite the spins precessing in a range of frequencies determined by the bandwidth of the RF pulse and the gradient strength. The selected plane refers to the field of view (FOV) whose characteristic features are the in-plane dimensions in square or rectangular geometry and the thickness.

The selection of the plane on which the FOV lays is carried out to assure that the region of interest is included in the image. Three main orientations are generally selected for the FOV. According to medical literature, slices that lie on the x–y plane are called axial or transversal, those lying on the y–z plane are called sagittal, and those on the x–z plane are called coronal. No further implications are involved unless the sample is in motion during image acquisition. Such conditions will be discussed later in this section.

Immediately after the first slice-selective RF pulse, a gradient along the y direction (phase encoding gradient, G_Φ) is switched on making rows within the FOV (and perpendicular to the gradient direction) to precess at different frequencies. After a time interval τ the gradient is switched off. The spins return to precess at the same frequency but the phase differences between rows persist. Next, a gradient along the x direction (frequency encoding or readout gradient, G_f) is applied during signal acquisition. This gradient makes each FOV column to precess at different frequencies, while the phase differences achieved in the phase-encoding direction are preserved. Thus, an echo is generated. The number of sampled points within the echo equals the number of encoded columns.

Spectroscopic Methods for Texture and Structure Analyses

The volume elements identified by pairs of row and column numbers are called voxels. Their phase will be imposed by their location within the FOV (x, y). The received signal over time is the sum throughout the FOV of the transverse magnetizations at each voxel with their corresponding encoded phases (Equation 5.3.12).

$$S(t) = \int \int m(x, y) e^{-i\gamma G_\Phi y \tau} e^{-i\gamma G_f x t} \, dx \, dy \qquad (5.3.12)$$

where
  $m(x, y)$ is the transverse magnetization of the voxel under spatial coordinates x and y
  $\gamma G_\Phi y \tau$ and $\gamma G_f x t$ refer to the acquired phase owing to the phase and frequency encoding, respectively (Figure 5.3.5)

This timing schedule is repeated over time at intervals termed repetition time (TR). After each RF pulse, the intensity of the phase encoding gradient is made to vary obtaining finally as much echoes as number of encoded rows. The image information is not encoded directly, the digitized values of each echo are sequentially stored in rows resulting in a 2D array, the so-called k-space (the k is by analogy wavenumber) or reciprocal domain, which can be regarded as a map of spatial frequencies (k_x and k_y) with the correspondence indicated in Equations 5.3.13 through 5.3.15. The k-space presents conjugate symmetry for positive and negative frequencies (in mathematical operations with complex notations, signed frequencies stand for rate and direction of rotation).

$$S(t) = S[k_x(t), k_y(t)] \qquad (5.3.13)$$

$$k_x(t) = \int_0^t \gamma G_f(\tau) \, d\tau \qquad (5.3.14)$$

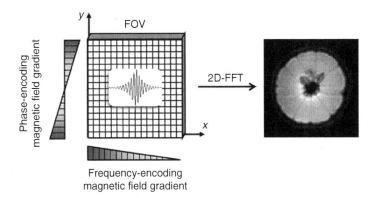

FIGURE 5.3.5 Schematic of signal spatial codification using linear magnetic field gradients along the x and y directions of the FOV and example of a reconstructed MR image after application of a 2D-FFT.

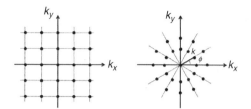

FIGURE 5.3.6 *k*-space sampling by Cartesian coordinates (left) and polar coordinates (right). For the latter data, interpolation is required before image reconstruction to obtain a Cartesian grid.

$$k_y(t) = \int_0^t \gamma G_\Phi(\tau)\,d\tau \tag{5.3.15}$$

For conventional MRI sequences, the magnetic field gradients are applied such that *k*-space is filled in a rectangular Cartesian way. In other sequences, the reciprocal domain data are collected using spiral or radial trajectories (Figure 5.3.6). Advantages and disadvantages are discussed later in the text. The total time required to complete data acquisition ranges from hours to tens of milliseconds depending on the type of acquisition parameters (number of phase cycles and number of cycles per TR) and hardware specifications of the equipment. To enhance the applicability of MRI technique for monitoring purposes, a compromise between total acquisition time, contrast, and spatial resolution has to be reached.

A 2D MR image is reconstructed by applying 2D fast Fourier transformation (2D-FFT) to the *k*-space. The resulting images are matrices, usually 2D (one or multiple) and 3D, consisting of an array of pixels whose intensity depends on acquisition parameters and different properties, such as the PD contained in the sample, diffusion, relaxation properties, etc. (Figure 5.3.5).

5.3.3.2 Effect of Movement on Image Quality

Internal quality evaluation under on-line conditions implies that the samples are conveyed through the magnet along the direction of the main magnetic field (*z* direction). When the volume to be excited by the RF pulses, the FOV, lies on the axial location (*x*–*y* plane in our scheme of magnet), that is, perpendicular to the motion axis, signal arising from adjacent tissue slices passing through it is registered during the acquisition time. Miscellaneous signal produces blurring artifact in the image that cannot be corrected. Such images only will be useful as long as the signal superimposition does not conceal the structures of interest within the sample under inspection owing to image blurring.

Coronal and sagittal locations of the FOV (*x*–*z* and *y*–*z* planes, respectively) assure the acquisition of the signal from the same slice of tissue for all the phase-encoding steps. In this case, the slice remains within the volume delimited by the FOV, although its position is changing during the acquisition since the sample moves forward. Such motion produces a measurable phase shift in the raw data that is susceptible of being corrected according to the Fourier shift theorem.

This theorem states that if a function $f(x)$ has an FT, $F(s)$, when $f(x)$ is shifted Δx, the resulting FT, $F'(s)$, will be $e^{-i2\pi\Delta xs}F(s)$ (Bracewell 2000). That is to say the signal from the moving sample is equivalent to that of the stationary sample $F(s)$ multiplied by a phase factor $e^{-i2\pi\Delta xs}$, which depends on the displacement of the object (Korin et al. 1989). Note that $F(s)$ is a complex number $[R + iI]$ and so is $F'(s)$ $[R' + iI']$.

The phase shift ($\Delta\Phi$) at any point of the k-space is mainly a consequence of the sample motion between phase-encoding steps. The expression for its computation is analogous for displacements along the phase-encoding direction and along the frequency-encoding direction (Equation 5.3.16). Such displacements derive from the velocity of the moving object along the corresponding direction and the time elapsed between consecutive phase encoding steps, v and t, respectively, in Equation 5.3.16.

$$\Delta\Phi = 2\pi \cdot vt \cdot \left[s - \left(\frac{N-1}{2} \right) \right] \cdot \frac{1}{N} \qquad (5.3.16)$$

For displacement in the phase-encoding direction, s is the number of the phase encoding step while N is the total number of steps. For displacement in the frequency-encoding direction, s is the point within the echo and N is the total number of echo points acquired. The equation establishes the reference of the phase shift in the central phase-encoding step and it increases gradually for the others. When the image is reconstructed from the corrected k-space, the sample will be positioned at the location occupied during the acquisition of that reference step.

The corrected values (R_c and I_c) corresponding to the real (R') and the imaginary (I') components of the complex points comprising the k-space acquired during sample motion are provided by implementing Equations 5.3.17 and 5.3.18.

$$R_c = R' \cdot \cos(\Delta\Phi) - I' \cdot \sin(\Delta\Phi) \qquad (5.3.17)$$

$$I_c = R' \cdot \sin(\Delta\Phi) + I' \cdot \cos(\Delta\Phi) \qquad (5.3.18)$$

Under given on-line conditions, that is, at a given belt speed, given acquisition settings such as TR and given number of pixels in the image along the motion direction, the induced phase shift is unique and straightforwardly computed. Therefore, correction algorithms can be generated for different experimental conditions so that the correction procedure could be standardized. Thus, its use could become routine to correct rectangular k-space of any MR image in the framework of on-line applications. The major requirement for the application of standard correction algorithms is the existence of a smooth and stable conveyor belt speed, since vibration and belt jerks would cause image blur that are more difficult to correct.

Even though motion correction is theoretically feasible for any belt speed, there are limiting FOV restrictions for increasing velocities. For coronal or sagittal images, the FOV has to be enlarged as to cover the sample displacement during signal acquisition, which increases at higher speeds. Consequently, distance between samples would have to be lengthened. For axial orientation, the distance covered by the sample while passing through the FOV thickness is shorter, which allows nearer placement of the fruits, and hence, raise the number of fruits inspected per unit time. However, as aforementioned, the image blurring cannot be corrected for this type of images. Therefore, a compromise between increasing performance and decreasing image quality should be achieved.

5.3.3.3 Sequence Parameters and Their Effects on MR Image Quality

The image quality with regard to the MRI techniques is defined by the contrast of the images, the ability to spatially resolve detail, and the SNR (Woodward 1995). This section provides a summary on several parameters that affect such quality image characteristics.

5.3.3.3.1 Effect on Contrast

In MRI, contrast is the difference in relative brightness between pixel units. Brightness is directly related to the signal intensity received from each voxel, which is translated into gray or false color scale ranges (Woodward 1995). Several sequence parameters affect the signal intensity differences between tissues, such as the TR, the echo time (TE), and the flip angle.

The TR is the TI between successive cycles of accumulations or phase steps. During this time, the longitudinal magnetization recovers and signal becomes available to be flipped into the transverse plane. Therefore, long TR implies higher signal. However, tissue contrast is poor if all tissues have recovered a similar amount of longitudinal magnetization. Therefore, to enhance tissue contrast through T_1 differences (T_1-weightening), TR needs to be shortened. Thus, tissues with short T_1 values recover faster and contribute with higher signal in the next excitation, while those with longer T_1 values will have varying recovering rates and so a variation in contrast.

The TE is the time elapsed between the excitation RF pulse and the maximum signal of the formed echo. TE controls the amount of dephasing between transverse magnetizations, and hence, the loss of signal. Long TE results in lower signal although enhanced T_2-weightening. Those tissues with long T_2 time take a longer time to dephase so that they will appear brighter in the image than those with faster loss of coherence. Moreover, the T_2 effect predominates when TR is lengthened as T_1 contrast minimizes.

Flip angle is the angle rotated by the longitudinal magnetization toward the transverse plane after RF excitation pulse. For short flip angles the T_2 contrast predominates. As flip angle increases, differences in longitudinal magnetization recovering between components intensify and those components with short T_1 will appear increasingly brighter. This parameter is in close relation with TR and usually their values are chosen accordingly.

There are other sequence-related parameters that affect tissue contrast that are not considered here, such as inversion time, diffusion parameters (diffusion gradient time and strength), number of accumulations, presence of magnetization transfer or

Spectroscopic Methods for Texture and Structure Analyses

magnetization saturation pulses, etc. Information on these can be found in other reviews or book chapters (e.g., Bernstein et al. 2004).

5.3.3.3.2 Effect on Spatial Resolution

The spatial resolution is an image characteristic related to the minimum size of an object that can be identified and to the clear definition of edges and boundaries between regions (Woodward 1995). The intensity of the magnetic field gradients [T/m] used for in-plane spatial encoding of the signal within the FOV alongside the FOV thickness defines the number and dimensions of a series of volume elements into which the FOV is divided, which are called voxels. The smaller the voxel, the higher the resolution appears, although a loss in the SNR is concomitant. For equal in-plane voxel dimensions, an increase in the slice thickness involves an increase in the signal as it arises from a higher excited volume. However, thicker slices may contain different tissues with different signal intensities so that their overlapped contributions lead to a signal misregistration during FT. This effect is called partial volume effect.

The FOV dimensions and the size of the raw data matrix also are involved in the spatial resolution. The former is selected to cover a particular tissue volume of interest. The latter is defined by the number of phase-encoding steps and the number of sampled points required to yield a desired spatial resolution. For a certain spatial resolution, the sampling size and FOV need to be chosen accordingly with the criteria given by the Nyquist rate to avoid image aliasing. The larger the matrix dimensions and the in-plane voxel dimensions, the larger the FOV. Changes in FOV while the matrix dimensions are the same affect the in-plane voxel size in such a way that increasing FOV means increasing voxel size.

5.3.3.3.3 Effect on Signal-to-Noise Ratio

The SNR is given by the relative contributions to a detected signal of the true signal and randomly superimposed signals (background noise). SNR is affected by many sequence parameters. Flip angles are related to the transversal magnetization after RF excitation pulse. Thus, increasing flip angles produce increasing transversal magnetization. One common method to enhance the SNR is to average several acquisitions or pulse sequence repetitions, as signal sums up and random contributions cancel out. The SNR increases with the square root of the number of acquisitions used for averaging. The SNR can also be improved by sampling larger volumes, that is, by increasing the FOV and slice thickness (with a corresponding loss of spatial resolution) or, within limits (e.g., relaxation properties, susceptibility artifacts, coils, etc.), by increasing the magnetic field strength of the NMR spectrometer. It is worthy to note that an improvement in SNR does not necessarily result in a net improvement of the image quality as SNR is inversely related to spatial resolution.

5.3.3.4 Fast and Ultrafast MRI Sequences

For food evaluation, speed is required under an economical interest as it increases the inspection performance in terms of inspected sample unit rate, whereas under a technical point of view there is a need for minimizing the sample motion effects, which are present whenever an on-line inspection is carried out.

The basic imaging time equation for a conventional MRI technique is given by Equation 5.3.19.

$$\text{Scan time} = \text{TR} \times \text{NPE} \times \text{NA} \qquad (5.3.19)$$

where
TR is the repetition time between successive excitation RF pulses
NPE stands for the number of phase encoding steps
NA refers the number of acquisitions to be averaged to enhance SNR

Therefore, the reduction of the total scan time may be approached with several strategies by managing the three parameters involved. The shortening of TR is favored by the use of short flip angles, although it is limited by gradient strength, which is normally an expensive engineering challenge. A different approach is using fractional or partial k-space filling to exclude redundant data. By acquiring fewer lines of k-space or, in other words, NPE is reduced, uncollected data need to be repopulated by performing a zero filling or by duplication of data taking advantage of the conjugate symmetry of the k-space. For zero filling, central lines of the k-space are acquired so that image contrast is not altered, whereas those corresponding to the highest phase-encoding steps, which present weak intensity signal, are substituted by strings of zeroes.

An important way of time reduction relies on the possibility of acquiring multiple k-space lines per TR, which is also called segmented k-space filling and has led to a family of sequences broadly recognized as echo train imaging. The scan time of these images is reduced by a factor that depends on the number of lines per segment obtained consecutively before the next excitation RF pulse. In addition, optimized k-space sampling techniques can contribute to speed acquisition times. Decrease in scan time may also be achieved by reducing the number of acquisitions used for signal averaging (NA). In fact, under on-line monitoring conditions NA is restricted to 1. In addition, more than one sample could be imaged within the same FOV.

The counterpart of the actions devoted to time shortening usually is associated with a decrease of the SNR and the spatial resolution or to an increase of susceptibility effects on signal decay.

5.3.3.4.1 One-Dimensional Image Profiles
One-dimensional image profiles are obtained by sampling with a single linear field readout gradient applied along the direction of interest. 1D FT provides a profile where each point corresponds to the signal average of the entire volume located at the position encoded by the applied gradient. Tissue differences are reflected in the profile, which may be useful for detecting the presence, or the absence, of internal structures such as pits in fruits or foreign bodies in fluids. This acquisition is faster than 2D or 3D imaging because it does not require repetitive sampling with intensity varying phase-encoding gradients. Therefore, motion-induced artifacts may be neglected, which enhances its applicability to dynamic measurements. In addition, the SNR is higher owing to the registration of signal from larger volumes. The main

disadvantage refers to the partial volume effect, which is greatly enhanced due to the average tissue signal so that small structures may be concealed by the surrounding tissue.

5.3.3.4.2 Fast Low-Angle Shot

The fast low-angle shot (FLASH) sequence combines a low flip angle RF excitation pulse and a reversed gradient instead of a 180° RF pulse, which refocuses the spins to form a gradient echo. As the excitation pulse angle is lower than 90°, the residual longitudinal magnetization permits shorter TR since it is not necessary to wait for its recovery by relaxation. The result of these arrangements is a significant scan time reduction up to 100 ms acquisition time. The k-space is filled linearly with one single k-space line being sampled after each RF pulse. As data are sampled on a rectangular grid, the reconstruction complexity is reduced. Such disposition facilitates the computation of the signal phase shift caused by sample motion and the application of correction procedures. However, the SNR is low. In addition, the T_2^*-weighting involves sensitivity to magnetic field distortions and to susceptibility effects, although internal quality inspection could take advantage of such susceptibility effects since tissues contrast may be enhanced, which would facilitate identification.

5.3.3.4.3 Echo Planar Imaging

The echo planar imaging (EPI) is the best example of echo train imaging (see above). The exceptional shortening in acquisition time that characterizes this sequence (up to tens of milliseconds) is achieved by the possibility of collecting part or even all the data points comprising the k-space after a single RF excitation pulse (compromising contrast, resolution, and image distortion). In the classical EPI pulse sequence, the image is acquired from one or several FIDs by applying a constant weak phase-encoding gradient and creating a series of echoes with strong, rapidly switched frequency-encoding gradient. As we have said, with modern gradient and RF hardware, EPI is capable of producing a 2D image in only a few tens of milliseconds (Bernstein et al. 2004). Such short times significantly overcome the problem of image degradation caused by the sample motion. However, the technical requirements in software and hardware such as those regarding the gradient speed, strength, eddy currents screening, controllers, amplifiers, etc. are significant. The total data acquisition must be performed within the T_2^* of the tissue, otherwise the signal is destroyed before sufficient information is acquired (Vlaardingerbroek and den Boer 1996). In these cases, image artifacts are considerable and the only alternative is to reduce the number of echoes in the echo train. Thus, the spatial resolution is limited by the number of gradient echoes that can be acquired during the T_2^* signal decay. Artifacts in EPI sequences are as k-space line carries a different T_2^*-weighting since the k-space lines are acquired at different times, which causes image blurring along the phase-encoded direction (Bernstein et al. 2004).

5.3.3.4.4 Rapid Acquisition Relaxation Enhanced

Rapid acquisition relaxation enhanced (RARE) is a good example of SE train. It was the first multiple SE based one to fill more than one k-space trajectory in a single excitation (Parikh 1992). This fact reduces the acquisition time, which diminishes the motion-induced artifacts. RARE employs a train of 180° pulses to refocus decaying echoes. The k-space filling can be accomplished in one shot by a long train of RF pulses or in multiple shots consisting of shorter trains of RF pulses. As EPI, the contrast is strongly dependent on T_2 relaxation because of the application of the 180° refocusing pulses providing an excellent T_2-weighted tissue contrast.

RARE images are less sensitive to main magnetic field inhomogeneities and tissue magnetic susceptibility variations than EPI sequences (Bernstein et al. 2004). The reconstruction is straightforward because of the rectilinear k-space.

Some disadvantages are related to the T_2 signal decay, which causes image blurring in the phase-encoding direction, and the presence of the 180° pulses that limit the minimum spacing between successive echoes (Parikh 1992). In addition, RARE sequence presents high requirements on software and hardware such as also found for EPI images since fast application of the 180° refocusing pulses is required.

5.3.3.4.5 Spiral or Radial k-Space Acquisition

Spiral k-space sampling is accomplished by the combination of two increasing, oscillating gradients. This modality completely eliminates rapid gradient switching required in sequences such as EPI, which means much less demanding requirements on gradient hardware. Here no echo procedure is required (Delpuech 1995) and there is no relevant differences between the k_x-axis and k_y-axis (Rodríguez et al. 2004). Data in k-space are identified by means of polar coordinates, that is, the angle φ with respect to the k_x-axis, and the distance k with respect to the origin of the (k_x, k_y) frame (Figure 5.3.6).

In spiral acquisitions, the trajectory of each readout step always starts at the origin of the k-space, and ends at its edge. Because of the geometry of the trajectory, the central region, where contrast information is contained, is rapidly acquired and oversampled. Such characteristics confer to the spiral and radial sequences robustness with respect to motion-induced phase errors, susceptibility, and T_2^* decay (Vlaardingerbroek and den Boer 1996), and for spiral acquisition allow ultrafast imaging with total acquisition time of tens of milliseconds with fine contrast.

As for EPI, traversing the complete k-space within the time T_2^*, while acquiring sufficient sampling points with one single spiral, is very difficult. Therefore, interleaved techniques are applied by acquiring a number of similar spiral arms after multiple shots, which are rotated with respect to each other (Vlaardingerbroek and den Boer 1996).

Radial k-space sampling can be considered a particular case of the spiral modality. Rodríguez et al. (2004) reported the development of a new imaging sequence called combined spiral and radial acquisition (COMSPIRA) that allows choosing between spiral and radial center-out k-space trajectories in two and three dimensions (Pérez-Sánchez et al. 2006). The new design includes two parameters that allow interleaved acquisition and sequence switching. The first is a measure of the angular difference between one readout step and the next, and the second one defines the angular difference between the beginning and the end of a readout step in a number of complete 2π radians, which is called spirality. For radial trajectories spirality is zero.

Data acquired along a spiral or radial trajectory entail special algorithms for image reconstruction, such as data interpolation to map the spiral data to a Cartesian grid.

5.3.4 Detailed View of Applications in Fruits

In this section, a number of applications of MRS, MRR, and MRI have been selected for a detailed view concerning a variety of fruits (avocados, cherries, olives, apples, pears, and citrus), as icons for the industrial sector.

Some of the applications are focused on the classification of fruits in a limited number of categories (seedless or seed-containing citrus, pitted and unpitted cherries), while others face the quantification of disorder degree (freezing injury, internal browning or breakdown, mealiness), or even the chemical composition of fruits (oil content in avocados). Also statistical features such as classification rates and correlation coefficient are provided for each case as a mean for the quantification of success.

Most of the applications that will be presented in detail and all the techniques (MRR, MRS, and MRI) have been investigated under an on-line configuration that is without doubt a main industrial demand, and a challenge for the practical interest of NMR in the field of horticultural quality assessment.

5.3.4.1 Maturity in Avocados

Chen and coworkers (1996) designed and built a conveying system to evaluate high-speed MRS technique in fruits and vegetables. Single-pulse spectra were obtained from moving avocados to determine the feasibility of the technique for quantifying the dry weight of the fruit at different conveyor speeds. The misalignment of the fruit with respect to the RF coil under both static and dynamic conditions, and the effect of the belt speed on the spectra were analysed.

The experiments were performed with a 2 T NMR spectrometer using a surface coil (Figure 5.3.7). The magnetic field was shimmed solely with the first sample. The static experiments were used to evaluate the effect of the fruit misalignment with respect to the center of the surface coil. The correlation between the oil/water resonance peak ratio and the percentage of dry weight of the avocados was 0.897 for zero misalignment. A simulation with 400 repetitions showed that a misalignment within ±4 mm from the surface coil would not significantly affect the peak ratio and hence, the correlation. Under motion conditions, the RF pulse was activated when the fruit was at position ranging from −10 to +10 mm. Results showed that the strongest signal was acquired when the sample was centered. The belt speed varied from 0 to 250 mm/s and the shape of the spectra as well as the line width showed very little change over mentioned speed range. This result supports that the field remains sufficiently homogeneous throughout the distance range even though shimming was only performed with the first sample. The correlation between dry matter and oil/water peak ratio showed an average value of 0.975 and a SD of 0.004. According to these results, authors concluded that the method was rapid and accurate and had a great potential for on-line sensing of avocados maturity.

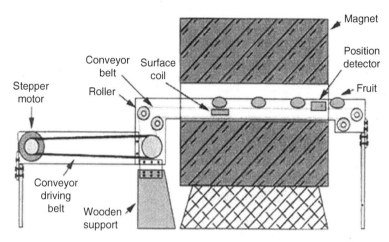

FIGURE 5.3.7 Schematic of an MR sensor with fruit conveying system for on-line fruit quality sensing. (Reprinted from Kim, S.M., Chen, P., McCarthy, M.J., and Zion, B., *J. Agric. Eng. Res.*, 74, 293, 1999. With permission.)

Kim and coworkers (1999) extended the previous work on avocados by using the same NMR equipment and conveyor belt and similar experimental procedure. The conveyor speed was varied from 0 to 250 mm/s and spectra were acquired at five different positions relative to the coil center, computing three different peak ratios. This study also showed that the maximum intensity and width of the peaks changed with relative location and belt speed. These authors explained that, for reduced coupling between sample and coil, the excitation of the nuclei of the sample decreases, leading to a decrease in the efficiency of signal recording and thus to a reduction of signal intensity together with an increase in the line width of the spectra. This fact was overcome by computing the peaks ratio, which revealed to be much less sensitive to belt speed and position. Increasing belt speed and sample displacement led to slight decreases in correlation coefficients between peaks ratio and dry weight, from 0.970 under the best conditions to 0.894 under the worst conditions. These results indicated, as in previous studies, the small effect of the motion of the sample on the acquired signal when single-shot sequences are used, as well as the great potential of the technique for this application.

Different studies have alerted the problems of using surface coils for signal acquisition on avocados, since these involve the excitation of the volume of tissue closest to the coil. As indicated by Pathaveerat and coworkers (2001), when such procedure is implemented for on-line systems either the composition of the sample should be uniform or the spatial variation in composition need to be known. To answer these questions, chemical shift images, consisting of a spatially resolved high-resolution NMR spectrum, were acquired with a Bruker Biospec 7 T spectrometer from avocados ranging from immature to very mature. It was found that the water distribution is higher near the skin and seed and lower in the middle of the flesh, and that the oil distribution increases toward the seed. As for the oil/water peaks ratio, water was found large along the middle ring of flesh near the seed and lower to the periphery. These results highlighted the importance of the knowledge on the distribution of the compound of interest within the sample that allows establishing optimum procedure settings for maximum accuracy.

5.3.4.2 Pit in Cherries and Olives

Zion and coworkers (1994) focused on real-time detection of pits in processed cherries by 1D MRI profiles. In this case, a superconductor magnet with a magnetic field strength of 2 T was used, together with a homemade volume coil, for the acquisition of NMR projections. A permanent magnet operating at 0.26 T was also used to determine the effective transverse relaxation times (T_2^*) of pits and flesh in cherries as prospecting study for enhancing the signal.

Based on the relative size of flesh and pits, a computer simulation of the projections of cherries with and without pits was performed with the aim of optimizing the shape of the projections for various slice thickness and various slice offsets.

For real applications, three cherries at a time were placed inside the magnetic field roughly aligned along the sagittal plane (Figure 5.3.8). In the first set of experiments, cherries were placed randomly, which is the most likely disposition owing to the difficulties in selecting orientation when conveying small-rounded fruits. A second set was obtained from pre-oriented cherries with the whole axis located perpendicular to the excited plane. The magnetic field was shimmed only once since, as explained by the authors, this operation is time consuming and could not be repeated under on-line specifications. An SE pulse sequence was used with no signal averaging. A devoted Matlab routine was implemented as to identify the peaks and valleys in the projections and to calculate their ratios enabling to overcome those problems arising from noise, shape distortions, and missing flesh parts. The SNR obtained with only one signal accumulation was enough to obtain reliable results.

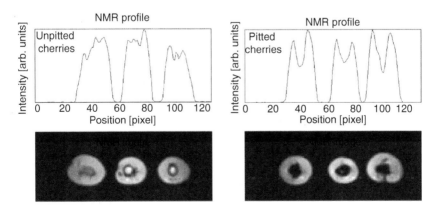

FIGURE 5.3.8 Examples of MR projections of a set of three cherries with pits (left) and a set of three pitted cherries (right) where pronounced valleys are observed for the latter samples. On the bottom, MR images of the corresponding cherries. (Courtesy of B. Zion, unpublished data.)

The classification routine yielded an 88% and 63% of correct classification for pit-containing and pitted cherries, respectively, corresponding to randomly placed samples. Logically, classification performance was improved for pre-oriented cherries, 96.7% of correct classified cherries either with or without pits. The simulation indicated that excitation of a narrow slice is preferable because it is less affected by shifts of the cherry from the excited plane, which is likely to happen under on-line conditions.

The short acquisition time of the projections ranged between 10 and 15 ms. Such fast data processing and the possibility of accommodating wider number of cherries in a longer volume coil reveal the high potential of this procedure to be transferred to industry application. However, the effect of the motion was not addressed in this study.

Kim and coworkers (1999) obtained 1D MRI profiles using the conveyor system presented in Chen et al. (1996) for studies on avocado, although making use of a volume coil instead of a surface coil. As before, only one acquisition was made for each projection. For comparison, cherries were placed with their axis both perpendicular and parallel to motion direction. Projections were acquired from 0 up to 250 mm/s belt speed. Sagittal imaging plane was chosen with the frequency-encoding gradient applied along the direction of the motion. Increasing motion again involved a decrease in signal intensity owing to relaxation effects. In contrast, there was little or no distortion in the shape of projections. The best results to discriminate fruit between pitted and unpitted cherries were obtained when the pit was orientated perpendicular to the motion direction. The classification rule was based on a simple threshold yielding no errors for whole cherries under motion conditions, whereas 71.7% error was found for pitted cherries placed with their axis parallel to the motion, and 1.7% for pitted cherries placed with their axis perpendicular to the motion. The classification performance was very similar to that of the static conditions (1%, 70%, and 2% for the same samples and placements). Despite these encouraging results, the authors pointed out the need for further improvements in the conveyor design and control, as well as in the sorting algorithms.

Zion and the working group applied the same procedure to the detection of pits in olives (Zion et al. 1997). Olives were conveyed at belt speeds up to 250 mm/s achieving classification errors in segregating pitted and non-pitted olives lower that 5%.

The use of profiles for the detection of pits in processed cherries and olives is justified by the large proportion of volume occupied by the pit within the piece, which involves significant pronounced valley appearing for pitted units in comparison with unpitted units. In contrast, the detection of small structures within larger samples (such as seeds in citrus) requires 2D imaging to avoid partial volume effect.

5.3.4.3 Internal Browning in Apples

Internal browning is a physiological disorder characterized by the development of brown discoloration areas throughout the cortex and core, and the formation of cavities without any external symptom. A high concentration of CO_2 and low levels of O_2 within the fruit are factors to induce this disorder. Thus, CA storage intensifies the occurrence of the internal browning, although it may also appear during the preharvest period. Other factors such as variety, growing regions, and cultural activities are also implicated in the development of this disorder.

Clark and Burmeister (1999) performed an MRI study on the development of browning in 'Braeburn' apples using a 1.5 T clinical instrument. Series of SE T_1-weighted images with an in-plane resolution of 0.35 mm/pixel (4 mm thick) were obtained from fruits during 28 days of CA storage under high CO_2 concentration. Nine images at different locations were acquired in each series with a total imaging time of 7.6 min.

In damaged apples, discrete areas of flesh appeared as pixels with higher intensity signal when compared to the background tissue. Their signal intensity increased over time, while the patches either enlarged or merged to form larger aggregates. When MR images were compared with photographs of dissected apples after the 28 days of storage, the highest MRI signal areas coincide with the brown damaged tissue. The increasing MRI signal was attributed to a decreasing mobility as a result of the concentration of metabolites such as acetaldehyde and ethanol produced by respiratory processes. However, authors also highlighted the difficulty of image contrast interpretation as changes in signal could also derive from changes in proton densities.

González and coworkers (2001) extended the study on internal browning in apples by identifying the individual contribution of the PD, longitudinal relaxation time (T_1), and transverse relaxation time (T_2) to the image contrast. Experimental work was carried out with a 0.6 T magnet. SE MR images of 'Fuji' apples were obtained along with T_1, T_2, and PD maps where a quantitative value of the corresponding parameter is displayed for each voxel.

MR images were compared to photographs captured after cutting in half the same apples. Brown-discolored areas in the pictures showed good correlation with the regions in the MR images presenting intensity signal different from that of the normal tissue. Different regions were identified on the basis of different ranges of intensities. Dark brown and light brown areas show high- and low-signal MRI intensity, respectively, when compared to sound tissue. Characteristic T_1, T_2, and PD values were computed for each region by averaging the results obtained from these areas in the corresponding maps. In damaged tissue, a reduction was observed in both the T_1 and the PD. As for T_2, light brown regions had shorter T_2, whereas the dark brown regions presented the highest relaxometric values. These values along with the experimental imaging parameters allowed calculating their relative contribution to the signal of each region. It was found that the transverse relaxation time contributed more to the contrast than PD and T_1.

To reduce the total acquisition time, resolution was diminished in the phase-encoding direction driving to a significant shortening in image acquisition time of two apples from 5.5 min to 20 s. Despite the increasing image blurring, the identification of affected regions was still feasible (Figure 5.3.9).

Jung and coworkers (1998) proposed a different approach for the detection of internal browning in apples. In this case, a global parameter, the transverse relaxation time (T_2), is computed to characterize the whole sample, instead of doing it regionally on each tissue disorder. Apples were placed in a 0.13 T permanent magnet and a

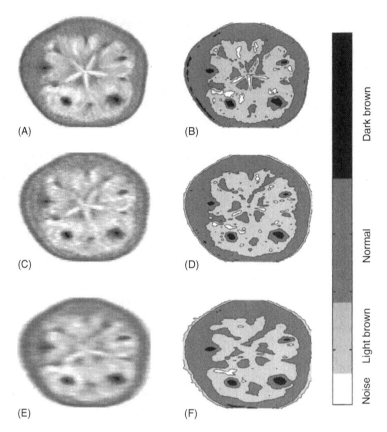

FIGURE 5.3.9 Example of SE MR images (A, C, and E with total acquisition times of 5.5 min, 40 s, and 20 s, respectively) acquired from an apple affected by internal browning. The corresponding contour maps (B, D, and F) were used for the identification of normal, light brown, and dark brown tissues. (Reprinted from González, J.J., Valle, R.C., Bobroff, S., Biasi, W., Mitcham, E.J., and MacCarthy, M.J., *Postharvest Biol. Technol.*, 22, 179, 2001. With permission.)

CPMG pulse sequence was used to compute the overall T_2 value. Results showed that the average T_2 value of the apples with internal browning decreased as the severity of the disorder increased. T_2–CPMG relaxation curves were fitted to a three-exponential model to associate T_2 values and proton relative populations with three different compartments, that is, the vacuole, the cytoplasm, and the extracellular space. Differences in proton spin density in the extracellular space were found between healthy apples and those presenting internal disorders. The properties or factors that could justify the differences observed in these results such as the diffusion through magnetic field gradients, the magnetic susceptibility discontinuities or the physiological changes during the development of this disorder were not investigated.

Chayaprasert and Stroshine (2005) carried out a study on the feasibility of global measurements under on-line conditions. Experimental work was performed with low-cost, low-field (0.13 T) equipment. The design of the NMR acquisition system for dynamic measurements was based on that proposed by Chen et al. (1996). The authors also claim that they have approached the concerns exposed by Hills and Clark (2003) on the transferability of the NMR systems toward on-line conditions. Experiments were conducted on 'Rome' and 'Red Delicious' apples. Transverse relaxation time (T_2) was computed for whole healthy and affected apples by means of a CPMG sequence (90°–180° pulses with pulse interspacing of 600 μs). Because of the sample motion, the CPMG sequence could not be completed before the apple left the sensitive area of the coil. Apparent decaying time T_2 (AT_2) was obtained after curve normalization, together with the sum of the areas of the normalized echo peaks (SUM). In this study both parameters AT_2 and SUM were found lower for apples with internal browning than those of the healthy fruit, even though the differences between the relaxation curves of healthy and affected apples reduced as conveyor speed increased.

To evaluate the effect of sample motion, CPMG signal acquisitions were tested at different belt speeds up to 250 mm/s at apples misalignments from −6 to +6 mm with respect to the coil center. The increasing conveyor belt speed induced a decreasing in the initial signal amplitude and a more rapid decay of the signal (Figure 5.3.10). The decreasing initial signal amplitude was attributed to the low degree of sample magnetization achieved during the shorter time inside the magnet. Normalization of different curves eliminated the variations in initial amplitudes (Figure 5.3.10). In addition, at increasing conveyor speed, samples get across the coil in less time, so that in some of the experimental conditions the pulse sequence is not fast enough to encode all data and thus the decaying curve is highly affected. Moreover, CPMG sequence does not compensate for the loss of phase coherence caused by magnetic field inhomogeneities along the direction of the sample motion. Thus, AT_2 decreases at higher conveyor speeds as larger distances are covered. Sample misalignments also exert an effect on the decay rate as consequence from the

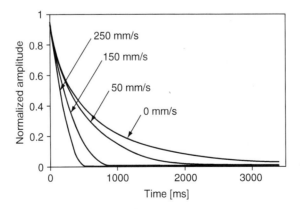

FIGURE 5.3.10 Normalized signal amplitude of the echo train in CPMG sequence acquired from an apple when conveyed at different speeds where increasing decaying rates are observed at increasing speeds. (Reprinted from Chayaprasert, W. and Stroshire, R., *Postharvest Biol. Technol.*, 36, 291, 2005. With permission.)

different times for each sample to pass across the RF coil and as the field inhomogeneities are not comparable.

The best classification performance was achieved at 50 mm/s by using the SUM parameter with classification error of 12% for 'Rome' apples and 0% for 'Red Delicious' apples. The extent of the damaged region was always higher than 40% of the whole tissue. Performance decreased to 20% of classification error at higher speeds. Misalignment also caused a significant increase of the errors. As a conclusion, from this study, it is possible to detect apples with internal browning at conveyor belt speed below 100 mm/s, as long as the misalignment of the samples is controlled. Recommended improvements to the system included a more robust conveyor system and modification of the magnet design to achieve a greater degree of pre-magnetization. Authors also highlighted the need of determining the effects of spacing between samples on the performance of the system under continuous operating conditions.

5.3.4.4 Mealiness in Apples and Wooliness in Peaches

Mealiness is a negative attribute of sensory texture that combines the sensation of a desegregated tissue with the loss of crispiness and a lack of juiciness without variation of the total water content in tissues. For mealy apples, these sensations are the result of the weakness of the middle lamella that involves cells separation instead of rupturing with releasing of aqueous content. Peach mealy textures are also known as wooliness. In this case, the characteristic lack of juiciness derives from a different origin. Under disorder-inducing conditions, cell membrane permeability is altered and pectin accumulates in cell walls and intercellular spaces. Those pectins interact with calcium, among other ions or molecules, and generate gel structures that retain the water molecules. In addition, cell adhesion is reduced, which also means a lack of crispiness. This physiological disorder is associated with inadequate overextended cold storage.

Barreiro and colleagues (1999) assessed mealiness in apples through spatially resolved MRR as T_2 maps corresponding to a centered tissue slice. Experiments were conducted in a 4.7 T magnet with stationary samples. Several parameters were extracted from the maps themselves such as the minimum, the average, and the maximum T_2 values with their associated SDs. The minimum T_2 value was the only parameter that showed significant differences between the non-mealy and the mealy texture stage of the fruit being shorter for affected apples. Differences were also found in the corresponding histograms. For mealy apples, histograms were skewed to shorter T_2 times, whereas the distribution of the T_2 values of sound apples was normal. In addition, the histogram of mealy apples showed a tail in the region of the highest values. A new study (Barreiro et al. 2000) also based on spatially resolved MRR reinforced previous results. In this case, more parameters were extracted from the T_2 histograms and an analysis of variance revealed those with the highest effect on texture categories, that is, fresh, intermediate, and mealy fruits. Such parameters entered to a stepwise discriminant analysis to create classification functions that segregate between categories. An 87% of well-classified apples was achieved by combining the mode height-to-interquartile range ratio, the maximum T_2 value, the number of pixels with T_2 value below 35 ms, the upper quartile, the mode height, and T_2 SD.

Mealiness (or wooliness) in peaches was studied on a set of fruits including crispy, intermediate (non-crispy but non-woolly), woolly, and very woolly peaches (Barreiro et al. 2000). Similar to apples, histograms extracted from the T_2 maps showed a decrease in T_2 values with disorder development (Figure 5.3.11).

The lack of applicability of spatially resolved MRR (T_2 maps) to on-line inspection owing to the long acquisition times redirected subsequent studies (Barreiro et al. 2002), which were focussed on rapid mealiness detection by non-spatially resolved MRR. Previous acquisition of T_2 maps (Figure 5.3.12) was complemented using CPMG experiments with a 90°–180° pulse spacing of 200 μs undertaken at 2.32 T on pieces of tissue extracted from apples presenting three levels of disorder, that is, non-mealy apples, medium stage apples, and mealy apples. The T_2 spectroscopic data showed three peaks. The dominant one, which also showed the longest T_2 value, was assigned to water in vacuole; the other two peaks were associated with cell wall and cytoplasm, respectively. The main relaxation time showed a significant decrease from 458.6 to 104.4 ms (repeatability of ±21.6 ms) for increasing mealiness level. These results highlighted a good prospect for on-line mealiness assessment.

In the same work, physical tissue properties related to tissue desegregation and dryness were analyzed with respect to T_2 maps parameters. Linear positive correlation ($r = 0.85$) was found between tissue hardness and minimum value at T_2 maps. Juiciness and SD of T_2 values in the corresponding maps showed nonlinear negative correlation ($r = -0.80$).

5.3.4.5 Internal Breakdown in Pears

Internal browning in pears is characterized by softening and browning of tissues and development of cavities. It is an important postharvest disorder that is observable only at the end of the commercial chain since the external appearance of the fruit is not altered, even when the characteristic brown colored tissue is widely spread from the core to the surrounding flesh. Many authors have identified the elevated CO_2 and decreased O_2 levels in the air composition during CA storage as the primary influence. Under such conditions, a decrease in antioxidant levels as well as in energy availability is induced so that membrane maintenance and free radical control are altered (Saquet et al. 2003; Veltman et al. 2003; Larrigaudière et al. 2004) in 'Conference' pears and in 'Blanquilla' pears, leading to cell decompartmentation and bringing browning reactions. Other characteristics favoring the development of the disorder include longer storage time along with overmaturity, heavy fruits, and hard tissue (Lammertyn et al. 2000) in 'Conference' pears. Thus, various names have been applied to call this disorder since it has been difficult to establish clear differences on the basis of similar symptoms. Among these are internal breakdown, brown heart, brown core, and core breakdown.

Several authors have used NMR techniques (MRI and MRR) to study breakdown development in 'Bartlett' and 'Conference' pears. Wang and Wang (1989) obtained SE images from 'Bartlett' pears with a 0.5 T NMR equipment. Bright

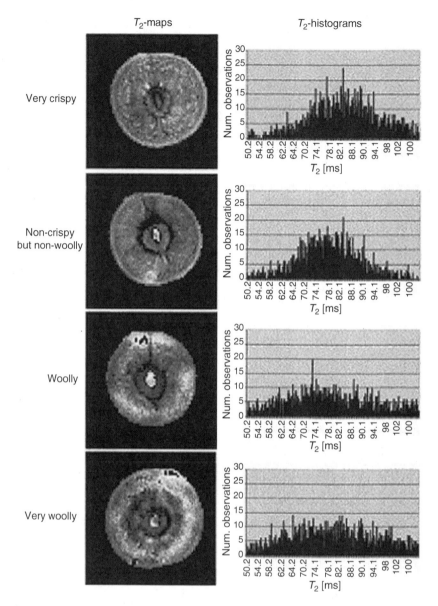

FIGURE 5.3.11 Examples of T_2 maps of a very crispy, a non-crispy but not woolly, a woolly, and a very woolly peaches, and the corresponding histograms where horizontal axis represents the T_2 values in milliseconds. (Reprinted from Barreiro, P., Ortiz, C., Ruiz-Altisent, M., et al., *Magn. Resonance Imaging*, 18, 1175, 2000. With permission.)

areas appeared in the MR images that corresponded to the affected region visually detected. The hyperintense signal was associated with an increase in the free water content. The presence of air spaces was revealed by dark pixel regions since PD decreases drastically in these spaces.

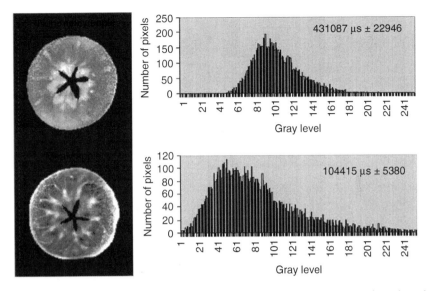

FIGURE 5.3.12 On the left, examples of MR images acquired from non-mealy and mealy apples. On the right, histograms from the corresponding T_2 map of each apple with increasing gray level corresponding to increasing T_2 values, and main T_2 peak. (Adapted from Barreiro, P., Moya, A., Correa, E., et al., *Appl. Magn. Resonance*, 22, 387, 2002.)

Lammertyn and coworkers studied the spatial distribution (2003a) as well as the time course (2003b) of the disorder in 'Conference' pears by means of MRI, spatially resolved MRR, and x-ray computer tomography. Series of T_1-weighted SE MR images were obtained, and PD and T_1 maps were constructed.

In contrast to results of Wang and Wang (1989), affected brown tissue appeared in T_1-weighted images as low-intensity signal regions, whereas healthy unaffected tissue showed high-intensity signal. Two patterns of browning, radial and local distributions, were observed.

The average PD values were found significantly higher for healthy tissue than for brown tissue, which indicated the occurrence of dehydration for the latter. In addition, shorter T_1 values were observed for the affected areas highlighting a more restricted motion, which involves higher pixel intensity on T_1-weighted images. Therefore, the net decrease in signal intensity for affected tissue pointed to the more important effect of the PD reduction on signal generation.

A hypothesis was proposed to explain differences with the results of Wang and Wang (1989). A higher free water level appears as cell membranes are affected by disorder-inducing conditions. When the moisture transport is faster than the cellular decompartmentation rate, the PD of the brown tissue is lower than that of the healthy tissue. Under faster membrane disintegration, more mobile protons lead to a signal intensity increase. The latter is prone to occur under extremely high CO_2 concentrations.

The study on the time course of this disorder (Lammertyn et al. 2003b) revealed that it does not grow spatially over time but only increases in severity and so in

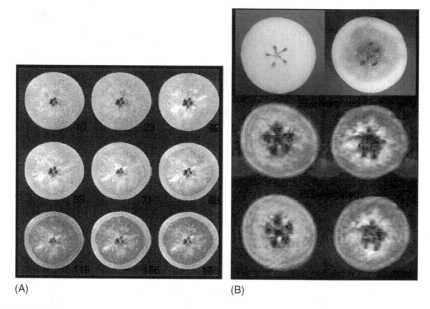

FIGURE 5.3.13 (A) T_1-weighted SE MR images from a stationary 'Conference' pear acquired along 178 days storage under core breakdown-inducing conditions. The numbers indicate days after harvest. (Reprinted from Lammertyn, J., T. Dresselaers, P. Van Hecke, P. Jacsók, M. Wevers, and B.M. Nicolaï, *Postharvest Biol. Technol.*, 29, 19–28, 2003b. With permission.) (B) Examples of T_2^*-weighted and PD-weighted FLASH MR images (703 ms and 484 ms acquisition time, respectively) after motion correction obtained from healthy (left) and affected (right) 'Blanquilla' pears while conveyed at 54 mm/s. (Reprinted from Hernández-Sánchez, N., Hills, B.P., Barreiro, P., and Marigheto, N., *Postharvest Biol. Technol.*, 44, 260, 2007. With permission.)

contrast magnitude (Figure 5.3.13A), and that cavities grow at the expense of the brown tissue. The authors of the study tentatively elucidated that the cavities formation is a result of the moisture transport toward the fruit boundary, which agrees with the decrease in PD and the subsequent image contrast obtained in their work on spatial distribution (Lammertyn et al. 2003a).

MRI offered higher sensitivity for detecting incipient browning than x-ray, since at the first stage of the disorder the cellular decompartmentation does not affect the density of the tissue but the mobility of the protons. In addition, a better contrast between affected and unaffected tissues was found for MRI inspection.

These earlier MRI and MRR studies were undertaken on stationary fruit using slow imaging sequences. Lately in 2007, Hernández-Sánchez and the working group performed a study on 'Blanquilla' pears where macroscopic dynamic MRI experiments were complemented with optical microscopy and with non-resolved MRR and NMR diffusion analyses to provide insight into the effect of the disorder on the microscopic tissue structure.

Optical microscopy confirmed that the cells lose their natural angular morphology and integrity in damaged tissues. This technique also showed that the whole tissue loses compactness and coherence. Accumulation of vesicles containing brown-colored compounds was also revealed by these microscopic images. A preliminary study performed on pieces of affected and healthy tissue, and juice from a healthy pear (Hernández Sánchez, 2006) demonstrated the great effect of the tissue microstructure on the transverse relaxation rate ($1/T_2$). CPMG sequences with varying pulse spacing pointed that such rate increases for altered tissue. This result was related to the water diffusion through magnified local gradients as a consequence of the magnetic susceptibility discontinuity across the phenolic vesicles and water interfaces. This effect was enhanced for 7 T compared to 2.35 T, as local gradients are proportional to the applied magnetic field. For juice, the water diffusivity effect on T_2 disappeared as tissue compartmentation is destroyed.

To obtain the probability density of different proton pools distinguished on the basis of their T_1 and T_2 values, 2D (T_1–T_2) cross-correlation data were obtained by acquiring CPMG sequences at different inversion times, and performing 2D inverse Laplace transformation. Proton pool assignment for tissue characterization was implemented by performing (T_2–D) correlation spectroscopy where the T_2 of the different water subcellular compartments were associated with water diffusion constants, D [m^2/s].

T_1–T_2 correlation data showed two main water proton pools distinguishable for fresh tissue (Figure 5.3.14), which were associated with water in vacuole and water in cytoplasm according to studies on parenchyma tissue of apple (Hills et al. 2004). These peaks merged when tissue was affected and a decrease in the T_2 values was

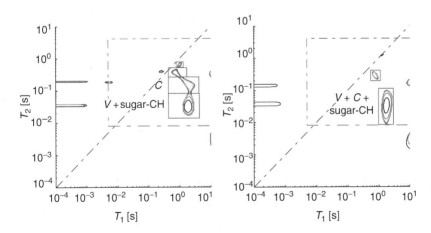

FIGURE 5.3.14 T_1–T_2 correlation spectra obtained from healthy pear tissue (left) and flesh presenting internal browning (right) at 7 T with a 90°–180° pulse spacing of 4000 μs. Contour lines delimit the probability of protons populations with peaks merging in affected tissue. C stands for water protons in cytoplasm, V stands for water protons in vacuole, and sugar-CH stands for protons in sugar molecules. (Reprinted from Hernández-Sánchez, N., Hills, B.P., Barreiro, P., and Marigheto, N., *Postharvest Biol. Technol.*, 44, 260, 2007. With permission.)

observed (Figure 5.3.14). Peak merging highlights the loss of membrane integrity and the enhancement of the diffusion exchange, which was confirmed by the higher diffusion coefficient computed by the T_2–D correlation spectroscopy for the vacuole compartment in affected tissue. The observed T_2 shortening was explained by the water diffusion through strong internal field gradients arising from susceptibility discontinuities within the affected tissue.

Such microscopic information was used to direct the macroscopic study. PD- and T_2*-weighted images were obtained with an in-plane resolution of 0.88 mm^2 per pixel (8.8 mm^3 per pixel volume resolution) using a 4.7 T magnet. The acquisition of 50% of the phase-encoding steps and the subsequent zero-filling yielded total acquisition times of 484 and 703 ms, respectively. Coronal images were acquired from the equatorial slice of intact fruits conveyed at 54 mm/s and phase shift, induced by sample motion, was corrected for image reconstruction (Figure 5.3.13B). Discriminant functions were obtained by selecting features, which were automatically extracted from the image histograms, among those showing significant differences between healthy and affected pears by ANOVA analysis. The achieved correct classifications were 98.4% for the former and 91% for the latter pears when using PD-weighted images; and 98% and 86% when using T_2*-weighted images. The final performance remains to be validated at low-magnetic field strengths and higher conveyor speeds.

5.3.4.6 Freeze Injury in Citrus

Freezing injury in oranges may appear whenever a temperature below the freezing point of the tissues is reached along the preharvest period. Injured fruits usually remain on the tree without any external symptoms. For injured fruit, the juice sacs dry out as the ice crystals burst the membranes and the cell walls. Also, the presence of water-soaking areas on the segment membranes becomes apparent for frozen fruits. Dehydrated tissues collapse at a severe stage, leading to the development of hollows between and within the segments. The occurrence of both dehydration and hollows in damaged fruits converts MRI as very suitable technique for freezing injury detection (Hernández Sánchez, 2006).

'Valencia' oranges were obtained after intense freezing conditions and measured in on-line MRI experiments using an NMR spectrometer of 4.7 T (Hernández-Sánchez et al. 2004). These samples along with nonexposed oranges were stabilized at room temperature. Then, oranges were placed in the conveyor belt with their stem–calyx axis along the z direction, and axial FLASH images with an in-plane resolution of 0.88 mm^2 per pixel (8.8 mm^3 per pixel volume resolution) obtained at 0, 50, and 100 mm/s belt speeds. To reduce the total acquisition time, half-Fourier gradient echo acquisitions with zero filling before image reconstruction were used to obtain sub-second scan times (780 ms). The loss of the image quality derived from lower k-space coverage (25%) did not compensate the significant reduction of the acquisition time (390 ms).

Affected tissue appeared as a region of hypointense signal, with different intensity levels according to the damage severity grade, while bright pixels related to non-affected tissue (Figure 5.3.15). Hypointense signal is derived from the

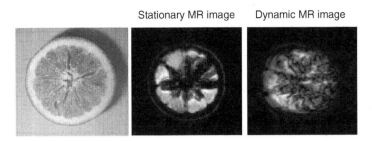

FIGURE 5.3.15 Example of axial FLASH images (780 ms acquisition time) obtained from a freeze injured orange while stationary (in the middle) and when conveyed at 54 mm/s (on the right). On the left, RGB image captured with a digital camera approximately corresponding to the center of the imaged FOV. (Adapted from Hernández-Sánchez, N., Barreiro, P., Ruiz-Altisent, M., Ruiz-Cabello, J., and Fernández-Valle, M.E., *Appl. Magn. Resonance*, 26, 431, 2004.)

reduction in PD as juice content decreases and from T_2^* reduction caused by the local field gradients at the interfaces between juicy and dehydrated structures. The central axis also showed hypointense signal as the mobility of the water is lower in this solid-like structure.

For the automatic analysis, two threshold segmentations were performed: one to separate the entire region of interest including healthy and affected tissue areas and the other to calculate the area of the healthy regions. This segmentation procedure uses an iterative process that is repeated until the number of pixels comprising each region reaches convergence. The signal hypointense fruit region was computed by subtracting those segmented areas. Thresholds were addressed at each belt speed from undamaged fruits as the ratio between the signal hypointense region and the region of interest. The threshold value increased from 10% under static conditions to 20% and to 30% at 50 and 100 mm/s belt speed, respectively. The lack of consistency was probably a consequence of the image blurring induced by the excitation of consecutive slices driven by the axial location of the FOV together with the sample motion. In fact, the resultant images were a rough projection of the entire orange volume. Authors indicated the major need of blurring reduction, although the use of coronal or sagittal images with motion correction procedures was not evaluated.

Gambhir and coworkers (2005) performed an MRR study on the effect of freezing ($-7°C$) and chilling ($5°C$) temperatures on the overall proton transverse relaxation time (T_2) of peel and juice sacs in 'Navel' oranges. Oranges were exposed to temperature treatment for 20 h. Samples warmed to room temperature and then were peeled. Measurements were conducted on a 0.235 T equipment. The exposure to chilling or freezing temperature did not affect the T_2 values of peel suggesting that the peel did not freeze under these conditions, whereas at $-20°C$ the T_2 reduced drastically. Freezing temperature caused an appreciable decrease in the T_2 values of flesh segments, which was associated with damage in the juice sacs membrane and subsequent leakage of juice.

5.3.4.7 Seed Identification in Citrus

One of the main concerns of citrus producers is the presence of seeds within oranges and mandarins as it greatly devalues the product quality. The acquisition of MR images where seeds are distinguishable from juice segments is one of the most promising procedures to identify seed-containing fruit. Seeds possess a solid-like structure that confers noticeable differences from the juicy pulp, which involve a straightforward contrast management in conventional NMR equipments, and subsequent image processing. The challenge is the achievement of practical image contrast by working procedures transferable to the industrial environment.

Blasco and colleagues (2003) evaluated the feasibility of obtaining enough MRI contrast to detect seeds in stationary oranges by using three types of sequences, that is, SE, half SE, and gradient echo, with varying TE, TR, and FOV. MR images were acquired with a medical NMR equipment of 0.2 T magnetic field strength. Three slices perpendicular to the stem-calyx axis of the fruit were imaged per sample. Among the analyzed sequences, SE with TR of 50 ms and TE of 18 ms provided the best discriminatory results. An algorithm developed to detect the seeds within the image achieved 100% of success. However, the acquisition time could not be set shorter than 7 s.

Hernández and colleagues (2005) used a faster MRI sequence, a gradient-echo FLASH with TR of 12.2 ms, TE of 3.8 ms, and flip angle 10° to detect seed-containing oranges when conveyed at 0, 54, and 90 mm/s through the 4.7 T magnetic field strength magnet of an NMR research equipment. Oranges were placed in the conveyor belt with their stem-calyx axis along the motion direction and axial images were acquired with an in-plane resolution of 0.88 mm^2 per pixel (8.8 mm^3 per pixel volume resolution). The region containing the seeds and central axis tissues appeared darker than the flesh pixels. As could be expected, the higher the belt speed the higher the blurring artifacts caused by the superimposition of signal arising from adjacent slices. Nevertheless, the contrast was still enough to allow a reliable automatic segmentation of the signal hypointense region and feature extraction when samples were conveyed at 50 mm/s belt speed with the acquisition of the 50% of k-space lines (zero filling was applied before reconstruction). At higher speeds, internal structures were completely masked. An unsupervised analysis on features extracted from the segmented regions revealed that seedless oranges and oranges with one seed were grouped into a single cluster. Taking into account that the discrimination between these two categories is critical for commercial purposes, it was concluded that on-line axial images could not be used for seed identification at current acquisition times (780 ms). In the same work (Hernández et al. 2005), coronal images from the equatorial slice of moving lemons were obtained, and the performance of a low time-consuming motion correction procedure was evaluated (Figure 5.3.16).

These promising results encouraged subsequent studies on mandarins (Hernández-Sánchez et al. 2006). Here, the NMR equipment also worked at 4.7 T and FLASH sequence was used. The imaging parameters were TR of 11 ms, TE of 3.8 ms, 10° flip angle, 50% of k-space lines, and zero filling, with a total acquisition time of 703 ms. Outstanding improvement in image quality was obtained since the features extracted from the segmented areas in motion-corrected images were not significantly different

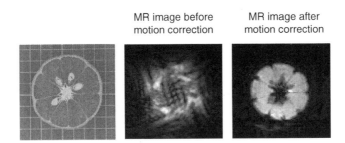

FIGURE 5.3.16 Example of motion-induced artifacts correction in a coronal MR image acquired during sample motion at 54 mm/s. On the left, RGB image captured with a digital camera approximately corresponding to the center of the imaged FOV; in the middle coronal, MR image reconstructed without artifact correction; on the right, same MR image reconstructed after artifact correction.

from those extracted from the static images (Figure 5.3.17). Those showing significant differences between seed-containing and seedless mandarins by ANOVA analysis entered a discriminant analysis. Finally, the perimeter of the region containing the

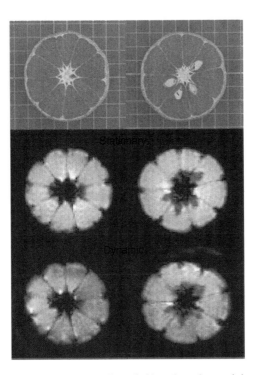

FIGURE 5.3.17 Examples of both seedless (left) and seed containing (right) mandarins. Upper line corresponds to RGB images, middle line to static FLASH MR images, and bottom line to motion-corrected MR FLASH images acquired at 54 mm/s conveyor belt speed. (Reprinted from Hernández-Sánchez, N., Barreiro, P., and Ruiz-Cabello, J., *Biosystems Eng.*, 95, 529, 2006. With permission.)

seeds and the central axis, and the maximum distance between the perimeter and the gravity center of this region were selected for classification. The number of samples correctly classified varied from 88.9% under static to 92.5% under dynamic conditions for seedless mandarins, and from 86.7% to 79.5% for fruits with seeds. A main conclusion of this study was that it is feasible to perform a straightforward export of models developed under static conditions toward on-line conditions.

Faster MRI sequences and more efficient segmentation procedures have been evaluated recently by Barreiro and colleagues (2007) with stationary mandarins. Two different types of fast MRI sequences were investigated using a 4.7 T magnet: a gradient-echo sequence (FLASH sequence) with 484 ms acquisition time and a spiral–radial sequence (COMSPIRA sequence) with 240 ms. Three segmentation techniques were applied for image postprocessing, that is, region-based, 1D histogram variance, and 2D histogram variance. These studies have demonstrated that the latter option provides the most promising results. Image features extracted from the signal hypointense region including perimeter, compactness, maximum distance to the gravity center, and aspect ratio are employed in a linear classification function, by which the seed identification can be achieved with 100% accuracy using spiral-radial [T2] sequence and 98.7% accuracy with gradient-echo images. These results need to be validated with a higher number of samples and using a low field magnet.

5.3.5 Concluding Remarks

We presented the technical background and survey of modern applications of NMR in fruits and vegetables. By analyzing the enormous available literature, we tried to provide an overview that could be interesting to those entering or to those who are currently applying other techniques in similar areas of agriculture and food. The high cost of commercial NMR instruments has probably prevented many authors to use NMR in their investigations. This is probably the reason why its use has not extended further. Being critical with the NMR work presented along these years, we can say that it has not yet gone beyond the consideration of being an interesting research tool. To go further with NMR applications, we should pay more attention to economy and efficiency, and break the vicious cycle that drives us to use only high-quality instruments, mainly designed to medical applications, whose quality standards logically are much more severe and whose results cannot in many cases be transferred to a final non-experience user working in a completely different environment such as that related to agricultural products. Joint efforts are worthy in view of the expansion of the NMR technique.

The basics of the NMR confer its exceptional capabilities to inspect internal quality parameters and to provide valuable insight on physiological processes as it is sensitive to the content, chemical environment, mobility, and diffusion among other phenomena, related to aqueous protons. These phenomena are influenced by cell compartmentation and tissue microstructure so that the NMR signal also acts as sensor of the tissue integrity.

The techniques derived from the NMR such as NMR relaxometry, NMR spectroscopy, and MRI greatly broaden the range of applications devoted to the inspection of food products and, particularly, of fruits and vegetables.

The different relaxation times obtained by means of MRR, longitudinal relaxation time T_1, and transverse relaxation time T_2 (including effective T_2^*) are sources of contrast that carry diverse underlying information for different types of tissue. T_1 distribution reveals differences between molecular motions within the tissue, which are related to their size and complexity, and the viscosity of the surroundings. As for T_2, the tumbling motions of the molecules and the chemical exchange between proton pools of different metabolites give rise to relaxation contrast. The diffusion through local magnetic field inhomogeneities induced by susceptibility variations at different interfaces and microstructure arrangements is also responsible for T_2 differences. The cell morphology and size, the membranes permeability, and hence, the membranes integrity affect such diffusion.

The MRS provides both qualitative and quantitative information on the composition of the sample with regard to a specific nucleus. Variations in electronic and chemical environment induce the groups of chemically equivalent nuclei of each molecule to precess at different frequencies. The qualitative information is provided by the different chemical shifts at which the resolved peaks are located in the NMR spectrum. The quantitative information is derived from the area of each peak.

MRI provides a picture that contains combined spectroscopy and relaxometry information both spatially resolved. An MR image is basically an indication of the PD contained in the sample enriched with the relaxation information. The management of the image acquisition parameters allows different weightings. In PD-weighted images, those regions with higher number of nuclei appear brighter, for T_1-weighting the bright pixels correspond to tissues with short T_1, whereas for T_2-weighted images the longer T_2 components present higher intensity signal.

The studies performed up to date are a demonstration of the potential of these techniques for the internal quality monitoring. Continuous technological advances in NMR hardware and computers allow more efficient explorations and are allies for the achievement of new correlations between quality aspects and NMR features as well as for the confirmation of those already found. However, there is additional intensive work that must be faced to optimize the NMR systems and the signal acquisition procedures, so that the transference of the acquired knowledge to commercial equipments is more realistic.

The implementation of the NMR techniques (MRR, MRS, and MRI) demands the use of appropriate technology at low-magnetic field strengths, efficient pulse sequences with ultrafast acquisition times (below 1 s per fruit), and the inspection under motion conditions to reach the typical conveyor velocities (up to 2 m/s). As aforementioned, one of the major hindrances for the transferability of the current applications to the horticultural industry is the high cost of the equipment used for research works, which is a consequence of the high field strength magnets and the high-performance hardware required for the application of some MRI sequences.

When comparing high- and low-field MRI applications, limits for the use at low-magnetic field strength mainly derive from the decrease in SNR (owing to the poorer sample polarization) for a similar pulse sequence, acquisition parameters, and coil. Other factors, such as relaxation times, magnetic susceptibility differences, and the loss of contrast arising from local field gradients (as local gradients decrease at lower magnetic field strength), need to be evaluated individually for each application, since the changes in these properties can be decisive for inspection performance. Hills and Clark (2003) proposed to overcome the loss of SNR at low field by using a prepolarizing unit. These authors also highlighted the field homogeneity within large volumes and the compatibility with commercial graders as major subjects to be considered.

For fast sample inspection with low hardware demand MRS, nonspatially resolved CPMG-T_2 measurements and 1D MRI provide the most efficient results. As for 2D MRI, high-quality images have been obtained with very short acquisition times, and motion-induced artifacts have been successfully corrected for coronal FOVs. However, it is a very demanding technique and validation of the results at low-magnetic field remains to be carried out. Nevertheless, the optimization and development of single-shot sequences along with alternative k-space sampling strategies bring optimistic expectations. In contrast, spatially resolved MRR experiments (e.g., relaxation maps) are often highly time consuming, although the provided information on inherent tissue characteristic is critical for the optimization of the more transferable experimental procedures.

With regard to the inspection under continuous motion, it seems to be a major requirement to reach the current conveyor velocities. Besides, the possible tumbling of the samples induced by the conveyor start and stop would bring artifacts that could spoil the measurement reliability. Additional time for sample stabilization would be needed. However, the industry could compromise the reduction in inspection rate as long as final results conform to significant improvements in quality standards.

The main applications that have been lately evaluated under on-line conditions are maturity in avocados, pit detection in cherries and olives, internal browning in apples, internal breakdown in pears, freeze injury in citrus, and seed in citrus, all of them related to internal quality aspects, which are not as easily detected with other techniques. Up to date, these applications have not been implemented by industry.

Collaborative works require a closer interaction between researchers and the food industry, as well as the implication of the NMR manufacturers to converge on realistic and reliable monitoring systems that exploit the great potential of the NMR techniques. There are some research groups that have already started, as those led by MJ McCarthy (University of California, United States) and BP Hills (Institute of Food Research, United Kingdom), who are better situated to materialize the encouraging expectations in the near future.

REFERENCES

Barreiro, P., J. Ruiz-Cabello, M.E. Fernández-Valle, C. Ortiz, and M. Ruiz-Altisent. 1999. Mealiness assessment in apples using MRI techniques. *Magnetic Resonance Imaging* 17: 275–281.

Barreiro, P., C. Ortiz, M. Ruiz-Altisent, J. Ruiz-Cabello, M.E. Fernandez-Valle, I. Recasens, M. Asensio. 2000. Mealiness assessment in apples and peaches using MRI techniques. *Magnetic Resonance Imaging* 18: 1175–1181.

Barreiro, P., A. Moya, E. Correa, M. Ruiz-Altisent, M. Fernández-Valle, A. Peirs, K.M. Wright, and B.P. Hills. 2002. Prospects for the rapid detection of mealiness in apples by nondestructive NMR relaxometry. *Applied Magnetic Resonance* 22: 387–400.

Barreiro, P., C. Zheng, D.W. Sun, N. Hernández-Sánchez, J.M. Pérez-Sánchez, and J. Ruiz-Cabello. 2007. Non-destructive seed detection in mandarins: Comparison of automatic threshold methods in FLASH and COMSPIRA MRIs. *Postharvest Biology and Technology*. doi:10.1016/j.postharvbio.2007.07.008.

Bernstein, M.A., K.F. King, and X.J. Zhou. 2004. *Handbook of MRI Pulse Sequences*. Elsevier Academic Press, United States of America.

Blasco, J., M.C. Alamar, and E. Moltó. 2003. Detección no destructiva de semillas en mandarinas mediante resonancia magnética. *Resúmenes del II Congreso Nacional de Agroingeniería*, Córdoba, Spain, 439–440.

Bracewell, R.N. 2000. *The Fourier Transform and its applications*, 3rd ed. McGraw-Hill Higher Education, Casson T, CA.

Brescia, M.A., T. Pugliese, E. Hardy, and A. Sacco. 2007. Compositional and structural investigations of ripening of table olives, Bella della Daunia, by means of traditional and magnetic resonance imaging analyses. *Food Chemistry* 105: 400–404.

Butz, P., C. Hofmann, and B. Tauscher. 2005. Recent developments in noninvasive techniques for fresh fruit and vegetable internal quality analysis. *Journal of Food Science* 70: 131–141.

Chayaprasert, W. and R. Stroshire. 2005. Rapid sensing of internal browning in whole apples using low-cost, low-field proton magnetic resonance sensor. *Postharvest, Biology and Technology* 36: 291–301.

Chen, P., M.J. McCarthy, S.M. Kim, and B. Zion. 1996. Development of a high-speed NMR technique for sensing maturity of avocados. *Transactions of the American Society of Agricultural Engineers* 39: 2205–2209.

Clark, C.J., P.D. Hockings, D.C. Joyce, and R.A. Mazzuco. 1997. Application of magnetic resonance imaging to pre- and post-harvest studies of fruits and vegetables. *Postharvest Biology and Technology* 11: 1–21.

Clark, C.J. and D.M. Burmeister. 1999. Magnetic resonance imaging of browning development in 'Braeburn' apple during controlled-atmosphere storage under high CO_2. *Horticultural Science* 34: 915–919.

Delpuech, J. 1995. *Dynamics of Solutions and Fluids Mixtures by NMR*. John Wiley & Sons, Chichester, United Kingdom.

Gambhir, P.N., Y.J. Choi, D.C. Slaughter, J.F. Thompson, and M.J. McCarthy. 2005. Proton spin–spin relaxation time of peel and flesh of navel orange varieties exposed to freezing temperature. *Journal of the Science of Food and Agriculture* 85: 2482–2486.

Goñi, O., M. Muñoz, J. Ruiz-Cabello, M.I. Escribano, and C. Merodio. 2007. Changes in water status of cherimoya fruit during ripening. *Postharvest Biology and Technology* 45: 147–150.

González, J.J., R.C. Valle, S. Bobroff, W. Biasi, E.J. Mitcham, and M.J. MacCarthy. 2001. Detection and monitoring of internal browning development in 'Fuji' apples using MRI. *Postharvest Biology and Technology* 22: 179–188.

Hernández-Sánchez, N., P. Barreiro, M. Ruiz-Altisent, J. Ruiz-Cabello, and M.E. Fernández-Valle. 2004. Detection of freeze injury in oranges by magnetic resonance imaging of moving samples. *Applied Magnetic Resonance* 26: 431–445.

Hernández, N., P. Barreiro, M. Ruiz-Altisent, J. Ruiz-Cabello, and M.E. Fernández-Valle. 2005. Detection of seeds in citrus using magnetic resonance imaging under motion conditions and improvement with motion correction. *Concepts in Magnetic Resonance Part B: Magnetic Resonance Engineering* 26B: 81–92.

Hernández Sánchez, N. 2006. Development of on-line NMR applications for the evaluation of fruit internal quality. Dissertation at the Polytechnic University of Madrid, Madrid, Spain.

Hernández-Sánchez, N., P. Barreiro, and J. Ruiz-Cabello. 2006. On-line identification of seeds in mandarins with magnetic resonance imaging. *Biosystems Engineering* 95: 529–536.

Hernández-Sánchez, N., B.P. Hills, P. Barreiro, and N. Marigheto. 2007. An NMR study on internal browning in pears. *Postharvest Biology and Technology* 44: 260–270.

Hills, B.P. and B. Remigereau. 1997. NMR studies of changes in subcellular water compartmentation in parenchyma apple tissue during drying and freezing. *International Journal of Food Science and Technology* 32: 51–61.
Hills, B.P. 1998. *Magnetic Resonance Imaging in Food Science*. John Wiley & Sons, New York.
Hills, B.P. and C.J. Clark. 2003. Quality assessment of horticultural products by NMR. *Annual Reports on NMR Spectroscopy* 50: 75–120.
Hills, B., S. Benamira, N. Marigheto, and K. Wright. 2004. T1-T2 correlation analysis of complex foods. *Applied Magnetic Resonance* 26: 543–560.
Jung, K.H., R. Stroshine, P. Cornillon, and P.M. Hirst. 1998. Low field proton magnetic resonance sensing of water core and internal browning in whole apples. *American Society of Agricultural Engineers Conference Paper* 98-6020.
Keener, K.M., R.L. Stroshine, and J.A. Nyenhuis. 1997. Proton magnetic resonance measurement of self-diffusion coefficient of water in sucrose solutions, citric acid solutions, fruit juices, and apple tissue. *Transactions of the American Society of Agricultural Engineers* 40: 1633–1641.
Kim, S.M., P. Chen, M.J. McCarthy, and B. Zion. 1999. Fruit internal quality evaluation using on-line Nuclear Magnetic Resonance Sensors. *Journal of Agricultural Engineering Research* 74: 293–301.
Korin, H.W., F. Farzaneh, R.C. Wright, and S.J. Riederer. 1989. Compensation for effects of linear motion in MR imaging. *Magnetic Resonance in Medicine* 12: 99–133.
Lammertyn, J., M. Aerts, B.E. Verlinden, W. Schotsmans, and B.M. Nicolaï. 2000. Logistic regression analysis of factors influencing core breakdown in 'Conference' pears. *Postharvest Biology and Technology* 20: 25–37.
Lammertyn, J., T. Dresselaers, P. Van Hecke, P. Jacsók, M. Wevers, and B.M. Nicolaï. 2003a. MRI and X-ray CT study of spatial distribution of core breakdown in 'Conference' pears. *Magnetic Resonance Imaging* 21: 805–815.
Lammertyn, J., T. Dresselaers, P. Van Hecke, P. Jacsók, M. Wevers, and B.M. Nicolaï. 2003b. Analysis of the time course of core breakdown in 'Conference' pears by means of MRI and X-ray CT. *Postharvest Biology and Technology* 29: 19–28.
Larrigaudière, C., I. Lentheric, J. Puy, and E. Pintó. 2004. Biochemical characterisation of core browning and brown heart disorders in pear by multivariate analysis. *Postharvest Biology and Technology* 31: 29–39.
Létal, J., D. Jirák, L. Šuderlová, and M. Hájek. 2003. MRI "texture" analysis of MR images of apples during ripening and storage. *Lebensmittel-Wissenschaft und-Technologie* 36: 719–727.
Marigheto, N., S. Duarte, and B.P. Hills. 2005. An NMR relaxation study of avocado quality. *Applied Magnetic Resonance* 29: 687–701.
McCarthy, M.J. 1994. *Magnetic Resonance Imaging in Foods*. Chapman and Hall.
Parikh, A.M. 1992. *Magnetic Resonance Imaging Techniques*. Elsevier, New York.
Pathaveerat, S., M.J. McCarthy, and P.P. Chen. 2001. Spatial distribution of avocado composition: Implications for on-line sorting by NMR spectroscopy. *VI International Symposium on Fruit, Nut, and Vegetable Production Engineering*. Potsdam (Germany), Conference Proceedings, 645–649.
Pérez-Sánchez, J.M., I. Rodríguez, R. Pérez De Alejo, M. Cortijo, and J. Ruiz-Cabello. 2006. COMSPIRA3D: A combined approach to radial and spiral 3D MRI. *Concepts in Magnetic Resonance Part B: Magnetic Resonance Engineering* 29B: 107–160.
Raffo, A., R. Gianferri, R. Barbieri, and E. Brosio. 2005. Ripening of banana fruit monitors by water relaxation and diffusion ^{1}H-NMR measurements. *Food Chemistry* 89: 149–158.
Rodríguez, I., R. Pérez de Alejo, J. Cortijo, and J. Ruiz-Cabello. 2004. COMSPIRA: A common approach to spiral and radial MRI. *Concepts in Magnetic Resonance Part B: Magnetic Resonance Engineering* 20B: 40–44.

Saquet, A.A., J. Streif, and F. Bangerth. 2003. Energy metabolism and membrane lipid alterations in relation to brown heart development in 'Conference' pears during delayed controlled atmosphere storage. *Postharvest Biology and Technology* 30:123–132.

Thybo, A.K., P.M. Szczypinski, A.H. Karlsson, S. Donstrup, H.S. Stodkilde-Jorgensen, and H.J. Andersen. 2004. Prediction of sensory texture quality attributes of cooked potatoes by NMR-imaging (MRI) of raw potatoes in combination with different image analysis methods. *Journal of Food Engineering* 61: 91–100.

Tu, S.S., J.C. Young, M.J. McCarthy, and K.L. McCarthy. 2007. Tomato quality evaluation by peak force and NMR spin-spin relaxation time. *Postharvest Biology and Technology* 44: 157–164.

Veltman, R.H., I. Lenthéric, L.H.W. Van der Plas, and H.W. Peppelenbos. 2003. Internal browning in pear fruit (*Pyrus communis L.* cv Conference) may be a result of a limited availability of energy and antioxidants. *Postharvest Biology and Technology* 28: 295–302.

Vlaardingerbroek, M.T. and J.A. den Boer. 1996. *Magnetic Resonance Imaging*. Springer-Verlag, Berlin/Heidelberg, Germany.

Wang, C.Y. and P.C. Wang. 1989. Nondestructive detection of core breakdown in 'Barlett' pears with nuclear magnetic resonance imaging. *Horticultural Science* 24: 106–109.

Woodward P. 1995. Contrast. In *MRI for Technologists*. McGraw-Hill Education, USA.

Zion, B., M.J. McCarthy, and P. Chen. 1994. Real-time detection of pits in processed cherries by magnetic resonance projections. *Lebensmittel-Wisenschaft und -Technologie* 27: 457–462.

Zion, B., S.M. Kim, M.J. McCarthy, and P.J. Chen. 1997. Detection of pits in olives under motion by nuclear magnetic resonance. *Journal of the Science of Food and Agriculture* 75: 496–502.

6 Process Monitoring

Ali Cinar and Sinem Perk

CONTENTS

6.1	Introduction	470
6.2	Multivariate Statistical Techniques for Process Monitoring, Quality Control, and Fault Diagnosis	471
	6.2.1 Principal Components Analysis and Partial Least Squares	471
	6.2.2 Contribution Plots	477
	6.2.3 Statistical Methods for Fault Diagnosis	479
	6.2.3.1 Clustering	480
	6.2.3.2 Discriminant Analysis	481
	6.2.4 Example: Analysis of the Growth of Fresh Apples	481
	6.2.4.1 PCA Analysis of the Spectral Data	481
6.3	SPM and Fault Diagnosis in Continuous Processes	485
	6.3.1 SPM Methods Based on PCA	485
	6.3.1.1 Hotelling's T^2 Charts	487
	6.3.1.2 SPE Charts	488
	6.3.2 SPM and SQC Methods Based on PLS	489
	6.3.3 Fault Diagnosis Using Contribution Plots	492
	6.3.4 Fault Diagnosis Using Statistical Methods	492
6.4	SPM Techniques for Multivariable Batch Processes	494
	6.4.1 SPM Charts for Batch Processes	496
	6.4.2 Multiway PLS-Based SPM for Postmortem Analysis	499
	6.4.3 MPCA-Based On-Line Monitoring of Batch Processes	505
	6.4.4 Example: MPCA-Based SPM of Fed-Batch Penicillin Fermentation	506
	6.4.5 Example: MPCA-Based SPM of the Growth of Fresh Apples	508
	6.4.6 MPLS-Based On-Line Estimation and Monitoring of Final Product Quality	515
6.5	Summary	519

Appendix 520
    Greek 520
    Subscripts 521
    Superscripts 521
    Abbreviations 521
References 521

6.1 INTRODUCTION

In food production and processing, new innovative technologies such as process-oriented multivariate data analysis, process monitoring, and quality control are needed for appropriate process management, to maintain nutritional product quality and diminish the product loss owing to the decaying along the supply chain of perishable agro-food products. Product quality in fresh produce is determined by the plant genome, environmental conditions, and the production system (Zude and McRoberts 2006).

Traditional quality control of fresh fruits is done using single-variable techniques, such as Shewhart charts. The availability of inexpensive sensors for typical process variables, such as temperature and humidity, and the development of advanced chemical analysis systems that can provide frequent reliable information on quality variables lead to new opportunities for powerful process monitoring and quality control. These sensors and analyzers increased the number of variables measured at high frequencies and enabled a data-rich food processing environment. Univariate statistical process monitoring (SPM) and statistical quality control (SQC) techniques have limited capabilities for monitoring multivariable processes. The critical limitation of univariate methods is their inability to use the correlation among various variables for developing powerful insight in the process (Cinar et al. 2007). The shortcomings of using univariate charts for monitoring multivariable processes include too many false alarms, too many missed alarms, and the difficulty of visualizing and interpreting the big picture about the process status. Plant personnel are expected to make decisions about product status by interpreting information from a large number of charts that ignore the correlation between the variables.

MSPM and SQC techniques are based on statistical methods such as principal components analysis (PCA) and partial least squares (PLS) regression. These methods capture the correlation information neglected by univariate monitoring techniques. The mathematical and statistical techniques used for developing MSPM and SQC charts are more complex, but most multivariate process monitoring software shield these computations from the user and provide easy-to-interpret graphs for monitoring a process (Cinar et al. 2003, 2007).

Section 6.2 presents the multivariate statistical techniques for process and quality monitoring, and fault diagnosis. Section 6.3 introduces SPM and SQC methods based on PCA and PLS for continuous processes. It also outlines fault diagnosis by using contribution plots and discriminant analysis. Section 6.4 presents SPM and SQC techniques for multivariable batch processes. Techniques for the analysis of a batch after its conclusion and for monitoring of the batch run during its development are described and illustrated by using data from the growth of apples and from simulations of batch fermentation. Section 6.5 provides a summary and conclusions.

6.2 MULTIVARIATE STATISTICAL TECHNIQUES FOR PROCESS MONITORING, QUALITY CONTROL, AND FAULT DIAGNOSIS

6.2.1 PRINCIPAL COMPONENTS ANALYSIS AND PARTIAL LEAST SQUARES

The PCA is a multivariate statistical technique that extracts the strong correlations in a data set and defines new orthogonal coordinate directions (Anderson 1984; Joliffe 1986; Jackson 1991; Johnson and Wichern 1998). The first practical application of PCA could be attributed to Pearson's work in biology (Pearson 1901).

PCA can be used as a visualization tool to describe correlation among variables, as a model-building platform, or as the foundation of monitoring techniques to describe the expected variation in a process under normal operation (NO). For a particular process, NO data provide information on satisfactory process performance under targeted operating conditions. PCA model is based on this data set and it can be used to detect outliers in data, data reconciliation, and deviations from NO that indicate significant variation from target values or unusual patterns of variation. PCA is an algebraic method of transforming the coordinate system of a data set for more efficient description of variability. The convenience of this representation is in the equivalence of data to measurable and meaningful physical quantities like temperatures, humidity, and compositions. In statistical analysis and modeling, quantification of data variance is of great importance. PCA provides a direct method of orthogonal decomposition onto a new set of basis vectors that are aligned with the directions of maximum data variance. Operation under various known upsets can also be modeled if sufficient historical data are available to develop automated diagnosis of source causes of abnormal process behavior (Raich and Cinar 1996).

Principal components (PCs) are a new set of orthogonal coordinate axes. The first PC indicates the direction of largest variation in data and the second PC indicates the largest variation unexplained by the first PC in a direction orthogonal to the first PC (Figure 6.2.1). The orthogonality assures that variation explained by all previous PCs is excluded in determining the next PC's direction. The number of PCs is usually less than the number of variables.

PCA involves the orthogonal decomposition of the set of process measurements along orthogonal directions that explain the maximum variation in the data. For a continuous process, the elements of the $n \times m$ data matrix X^* are x_{ij}^* where $(i = 1, \ldots, n)$ indicates the number of samples and $(j = 1, \ldots, m)$ indicates the number of variables. To remove magnitude and variance biases in data, X^* is mean-centered and variance-scaled to get X. Each row of X represents the time series of a process measurement with mean 0 and variance 1 reflecting equal importance of each variable. Specific variables can be given a higher scaling weight than that corresponding to unit variance scaling if a priori knowledge about the relative importance of the variables is available (Gurden et al. 2001; Bro and Smilde 2003). The directions

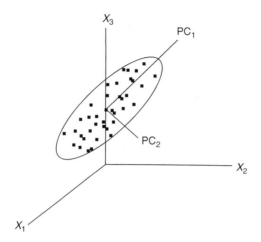

FIGURE 6.2.1 The first two PCs of a three-dimensional data set projected on a single plane (the PC_1–PC_2 plane).

extracted by the orthogonal decomposition of X are the eigenvectors $\vec{p}_i$ of $X^T X$ or the PC loadings,

$$X = \vec{t}_1 \vec{p}_1^T + \vec{t}_2 \vec{p}_2^T + \cdots + \vec{t}_a \vec{p}_a^T + E \qquad (6.2.1)$$

where E is the $n \times m$ matrix of residuals and the superscript T denotes the transpose of a matrix. The dimension a is chosen such that most of the significant process information is extracted from E and E represents mostly random error. The first eigenvector is in the direction of maximum variation in data and the second one, while being orthogonal to the first, is in the direction of maximum variance of the residual. The same pattern is followed in determining additional PCs. The residual is obtained at each step by subtracting the variance already explained by the PC loadings selected before, and E is used as the data matrix for the computation of the next PC loading.

The eigenvalues of the covariance matrix of X define the corresponding amount of variance explained by each eigenvector. The projections of the measurements (observations) onto the eigenvectors provide new coordinate values, which constitute the scores matrix T whose columns are $\vec{t}_i$ given in Equation 6.2.1. The relationship between T, P, X can also be expressed as

$$T = XP, \quad X = TP^T + E \qquad (6.2.2)$$

where
  P is an $m \times a$ matrix whose jth column is the jth eigenvector of $X^T X$
  T is an $n \times a$ score matrix

The PCs can be computed by spectral decomposition (Johnson and Wichern 1998), computation of eigenvalues and eigenvectors, or singular value decomposition.

Process Monitoring

The covariance matrix S ($S = X^T X/(m-1)$) of data matrix X can be decomposed by spectral decomposition as

$$S = PLP^T \tag{6.2.3}$$

where P is a unitary matrix (a unitary matrix A is a complex matrix, in which the inverse is equal to the conjugate of the transpose: $A^{-1} = A^*$. Orthogonal matrices are unitary. If A is a real unitary matrix, then $A^{-1} = A^T$, whose columns are the normalized eigenvectors of S and L is a diagonal matrix that contains the ordered eigenvalues $\vec{l}_i$ of S. The scores T are computed by using the relation $T = XP$.

Singular value decomposition of X is carried out as

$$X = U\Lambda V^T \tag{6.2.4}$$

where
- the columns of U are the normalized eigenvectors of XX^T
- the columns of V are the normalized eigenvectors of $X^T X$
- Λ is a diagonal matrix having the positive square roots of the magnitude ordered eigenvalues of $X^T X$ as its elements

For an $n \times m$ matrix X, U is $n \times n$, V is $m \times m$, and Λ is $n \times m$. Let the rank of X be denoted as r, $r \leq \min(m, n)$. The first r rows of Λ make an $r \times r$ diagonal matrix, the remaining $(n - r)$ rows are filled with zeros. Term-by-term comparison of the last two equations yields

$$P = V \quad \text{and} \quad T = U\Lambda \tag{6.2.5}$$

For a data set that is described satisfactorily by two PCs, the data can be displayed in a plane (biplot). The data with random variation around the mean values (NO data) are scattered as an ellipse, whose axes are in the direction of PC loadings in Figure 6.2.1. For higher number of variables data will be scattered as an ellipsoid.

The selection of appropriate number of PCs or the maximum significant dimension $\vec{a}$ is critical for developing a parsimonious PCA model (Jackson 1980; Runger and Alt 1996; Johnson and Wichern 1998). A quick method for computing an approximate value for a is to add PCs to the model until the percent of the variation explained by adding additional PCs becomes small. Inspect the ratio where L is the diagonal matrix of ordered eigenvalues of S, the covariance matrix. The sum of the variances of the original variables is equal to the trace (tr(S)), the sum of the diagonal elements of S:

$$S_1^2 + S_2^2 + \cdots + S_r^2 = \text{tr}(S) \tag{6.2.6}$$

where $\text{tr}(S) = \text{tr}(L)$. A more precise method that requires large computational time is cross-validation (Wold 1978; Krzanowski 1987). Cross-validation is implemented by excluding part of the data, performing PCA on the remaining data, and computing the prediction error sum of squares (PRESS) using the data retained (excluded from model development). The process is repeated until every observation is left out once. The order a is selected as the number of PCs that minimizes the overall PRESS. Other criteria for choosing the optimal number of PCs have also been proposed. Wold (1978) proposed checking the ratio

$$R = \frac{\text{PRESS}_a}{\text{RSS}_{a-1}} \qquad (6.2.7)$$

where RSS_a is the residual sum of squares using the PCA model after ath PC is added to the model. When R exceeds unity upon addition of another PC, it suggests that the ath component did not improve the prediction power of the model and it is better to use $(a-1)$ components. Another approach is based on the SCREE plots that indicate the dimension at which the smooth decrease in the magnitude of the covariance matrix eigenvalues appear to level off to the right of the plot (Romagnoli and Palazoglu 2005).

Principal components regression (PCR) is one of the techniques to deal with ill-conditioned regressor data matrices by regressing the dependent variables such as quality measurements on the principal components scores of regressor variables such as the measured variables (temperature, humidity) of the process. The implementation starts by representing the data matrix X with its scores matrix T using the transformation $T = XP$. The regression equation becomes

$$Y = T\beta + E \qquad (6.2.8)$$

where the optimum matrix of regression coefficients β is obtained as

$$\beta = (T^{\text{T}}T)^{-1}T^{\text{T}}Y \qquad (6.2.9)$$

In contrast to the inversion of $X^{\text{T}}X$ when some of the $\vec{x}$ are collinear, the inversion of $T^{\text{T}}T$ does not cause any problems due to the mutual orthogonality of the PCs. Score vectors corresponding to small eigenvalues can be left out to avoid collinearity problems. Since PCR is a two-step method, there is a risk that useful predictive information would be discarded with a PC that is excluded. Hence caution must be exercised, while leaving out vectors corresponding to small eigenvalues. If regression based on the original variables $\vec{x}$ is preferred, inspecting the variables that contribute to the first few loadings and avoiding those that provide duplicate information can select the most important variables.

To include information about process dynamics, lagged variables can be included in X. The autocorrelograms of all $\vec{x}$ variables are plotted first to determine how many lagged values are relevant for each variable. Then the data matrix is augmented accordingly and used to determine the PCs for the regression step.

Nonlinear extensions of PCA have been proposed by using autoassociative neural networks (NNs) (Kramer 1992) or by using principal curves and surfaces (Hastie and Stuetzle 1989; LeBlanc and Tibshirani 1994).

PLS, also called projections to latent structures, develops a biased model between two blocks of variables X and Y. PLS selects latent variables so that

Process Monitoring

variation in X, which is most predictive of the Y, is extracted. The PLS approach was developed in the 1970s by Herman Wold for analyzing social sciences data by estimating model parameters using the nonlinear iterative partial squares (NIPALS). PLS works on the sample covariance matrix $(X^T Y)(Y^T X)$ (Wold et al. 1984). PLS regression develops a biased linear regression model between X and Y. In the context of process operations, usually X denotes the process variables and Y the quality variables. PLS selects latent variables so that variation in X that is most predictive of the product quality data Y is extracted. PLS works on the sample covariance matrix $(X^T Y)(Y^T X)$ (Wold et al. 1984; Geladi and Kowalski 1986a,b; Lorber et al. 1987; Wold et al. 1987; Hoskuldsson 1988; Martens and Naes 1989). Measurements of m process variables taken at n different times are arranged into an $n \times m$ process data matrix X. The q quality variables are given by the corresponding $n \times q$ matrix Y. Both X and Y data blocks are usually centered and scaled to unit variance because in PLS the influence of a variable on model parameters increases with the variance of the variable. The PLS model consists of outer relations (X and Y blocks individually) and an inner relation (linking both blocks). The outer relations for the X and Y blocks are, respectively,

$$X = TP^T + E = \sum_{i=1}^{a} \vec{t}_i \vec{p}_i^T + E \qquad (6.2.10)$$

$$Y = UQ^T + F = \sum_{i=1}^{a} \vec{u}_i \vec{q}_i^T + F \qquad (6.2.11)$$

where E and F represent the residuals matrices. Linear combinations of $\vec{x}$ vectors are calculated from the latent variable scores $\vec{t}_i = \vec{w}_i^T \vec{x}$ and those for the $\vec{y}$ vectors from $\vec{u}_i = \vec{q}_i^T \vec{y}$ so that they maximize the covariance between X and Y explained at each dimension. $\vec{w}_i$ and $\vec{q}_i$ are the weight vectors and $\vec{p}_i$ are the loading vectors of X. The number of latent variables can be determined by cross-validation (Wold 1978) or pragmatic techniques discussed earlier in the chapter.

T and U are PLS scores matrices of X and Y blocks, respectively, P contains the X loadings, W and Q are weight matrices for X and Y blocks, respectively, and E and F are residual matrices of X and Y blocks (Figure 6.2.2).

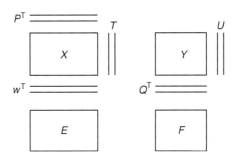

FIGURE 6.2.2 The matrix relationships in PLS.

For the first latent variable, PLS decomposition is started by selecting $\vec{y}_j$, an arbitrary column of Y as the initial estimate for $\vec{u}_1$. Usually, the column of Y with greatest variance is chosen. Starting in the X data block, for the first latent variable

$$\vec{w}_1^T = \frac{\vec{u}_1^T X}{\|\vec{u}_1^T \vec{u}_1\|}, \quad \vec{t}_1 = \frac{X\vec{w}_1}{\|\vec{w}_1^T \vec{w}_1\|} \tag{6.2.12}$$

In the Y data block

$$\vec{q}_1^T = \frac{\vec{t}_1^T Y}{\|\vec{t}_1^T \vec{t}_1\|}, \quad \vec{u}_1 = \frac{Y\vec{q}_1}{\|\vec{q}_1^T \vec{q}_1\|} \tag{6.2.13}$$

Convergence is checked by comparing $\vec{t}_1$ in Equation 6.2.12 with the $\vec{t}_1$ from the previous iteration. If their difference is smaller than a prespecified threshold, one proceeds to Equation 6.2.14 to calculate X data block loadings $\vec{p}_1$ and weights $\vec{w}_1$ are rescaled using the converged $\vec{u}_1$. Otherwise, $\vec{u}_1$ from Equation 6.2.13 is used for another iteration. If Y is univariate, Equation 6.2.13 can be omitted, and $\vec{q}_1 = 1$. The loadings of the X data block and are computed and the scores and weights are rescaled:

$$\vec{p}_1^T = \frac{\vec{t}_1^T X}{\|\vec{t}_1^T \vec{t}_1\|}, \quad \vec{p}_{1n} = \frac{\vec{p}_{1o}}{\|\vec{p}_{1o}\|} \tag{6.2.14}$$

$$\vec{t}_{1n} = \vec{t}_{1o}\|\vec{p}_{1o}\|, \quad \vec{w}_{1n} = \vec{w}_{1o}\|\vec{p}_{1o}\| \tag{6.2.15}$$

where the subscript o refers to old and n to new values. The regression coefficient b_i for the inner relation ($\vec{u}_i = b_i \vec{t}_i$) is computed using

$$b_1 = \frac{\vec{t}_1^T \vec{u}_1}{\|\vec{t}_1^T \vec{t}_1\|} \tag{6.2.16}$$

When the scores, weights, and loadings have been determined for a latent variable (at convergence), X- and Y-block matrices are adjusted to exclude the variation explained by that latent variable. Equations 6.2.17 and 6.2.18 illustrate the computation of the residuals after the first latent variable and weights have been determined:

$$E_1 = X - \vec{t}_1 \vec{p}_1^T \tag{6.2.17}$$

$$F_1 = Y - b_1 \vec{t}_1 \vec{q}_1^T \tag{6.2.18}$$

The entire procedure is repeated for finding the next latent variable and weights starting with Equation 6.2.12. The variations in data matrices X and Y explained by the earlier latent variables are excluded from X and Y by replacing them in the next iteration with their residuals that contain unexplained variation. After the convergence of the first set of latent variables to their final values, X and Y are replaced with the residuals E_1 and F_1, respectively, and all subscripts are incremented by 1.

Several enhancements have been made to the PLS algorithm (Wold et al. 1989; Wold 1992; Lindgren et al. 1994; Dayal and MacGregor 1997; Goutis 1997; Malthouse et al. 1997; Wold et al. 2001). Commercial software is available for developing PLS models (PLS Toolbox for MATLAB (Version 1.2) 2001, Seattle, Washington; SIMCA (Version 6.0) 2001, Umeaa, Sweden).

To model nonlinear relationships between X and Y, their projections should be nonlinearly related to each other (Wold et al. 1989). One alternative is the use of a polynomial function such as

$$\vec{u}_i = c_{0i} + c_{1i}\vec{t}_i + c_{2i}\vec{t}_i^{\,2} + \vec{\varepsilon}_i \qquad (6.2.19)$$

where
 i represents the model dimension
 c_{0i}, c_{1i}, and c_{2i} are constants
 $\vec{\varepsilon}_i$ is a vector of errors (innovations)

This quadratic function can be generalized to other nonlinear functions of $\vec{t}_i$:

$$\vec{u}_i = f(\vec{t}_i) + \vec{\varepsilon}_i \qquad (6.2.20)$$

where $f(\bullet)$ may be a polynomial, exponential, or logarithmic function.

Another structure for expressing a nonlinear relationship between X and Y is splines (Wold 1992) or smoothing functions (Frank 1990). Splines are piecewise polynomials joined at knots (denoted by z_j) with continuity constraints on the function and all its derivatives except the highest. Other nonlinear PLS models that rely on nonlinear inner relations have been proposed (Taavitsainen and Korhonen 1992; Haario and Taavitsainen 1994; Doymaz et al. 2003). Nonlinear relations within X or Y can be modeled by using variable transformations such as defining a new variable that is the square root of the original variable.

6.2.2 CONTRIBUTION PLOTS

MSPM techniques detect significant deviations in process operation from the desired or NO and trigger the need to determine causes affecting the process. MSPM charts such as T^2 and squared prediction error (SPE) charts (Section 6.3) indicate when the process goes out of control, but they do not provide information on the source causes of abnormal process operation. The engineers and plant operators need to determine the actual problem once an abnormality is indicated. Miller and coworkers (1993, 1998) proposed variable contributions and contribution plots concept to address this need. Contribution plots indicate the process variables that have contributed significantly to inflate the T^2-statistic, the SPE-statistic (or Q), and/or the scores. The fault diagnosis activity is completed by using process knowledge of operation personnel or a knowledge-based system to relate these process variables to various equipment failures and disturbances.

Two different approaches for calculating variable contributions to T^2-statistic have been proposed. The first approach calculates the contribution of each process variable to a separate score (MacGregor et al. 1994; Kourti and MacGregor 1996; Miller et al. 1998). T^2 can be written as

$$T^2 = \sum_{i=1}^{m} \frac{\vec{t}_i^2}{\lambda_i} = \sum_{i=1}^{m} \frac{\vec{t}_i^2}{S_i^2} \tag{6.2.21}$$

where
$\vec{t}_i$ are the scores
λ_i are the eigenvalues of S
m is the number of variables
S_i^2 is the variance of t_i (the ith ordered eigenvalue of S)

Each score can be written as

$$\vec{t}_i = \vec{p}^T (\vec{x} - \vec{\bar{x}}) = \sum_{j=1}^{m} p_{i,j}(x_j - \bar{x}_j) \tag{6.2.22}$$

where
$\vec{p}_i$ is the loading
the eigenvector of S corresponding to λ
$p_{i,j}$, x_j, and $\bar{x}_j$ are associated with the jth variable

The contribution of each variable x_j to the score of PC i is given by Equation 6.2.22 as $p_{i,j}(x_j - \bar{x}_j)$. Considering that variables with high levels of contribution that are of the same sign as the score are responsible for driving T^2 to higher values, only those variables are included in the analysis (Kourti and MacGregor 1996). For example, only variables with negative contributions are selected if the score is negative.

The overall contribution of each variable is computed by summing over all scores with high values. For each score with high values (e.g., using a threshold value of 2.5) the variable contributions are calculated (Kourti and MacGregor 1996). Then, the values over all the l high scores are summed for contributions that have the same sign as the score:

1. For all l high scores ($l \leq m$):
   Compute the contribution (cont) of variable x_j to the normalized score $(\vec{t}_i/S_i)^2$:

$$\text{CONT}_{i,j}^{T^2} = \frac{\vec{t}_i^2}{S_i^2} p_{i,j}(x_j - \vec{\bar{x}}_j) = \frac{\vec{t}_i^2}{\lambda_i} p_{i,j}(x_j - \bar{x}_j) \tag{6.2.23}$$

   Set $\text{CONT}_{i,j}^{T^2}$ to zero, if it is negative (sign opposite to the score $\vec{t}_i$)
2. Calculate the total contribution of variable x_j:

$$\text{CONT}_j^{T^2} = \sum_{i=1}^{l} \text{CONT}_{i,j}^{T^2} \tag{6.2.24}$$

Process Monitoring

The second approach calculates contributions of each process variable to the T^2 statistic rather than contributions of separate scores (Nomikos 1996).

$$\text{CONT}_j^{T^2} = \sum_{i=1}^{m} \frac{\vec{t}_i^{\,2}}{S_i^2} p_{i,j}(x_j - \bar{x}_j) \qquad (6.2.25)$$

Contribution to SPE-statistic is calculated by using the individual residuals,

$$\text{CONT}_j^{\text{SPE}} = \left(\vec{x}_j - \hat{\vec{x}}_j\right)\left(\vec{x}_j - \hat{\vec{x}}_j\right)^T = \sum_{i=1}^{n} e_{i,r}^2 \qquad (6.2.26)$$

where
  $\hat{x}_j$ is the vector of predicted values of the (centered and scaled) measured variable j (with n observations)
  $\vec{e}_j$ denotes the residuals

It is always a good practice to check individual process variable plots for those variables diagnosed as responsible for flagging an out-of-control situation. When the number of variables is large, analyzing contribution plots and corresponding variable plots to reason about the source cause of the abnormality may become tedious and challenging. This analysis can be automated and linked with real-time diagnosis (Norvilas et al. 2000; Undey et al. 2000) by using knowledge-based systems.

6.2.3 STATISTICAL METHODS FOR FAULT DIAGNOSIS

Fault diagnosis determines the source causes of abnormal process operation. The fault may be one of many that are already known because of previous experience or a new one. Fault diagnosis activity usually compares the performance of the process (trajectories of process variables) under the current fault to process behavior under various faults (fault signatures) to determine the current fault. A combination of statistical techniques and process knowledge should first be used to catalog process behaviors (fault signatures) from historical data. Pattern-matching methods for this activity have been proposed (Singhal and Seborg 2002a,b, 2006). The identification of fault signatures for faults that have not been determined by operation personnel may necessitate unsupervised learning. This can be achieved by clustering. Once data clusters with various faults have been determined, discrimination and classification are used for fault diagnosis (Fukunaga 1990; Duda et al. 2001). Two linear statistical techniques, discriminant analysis and Fisher discriminant analysis, provide powerful fault diagnosis tools. Neural networks (NNs) have also been used for fault classification and diagnosis. NN-based classification is useful when a small number of faults in a closed set are to be diagnosed, but for more complex cases with multiple faults or new faults, NN does not provide a reliable framework and they may converge to local optima during training. Support vector machines (SVM) provide another nonlinear technique for event classification and fault diagnosis (Cinar et al. 2007).

6.2.3.1 Clustering

Cluster analysis groups data on the basis of similarity measures (Johnson and Wichern 1998). Items and cases are usually clustered by indicating proximity using some measure of distance or angle. Variables are usually grouped on the basis of measures of association such as correlation coefficients.

The distance $d(\vec{x}, \vec{y})$ between two items $\vec{x} = [x_1, x_2, \ldots, x_m]^T$ and $\vec{y} = [y_1, y_2, \ldots, y_m]^T$ can be expressed as the Euclidian distance:

$$d(\vec{x}, \vec{y}) = \sqrt{(\vec{x} - \vec{y})^T (\vec{x} - \vec{y})} \qquad (6.2.27)$$

or the statistical distance (or Mahalanobis distance),

$$d(\vec{x}, \vec{y}) = \sqrt{(\vec{x} - \vec{y})^T S^{-1} (\vec{x} - \vec{y})} \qquad (6.2.28)$$

where S is the covariance matrix. Clustering can be hierarchical such as grouping of species and subspecies in biology or nonhierarchical such as grouping of items. For fault diagnosis, nonhierarchical clustering is used to group data to k clusters corresponding to k known faults.

k-means clustering is a popular nonhierarchical clustering method that assigns each item to the cluster having the nearest centroid (mean) (MacQueen 1967). It consists of

1. Partitioning the items into k initial clusters or specifying k initial mean values as seed points
2. Proceeding through the list of items by assigning an item to the cluster, whose mean is nearest (using a distance measure, usually the Euclidian distance)
3. Recalculation of the mean for the cluster receiving the new item and the cluster losing the item
4. Repeating steps 2 and 3 until no more reassignments take place

The k-means algorithm requires the estimate of the number of clusters, i.e., k, and its solution depends on the initial assignments as the optimization can get stuck in local minima. Furthermore, time series data are inherently autocorrelated that violates the key assumption of independent data elements for traditional clustering algorithms. Beaver and Palazoglu (2006a,b) proposed an agglomerative k-means algorithm that overcomes these drawbacks.

Displaying multivariate data in low-dimensional space can be useful for visual clustering of items. For example, plotting the scores of the first few pairs of PCs as biplots of the first versus the second or the third PC can cluster normal process operation and operation under various faults.

Pattern-matching methods to catalog process behaviors (fault signatures) from historical data have been proposed (Singhal and Seborg 2002a,b, 2006). For high-dimensional data, distance measures may not be enough to describe the locations of specific clusters with respect to one another. Angle measures provide additional information (Krzanowski 1979; Raich and Cinar 1997).

6.2.3.2 Discriminant Analysis

Statistical discrimination separates distinct sets of objects or events and classification allocates new objects or events into previously defined groups (Johnson and Wichern 1998). Discrimination focuses on discrimination criteria, called discriminants, for converting salient features of objects from several known populations to quantitative information separating these populations as much as possible. Classification sorts new objects or events into previously labeled classes by using rules derived to optimally assign new objects to the labeled classes. A good classification rule takes the "prior probabilities of occurrence" into consideration, accounts for the costs associated with misclassification, and yields few misclassifications.

Fisher suggested the transformation of multivariate observations $\vec{x}$ to another coordinate system that enhances the separation of the samples belonging to each class (Fisher 1938). Fisher's discriminant analysis (FDA) is optimal in terms of maximizing the separation among the set of classes. Discriminant analysis and FDA are discussed in most multivariate statistics books (Johnson and Wichern 1998) and books on SPM and fault diagnosis (Chiang and Braatz 2001; Cinar et al. 2007).

6.2.4 EXAMPLE: ANALYSIS OF THE GROWTH OF FRESH APPLES

Site-specific monitoring of Glindow apples (apple growing location in Germany) from three different cultivars, namely 'Elstar', 'Topaz', and 'Pinova', was carried out over a harvesting period to illustrate PCA techniques. Nondestructively recorded apple data were obtained during fruit development on tree. The data set is open for further calculation at www.atb-potsdam.de/fruitloc with login as user: gast2006 using the password: apple2006 +. There have been slightly abnormal developments in growing location Glindow. Trees denoted as 'Elstar' DS suffered drought stress. Resulting from the water deficiency, these apples ripened earlier, but, however, were picked on the same days with the others and therefore had a shorter shelf life. In this study, nondestructive optical sensing of fresh apples as well as conventional laboratory analysis were used to study the fruit ripening and quality changes in time and to monitor new growing samples to detect any abnormality in their growth conditions.

6.2.4.1 PCA Analysis of the Spectral Data

Nondestructive spectral data are available for 10 fruit samples from each cultivar. The measurements were made once a week for a total of 8 weeks, starting on August 7, 2006.

TABLE 6.2.1
Percent Variation Explained by Each PC Used in the Model

Principal Component Number	Eigenvalue of Cov(X)	Percent Variance Captured by This PC	Percent Variance Captured by Total
1	5.72e+008	64.90	64.90
2	1.42e+008	16.15	81.04
3	1.14e+008	12.87	93.92

First, data from all cultivars measured on 7/8/2006 are modeled using PCA. Spectral data for equally spaced wavelengths ranging from 403 to 1098 nm, for each fruit sample, constitute the rows of the PCA data matrix. After mean centering, a PCA model is built using three PCs on the (30 × 211) data matrix. The number of PCs is found using the broken stick rule. The first PC explains 64.90% and the first three explain 93.92% of the total variation in the data (Table 6.2.1).

The score biplots show how the samples are related to each other and also reveal the strong outliers in the data. The score biplot of the first two PCs is shown in Figure 6.2.3. The scores corresponding to different cultivars are shown using different markers, to reveal the clustering in the data. Fruit samples #6 and 8 in 'Elstar' cluster, sample #1 in 'Topaz', and, finally, samples #2, 3, and 6 in 'Pinova' are classified as outliers and will be excluded from the model data.

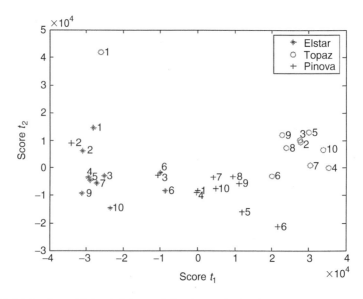

FIGURE 6.2.3 Score biplot (t_2/t_1) reveal the clusters in the data.

Process Monitoring 483

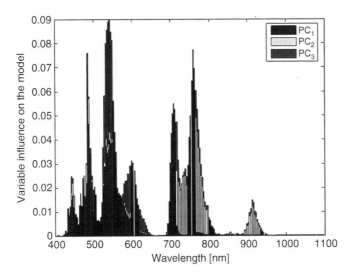

FIGURE 6.2.4 (See color insert following page 376.) Explained variation of the variables by each PC.

The PCA model was rebuilt after excluding the outlier samples. The first three PCs explain 95.87% of the total variation.

Figure 6.2.4 shows the variable influences on the model, in other words, explained variation of the variables by each PC. The first PC captures variations in wavelengths 400–650 nm, the second PC explains variations in wavelengths 700–1100 nm, and the third PC explains variations in wavelengths 550–650 nm. Wavelengths 650–700 nm and 800–900 nm are not explained by any of the components since there is no significant change in those ranges for any of the cultivars. From the spectra, it is concluded that Pinova and Elstar cultivars exhibit similar landmark curves, whereas, Topaz is different from them, especially for low wavelengths. As shown in Section 3.2, such result is reasonable, since the profile of anthocyanins absorbing in the wavelengths 400–550 nm present in the different cultivars depend on the genome. However, the lack of variations in the chlorophyll absorption range 650–700 nm can be explained by complete extinction of all photos injected in the fruit. The score plot of the new model is given in Figure 6.2.5. In this figure, the clusters are easily detectable and there are no strong outliers present.

In the next step, the time rate of change of apple spectra over several days is considered. Figure 6.2.6 shows the score biplot of a three PC model, which is built using spectral data of six fruit samples taken from the Elstar cultivar on five different days. In the figure, the first 5 days are plotted. The initial samples are more similar to each other and cluster very closely. They deviate some more from each other everyday; this is reflected as the increasing distance between samples in every cluster. Moreover, the deviation between days is also reflected as the increasing distance between the clusters themselves. The second cluster showing the measurements

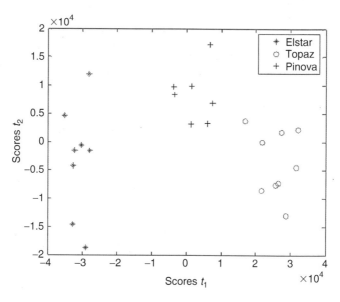

FIGURE 6.2.5 Score biplot show the three clusters, no strong outliers.

taken on the second day falls close to the first cluster showing the initial measurements. The inter-distance increases for consecutive days, reflecting the time rate of change in the samples' spectral properties.

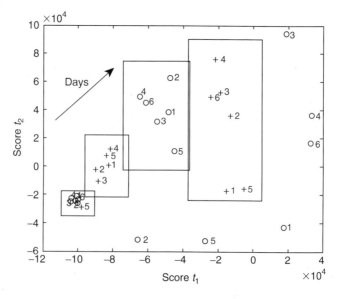

FIGURE 6.2.6 Score biplot showing data taken on different days for six fruit samples taken from 'Elstar'.

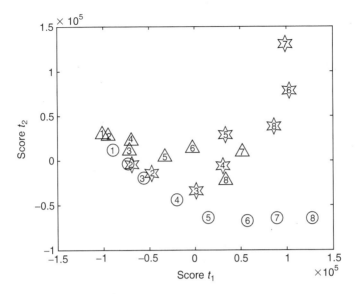

FIGURE 6.2.7 Score biplot showing data taken on different days for each cultivar.

The mean spectrum for these samples is calculated for each day and for each cultivar, to see if a time dependency between observations occurs and how they are related to each other (Figure 6.2.7).

6.3 SPM AND FAULT DIAGNOSIS IN CONTINUOUS PROCESSES

6.3.1 SPM Methods Based on PCA

MSPM methods with PCs can use various types of charts. If only a few PCs can describe the process behavior satisfactorily, biplots could be used as visual aids that are easy to interpret. Biplots can be generated by projecting the data to two-dimensional surfaces such as PC_1 versus PC_2, PC_1 versus SPE, and PC_2 versus SPE (Figure 6.3.1). A shortcoming of these plots is the loss of the visual inspection of trends over time. This is alleviated by including the time stamp for each data point.

Data representing NO and various faults are clustered in different regions, providing the opportunity to diagnose source causes as well (Kresta et al. 1991). Score biplots are used to detect any departure from the in-control region defined by the confidence limits calculated from the reference set. The length of the confidence ellipsoid's axis in the direction of ith PC is given by Johnson and Wichern (1998)

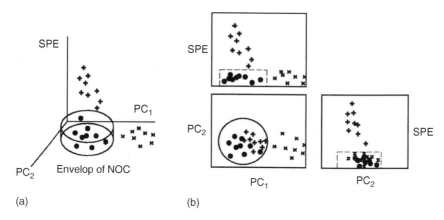

FIGURE 6.3.1 Multivariate monitoring space. (a) Three-dimensional representation, (b) two-dimensional representation. (From Cinar, A., Palazoglu, A., and Kayihan, F., *Chemical Process Performance Evaluation*, CRC Press, 2007. With permission.)

$$\pm \left[S_i F_{a,n-a,\alpha} \frac{a(n^2-1)}{n(n-a)} \right]^{\frac{1}{2}} \quad (6.3.1)$$

where
  S is the estimated covariance matrix of scores
  $F_{a,n-a,\alpha}$ is the F-distribution value with a and $n-a$ degrees of freedom in α significance level
  a is the number of PCs retained in the model
  n is the number of samples in the reference set

Inspection of many biplots becomes difficult to interpret when a large number of PCs are needed to describe the process. Monitoring charts based on SPEs and T^2 become more useful. An MSPM chart is easy to interpret as a Shewhart chart is obtained by plotting SPE and T^2 statistics and their confidence intervals.

Sometimes, plots of individual PC scores can be used for preliminary analysis of variables that contribute to an out-of-control signal. The control limits for new $\vec{t}$ scores under the assumption of normality at significance level α at any time k is (Hahn and Meeker 1991)

$$\pm t_{n-1,\alpha/2} s_{\text{ref}} \left(1 + \frac{1}{n} \right)^{1/2} \quad (6.3.2)$$

where
  n and s_{ref} are the number of observations and the estimated standard deviation of the $\vec{t}$-score sample at sampling time k (mean is always 0)
  $t_{n-1,\alpha/2}$ is the critical value of the Studentized variable with $n-1$ degrees of freedom at significance level $\alpha/2$

6.3.1.1 Hotelling's T^2 Charts

Hotelling's T^2 plot detects the small shifts and deviations from NO better than charts of individual variables since it includes contributions of all variables and can become significantly faster than the deviation of an individual variable. The T^2 statistic based on process variables at sampling time k is

$$T^2(k) = (\vec{x}(k) - \bar{\vec{x}})^T S^{-1} (\vec{x}(k) - \bar{\vec{x}}) \quad (6.3.3)$$

where $\bar{\vec{x}}$ and S are estimated from process data. If the individual observation vector $\vec{x}(k)$ is independent of $\bar{\vec{x}}$ and S, then T^2 follows an F-distribution with m and $n - m$ (m measured variables, n sample size) degrees of freedom (Mason and Young 2002):

$$T^2 \sim \left[\frac{m(n+1)(n-1)}{n(n-m)} \right] F_{m,n-m} \quad (6.3.4)$$

If the observation vector $\vec{x}$ is not independent of the estimators $\bar{\vec{x}}$ and S, but is included in their computation, then T^2 follows a beta distribution with $m/2$ and $(n - m - 1)/2$ degrees of freedom (Mason and Young 2002):

$$T^2 \sim \left[\frac{(n-1)^2}{n} \right] B_{m/2, (n-m-1)/2} \quad (6.3.5)$$

The T^2 charts based on PCs use

$$T^2(k) = \vec{t}_a^T(k) S^{-1} \vec{t}_a(k) \quad (6.3.6)$$

and follow an F- or a beta distribution for the same conditions leading to Equations 6.3.4 and 6.3.5, with a and $n - a$ degrees of freedom for the F-distribution, and $a/2$ and $(n - a - 1)/2$ degrees of freedom for the beta distribution, for data that follow a multivariate normal distribution (Jackson 1980, 1991). As before, a denotes the number of PCs, $\vec{t}_a$ is a vector containing the scores from the first a PCs (Jackson 1991), and S is the $(a \times a)$ estimated covariance matrix, which is diagonal due to the orthogonality of the $\vec{t}$ scores (Tracy et al. 1992). The T^2 based on PCs can also be calculated at each sampling time k as (Jackson 1991)

$$T^2(k) = \sum_{i=1}^{a} \frac{t_i^2(k)}{\lambda_i} = \sum_{i=1}^{a} \frac{t_i^2(k)}{S_i^2} \quad (6.3.7)$$

where the PC scores $\vec{t}_i$ have variance λ_i (or estimated variance S_i^2 from the scores of the reference set) which is the ith largest eigenvalue of the covariance matrix S. The term k that indicates the explicit dependence on sampling time will be omitted from the T^2 equations in the remainder of the section. If tables for the beta distribution are not readily available, this distribution can be approximated by using (Tracy et al. 1992)

$$B_{a/2,(n-a-1)/2,\alpha} = \frac{[a/(n-a-1)]F_{a,(n-a-1),\alpha}}{1 + [a/(n-a-1)]F_{a,(n-a-1),\alpha}} \quad (6.3.8)$$

Another chart based on the T^2 statistic is the D chart:

$$D(k) = \frac{\vec{t}_a^T S^{-1} \vec{t}_a n}{(n-1)^2} \sim B_{a/2,(n-a-1)/2} \qquad (6.3.9)$$

6.3.1.2 SPE Charts

The SPE charts show deviations from NO based on variations that are not captured by the PC model. Since the model does not describe variations that are not captured in it, SPE rather than T^2 is inflated. Equation 6.2.2 can be rearranged to compute the prediction error (residual) E

$$X = TP^T + E, \quad E = X - \hat{X} \qquad (6.3.10)$$

where $\hat{X} = TP^T$ denotes the estimates of the data X. The location of the projection of an observation at time k on the a-dimensional PC space is given by its score $t_a(k)$. The orthogonal distance of the observation $\vec{x}(k)$ from the projection space is the prediction error $\vec{e}(k)$. SPE(k) is the sum for all times k of the square of $\vec{e}(k)$, which indicates how close the observation at time k is to its description in the a-dimensional space:

$$\text{SPE}(k) = \vec{e}(k)^T \vec{e}(k) = \sum_{j=1}^{m} e_j^2(k) = \sum_{j=1}^{m} [x_j(k) - \hat{x}_j(k)]^2 \qquad (6.3.11)$$

where $\hat{x}_j(k)$ is computed from the PCA model. SPE is also called the Q-statistic.

Statistical limits on the Q-statistic are computed by assuming that the data have a multivariate normal distribution (Jackson 1980, 1991). The control limits for Q-statistic are given by Jackson and Mudholkar (1979) based on Box's formulation (Equation 6.3.12) for quadratic forms with significance level of α given in the following equations as

$$Q_\alpha = g\chi^2_{h,\alpha} \qquad (6.3.12)$$

$$Q_\alpha = \theta_1 \left[1 - \frac{\theta_2 h_0 (1 - h_0)}{\theta_1^2} + \frac{z_\alpha (2\theta_2 h_0^2)^{1/2}}{\theta_1} \right]^{\frac{1}{h_0}} \qquad (6.3.13)$$

where
χ^2_h is the chi-squared variable with h degrees of freedom
z is the standard normal variable corresponding to the upper $(1 - \alpha)$ percentile with z_α having the same sign as h_0. θ values are calculated using unused eigenvalues of the covariance matrix of observations that are not included in the model as (Wise and Gallagher 1996)

Process Monitoring

$$\theta_i = \sum_{j=a+1}^{m} \lambda_j^i, \quad \text{for } i = 1, 2, \text{ and } 3 \tag{6.3.14}$$

The other parameters are

$$g = \frac{\theta_2}{\theta_1}, \quad h = \frac{\theta_1^2}{\theta_2}, \quad h_0 = 1 - \frac{2\theta_1\theta_3}{3\theta_2^2} \tag{6.3.15}$$

$\theta_i's$ can be estimated from the estimated covariance matrix of prediction errors (residuals) (residual matrix used in Equation 6.3.11) for use in Equation 6.3.13 to develop control limits on Q for comparing prediction errors. A simplified approximation for Q-limits has also been suggested in the text book of Evans and coworker (1993) by rewriting Box's equation (Equation 6.3.12) by setting $\theta_2^2 \approx \theta_1\theta_3$.

$$Q_\alpha \cong gh\left[1 - \frac{2}{9h} + z_\alpha\left(\frac{2}{9h}\right)^{1/2}\right]^3 \tag{6.3.16}$$

SPE values for new data at time k are calculated using

$$\text{SPE}(k) = \sum_{j=1}^{m} \left[x_j(k) - \hat{x}_j(k)\right]^2 \tag{6.3.17}$$

These SPE(k) values computed using Equation 6.3.17 follow the χ^2 (chi-squared) distribution (Box 1954). This distribution can be well approximated at each time interval using Box's equation in Equation 6.3.12 (or its modified version in Equation 6.3.16).

T^2 statistic is used to monitor the systematic variation and SPE statistic is used to monitor the variation of residuals. Hence, in the case of a process disturbance, either of these statistics will exceed the control limits. If only the T^2 statistic is out of control, the model of the process is still valid but the contributions of each process variable to this statistic should be investigated to find a cause for the deviation from NO. If SPE is out of control, a new event that is not described by the process model is found in the data. Contributions of each variable to SPE will unveil the responsible variables to that deviation.

6.3.2 SPM AND SQC METHODS BASED ON PLS

Large amounts of process data, such as temperatures, humidity, and flow rates, are collected at high frequency by process data collection systems. Information on product quality variables is collected less frequently. Although it is possible to measure some quality variables on-line by means of sophisticated devices, measurements are generally made off-line in the quality control laboratory and often involve time lags between data collection and receiving the analysis results. Process data contain valuable information about both the quality of the product and the performance of the process operation. Partial least square (PLS) models provide the quantitative relations for estimating product quality from process data. They can also be used to quickly detect process upsets and unexpected behavior.

Cross correlations and collinearity among process variables severely limit the use of traditional linear regression techniques. PLS offers a suitable solution for modeling such data.

The first step in the development of a PLS model is to group the process variables as X and the product quality variables as Y (Figure 6.3.2). This selection is dependent on the measurements available and the objectives of monitoring. The reference set used to develop the multivariate monitoring chart will determine the variations considered to be part of NO and ideally include all variations leading to desired process performance and product quality. If variations in the reference set are too small, the resulting model that is used for process monitoring will cause frequent alarms. If the reference data set contains large variations beyond the acceptance tolerance, the sensitivity for detecting abnormal operation will be poor.

Since PLS technique is sensitive to outliers and scaling, outliers should be removed and data should be scaled before modeling. After data pretreatment, the number of latent variables (PLS dimensions) to be retained in the model is determined. Cumulative prediction sum of squares (CUMPRESS) versus the number of latent variables or prediction sum of squares (PRESS) versus the number of latent variables plots are used for this purpose. It is usually enough to consider the first few PLS dimensions for monitoring activities, while more PLS dimensions are needed for prediction to improve the accuracy of predictions.

The SPE can be calculated for the X and the Y block models.

$$\text{SPE}_{X,k} = \sum_{j=1}^{m} \left(x_{kj} - \hat{x}_{kj} \right)^2 \quad (6.3.18)$$

$$\text{SPE}_{Y,k} = \sum_{j=1}^{q} \left(y_{kj} - \hat{y}_{kj} \right)^2 \quad (6.3.19)$$

where

$\vec{\hat{x}}$ and $\vec{\hat{y}}$ are predicted observations in X and Y using the PLS model, respectively
k and j are the indexes for observations and variables in X or Y, respectively

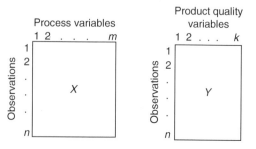

FIGURE 6.3.2 Arrangement of data in PLS for SPM. (From Cinar, A., Parulekar, S.J., Undey, C., and Birol, G., *Batch Fermentation: Modeling, Monitoring and Control*, Marcel Dekker, 2003. With permission.)

$\hat{x}_{kj}$ and $\hat{y}_{kj}$ in Equations 6.3.18 and 6.3.19 are calculated for new observations by using

$$t_{i,\text{new}} = \sum_{j=1}^{m} x_{\text{new},j} w_{i,j} \qquad (6.3.20)$$

$$\hat{x}_{\text{new},j} = \sum_{i=1}^{a} t_{i,\text{new}} p_{i,j} \qquad (6.3.21)$$

$$\hat{y}_{\text{new},j} = \hat{x}_{\text{new},j} b \qquad (6.3.22)$$

where
 $w_{i,j}$ denotes the weights
 $p_{i,j}$ the loadings for the X block (process variables) of the PLS model
 $t_{i,\text{new}}$ the scores of new observations
 b the regression coefficient for the inner relations

Multivariate monitoring charts based on T^2 and squared prediction errors (SPE_X and SPE_Y) are constructed using the PLS models. The T^2 statistic for a new independent $\vec{t}$ vector is (Tracy et al. 1992)

$$T^2 = \vec{t}_{\text{new}}^T S^{-1} \vec{t}_{\text{new}} \sim \frac{a(n^2-1)}{n(n-a)} F_{a,n-a} \qquad (6.3.23)$$

where
 S is the estimated covariance matrix of PLS model scores
 a the number of latent variables retained in the model
 $F_{a,n-a}$ the F-distribution value

The control limits on SPE charts can be calculated by an approximation of the χ^2 distribution given as $\text{SPE}_\alpha = g\chi^2_{h\alpha}$ (Box 1954). This equation is well approximated as (Jackson and Mudholkar 1979; Evans et al. 1993; Nomikos and MacGregor 1995a)

$$\text{SPE}_\alpha \cong gh\left[1 - \frac{2}{9h} + z_\alpha \left(\frac{2}{9h}\right)^{1/2}\right]^3 \qquad (6.3.24)$$

where
 g is a weighting factor
 h is the degrees of freedom for the χ^2 distribution

These can be approximated as $g = v/(2m)$ and $h = 2m^2/v$, where v is the variance and m the mean of the SPE values from the PLS model.

Biplots of scores ($\vec{t}_i$ versus $\vec{t}_{i+1}$, for $i = 1, \ldots, a$) can also be developed. The control limits at significance level $\alpha/2$ for a new independent t score under the assumption of normality at any sampling time are

$$\pm t_{n-1,\alpha/2} s_{est} \left(1 + \frac{1}{n}\right)^{1/2} \qquad (6.3.25)$$

where

n and s_{est} are the number of observations and the estimated standard deviation of the score sample at the chosen time interval, respectively

$t_{n-1,\alpha/2}$ is the critical value of the t test with $n-1$ degrees of freedom at significance level $\alpha/2$ (Hahn and Meeker 1991; Nomikos and MacGregor 1995a).

6.3.3 FAULT DIAGNOSIS USING CONTRIBUTION PLOTS

When T^2 or SPE charts exceed their control limits to signal abnormal process operation, variable contributions can be analyzed to determine the variables causing the inflation of the monitoring statistic and initiating the alarm. The variables identified provide valuable information to operation personnel, who are responsible for associating these process variables with equipment malfunctions or external disturbances, and diagnosing the source causes for the abnormal process behavior.

Analysis of contribution plots can be automated and linked with fault diagnosis by using real-time knowledge-based systems (KBS). The integration of statistical detection tools and contribution plots with fault diagnosis by using a supervisory KBS has been illustrated for both continuous (Norvilas et al. 2000) and batch processes (Undey et al. 2003a, 2004).

6.3.4 FAULT DIAGNOSIS USING STATISTICAL METHODS

Contribution plots provide an indirect approach to fault diagnosis by first determining the process variables that have inflated the detection statistics. These variables are then related to equipment malfunctions and disturbances. A direct approach would associate the trends in process data to faults explicitly. Statistical classification and discriminant analysis techniques provide methods for implementing fault diagnosis directly.

When a process can be represented by a few PCs, the biplots of PCs and SPE provide a visual aid to identify data clusters that indicate NO or operation under a specific fault (Figure 6.3.1). An integrated statistical method was developed by utilizing PCA and discriminant analysis techniques for processes that need to be described by a higher number of PCs or for automation of diagnosis activities (Raich and Cinar 1996). PCA is used to develop a model describing NO. This PC model is used to detect outliers from NO, as excessive variation from normal target or unusual patterns of variation. Operation under various known upsets is also modeled using PCA, provided that sufficient historical data are available. These fault models are then used to isolate source causes of abnormal operation based on the proximity of

Process Monitoring

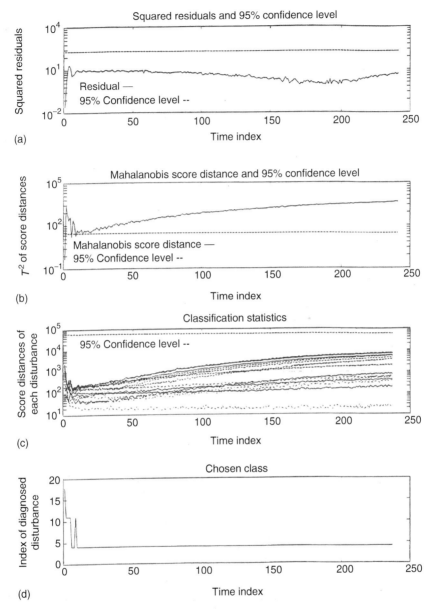

FIGURE 6.3.3 Detection and diagnosis of process upsets. (a) Detection of fault based on residuals, (b) detection based on T^2 test of scores, (c) diagnosis statistics considering each possible disturbance, and (d) index of chosen disturbance for each observation. (From Raich, A. and Cinar, A., *AIChE J.*, 42, 995, 1996. With permission.)

current process operation to one of the data clusters indicating a specific source cause. Using PCs for several sets of data under different operating conditions (NO and with various upsets) statistics can be computed to describe distances of the current operating point to regions representing other conditions of operation. Both score distances and model residuals are used to measure such distance-based statistics. In addition, angle-based discrimination criteria can also be used. The diagnosis system design includes the development of PC models for NO and abnormal operation with specific malfunctions and disturbances, and the computation of threshold limits using historical data sets collected during NO and operation under specific causes leading to abnormalities.

At each data collection, the implementation of the abnormality detection and diagnosis system starts with monitoring. The model describing NO is used with new data to decide if the current operation is in control. If there is no significant evidence that the process is out of control, further analysis is not necessary and the procedure is concluded for that time. If scores or residuals tests exceed their statistical limits, there is significant evidence that the process is out of control. Then, the PC models for all faults are used to carry out the score and residuals distance tests and/or angle tests, and discriminant analysis is performed by using PC models for various faults to diagnose the source cause of abnormal behavior.

The diagnosis system is implemented by plotting and inspecting the diagnosis statistics visually. It can be automated using software to compare the magnitudes of various diagnosis statistics and inferences, and to report the results to operation personnel. The test statistics using different types of discriminants can be plotted versus the sample number. An example for fault diagnosis from a continuous chemical process illustrates the plots generated and the diagnosis.

Figure 6.3.3a and b shows residuals and scores, respectively (plotted as the negative of the discriminant), at each sampling time during a run with a disturbance (disturbance number 3) in the Tennessee–Eastman industrial challenge problem, a benchmark problem for continuous processes (Downs and Vogel 1993). The figure illustrates the fault isolation process when disturbance 3 (step change in feed temperature) is introduced. Score discriminants are calculated using PC models for the various known faults (Figure 6.3.3c); this semilog plot shows the negative of the discriminant. The most likely fault is chosen over time by selecting the fault corresponding to the maximum discriminant (curve with the lowest magnitude). Figure 6.3.3d reports the fault selected at each sampling time. Fault 3, which is the correct fault, is reported consistently after the first 10 sampling times.

6.4 SPM TECHNIQUES FOR MULTIVARIABLE BATCH PROCESSES

Batch processes are used frequently in food processing. They often exhibit some batch-to-batch variation. Variations in charging the production recipe, differences in types and levels of impurities in raw materials, shift changes of operators, and disturbances during the progress of the batch are some of the reasons for this behavior. Growth and preservation of fresh fruits and vegetables are also batch processes that can benefit from MSPM and quality control techniques.

Monitoring the trajectories of the process variables provides four different types of batch process monitoring and detection activities:

1. End of batch quality control: This is similar to the traditional quality-control approach. The ability to merge information from quality variables and process variable trajectories by multivariate statistical tools enables more accurate decision making. Since process variable trajectories are available immediately at the conclusion of the batch, product quality can be inferred from them without any time delay.
2. Analysis of process variable trajectories after the conclusion of the batch: This postmortem analysis of the batch progress can identify major deviations in process variable trajectories. Since the analysis is carried out after the conclusion of the batch, it cannot be used to improve the product of that batch, but of future products.
3. Real-time batch process monitoring: Real-time monitoring provides information about the progress of the batch, while the physical, physiological, and chemical changes are occurring during the run. This enables the observation of deviations from desired trajectories, implementation of interventions to eliminate the effects of disturbances, and abortion of the batch if corrections are not feasible.
4. Real-time quality control: This is the most challenging and valuable case. During the progress of the batch, frequently measured process variables can be used to estimate product quality at the end of the batch run. This provides an opportunity to foresee, if there is a tendency toward inferior product quality and take necessary actions to prevent final product quality.

Batch process data form a three-way array with dimensions: batches (I), variables (J), and time (K). PCA and PLS are extended to multiway PCA (MPCA) and multiway PLS (MPLS) to handle the added data dimension. There will be a temporal variation, in addition to magnitude variation, in process trajectories for each batch run resulting in unequal and unsynchronized data. Before model development, it is crucial to apply a batch data length equalization/synchronization technique. Three techniques are available in the literature: indicator variable (IV), dynamic time warping (DTW), and curve registration (Cinar et al. 2003). Equalized/synchronized data form a three-way array. Unfolding the three dimensional (3D) array into a matrix enables the use of simpler computational tools. Subtracting the mean trajectory set from each batch trajectory set removes most of the nonlinearity and improves the accuracy of MPCA and MPLS models.

Developing empirical models and MSPM charts requires a reference database that contains past successful batches representing NO. The historical database containing only common cause variations will provide a reference distribution against which future batches can be compared. Selection of the reference batch records from a historical database depends on the objective of the monitoring paradigm that will be implemented. MPCA-based modeling is suitable if only the process variables are of interest. The MPLS model will allow inclusion of quality variables in the monitoring scheme. The initial set of NO reference set may contain outlying batches. These batches must be identified and removed before MPCA or MPLS modeling. Once the reference model is developed, multivariate monitoring chart limits are constructed.

6.4.1 SPM Charts for Batch Processes

Various multivariate charts are used as visual aids for interpreting multivariate statistics calculated based on empirical models. Each chart can be constructed to monitor batches at the end of the run or during the progress of a batch.

Score biplots or 3D plots are used to detect any departure from the in-control region defined by the confidence limits calculated from the reference set. The score plots provide a summary of process performance from one batch to the next. Assuming normal distribution, the control limits for new independent $\vec{t}$ scores at significance level α at any time step k is (Hahn and Meeker 1991)

$$\pm t_{n-1,\alpha/2} s_{\text{ref}} (1 + 1n)^{1/2} \qquad (6.4.1)$$

where

n and s_{ref} are the number of observations and the estimated standard deviation of the $\vec{t}$-score sample at time k (the mean is always zero)

$t_{n-1,\alpha/2}$ is the critical value of t with $n-1$ degrees of freedom at significance level $\alpha/2$ (Nomikos and MacGregor 1995a)

The axis lengths of the confidence ellipsoids in the direction of ath PC are given by (Johnson and Wichern 1998)

$$\pm \left\{ S_a F_{A,I-A,\alpha} A(I^2 - 1)/[I(I - A)] \right\}^{1/2} \qquad (6.4.2)$$

where

S is the estimated covariance matrix of scores

$F_{A,I-A,\alpha}$ is the F-distribution value with A and $I-A$ degrees of freedom in α significance level

I is the number of batches in the reference set

A is the number of PCs retained in the model

Hotelling's T^2 plot detects the shifts and deviations from NO described by the model. The T^2-statistic is defined in Equation 6.3.6 and the D-statistic in Equation 6.3.9. Statistical limits on the T^2- and D-statistic are computed by assuming that the data follow at multivariate normal distribution (Jackson 1980, 1991). D-statistic for end-of-batch SPM for batch i is

$$D_i = \frac{\vec{t}_a^T S^{-1} \vec{t}_a I}{(I-1)^2} \sim B_{A/2,(I-A-1)/2} \qquad (6.4.3)$$

Process Monitoring

where
$\vec{t}_a$ is a vector of A scores (Jackson 1991)
S is the $(A \times A)$ estimated covariance matrix, which is diagonal because of the orthogonality of the $\vec{t}$ scores (Tracy et al. 1992)

The T^2-statistic follows the beta distribution. It can also be calculated for each batch as (Jackson 1991)

$$T_i^2 = \sum_{a=1}^{A} \frac{t_{ia}^2}{\lambda_a} = \sum_{a=1}^{A} \frac{t_{ia}^2}{S_a^2} \qquad (6.4.4)$$

where the PCA scores $\vec{t}$ in dimension a have variance λ_a (or estimated variance S_a^2 from the scores of the reference set), which is the ath largest eigenvalue of the scores covariance matrix S.

If tables for the beta distribution are not readily available, this distribution can be approximated using Equation 6.4.5 (Tracy et al. 1992).

$$B_{A/2,(I-A-1)/2,\alpha} = \frac{[A/(I-A-1)]F_{A,I-A-1,\alpha}}{1+[A/(I-A-1)]F_{A,I-A-1,\alpha}} \qquad (6.4.5)$$

For real-time monitoring and for analysis of variable trajectories after the conclusion of the batch run, T^2 values are calculated as batch progresses. As soon as the batch is complete, Equation 6.4.6 is applied for each observation at time k.

$$T_{ik}^2 = \vec{t}_{iAk}^T S^{-1} \vec{t}_{iAk} \qquad (6.4.6)$$

T^2 values for each time k for a new batch can also be calculated similar to Equation 6.4.4 as

$$T_k^2 = \sum_{a=1}^{A} \frac{t_{ak}^2}{\lambda_a} = \sum_{a=1}^{A} \frac{t_{ak}^2}{S_{ak}^2} \qquad (6.4.7)$$

T^2 values follow F-distribution (Tracy et al. 1992)

$$T_k^2 \sim \frac{A(I^2-1)}{I(I-A)} F_{A,I-A} \qquad (6.4.8)$$

where
A denotes the number of PCs
I denotes the number of batches in the reference set

SPE plot shows deviations from NO that are not described by the model. The ith elements of the $\vec{t}$-score vectors correspond to the ith batch with respect to the other batches in the database over the entire history of the batch. The P loading matrices summarize the time variation of the measured variables about their average trajectories. If a new batch is good and consistent with the normal batches (used to develop MPCA model), its scores should fall within the normal range and the sum of the squared residuals (Q-statistic) should be small. The Q-statistic for end-of-batch SPM for batch i is

$$Q_i = \vec{e}_i \vec{e}_i^T = \sum_{c=1}^{KJ} E(i,c)^2 \tag{6.4.9}$$

where
$\vec{e}_i$ is the ith row of E
K is the time of duration of the batch
J is the number of variables

Statistical limits on the Q-statistic are computed by assuming that the data have a multivariate normal distribution (Jackson 1980, 1991). The control limits for Q-statistic are given by Jackson and Mudholkar (1979) based on Box's (1954) formulation (Equation 6.3.12) for quadratic forms with significance level of α given in Equations 6.3.12 and 6.3.13.

Since the covariance matrices $E^T E \sim (JK \times JK)$ and $EE^T \sim (I \times I)$ have the same nonzero eigenvalues (Nomikos and MacGregor 1995a), EE^T can be used in estimating θ_i's owing to its smaller size for covariance estimation as

$$V = \frac{EE^T}{I-1}, \quad \theta_i = \text{trace}(V^i), \quad \text{for } i = 1, 2, \text{ and } 3 \tag{6.4.10}$$

Equation 6.3.13 can be used together with Equation 6.4.10 to calculate control limits for SPE when comparing batches (Q_i in Equation 6.4.9).

SPE values for batch I at measurement time k are computed by using (Nomikos and MacGregor 1995a)

$$\text{SPE}_{jk} = \sum_{j=1}^{J} (x_{ijk} - \hat{x}_{ijk})^2 = \sum_{j=1}^{J} e_{ijk}^2 \tag{6.4.11}$$

Calculated SPE values follow χ^2 (chi-squared)-distribution (Box 1954) approximated at each time k using Box's equation in Equation 6.3.12 (or its modified version in Equation 6.3.16). Parameters g and h can be approximated by matching moments of the $g\chi_h^2$ (Nomikos and MacGregor 1995a):

$$g = \frac{\nu}{2m}, \quad h = \frac{2m^2}{\nu} \tag{6.4.12}$$

where m and ν are the estimated mean and variance of the SPE at time k, respectively. These matching moments were susceptible to error in the presence of outliers in the data or when the number of observations was small.

Contribution plots are used for fault diagnostics. Both T^2 and SPE charts produce an out-of-control signal when a fault occurs, but they do not provide any information about the cause. Variable contributions to T^2 and SPE values indicate that variables are responsible for the deviation from NO.

Explained variance, loadings, and weight plots indicate the variability of batch profiles. Explained variance is based on the ratio of the variance of the model estimate to the variance of the real process data. This can be calculated as a function of batch number, time, or variable number. The value of explained variance increases as the model accounts for more variability in the data and for the correlation that exists among the variables. Variance plots over time can be used as an indicator of the phenomenological/operational changes that occur during the process evolution. This measure can be computed as

$$\text{SS explained } [\%] = \frac{\hat{\sigma}^2}{\sigma^2} \times 100 \quad (6.4.13)$$

where
SS stands for the sum of squares
σ^2 and $\hat{\sigma}^2$ are the true and estimated sum of squares, respectively

Loadings also represent variability across the entire data set. Although the loadings look like contributions, a practical difference occurs when some of the contributions of the process variables have values much smaller than their corresponding loadings and vice versa.

In MPLS-based empirical modeling, variable contributions to weights (W) carry valuable information, since these weights summarize information about the relationship between X and Y blocks. The overall effect of all process variables on quality variables over the course of process can be plotted, or this can be performed for a specific period of the process to reflect the change of the effect of the predictor block (X).

6.4.2 MULTIWAY PLS-BASED SPM FOR POSTMORTEM ANALYSIS

MPLS (Wold et al. 1987) extends PLS to predict final product quality during or at the end of a batch run (Kourti et al. 1995; Nomikos and MacGregor 1995b; Wold et al. 1998). When a batch is finished, a block of recorded process variables X_{new} ($K \times J$) and a vector of quality measurements $\vec{y}_{\text{new}}$ ($1 \times M$) are obtained. $\vec{y}_{\text{new}}$ is usually measured with a delay owing to off-line quality analysis. X_{new} ($K \times J$) is unfolded to $\vec{x}_{\text{new}}$ ($1 \times KJ$), and both $\vec{x}_{\text{new}}$ and $\vec{y}_{\text{new}}$ are scaled using the reference set scaling factors. Then, they are processed with MPLS model loadings and weights that contain structural information on the behavior of NO data set as

$$\vec{t}_{\text{new}} = \vec{x}_{\text{new}} W \left(P^T W \right)^{-1}, \quad \vec{e}_{\text{new}} = \vec{x}_{\text{new}} - \vec{t}_{\text{new}} P^T \quad (6.4.14)$$

$$\vec{\hat{y}}_{\text{new}} = \vec{t}_{\text{new}} Q^T, \quad \vec{f}_{\text{new}} = \vec{y}_{\text{new}} - \vec{\hat{y}}_{\text{new}} \quad (6.4.15)$$

where
$\vec{t}_{\text{new}}$ ($1 \times A$) denotes the predicted $\vec{t}$-scores
$\vec{\hat{y}}_{\text{new}}$ ($1 \times M$) denotes the predicted quality variables
$\vec{e}$ and $\vec{f}$ denote the residuals

Similar to MPCA framework, the MPLS framework has two main stages: model development using a historical reference batch database that defines NO and process monitoring and quality prediction that uses the model developed. The latter includes prediction of the product quality, which is the main difference between MPCA and MPLS. Quality prediction is made at the end of the batch run, while waiting to receive quality analysis laboratory results. MPLS can also be implemented on-line by predicting the final product quality as the batch progresses (Section 6.4.6).

Example: A set of data are produced using a simulator for fed-batch penicillin production, PenSim (Birol et al. 2002). To generate data representing good process behavior under normal operating conditions, the values of the initial conditions and set points of input variables are slightly varied for each batch, resulting in unequal and unsynchronized batch trajectories that are typical in most experimental cases. Batch lengths varied between 375 and 390 h. One of the batches that have a batch length of 382 h, close to the median batch length is chosen as a reference batch. Data of the other batches are equalized using the multivariate DTW algorithm (Cinar et al. 2003).

The reference set includes 41 batches containing 14 variables (sampled at 0.5 h). A three-way array of size (41 × 14 × 764) is formed based on this initial analysis. The variables are listed in Table 6.4.1. Although on-line real-time measurement availability of some of the product related variables such as biomass and penicillin concentrations is somewhat limited in reality, it is assumed that these can be measured along with frequently measurable variables such as feed rates and temperature. If the sampling rates are different, an estimator such as Kalman filter can be used to estimate these variables from the measured values of frequently measured variables. A number of product quality variables are also recorded at the end of the batch (Table 6.4.1).

TABLE 6.4.1
List of Process Variables and Product-Quality Variables for the Penicillin Fermentation Process

	Process Variables Measured during the Run		Quality Variables Measured after the Completion of Batches	
	Variable Name	Unit	Variable Name	Unit
1	Aeration rate	L/h	Final penicillin concentration	g/L
2	Agitation power	W	Overall productivity	g/h
3	Glucose feed rate	L/h	Terminal yield of penicillin on biomass	
4	Glucose feed temperature	K	Terminal yield of penicillin on glucose	
5	Glucose concentration	g/L	Amount of penicillin produced	g
6	Dissolved oxygen concentration	mmol/L		
7	Biomass concentration	g/L		
8	Penicillin concentration	g/L		
9	Culture volume	L		
10	Carbon dioxide concentration	g/L		
11	pH			
12	Fermenter temperature	K		
13	Generated heat	kcal		
14	Cooling water flow rate	L/h		

TABLE 6.4.2
Percent Variance Captured by MPLS Model

LV Number	X-block		Y-block	
	This LV	Cumulative	This LV	Cumulative
1	16.11	16.11	57.58	57.58
2	9.55	25.66	26.26	83.84
3	5.84	31.50	11.78	95.62
4	6.89	38.39	1.82	97.44

Additional batches were simulated to illustrate detection, diagnosis, and prediction capabilities of the MPLS models used for both end-of-batch and on-line SPM. Fault scenarios are chosen such that they resemble the ones generally encountered in industry. The fault reported here is a temporary small downward drift in the glucose feed rate (variable 3) right after the start of feeding in fed-batch operation. The abnormal operation develops slowly and none of the individual measurements reveal it clearly when their univariate charts are examined. The faulty batch is of length 380 h (760 samples).

MPLS model development stage: MPLS model is developed from the data set of equalized/synchronized, unfolded, and scaled 38 good batches (14 process variables 764 measurements) resulting in a three-way array of size $\underline{X}$ (38 × 14 × 764). The underlined X denotes a three-dimensional array. After unfolding by preserving the batch direction (I), the unfolded array becomes X (38 × 10,696). Three batches in the original 41 batches of data are excluded from the reference set because of their high variation. In addition, there is a Y (38 × 5) block with five quality variables measured at the end of each batch (Table 6.4.1). MPLS is performed between the unfolded and scaled X and Y with four latent variables (LVs) resulting in scores T (38 × 4), U (38 × 4), weights W (10,696 × 4), Q (5 × 38), and loading matrices P (10,696 × 4). The variability explained on both X and Y blocks by MPLS model is summarized in Table 6.4.2. 38.39% of X explains 97.44% of Y with four LV MPLS model. Cumulative percentage of sum of squares explained by four LVs on each $\bar{y}$ in Y block is also tabulated in Table 6.4.3. All control limits are developed using the expressions given earlier.

TABLE 6.4.3
Cumulative Percent Variance Captured by MPLS Model on each Quality Variable

LV Number	X	Y	y_1	y_2	y_3	y_4	y_5
1	16.11	57.58	48.08	51.24	45.42	91.89	51.24
2	25.66	83.84	85.62	87.23	62.18	96.89	87.23
3	31.50	95.62	95.64	95.98	92.38	98.08	95.98
4	38.39	97.44	97.41	97.23	97.20	98.10	97.23

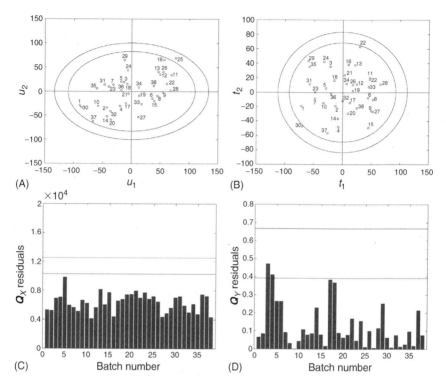

FIGURE 6.4.1 MPLS-based end-of-batch statistics for the reference batches. (From Cinar, A., Parulekar, S.J., Undey, C., and Birol, G., *Batch Fermentation: Modeling, Monitoring and Control*, Marcel Dekker, 2003. With permission.)

NO region is defined by the ellipses in Figure 6.4.1A and B based on the MPLS model. As expected, all reference batches fall inside the 99% limits. All of the reference set batches are also inside the control limits of the sum of squared residuals as shown in Figure 6.4.1C and D in both process and quality spaces. Hence, MPLS model can be used to discriminate between the acceptable and poor batches at the end of the batch.

The MPLS model developed is used to monitor completed batch runs to classify them as good or poor based on how well they follow similar trajectories to achieve good quality product. The fault scenario with a small downward drift on glucose feed is used to illustrate the performance of the MPLS-based end-of-batch SPM framework. The MPLS model is used to predict product quality as soon as the batch finishes, providing information ahead of time for initial fast assessment of batch quality before the real Y is available. New batch data are processed with MPLS model at the end of the batch using Equations 6.4.14 and 6.4.15 after proper equalization, synchronization, unfolding, and scaling, resulting in MSPM charts (Figures 6.4.2 and 6.4.3), for detection and diagnosis.

Figure 6.4.2 summarizes several statistics to compare new batch (batch 39) with the reference batches (the first 38 batches). Figure 6.4.2A and B indicates that there is a significant difference between the new batch and the NO batches in both process

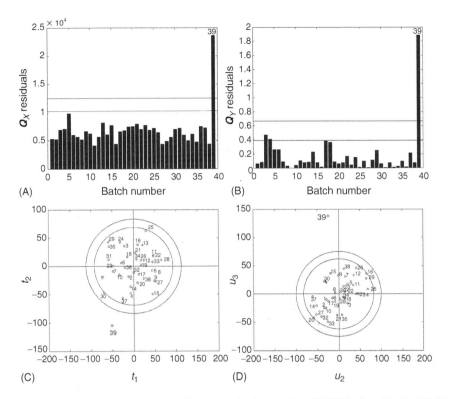

FIGURE 6.4.2 MPLS-based end-of-batch monitoring results: (A) X block residuals, (B) Y block residuals, (C) X block biplot, and (D) Y block biplot. Batch 39 is out of control. (From Cinar, A., Parulekar, S.J., Undey, C., and Birol, G., *Batch Fermentation: Modeling, Monitoring and Control*, Marcel Dekker, 2003. With permission.)

and quality spaces. Scores of the new batch in both spaces fall outside of the in-control regions defining NO in Figure 6.4.2C and D. These charts suggest that an unusual event occurred in new batch and should be investigated further. To find out when the process went out of control and which variables were responsible for this abnormality, the SPE_X chart and a variety of contribution plots are used (Figure 6.4.3). SPE_X chart of process space in Figure 6.4.3A reveals a deviation from NO and process goes out of control around 570th observation. The overall variable contributions to SPE_X in Figure 6.4.3B over the course of batch run indicate that variable 9 (culture volume) has changed unexpectedly, causing the deviation. Variable contributions to SPE_X for a specified time interval can also be calculated to zoom at the interval where the abnormal situation is observed.

Figure 6.4.3D shows average variable contributions to SPE_X between 570th and 690th measurements. Variables 3, 6, and 9 (Table 6.4.1) have the highest contributions to the deviation during this time. Further analysis can be performed by calculating contributions to process variable weights. Since weights (W) bear information about the relationship between process and product variables, variable contributions to weights will reveal the variables responsible for causing the

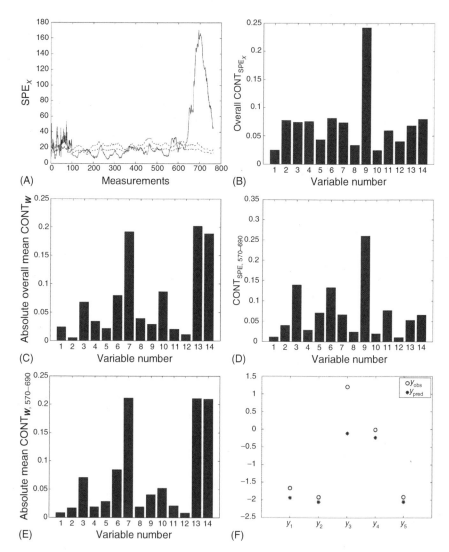

FIGURE 6.4.3 MPLS-based end-of-batch monitoring results: (A) SPE plot for X to conduct SPM and check the validity of the model, (B) contributions to SPE_x, (C) contributions to W, (D) contributions to SPE_x during the period 570–690, (E) contributions to W during the period 570–690, and (F) observed and predicted values of quality variables. (From Cinar, A., Parulekar, S.J., Undey, C., and Birol, G., *Batch Fermentation: Modeling, Monitoring and Control*, Marcel Dekker, 2003. With permission.)

abnormality in product quality. These contributions are calculated similar to SPE_X contributions. Figure 6.4.3C shows overall absolute variable contributions to the weights over the course of the batch run. Variables 3, 6, 7, 10, 13, and 14 have high levels of contributions compared to other variables. These contributions are also calculated between 570th and 690th measurements where the out-of-control occurs.

Variables 3, 6, 7, 13, and 14 are found to be significant in that case (Figure 6.4.3E). Since the original disturbance was introduced into variable 3 as a small downward drift, its effect on the other structurally important process variables such as dissolved oxygen concentration (variable 6) and biomass concentration (variable 7) becomes more apparent as the process progresses in the presence of that disturbance. Culture volume is directly affected by this disturbance as confirmed by SPE_X contribution plots. The weight contributions indicate the effect of this change on process variables that are influential in the quality space. Variables 13 and 14 (heat generated and cooling water flow rate, respectively) being highly correlated with variable 7 also show high contribution. The MPLS model is used to predict end-of-batch quality, and model predictions are compared with actual measurements in Figure 6.4.3F. Quality variable 3 is predicted somewhat poorly. This is due to model order, and the prediction can be improved by increasing the number of LVs retained in the MPLS model.

6.4.3 MPCA-Based On-Line Monitoring of Batch Processes

The on-line evolution of a new batch is monitored in the reduced space defined by the PCs of the MPCA model. MPLS enables the prediction of product properties during the progress of the batch. The problem with applying MPCA and MPLS techniques for on-line statistical process and product quality monitoring is the necessity to have the complete $\vec{x}_{new}$ vector as in Equation 6.4.14 until the end of the batch run for computing the monitoring statistics. At time k, the matrix X_{new} has only its first k rows complete and all the future observations [$(K-k)$ rows] are missing. The loadings of the reference data set cannot be used with incomplete data because the vector dimensions do not match. Several approaches have been proposed to overcome this problem for MPCA- and MPLS-based on-line monitoring. One alternative is to estimate the remaining portions of variable trajectories to the end of the run (Nomikos and MacGregor 1994, 1995a,b). The future portions of variable trajectories are estimated by making various assumptions (Nomikos and MacGregor 1994). Three approaches are suggested to fill in the missing values in $\vec{x}_{new}$ (Nomikos and MacGregor 1994, 1995a):

1. Assume that future observations are in perfect accordance with their mean trajectories
2. Assume that future values of disturbances remain constant at their current values over the remaining batch period
3. Treat unknown future observations as missing values from the batch in MPCA model and use the MPCA model of the reference batches for predicting the missing values

All three assumptions introduce some arbitrariness in the estimates of variable trajectories. Selecting which approach to use depends on the inherent characteristics of the process being monitored and information about disturbances. If process measurements do not contain discontinuities or early deviations, the third approach may be used after some data have been collected. If it is known that the disturbances in a given process are persistent, it is reported that the second approach works well (Nomikos and MacGregor 1995a). When no prior knowledge exists about the process, the first or second estimation technique can be used, until some process data have been collected and then the third method can be used.

As the new vector of variable measurements is obtained at each time k, the future portions of the variable trajectories are estimated for use in regular MPCA-based SPM framework. Then the scores and the residuals can be computed.

$$\vec{t}_{new,k} = \vec{x}_{new}^{est} P, \quad \vec{e}_{new,k} = \vec{x}_{new}^{est} - \sum_{a=1}^{A} \vec{t}_{new,ak} \vec{p}_a \quad (6.4.16)$$

where
  $\vec{x}_{new}^{est}$ denotes the full variable measurements vector $(1 \times KJ)$ that is estimated at each k onward to the end of the batch run
  $\vec{t}_{new,k}$ $(1 \times A)$ denotes the predicted scores at sampling time k from the P loadings
  $\vec{e}_{new,k}$ $(1 \times KJ)$ denotes the residuals vector at time k

To construct the control limits for on-line monitoring of new batches, each reference batch is passed through the on-line monitoring algorithm above, as if they are new batches, and their predicted scores $\vec{t}_{new,k}$ and squared prediction errors (SPE_k) are stored at each sampling time k.

An alternate unfolding of the data matrix eliminates the need for estimating future portions of variable trajectories. The $(I \times J \times K)$ data matrix can be unfolded by preserving the variable direction J to yield an $(IK \times J)$ matrix. The algorithm for developing monitoring charts with variable-wise unfolding and adding a second MPLS model based on batch-wise unfolding for final quality prediction is presented by Undey et al. (2003b).

6.4.4 EXAMPLE: MPCA-BASED SPM OF FED-BATCH PENICILLIN FERMENTATION

MPCA-based on-line SPM framework is illustrated with the same simulated data set of fed-batch penicillin production used earlier. A downward drift fault in glucose feed rate starting the fed-batch stage is used as a case study. The MPCA model development follows the general procedure and the construction of control limits is performed by passing each batch data in the reference set through the estimation-based on-line SPM procedure. Process monitoring results are affected by the estimation method used. All three methods are implemented in this example. Greater difference caused by the data estimation method used is observed in the T^2 chart in Figure 6.4.4A. The out-of-control signal is first detected by the second technique (the future values of disturbances remain constant at their current values over the remaining batch period) at the 325th measurement in the T^2 chart. The SPE chart detected the fault around the 305th measurement in all of the techniques. Variable contributions to SPE and T^2 and score biplots are presented for method 2. Contribution plots revealed the variables responsible for the deviation from NO when the

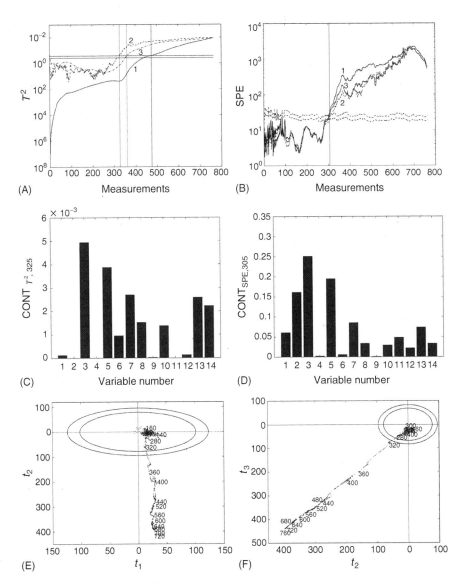

FIGURE 6.4.4 MPCA-based on-line SPM results of a faulty batch. In (A) and (B) method 1 (solid curve), method 2 (dashed curve), and method 3 (dash-dotted curve); (C) and (D) variable contributions to T^2 and SPE at 325th and 305th measurements, respectively. Score biplots based on method 2 (E) first versus second PC and (F) second versus third PC. (From Cinar, A., Parulekar, S.J., Undey, C., and Birol, G., *Batch Fermentation: Modeling, Monitoring and Control*, Marcel Dekker, 2003. With permission.)

abnormality is detected. Variables 3 and 5 (glucose concentration in the fermentes) in SPE contributions (Figure 6.4.4D) at measurement time 305 and variables 3 and 5 (and 7, 13, 14 to a lesser extent) in T^2 contribution plot (Figure 6.4.4C) at measurement time 325 are identified as responsible for the out-of-control situation. Variable

3 is the main problematic variable affecting the other variables gradually. Variable 5 is the first variable directly affected by the drift in variable 3. Since T^2 detects the out-of-control state later, the effect of the drift develops significantly on variables such as 7, 13, and 14 that are signaled by the T^2 contribution plot (Figure 6.4.4C). Score biplots also show a clear deviation from NO region defined by confidence ellipses of the reference model (Figure 6.4.4E and F). The numbers on the biplots stamp the measurement time.

6.4.5 EXAMPLE: MPCA-BASED SPM OF THE GROWTH OF FRESH APPLES

The MPCA-based methods are illustrated by using the apple growth data (www.atb-potsdam.de/fruitloc, August, 2007).

MPCA analysis of the spectral data: In the previous PCA analysis of the spectral data of apples, the fruit samples' spectral data are modeled without considering the time rate of change of spectral data explicitly. With MPCA, the data can be modeled including the time as a separate dimension in the model data. So that, time evolution of spectral properties for different samples can be investigated.

The multiway array is made up of fruit samples, spectral data, and measurement dates, which will be denoted as I, J, and K, respectively. The array will be unfolded conserving the sample number, in which case the resulting 2D array will be size I by JK. With this unfolding technique, the change in the spectral data in time around the mean trajectory for each sample can be investigated.

After unfolding, the data are mean centered and then a PCA model is built. There are 20 fruit samples used in the model building, six samples from Elstar, eight samples from Topaz, and finally, six from Pinova cultivars. These samples are selected after the outliers are removed. For these 20 samples, spectral data are measured for equally spaced wavelengths from 403 to 1098 nm, generate 211 variables. The measurements were done using nondestructive techniques on site on eight different dates 1 week spaced apart. The size of the data to be model is ($20 \times 211 \times 8$). After unfolding, the size reduces to (20×1688). Three PCs explain 82.81% of the total variation, where the first two individually explain 44.67% and 31.76%, respectively.

Figure 6.4.5 shows the score biplot for this model. The score plot shows how the samples are related to each other. Elstar, Topaz, and Pinova cultivars make distinct clusters in this plot. This model can be used to test the normality of the growing conditions on the growing site. After a good statistical model is built using data from normal growing samples, new data obtained from a new sample can be projected onto this model, and any abnormality that is within this new sample is detected.

New data can be monitored in two different ways: off-line and on-line. In off-line monitoring, new data is collected on each measurement date, and monitored at the end of the growing time. In on-line monitoring, data taken from the new sample is monitored each time it becomes available.

The MPCA model is used to test samples with slightly abnormal growth conditions owing to a mild water deficiency. Resulting from the drought stress, the apples ripened slightly earlier; however, they were picked on the same dates. Using the multivariate monitoring charts such as the SPE and Hotelling's T^2 charts, an abnormality captured in the data can be detected.

Process Monitoring 509

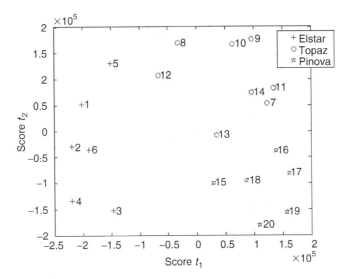

FIGURE 6.4.5 Score biplot shows how the fruit samples are related.

The spectral data for the new sample measured each time is arranged side by side and plotted in Figure 6.4.6B. The same thing is done for the mean of the normal model data and is shown in Figure 6.4.6A. From these two plots, there is a visible difference between the two spectra. For the abnormal sample, the number of peaks, their amplitude, and spacing are very different than the model (reference) data. The difference is especially noticeable for days 5–7.

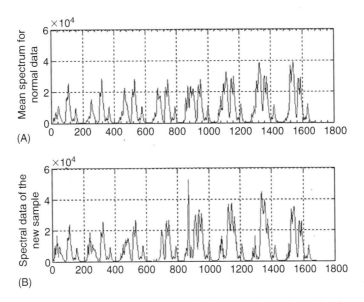

FIGURE 6.4.6 Comparison of mean spectrum for the normal growth and the abnormal new sample.

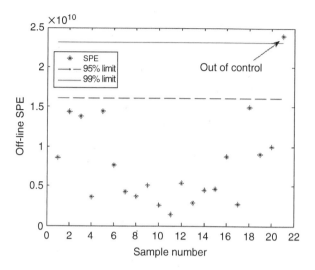

FIGURE 6.4.7 SPE statistic for off-line MPCA.

Figure 6.4.7 shows the SPE statistics for off-line MPCA monitoring, where the 21st sample is the new fruit sample taken from 'Elstar' and was object to drought stress. The abnormality is detected in the SPE graph, since it exceeds both the 95% and the 99% confidence limits.

Figure 6.4.8 shows the SPE statistic calculated for the time of growth of the new sample. And this figure reveals that the difference of the new sample and the model is highest on day 5.

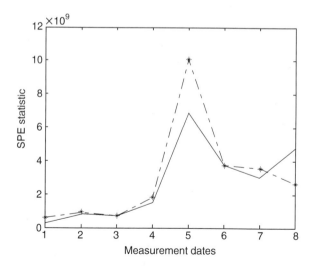

FIGURE 6.4.8 SPE statistic for the growth of the new sample.

Process Monitoring

In on-line monitoring, the new data are tested for normal process operation every time new measured values become available. In this case, all the measurements are taken using nondestructive techniques on site. Hence, on-line monitoring can be used for the timely detection of any abnormality in the growth of the new sample.

When abnormalities are detected, necessary actions can be taken without delay to save the product. However, statistical on-line monitoring is more complicated than off-line monitoring. Both the reference and the new data sets should have the same lengths for implementing on-line monitoring. Since the statistical model is built using the reference data set for the whole growth duration, new samples taken daily cannot be projected onto the model as they are, because of size mismatch of the new data. In the on-line monitoring calculations presented, PCA-based estimates are used to estimate the future values for the new sample.

Each time new data are available, unknown future values are estimated and the whole data are projected onto the model, to calculate the monitoring chart statistics. Figure 6.4.9 displays the SPE graph for the time duration of the growth of a new 'Elstar' sample, which was exposed to drought conditions during growth. The figure shows for comparison a normal sample that remains within the confidence limits during the entire period (denoted by hollow circles) and also the abnormal sample with drought stress.

The abnormal sample is different from the beginning. And the abnormality increases in time. When these statistics are available during growth of a sample, necessary actions can be taken at the early steps. For example, irrigation can be considered to reduce the effects of the drought.

MPCA analysis of individual data: Individual data for each fruit sample from each cultivar is available from laboratory analysis. These variables are listed in Table 6.4.4, with their respective scientific units. Instead of the spectral data, these variables can be used to build an MPCA model and to monitor a new sample.

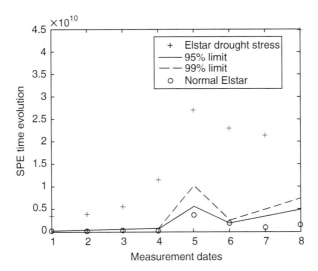

FIGURE 6.4.9 SPE statistic for the new abnormal sample during its growth.

TABLE 6.4.4
Variables Calculated through Laboratory Analysis

1	Normalized difference vegetation index (NDVI) [1;−1]
2	Normalized anthocyanin index (NAI) [−1;1]
3	Dry matter [%]
4	Respiration rate [mg CO_2/kg h]
5	Fruit flesh firmness [N/cm^2]
6	Soluble solids content [%Brix]
7	Starch index [1–10]
8	Chlorophyll [mg/m^2]
9	β-Carotene [mg/m^2]
10	Titratable acidity [%]

A new MPCA model is built using three sample fruits from each of the three cultivars. The dimensions of the 3D data matrix are (9 × 10 × 8), since 10 variables are recorded on every measurement day for nine samples. After unfolding, the data matrix becomes (9 × 80). The data matrix is mean centered and unit variance scaled, before statistical analysis. The first two PCs explain 96.69% of the total variation, where the first PC itself explains 61.17%. The model is built using two PCs and the model scores are shown in Figure 6.4.10. The clustering of the different cultivars is also visible with this model. As before, both off-line and on-line analyses are done on the data. And both analyses prove the effectiveness of MPCA in detecting an outlier.

Figure 6.4.11 shows the off-line SPE values for the nine samples used to build the model and also the 10th sample, which was monitored. The new sample is selected from 'Elstar' drought stress (DS), and was subjected to drought stress during

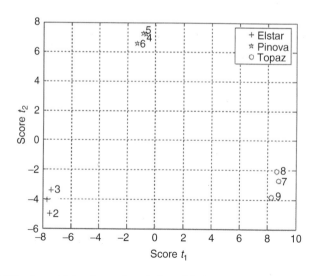

FIGURE 6.4.10 Score biplot shows the relation of fruit samples.

Process Monitoring

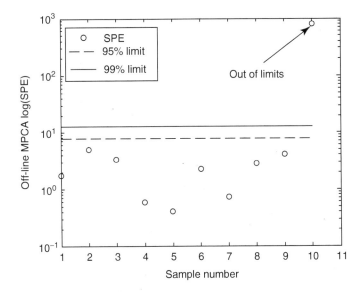

FIGURE 6.4.11 Off-line SPE plot shows sample number 10 as out of control.

growth. The new sample is well beyond the 99% confidence limit and it is declared to be significantly different than the apples in the reference set.

Figure 6.4.12 shows the online SPE values for the growth of a new abnormal sample and also a normal sample. From the figure, it is clear that the abnormal sample, denoted with "o," is beyond the confidence limits from day 1. However, the normal sample is within the limits during its entire growth.

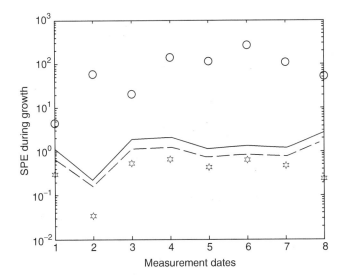

FIGURE 6.4.12 Online SPE for the growth of an abnormal and a normal sample.

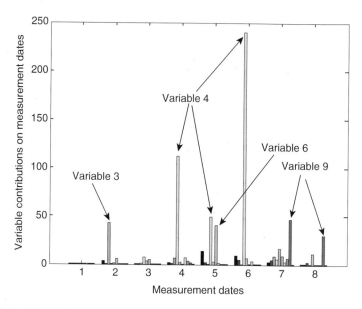

FIGURE 6.4.13 Variable contributions to SPE during the growth period.

The variable contribution graphs reveal which variables were affected from the abnormality the most and help identify the source cause of the abnormality. The variable contributions to SPE during the growth period (Figure 6.4.13) indicate that variable 4 (respiration rate) dominates the contributions to the SPE statistics, especially on days 4, 5, and 6. This is followed by variable 6 (soluble solids content) around day 5 and variable 9 (β-carotene) on days 7 and 8. Figure 6.4.14 shows the

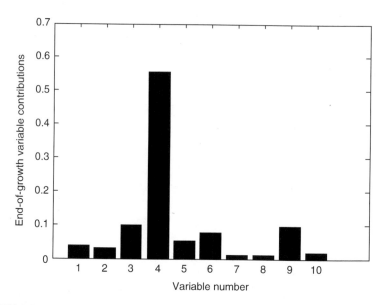

FIGURE 6.4.14 Sum of variable contributions to SPE over the entire growth.

Process Monitoring

sum of contributions of variables to SPE over the entire growth period. And variable 4 followed by variables 9, 3, and 6 seems to have been affected from the drought stress more than other variables.

6.4.6 MPLS-Based On-Line Estimation and Monitoring of Final Product Quality

Another approach to deal with missing future portions of the trajectories when implementing MPLS on-line is to use the ability of PLS to handle missing values (Nomikos and MacGregor 1995b). Measurements available up to time k are projected onto the reduced space defined by the W and P matrices of the MPLS model in a sequential manner for all A LVs.

$$\vec{t}_{\text{new},k}(1, a) = \vec{x}_{\text{new},k} \frac{W(1:kJ, a)}{W(1:kJ, a)^T W(1:kJ, a)} \qquad (6.4.17)$$

$$\vec{x}_{\text{new},k} = \vec{x}_{\text{new},k} - \vec{t}_{\text{new},k}(1,a) P(1:kJ, a)^T \qquad (6.4.18)$$

where (1:kJ, a) indicates the elements of the ath column from the first row up to the kJth row. The missing values are predicted by restricting them to be consistent with the values already observed, and with the correlation structure that exists between the process variables as defined by the MPLS model. This approach is reported to give $\vec{t}$-scores very close to their final values as $\vec{x}_{\text{new}}$ is getting filled with measured data (k increases) and it works well after 10% of the batch evolution is completed (Nomikos and MacGregor 1994, 1995a,b).

When a new variable measurement vector is obtained and k is incremented, scores $\vec{t}_{\text{new},k}(1, a)$ can be estimated and used in MPLS (Equations 6.4.14 and 6.4.15). There are no residuals $\vec{f}$ on quality variable space during on-line monitoring, since the actual values of the quality variables will be known only at the end of the batch. Each batch in the reference database is passed through the on-line MPLS algorithm as if they were new batches to construct control limits. Since MPLS provides predictions for the final product qualities at each sampling interval, the confidence intervals for those can also be developed (Nomikos and MacGregor 1995b). The confidence intervals at significance level α for an individually predicted final quality variable $\hat{y}$ are (Nomikos and MacGregor 1995b)

$$\hat{y} \pm t_{I-A-1,\alpha/2}(\text{MSE})^{1/2}\left[1 + \vec{t}(T^T T)^{-1}\vec{t}^T\right]^{1/2} \qquad (6.4.19)$$

where
    T is the scores matrix
    $t_{I-A-1,\alpha/2}$ is the critical value of the studentized variable with $(I-A-1)$ degrees of freedom at significance level $\alpha/2$

Mean squared errors (MSE) on prediction are given as

$$\text{SSE} = (y - \hat{y})^{\text{T}}(y - \hat{y}), \quad \text{MSE} = \text{SSE}/(I - A - 1) \qquad (6.4.20)$$

where
 I refers to the number of batches
 A refers to the number of LVs retained in the MPLS model
 SSE refers to the sum of squared errors in prediction

Example: To illustrate on-line implementation of MPLS for monitoring and prediction of end-product quality, the same reference set and MPLS model are used. All batches in the reference set are passed through the on-line algorithm to construct the control limits. Figure 6.4.15 presents predictive capability of the model. The solid curves indicate values at the end of the batch for all quality variables estimated at the corresponding measurement times. The dashed curves are the 95% and 99% control limits on end-of-batch estimates. End-of-batch values of all five quality variables are predicted reasonably while the batch is in progress. The fault with a significant downward drift on substrate feed rate is used to illustrate MPLS-based on-line SPM.

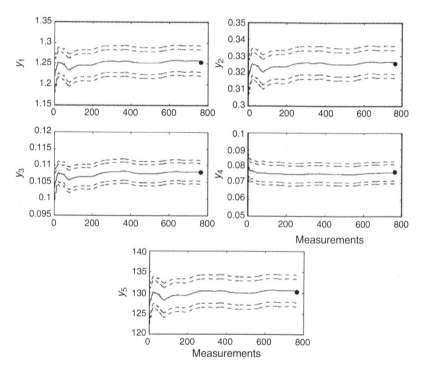

FIGURE 6.4.15 MPLS-based on-line predictions of end-of-batch product quality of an in-control batch. (•) represents the actual value of the end-of-batch product quality measurement. (From Cinar, A., Parulekar, S.J., Undey, C., and Birol, G., *Batch Fermentation: Modeling, Monitoring and Control*, Marcel Dekker, 2003. With permission.)

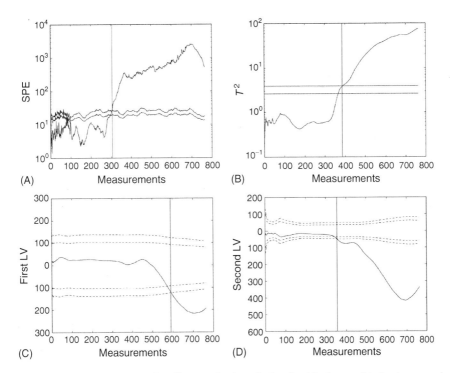

FIGURE 6.4.16 MPLS-based on-line monitoring of a batch with abnormal behavior caused by drift in feed properties (glucose feed rate). (From Cinar, A., Parulekar, S.J., Undey, C., and Birol, G., *Batch Fermentation: Modeling, Monitoring and Control*, Marcel Dekker, 2003. With permission.)

The first out-of-control signal is generated by the SPE chart at measurement 305 (Figure 6.4.16A), followed by the second LV plot at measurement 355 (Figure 6.4.16D), the T^2 chart at measurement 385 (Figure 6.4.16B), and finally by the first LV plot at measurement 590 (Figure 6.4.16C).

The corresponding contribution plots are plotted when out-of-control status is detected on these charts. Variable contributions to SPE in Figure 6.4.17A reveal the root cause of the deviation that is variable 3. Second highest contribution in this plot is from variable 5, which is directly related to variable 3. The rest of the corresponding contribution plots reveal variables that are affected sequentially as the fault continues. For instance, the second LV signals the fault later than SPE, hence there is enough time to see the effect of the fault on other variables such as variables 12 (temperature in the fermenter) and 13, while variable 3 is still having the maximum contribution (Figure 6.4.17E). T^2 chart signals out-of-control a little later than the second LV and at that point variables affected include variables 7, 13, and 14 (Figure 6.4.17B). An upward trend toward the out-of-control region is progressing in T^2 charts in Figure 6.4.16C when SPE chart detects the out-of-control situation. Variable contributions at measurement 305 in Figure 6.4.17C

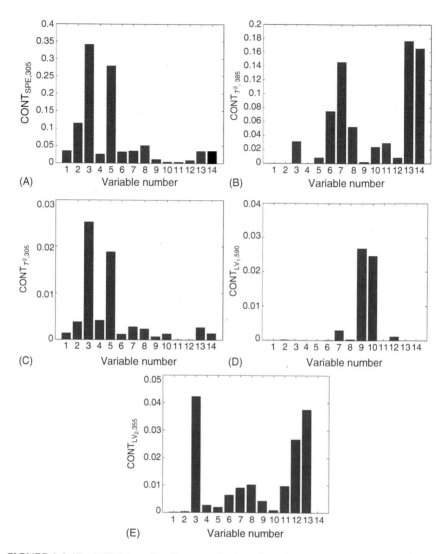

FIGURE 6.4.17 MPLS-based on-line contribution plots of a batch with abnormal behavior caused by drift in glucose feed rate. (From Cinar, A., Parulekar, S.J., Undey, C., and Birol, G., *Batch Fermentation: Modeling, Monitoring and Control*, Marcel Dekker, 2003. With permission.)

reveal that variables 3 and 5 are responsible for that upward trend toward the out-of-control region.

End-of-batch product quality is also predicted (Figure 6.4.18). Significant variation is predicted from desired values of product quality variables (compare Figures 6.4.18 and 6.4.15). The confidence intervals are plotted only until the

Process Monitoring

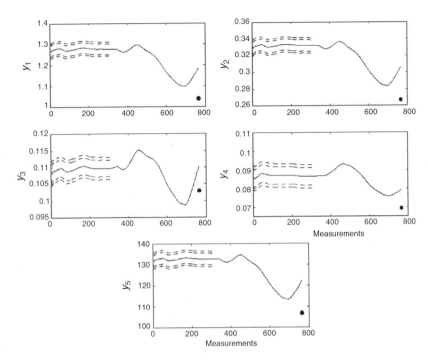

FIGURE 6.4.18 MPLS-based on-line predictions of end-of-batch product quality for a batch with drift in feed flow rate. (•) represents the actual value of the end-of-batch product quality measurement. (From Cinar, A., Parulekar, S.J., Undey, C., and Birol, G., *Batch Fermentation: Modeling, Monitoring and Control*, Marcel Dekker, 2003. With permission.)

SPE signals out-of-control status at measurement 305 because the model is not valid after that point.

6.5 SUMMARY

Multivariate statistical methods provide a powerful foundation for developing very useful SPM and SQC techniques that can be integrated with fault diagnosis. PCA and PLS and their extensions to batch process operations, MPCA and MPLS, leverage the correlation among variables and provide abnormality detection when many process or product variables make small but coherent changes from NO. This enables earlier detection that uses univariate SQC charts where larger deviations in a variable would trigger the detection of abnormal process operation or product quality. Data from growth of fresh apples and simulation data from batch fermentation illustrate the use of the techniques described in this section. Several extensions of these monitoring and quality control techniques have been suggested in the References, to reduce the number of false alarms and missed alarms, for earlier detection of abnormalities, and to monitor multistage processes.

APPENDIX

a, A	number of PCs retained in the PCA model
$B_{m/2,(n-m-1)/2}$	beta distribution with $m/2$ and $(n-m-1)/2$ degrees of freedom
CONT	contributions to T^2 or SPE
$d(\vec{x}, \vec{y})$	distance between two vectors $\vec{x}$ and $\vec{y}$
E	residuals matrix for X
F	residual matrix for Y
$F_{a,n-a,\alpha}$	F-distribution with a and $n-a$ degrees of freedom in α significance level
I	total number of batches in X ($i=1,\ldots,I$) (Section 6.4)
J	total number of variables in X ($j=1,\ldots,J$) (Section 6.4)
k	number of clusters
K	total time of duration of the batch ($k=1,\ldots,K$) (Section 6.4)
m	number of process variables
MSE	mean squared errors
n	number of samples or observations
P	PCA loading matrix, whose column are the eigenvectors of $X^T X$
PRESS	prediction error sum of squares
q	number of quality variables
Q	PLS weight matrix for Y
RSS	residual sum of squares
S	covariance matrix
SPE	squared prediction error (also called the Q-statistic)
SS	sum of squares
SSE	sum of squared errors
$t_{n-1,\alpha/2}$	studentized variable with $n-1$ degrees of freedom at significance level $\alpha/2$
T	PCA scores matrix, whose columns are $\vec{t}$ vectors
U	PLS score matrix for Y
W	weight matrix for X
X	mean-centered and scaled matrix of process measurements, with row vectors $\vec{x}$
X^*	matrix of process measurements
$\underline{X}$	three-way (I, J, K) batch process measurements matrix
Y	matrix of quality variables
z	standard normal variable corresponding to the upper $(1-\alpha)$ percentile in Q-statistic

GREEK

α	significance level in confidence limits
β	matrix of regression coefficients in PCR
χ_h^2	chi-squared variable with h degrees of freedom
λ	eigenvalues of S

σ^2	true sum of squares
θ	eigenvalues not used in the model, required in Q-statistic limit calculation

SUBSCRIPTS

new	belongs to the new measurement

SUPERSCRIPTS

T	transpose of a variable
$\wedge$	estimate of a variable

ABBREVIATIONS

CUMPRESS	cumulative prediction sum of squares
DTW	dynamic time warping
IV	indicator variable
LV	latent variable
MPCA	multiway principal components analysis
MPLS	multiway partial least squares
MSPM	multivariate statistical process monitoring
NIPALS	nonlinear iterative partial squares
NN	neural network
NO	normal operation
PC	principal component
PCA	principal components analysis
PCR	principal components regression
PLS	partial least squares regression (or projections to latent structures)
SPM	statistical process monitoring
SQC	statistical quality control
SVM	support vector machines
UCL	upper confidence limit

REFERENCES

Anderson, T.W. 1984. *Introduction to Multivariate Statistical Analysis*. Wiley, New York.

Beaver, S. and A. Palazoglu. 2006a. A cluster aggregation scheme for ozone episode selection in the San Francisco, CA Bay area. *Atmospheric Environment* 40: 713–725.

Beaver, S. and A. Palazoglu. 2006b. Cluster analysis of hourly wind measurements to reveal synoptic regimes affecting air quality. *Journal Applied Meteorology and Climatology* 45: 1710–1726.

Birol, G., C. Undey, and A. Cinar. 2002. A modular simulation package for fed-batch fermentation: Penicillin production. *Computers and Chemical Engineering* 26: 1553–1565.

Box, G.E.P. 1954. Some theorems on quadratic forms applied in the study of analysis of variance problems: Effect of inequality of variance in one-way classification. *The Annals of Mathematical Statistics* 25: 290–302.

Bro, R. and A.K. Smilde. 2003. Centering and scaling in component models. *Journal of Chemometrics* 17: 16–33.

Chiang, L.H. and R.D. Braatz. 2001. *Fault Detection and Diagnosis in Industrial Systems*. Springer-Verlag, London, United Kingdom.

Cinar, A., S.J. Parulekar, C. Undey, and G. Birol. 2003. *Batch Fermentation: Modeling, Monitoring and Control*. Marcel Dekker, New York.

Cinar, A., A. Palazoglu, and F. Kayihan. 2007. *Chemical Process Performance Evaluation*. CRC Press, Boca Raton.

Dayal, B.S. and J.F. MacGregor. 1997. Improved PLS algorithms. *Journal of Chemometrics* 11: 73–85.

Downs, J.J. and E.F. Vogel. 1993. Plant-wide industrial process control problem. *Computers and Chemical Engineering* 17: 245–255.

Doymaz, F., A. Palazoglu, and J.A. Romagnoli. 2003. Orthogonal nonlinear partial least-squares. *Industrial and Chemistry Engineering Research* 42: 5836–5849.

Duda, R.O., P.E. Hart, and D.G. Stork. 2001. *Pattern Classification*, 2nd edn. John Wiley, New York.

Evans, M., N. Hastings, and B. Peacock. 1993. *Statistical Distributions*. John Wiley, New York.

Fisher, R.A. 1938. The statistical utilization of multiple measurements. *Annals of Eugenics* 8: 376–386.

Frank, I. 1990. A nonlinear PLS model. *Chemometrics Intelligent Laboratory* 8: 109–119.

Fukunaga, K. 1990. *Introduction to Statistical Pattern Recognition*. Academic Press, San Diego, CA.

Geladi, P. and B.R. Kowalski. 1986a. Partial least squares regression: A tutorial. *Analytica Chimica Acta* 185: 1–17.

Geladi, P. and B.R. Kowalski. 1986b. An example of 2-block predictive partial least squares regression with simulated data. *Analytica Chimica Acta* 185: 19–32.

Goutis, C. 1997. A fast method to compute orthogonal loadings partial least squares. *Journal of Chemometrics* 11: 33–38.

Gurden, S.P., J.A. Westerhuis, R. Bro, and A.K. Smilde. 2001. A comparison of multiway regression and scaling methods. *Chemometrics Intelligent Laboratory* 59: 121–136.

Haario, H. and V.M. Taavitsainen. 1994. Nonlinear data analysis. II. Examples on new link functions and optimization aspects. *Chemometrics Intelligent Laboratory* 23: 51–64.

Hahn, G.J. and W.Q. Meeker. 1991. *Statistical Intervals: A Guide to Practitioners*. John Wiley, New York.

Hastie, T. and W. Stuetzle. 1989. Principal curves. *Journal of the American Statistical Association* 84: 502–516.

Hoskuldsson, A. 1988. PLS regression methods. *Journal of Chemometrics* 2: 211–228.

Jackson, J.E. 1980. Principal components and factor analysis: Part 1—Principal components. *Journal of Quality Technology* 12: 201–213.

Jackson, J.E. 1991. *A User's Guide to Principal Components*. Wiley, New York.

Jackson, J.E. and G.S. Mudholkar. 1979. Control procedures for residuals associated with principal components analysis. *Technometrics* 21: 341–349.

Johnson, R.A. and D.W. Wichern. 1998. *Applied Multivariate Statistical Analysis*. Prentice-Hall, Englewood Cliffs, NJ.

Joliffe, I.T. 1986. *Principal Component Analysis*. Springer Verlag, New York.

Kourti, T. and J.F. MacGregor. 1996. Multivariate SPC methods for process and product monitoring. *Journal of Quality Technology* 28: 409–428.

Kourti, T., P. Nomikos, and J.F. MacGregor. 1995. Analysis, monitoring and fault diagnosis of batch processes using multiblock and multiway PLS. *Journal of Process Control* 5: 277–284.

Kramer, M.A. 1992. Autoassociative neural networks. *Computers and Chemical Engineering* 16: 313–328.
Kresta, J.V., J.F. MacGregor, and T.E. Marlin. 1991. Multivariate statistical monitoring of process operating performance. *Canadian Journal of Chemical Engineering* 69: 35–47.
Krzanowski, W.J. 1979. Between-groups comparison of principal components. *Journal of the American Statistical Association* 74: 703–707.
Krzanowski, W.J. 1987. Cross-validation choice in principal component analysis. *Biometrics* 43: 575–584.
LeBlanc, M. and R. Tibshirani. 1994. Adaptive principal surfaces. *Journal of the American Statistical Association* 89: 53–64.
Lindgren, F., P. Geladi, S. Rannar, and S. Wold. 1994. Interactive variable selection (IVS) for PLS. Part I. Theory and algorithms. *Journal of Chemometrics* 8: 349–363.
Lorber, A., L. Wangen, and B. Kowalski. 1987. A theoretical foundation for the PLS algorithm. *Journal of Chemometrics* 1: 19–31.
MacGregor, J.F., C. Jaeckle, C. Kiparissides, and M. Koutoudi. 1994. Process monitoring and diagnosis by multiblock PLS methods. *AIChE Journal* 40: 826–838.
MacQueen, J.B. 1967. Some methods for classification and analysis of multivariate observations. *Proceeding 5th Berkeley Symposium on Mathematical Statistics and Probability* 1: 281–297.
Malthouse, E.C., A.C. Tamhane, and R.S.H. Mah. 1997. Nonlinear partial least squares. *Computers and Chemical Engineering* 21: 875–890.
Martens, H. and T. Naes. 1989. *Multivariate Calibration*. Wiley, New York.
Mason, R.L., and J.C. Young. 2002. *Multivariate Statistical Process Control with Industrial Applications*. ASA_SIAM, Philadelphia.
Miller, P., R.E. Swanson, and C.F. Heckler. 1993. Contribution plots: The missing link in multivariate quality control. *Proceeding of the 37th Annual Fall Technical Conference of ASQC* Rochester, New York.
Miller, P., R.E. Swanson, and C.F. Heckler. 1998. Contribution plots: The missing link in multivariate quality control. *International Journal of Applied Mathematics and Computer Science* 8: 775–792.
Nomikos, P. 1996. Detection and diagnosis of abnormal batch operations based on multiway principal components analysis. *ISA Transactions* 35: 259–266.
Nomikos, P. and J.F. MacGregor. 1994. Monitoring batch processes using multiway principal component analysis. *AIChE Journal* 40: 1361–1375.
Nomikos, P. and J.F. MacGregor. 1995a. Multivariate SPC charts for monitoring batch processes. *Technometrics* 37: 41–59.
Nomikos, P. and J.F. MacGregor. 1995b. Multi-way partial least squares in monitoring batch processes. *Chemometrics Intelligent Laboratory* 30: 97–108.
Norvilas, A., A. Negiz, J. DeCicco, and A. Cinar 2000. Intelligent process monitoring by interfacing knowledge-based systems and multivariate statistical monitoring. *Journal of Process Control* 10: 341–350.
Pearson, K. 1901. On lines and planes of closest fit to systems of points in space. *Philosophical Magazine* 2: 559–572.
PLS Toolbox for MATLAB (Version 1.2). 2001. *User's Guide*. Eigenvector Research, Seattle, WA. Available at http://www.eigenvector.com, October, 2007.
Raich, A. and A. Cinar. 1996. Statistical process monitoring and disturbance diagnosis in multivariable continuous processes. *AIChE Journal* 42: 995–1009.
Raich, A. and A. Cinar. 1997. Diagnosis of process disturbances by statistical distance and angle measures. *Computers and Chemical Engineering* 21: 661–673.
Romagnoli, J.A. and A. Palazoglu. 2005. *Introduction to Process Control*. Taylor & Francis, Boca Raton, FL.

Runger, G.C. and F.B. Alt. 1996. Choosing principal components for multivariate statistical process control. *Communications in Statistics—Theory and Methods* 25: 909–922.

SIMCA (Version 6.0). 2001. *User's Guide*. UMETRICS AB, Umeaa, Sweden. Available at http://www.umetrics.com, October, 2007.

Singhal, A. and D.E. Seborg. 2002a. Pattern matching in historical batch data using PCA. *IEEE Control Systems Magazine* 22: 53–63.

Singhal, A. and D.E. Seborg. 2002b. Pattern matching in multivariate time series databases using a moving window approach. *Industrial and Engineering Chemistry Research* 41: 3822–3838.

Singhal, A. and D.E. Seborg. 2006. Evaluation of a pattern matching method for the Tennessee Eastman challenge process. *Journal of Process Control* 16: 601–613.

Taavitsainen, V.M. and P. Korhonen. 1992. Nonlinear data analysis with latent variables. *Chemometrics Intelligent Laboratory* 14: 185–194.

Tracy, N.D., J.C. Young, and R.L. Mason. 1992. Multivariate control charts for individual observations. *Journal of Quality Control* 24: 88–95.

Undey, C., E. Tatara, B. Williams, G. Birol, and A. Cinar. 2000. A hybrid supervisory knowledge-based system for monitoring penicillin fermentation. *Proceeding American Control Conference*, Chicago, IL.

Undey, C., E. Tatara, and A. Cinar. 2003a. Real-time batch process supervision by integrated knowledge-based systems and multivariate statistical methods. *Engineering Applications of Artificial Intelligence* 16: 555–566.

Undey, C., S. Ertunc, and A. Cinar. 2003b. Online batch/fed-batch process performance monitoring, quality prediction, and variable contribution analysis for diagnosis. *Industrial and Engineering Chemistry Research* 42: 4645–4658.

Undey, C., E. Tatara, and A. Cinar. 2004. Intelligent real-time performance and monitoring and quality prediction for batch-fed-batch cultivations. *Journal of Biotechnology* 108: 61–77.

Wise, B.M. and N.B. Gallagher. 1996. The process chemometrics approach to process monitoring and fault detection. *Journal of Process Control* 6: 329–348.

Wold, S. 1978. Cross-validatory estimation of the number of components in factor and principal components models. *Technometrics* 4: 397–405.

Wold, S. 1992. Nonlinear partial least squares modeling: II. Spline inner relation. *Chemometrics Intelligent Laboratory* 14: 71–84.

Wold, S., A. Ruhe, H. Wold, and W.J. Dunn. 1984. The collinearity problem in linear regression. Partial least squares (PLS) approach to generalized inverses. *SIAM Journal on Scientific and Statistical Computing* 3: 735–743.

Wold, S., P. Geladi, K. Esbensen, and J. Ohman. 1987. Multiway principal component and PLS analysis. *Journal of Chemometrics* 1: 41–56.

Wold, S., N. Kettaneh-Wold, and B. Skagerberg. 1989. Nonlinear PLS modeling. *Journal of Chemometrics* 7: 53–65.

Wold, S., N. Kettaneh, H. Friden, and A. Holmberg. 1998. Modeling and diagnostics of batch processes and analogous kinetic experiments. *Chemometrics Intelligent Laboratory* 44: 331–340.

Wold, S., M. Sjostrom, and L. Eriksson. 2001. PLS regression: A basic tool of chemometrics. *Chemometrics Intelligent Laboratory* 58: 109–130.

Zude, M. and N. McRoberts. 2006. Product monitoring and process control in the crop supply chain. *Agricultural Engineering* 61: 2–3.

Index

A

AAC, *see* Amylose content
Absorption, 394
Accusto-optic tunable filters, 115, 204, 269, 399
Advanced glycated end products (AGEs), 337
Agricultural crops and fruits
   NIRS application in
      fresh fruit, 222–225
      grains and grain products, 225–228
      on-line quality monitoring of, 229–234
   quality monitoring parameters of
      assessment of quality, 2–7
      harvest maturity of, 7–11
      nondestructive methods, advantages of, 13–14
      post-harvest processing, maintenance of quality in, 11–13
   VIS spectroscopy in monitoring and mapping of
      in apple fruit spectral measurements, 175–180, 185–189
      mapping, GPS technique and data management in, 170–175
      in potato tuber spectral signature measurement, 184–185
      product pigment contents and spectral signature and indices, 157–160
      in spectral measurements of carrot root, 181–184
      in sweet cherry spectral measurements, 181
      wavelength range of, 160–170
Agricultural crops, machine vision system in, 84–85
   commercial machine vision
      CCD cameras and video signals, 92–95
      lighting sources, 85–91
      machine vision applications, 95–105
   commercial sorters
      elements of packing line, 105–112
      limitations of, 113–114
   future of
      exploitation of fluorescent properties, 114–115
      hyperspectral images, 115–117
      internal quality, 117–120
      quality inspection system, 120
Agricultural supply chains
   definition and problems of, xxix–xxx
   ethical management measurement, xxxiii–xxxv
   information hierarchy measurement, xxxv–xxxviii
   quality measurements and models, xxx–xxxiii
Agrofoods
   chlorophyll-a role in, 278
   fluorescence ratio imaging in, 297–304
   monitoring, LIFS in, 319–320
      applications in, 326–331
      blue-green lif spectra of, 323–326
      fluorescence mechanism, 320–322
      future prospects of, 331–332
      instrumentation for, 322–323
   products, quality controls of, 126–127
      applications of, 132–137
      image analysis, 127–130
      machine learning, 130–132
Agro-products, integrated system design in
   field server application, 365–369
   multiband spectroscopy, 359–365
   tasting robot, 369–373
Alkaloids role, 29
Allium sp., 28
Amylose content, 226, 421
Analogue-to-digital converter, 432
Anisotropy, measurement of, 261–262
ANN, *see* Artificial neural nets
AOTFs, *see* Accusto-optic tunable filters
Apples
   firmness, evaluation of, 409
   hyperspectral scattering image, 402–403
   internal browning in, 448–452
   light propagation in, 411–412
      Monte carlo simulation, 411–412
   mealiness in, 452–453
   multicolor fluorescence images of, 304
   spectral measurements of, 175–180
      maturity and development of
         fruit yield and climatic condition, 186–187
         and geographical site, 185–186
         mapping of, 187–189
   spectra, time rate of change of, 483–484
   varieties of, 407
Approximate diffusion theory model, 396–397
Argon-ion laser, 148–149
Artificial neural nets, 246
Avocados, maturity in, 445–447

B

Background limited induced power, 155
Batch processes
  food processing, used in, 494
  MPCA-based on-line
    abnormal behavior, in feed properties, 517
    contribution plots, abnormal behavior in glucose feed rate, 518
    end-of-batch product quality, predictions, 519
    estimation of final product quality, 515
    monitoring of, 505–506
    predictions, 516
    SPM results of faulty batch, 507
    variable trajectories, 505
  real-time monitoring, 497
  SPM charts for, 496–499
  types of, 495
β-Barium borate, β-BaB$_2$O$_4$ (BBO), 151
Beer–Lambert law, 44, 144, 158, 160, 392, 394
Beta distribution, 497
Beta vulgaris L., 309
BG-LIF spectra, in agrofoods, 326–327
BHT, *see* Butylated hydroxytoluene
Biased linear regression model, 475
Biased model, 474
Bio information exchange, 368–369
Biological materials
  light propagation
    diffusion theory model, 395–396
    scattering and absorption, 393–395
    steady-state solutions, 396–398
  optical properties
    steady-state spatially resolved technique, 397
    photon distributions, 396
Bismuth triborat, BiB$_3$O$_6$ (BiBo), 151
BIX, *see* Bio information exchange
BIX-IBS, *see* BIX image broker system
BIX image broker system, 369
Blackbody radiation, 145–146
BLIP, *see* Background limited induced power
Block residuals, 503
Blue-green and red+far-red fluorescence wavelength, in plant tissues, 285–286
Blue-green fluorescence, in plants, 275–276
Blue-green lif spectra, 323–326
Boltzmann equation, 395
Botrytis cinerea, xxxvii
Box's equation, 489, 498
Brassica rapa L., 326
Butylated hydroxytoluene, 27

C

Carboxymethyllysine, 31, 338
Carotenoids, source and role of, 27
Carrot root, spectral measurements of, 181–184
Carrot xylem, optical properties and spectra of, 51–53
CCD, *see* Charge-coupled device; CCD detector
CCD cameras, 90, 313, 322
  and video signals, in agricultural crops, 92–95
  (*see also* Agricultural crops, machine vision system in)
CCD detector, 399
CCIR, *see* Comite Consultatif International des RadioCommunications
Cell wall
  softening, measurement of, 386
  and texture, 380–381
CFD, *see* Constant fraction discriminator
Charge-coupled device, 90, 156, 263, 313, 322, 399
Cheese, *fromage frais*, 385
Chemical shift phenomenon, 434
Cherries pits in, NMR of, 447–448
Chilling injury, 383
Chlorophyll-a
  agrofood assessment by, 278
  fluorescence dependence on, 287–290
  red+far-red chlorophyll fluorescence in, 276
Chlorophyll, absorption of light, 407, 414
Chlorophyll fluorescence, 272
Citrus
  freezing injury, detection of, 458–459
  seed identification in, 460–462
Citrus fruit, algorithm for, 105
Climacteric fruits and vegetables, 7–8
Cluster analysis groups data, 480
CM, *see* Convolution masks
CML, *see* Carboxymethyllysine
CMOS, *see* Complementary metal oxide semiconductor
Combined spiral and radial acquisition, 444
Comite Consultatif International des RadioCommunications, 94
Commelina communis L., 272, 275
Commercial machine vision; *see also* Agricultural crops, machine vision system in
  CCD cameras and video signals, 92–95
  lighting sources, 85–91
  machine vision applications, 95–105
Complementary metal oxide semiconductor, 94
Computer vision, in agro-food products, 126–127; *see also* Agrofoods
COMSPIRA, *see* Combined spiral and radial acquisition
Constant fraction discriminator, 68

Index

Continuous-wave, 57, 149
   and modulation, 156
Contrast, 440–441
   echo time, 440
   flip angle, 440
   repetition time, 440
Convolution masks, 133
CUMPRESS, *see* Cumulative prediction sum of squares
Cumulative prediction sum of squares, 490
Curie's law of magnetization, 427
Curve-fitting algorithm, nonlinear, 418
CW, *see* Continuous-wave
Cyanogenic glycosides, sources and role of, 28

D

Dairy products
   cheese, 385
   milk, 385–386
   texture, measurement of, 385–386
   texture of
      NIR spectra, 385
      two-dimensional spin warp MRI, 385
Data block loadings, 476
DDS, *see* Direct digital synthesizer
Degree of milling, 226, 228
Delay line, 68
Destructive methods, for food quality determination, 13–14
2D-Fast Fourier transformation (2D-FFT), 438
Diffuse illumination, in agricultural crops, 87
Diffuse photon density wave, 58–60
Diffuse reflectance, 48–51, 66, 398
Diffuse reflectance profiles, measurement of
   distortions, 405
   fruit size and scattering, effect of, 406
Diffuse transmission, 51
Diffusion theory model, 395–396
Diffusion, water molecules, 431
Digital signal processors, 104–105
Diode laser, 149–151
Diode pumped solid-state, 268
Diphenylamine, 6
Direct digital synthesizer, 365
Direct lighting, in agricultural crops, 86–87
Direct standardization, 243
DL, *see* Delay line
DOM, *see* Degree of milling
DPA, *see* Diphenylamine
DPDW, *see* Diffuse photon density wave
DPSS, *see* Diode pumped solid-state
DS, *see* Direct standardization

D/8 sphere geometry, in optical reflectance measurement, 161, 166
DSPs, *see* Digital signal processors
Dynamic time warping (DTW), 495, 500

E

EBC, *see* Extrapolated boundary condition
Echo planar imaging, 443–444
EDA, *see* Emitting diode array
EEM, *see* Excitation–emission matrix
Eigenvalues and eigenvectors, computation of, 472
Emitting diode array, 163
Enzymatic-induced changes, in plant food processing, 29–30
EPI, *see* Echo planar imaging
Escherichia coli, 7
Excitation–emission matrix, 313, 337–340, 343–344, 347–348, 352, 356
Extrapolated boundary condition, 64–67

F

Fabry–Perot interferometer, 153
Faraday's law of induction, 432
Far-infrared region, 143–144
Farrell's diffusion theory model, 397–398, 404
Farrell's scattering profiles, 404
Fast low-angle shot, 443, 456, 458–462
Fault diagnosis
   abnormality detection, 494
   contribution plots, 498
   detection of, 493
   diagnosis system, plotting and inspecting, 494
   using real-time knowledge-based systems, 492
Fault signatures, 479
Faulty batch, MPCA-based on-line SPM results of, 507
FBGs, *see* Fiber bragg gratings
FDA, *see* Fisher's discriminant analysis; Food and Drug Administration
Fed-batch penicillin fermentation, MPCA-based on-line SPM framework, 506–508
Ferulic acid, fluorescence dependence on, 291–292
FFFS, *see* Front-face fluorescence spectroscopy
FHG, *see* Fourth harmonic generation
FIA, *see* Fluorescence immunoassay
Fiber bragg gratings, 269
Fiber optical chemical sensors, 269–270
FID, *see* Free induction decay
Field of view (FOV), magnetic, 436–437, 441
Field server engine, 365
Field server (FS), in plant monitoring, 365–367
FIR, *see* Far-infrared region

Firmness; see also Apples
 of apple, 409
 defined, 392–393
First-order statistics, 133
FIS, see Flashlamp imaging system; Fluorescence imaging system
Fisher's discriminant analysis, 479, 481
FLASH, see Fast low-angle shot
Flashlamp imaging system, 313
Fluorescence
 characteristics, in plants, 280–282
 emission spectra, in plant tissues
  blue-green and red+far-red fluorescence, wavelength behavior of, 285–286
  fluorescence ratios, 286–287
  shape of emission spectra, 282–284
  wavelength choice, 284–285
 excitation spectra, 280
 extraction of, 344–345
 imaging technique, advantage of, 273–274
 instrumentation for, 322–323
  light sources, 268–269
  steady-state spectrophotometers, 262–265
  time-resolved spectrophotometers, 265–268
 mechanism of, 320–322
 in plant tissues, 286–287
 quantum yield, 274
 ratios, determination of, 274
 signatures, in plant tissues
  chlorophyll a and ferulic acid concentration, 287–292
  excitation light and tissue structure, 292–294
  photosynthetic activity, 295–296
  tissue structure, 294–295
Fluorescence imaging system, 296, 323
 in plant tissues, 310–314
Fluorescence immunoassay, 258, 267
Fluorescence spectroscopy, 253, 386
 of milk products, 385
 role of, 337
Fluorophores, in plants
 blue-green fluorescence, 275–276
 fluorescence characteristics, 280–282
 fluorescence excitation spectra, 280
 plant fluorescence, 274–275
 red+far-red chlorophyll fluorescence, 276–280
FOCS, see Fiber optical chemical sensors
Food and Drug Administration, 21
Food processing
 analysis of, 24–26
 and preservation, 21–24
 steps of, 20–21

Food processing and neoformed contamination, 337–338
 front-face fluorescence signal, 345–347
 methodology, 347–352
 multiway parafac decomposition, 352–354
 parafac scores and NFC content, 354–355
 process-induced food physicochemical changes
  extraction of pure fluorescence, 344–345
  matrix-induced distortions, 338–341
  transfer function assessment, 342–344
Food refrigeration and computerized food technology, 135
Food texture
 fruit and vegetable cell walls, 380–381
 nondestructive sensing
  mastication, 379
  MRI spectroscopy, 382
  NIR spectroscopy, 381–382
  NMR spectroscopy, 381
  time-domain NMR, 386
  TRS, 383
  VIS-NIR spectra, 383
 sensory perception, 379–380
Förster or fluorescence resonance energy transfer, 257–258, 264
FOS, see First-order statistics
FOSS Infratec grain network, 246–247
Fourier shift theorem, 439
Fourier transformation, 60, 435, 439, 441–442
Fourier transform infrared (FT-IR) spectroscopy, 362, 384–385
Fourier transform near-infrared spectroscopy, 206, 210
Fourier-transform spectrophotometer, 153
Fourth harmonic generation, 151
Fractal theory, 102; see also Machine vision system
Franck–Condon principle, in photoluminescence, 260
FRCFT, see Food refrigeration and computerized food technology
Free induction decay, 432, 435
Free spectral range, 153
FRET, see Förster or fluorescence resonance energy transfer
Front-face fluorescence analysis, 337–338
 front-face fluorescence signal, 345–347
 methodology, 347–352
 multiway parafac decomposition, 352–354
 parafac scores and NFC content, 354–355
 process-induced food physicochemical changes
  extraction of pure fluorescence, 344–345
  matrix-induced distortions, 338–341
  transfer function assessment, 342–344
Front-face fluorescence spectroscopy, 36, 346

Index

Fruit and vegetable monitoring and mapping
    NIRS application in, 222–228
    on-line quality monitoring of, 229–234
    parameters for quality monitoring of
        assessment of quality, 2–7
        harvest maturity of, 7–11
        nondestructive methods, advantages of, 13–14
        postharvest processing, maintenance of quality in, 11–13
    VIS spectroscopy in
        apple fruit, 175–180, 185–189
        carrot root, 181–184
        mapping, GPS technique and data management in, 170–175
        potato tuber spectral measurement, 184–185
        product pigment contents and spectral signature and indices, 157–160
        sweet cherry spectral measurements, 181
        wavelength range of, 160–170
Fruit firmness, light-scattering technique, 413–414
Fruits
    cell walls, 380–381
    fluorescence signatures changes in, 306–307
    light penetration estimation, 409–411
    optical properties
        absorption coefficient, 408
        absorption peak, 408
        diffuse reflectance profiles, 405–407
        estimation process, 403–404
        light–tissue interactions, 405
        scattering coefficient, 408
    quality inspection system, 120 (see also Machine vision system)
    ripening, chlorophyll in, 283–284
    texture, apples
        Multi-slice MRI, 382
        NIRS, 383
        NMR, 384
        TRS, 383
    and vegetables
        optical properties of, 405–409
        quality attributes and disorders in, 424–425
    and vegetables, quality control of, 132–134
FSE, see Field server engine
fsr, see Free spectral range
FT, see Fourier transformation
FTNIR, see Fourier transform near-infrared spectroscopy
FTS, see Fourier-transform spectrophotometer
Full width at half maximum, 144, 147, 183, 202, 206
Functional food, definition and importance of, 5–6
Fuzzy logic, in food quality evaluation, 131–132; see also Agrofoods
FWHM, see Full width at half maximum

G

Generalized linear models, 352, 354–357
Generally recognized as safe, 21
45/0 Geometry, in optical reflectance measurement, 161–162
GLMs, see Generalized linear models
Global positioning system, 175, 369
Glucosinolates, sources and role for, 26–27
'Golden Delicious' apples, 407
    standard deviation (SD) spectra for, 409
Gompertz functions, 417
GPS, see Global positioning system
Grain analysis, filter instruments in, 246
Grain, quality control of, 134–135
Grapes, multicolor fluorescence images of, 298–300
GRAS, see Generally recognized as safe
Grating monochromator, in spectral selection, 152
Green bell peppers, multicolor fluorescence images of, 300–304

H

Harvesting of fruits and vegetables, importance of, 19–20
Harvest maturity, of fruit and vegetable, 7–9
    maturity measurements, 9–11
    variation of, 11
Helium:neon laser, 149
HHP, see Hydrostatic pressure processing
High-performance liquid chromatography, 247, 330, 373
High-pressure processing, 22–24
HMF, see Hydroxymethylfurfural
Horticultural maturity, of fruits and vegetables, 8–11
Horticultural products
    firmness, 391–392
    hyperspectral imaging technique
        calibrations of, 401
        for optical properties measurement, 399–400
        sensing modes, 399
        spatially resolved (SR) spectroscopic principle, 397, 399
    hyperspectral scattering images
        optical properties, determination procedure, 403
    optical properties, 393
        light–tissue interactions, 405
    quality inspection of, 392
    scattering and absorption properties, 393
Horticultural products texture
    food sensory, 391
    Magness–Taylor firmness tester, 391–392
    NIR spectroscopy, 392

Hotelling's T^2 charts, 487
HPLC, see High-performance liquid chromatography
HPP, see High-pressure processing
Hydrostatic pressure processing, 22
Hydroxymethylfurfural, 30, 348, 357
Hygienic quality, of fruits and vegetables, 6–7
Hyperspectral images, of fruits, 115–117; see also Machine vision system
Hyperspectral imaging system, 398
    for determining optical properties, 402–405
    nonuniform response curves for, 402
    principle and instrumentation, 399–402
    schematic of, 400

I

ICA, see Independent component analysis
ICCD, see Image-intensified CCD; Intensified charge-coupled device
I_{ChlD}, see Index of chlorophyll decrease
ICNIRS, see International Council for Near Infrared Spectroscopy
ICT, see Information Communications Technology
IEEE, see Institute of Electrical and Electronics Engineers
Image analysis, in agro-food products, 127–130; see also Agrofoods
Image-intensified CCD, 323
Image quality, and effect of movement on, 438–440
iMCS, see Inverse Monte Carlo simulation
Impact and sonic techniques, 392
Independent component analysis, 132
Index of anthocyanins changes (I_{Anl}), measurement of, 178
Index of chlorophyll decrease, 177–180, 188
Indium gallium arsenide, 155, 165, 204–205
Information Communications Technology, 332–333
Infrared and THz Spectroscopy, 362–364; see also Integrated system design, in agro-products
Infrared imaging, 127
Infrared (IR) spectroscopy, 193
InGaAs, see Indium gallium arsenide
Institute of Electrical and Electronics Engineers, 95
Instrument response function, 71–73
Integrated system design, in agro-products
    field server application, 365–369
    multiband spectroscopy, 359–365
    tasting robot, 369–373
Intensified charge-coupled device, 67, 69–71, 263, 267
Interference filters, in spectral selection, 152–153

International Council for Near Infrared Spectroscopy, 194
Inverse Monte Carlo simulation, 47, 51
    principle of, 48–50
IRF, see Instrument response function
IRI, see Infrared imaging
Isoflavones, role of, 28
Isothiocyanates (ITC), 34

J

Journal of Near Infrared Spectroscopy (JNIRS), 195

K

Kiwifruit, 407
k-means algorithm, 480
k-space
    reciprocal domain, 437–438
    sampling, spiral and radical, 444–445
Kubelka–Munk theory, 47, 160–161

L

Lactuca sativa L., 326
Lambert–Beer law, 383
Lambertian cosine law, 406
Larmor frequency, defined, 428
Laser-induced fluorescence, 273, 296, 319
Laser-induced fluorescence spectroscopy (LIFS), in agrofoods monitoring, 319–320
    applications of, 326–331
    blue-green lif spectra of, 323–326
    fluorescence mechanism, 320–322
    future prospects of, 331–332
    instrumentation for, 322–323
Laser operations, 147–151
Laser spectrophotometer, 153–154
LCTF, see Liquid crystal tunable filter
Leaf, fluorescence emission by, 279
Least square fit formula, 243
LEDs, see Light-emitting diodes
LIDAR, see Light detection and ranging
LIF, see Laser-induced fluorescence
Light detection and ranging, 149, 332
Light-emitting diodes, 36, 146–147, 163, 166–167, 169, 183, 225, 265, 313, 322
Light intensity, 394
Light penetration depths, in fruit, 409–411
Light-scattering technique
    fruit firmness assessment, 418–419
    instrumentation setup, 414–416

Index

mathematical methods and procedures, 416–418
wavelengths selection, 414
Liquid crystal tunable filter, 115, 117, 322–323, 399, 414–415, 418
Lithium niobate (LiNbO$_3$), 151
Lithium triborate, LiB$_3$O$_5$ (LBO), 151
Look-up table, 104
Lorentzian distribution functions, 417
LSF, *see* Least square fit formula
Luminescence probes, 270–271
LUT, *see* Look-up table

M

Machine learning, in agro-food products, 130–132; *see also* Agrofoods
Machine vision system, 84–85; *see also* Agricultural crops, machine vision system in commercial sorters
    elements of packing line, 105–112
    limitations of, 113–114
    future of
        exploitation of fluorescent properties, 114–115
        hyperspectral images, 115–117
        internal quality, 117–120
        quality inspection system, 120
    steps of
        algorithms and applications in real time, 104–105
        feature extraction, 100–103
        image preprocessing, 95–98
        image segmentation, 98–100
        object classification, 103
Magness–Taylor (MT) firmness tester, 391
Magnetic fields
    field of view, 436–437
    in liquid molecules, 430–431
    in solid molecules, 430
Magnetic resonance imaging, 36, 110, 380
    in fruit inspection, 118–120 (*see also* Machine vision system)
    gradient-echo sequence of, 460, 462 (*see also* Fast low-angle shot)
    for image acquisition and reconstruction, 436–438
    quality attributes and disorders, detection of, 424–425
    scan time, 442
    spiral–radial sequence, 462 (*see also* Combined spiral and radial acquisition)
    viscosity, measurement of, 385
Magnetic resonance relaxometry
    quality attributes and disorders in fruits and vegetables, 424–425

Magnetic resonance spectroscopy
    quality attributes and disorders in fruits and vegetables, 424–425
Magnetization, evaluation of, 427
    longitudinal, 428
    transverse, 429
        decay in, rate of, 430
        field of view, 437
Maillard reaction, 30–31
MAP, *see* Modified atmosphere packaging
Marketing chain of fruit and vegetable, in developed countries, 3–4
Matrix cameras, in agricultural crops, 92
Matrix-induced distortions, 338–341; *see also* Food processing and neoformed contamination
MCA, *see* Multi channel analyzer
MCP, *see* Microchannel plate
MCP-PMT, *see* Microchannel plate photomultiplier tube
MCS, *see* Monte Carlo simulation
Melons, water-soaking disorder in, 384
MEMS, *see* Micromechanical systems
Michelson interferometer, 153
Microchannel plate, 69–70, 311, 313
Microchannel plate photomultiplier tube, 69–70
Micromechanical systems, 205, 224
Mid-infrared (MIR) spectroscopy, 359
Mid-infrared reflectance (MIR) spectrum, 385
Mid-infrared region, 143–144, 149, 155, 209, 359–362
Milk, viscosity of, 385
MIR, *see* Mid-infrared region
MLR, *see* Multiple linear regression
Modified atmosphere packaging, 20, 29
Molecular quenching, mechanisms of, 256
Monochromatic cameras, in agricultural crops, 92–93
Monte Carlo method, in fluorescence extraction, 344–345
Monte Carlo simulation, 49–50, 52, 64, 341, 344, 411–412
    for agriculture products, 48
MPCA, *see* Multiway PCA
MPCA analysis
    of individual data, for each fruit, 511
    spectral data, 508
MPCA model
    abnormal growth conditions, sample test, 508
MPLS-based empirical modeling, 499
MPLS model
    cumulative percent variance, 501
    development stage, 501
    end-of-batch monitoring results, 503

end-of-batch statistics for reference batches, 502
on-line estimation and monitoring of final product quality, 515–519
MRI, *see* Magnetic resonance imaging
MRR, *see* Magnetic resonance relaxometry
MRS, *see* Magnetic resonance spectroscopy
MSC, *see* Multiplicative scatter correction
MSPM methods, with PCs, 485
MSPM techniques, 477
MT, *see* Magness-Taylor
Multi channel analyzer, 68
Multicolor fluorescence imaging, in plant tissues, 274, 296–297
   developmental stage, 304–306
   fluorescence ratio imaging, 297–304
   preventive measures in, 308–310
   stress or strain effects, 306–308
Multiple linear regression, 193, 217, 219, 224, 245, 354, 373
Multiplicative scatter correction, 348, 354–355
Multi-spectral cameras, in agricultural crops, 93–94
Multivariate monitoring charts, 491
Multivariate monitoring space, 486
Multivariate statistical techniques
   fault diagnosis, statistical methods for
      cluster analysis groups data, 480
      discriminant analysis, 481
      pattern-matching methods, 479
   fresh apples, growth analysis
      spectral data, PCA analysis, 481–485
   process and quality monitoring
      eigenvalues and eigenvectors, computation of, 472–473
      PLS, 474–477
      PLS, matrix relationships, 475
      principal components analysis, 471
      principal components regression, 474
      squared prediction error, 477–479
Multiway parafac decomposition, 352–354; *see also* Food processing and neoformed contamination
Multiway parallel factor analysis, 338
Multiway PCA, 495–497, 500, 505–508, 510–513, 519

N

NDVI, *see* Normalized difference vegetation index
Nd:YAG laser, 149, 151
Near infrared (NIR) spectroscopy, 36, 47, 192–194; *see also* Aricultural crops and fruits
   application in food and feed industries
      fresh fruit, 222–225
      grains and grain products, 225–228

instruments used in
   ambient light and wavelength range in, 204
   light sources of, 201–204
   reference measurement in, 200–201
   source–sample–detector optical geometry, 199–200
   spectrograph dispersive system, 204–205
   wavelength resolution and signal-to-noise ratio, 206–207
national and regional societies for, 194–195
network of instruments of
   calibration transfer, 239–240
   prediction correcting strategies for, 240–246
sampling strategies in, 196–198
in sucrose concentration assessment, 195–196
Near-infrared reflectance spectroscopy (NIRS), 199, 381–382
   of cooked carrots, 384
   horticultural products, quality control of, 392
   of potatoes, 384
Near-infrared (NIR) region, 143
NEDO, *see* New Energy and Industrial Technology Development Organization
Neoformed compounds, 18, 35, 344, 352, 354
Neoformed contaminants, 338
NEP, *see* Noise equivalent power
Neural networks, in food quality evaluation, 130–131; *see also* Agrofoods
New Energy and Industrial Technology Development Organization, 369
NFC, *see* Neoformed compounds; Neoformed contaminants
NIM, *see* Nuclear instrumentation module
NIPALS, *see* Nonlinear iterative partial squares
NIR transmittance (NIRT) spectroscopy, 199
Nitrogen laser, 148
NMF, *see* Nonnegative matrix factorization
NMR, *see* Nuclear magnetic resonance
NMR relaxometry, 432–434
NMR spectroscopy, 434–435
Noise equivalent power, 154–155
Non-climacteric fruits and vegetables, 7–8
Nondairy processed foods, texture of SAXS, 386
Nondestructive methods, food quality determination, 13–14
Nonhierarchical clustering method, k-means clustering, 480
Nonlinear iterative partial squares, 475
Nonnegative matrix factorization, 132
Nonthermal food processing technology, 22–23, 33–35
Nonuniform instrument response, 401–402
Normalized difference vegetation index, 160, 512
Nuclear instrumentation module, 69

Index

Nuclear magnetic resonance, 36, 247, 380–382, 384–386, 423–424, 426–430, 432, 434–436, 441, 445–447, 451, 453, 456, 458, 460, 462–464
Nucleus
    chemical shift of, 434
    magnetic moment of, 426–429
    relaxation time of, 429–432
Nutritional quality, of fruit and vegetables, 5–6

O

OEM, see Original equipment manufacturer
Off-line MPCA, SPE statistic for, 510
Off-line quality analysis, 499
Off-line SPE plot, 513
Olives pits in, NMR of, 447–448
One-dimensional Fourier transformation (1D-FT), 442–443
Online SPE values, growth of abnormal and normal sample, 513
OPOs, see Optical parametric oscillators
Optical farming, 368–369; see also Integrated system design, in agro-products
Optical parametric oscillators, 268
Optical properties, of food and agricultural products, 393
Optical radiation, definition of, 143
Optical reflectance measurements, optical geometries for, 161–162
Optical sensing in turbid media, determination of
    continuous wave approach
        Beer–Lambert law, 44–45
        horticultural product measurement, experimental setup, 50–51
        Kubelka–Munk theory and diffusion theory, 47–48
        Monte Carlo simulation and inverse Monte Carlo simulation, 48–50
        optical properties and spectra of, 51–53
        radiation transport theory, 45–47
    time-resolved approach
        fresh fruit and vegetable, time-resolved reflectance spectroscopy in, 73–78
        light propagation in diffusive media, 55–60
        photon migration, mathematics for, 60–67
        time-resolved instrumentation for photon migration, 67–73
Orange-red bell peppers, multicolor fluorescence images of, 300–301
Oranges
    freezing injury in, 458–459
    seed identification in, 460–462

Organoleptic characteristics, of fruit and vegetables, 4–5
Original equipment manufacturer, 170
Orthogonal signal correction before n-way partial least squares regression (OSC-NPLS), 355

P

PAH, see Polycyclic aromatic hydrocarbon
PaPeRo robot, 371
Parafac scores and NFC content, 352, 354–355; see also Food processing and neoformed contamination
Parenchyma tissue
    for food consumption, 380
Partial current boundary, 64–65
Partial least squares, 36, 193
    biased model, 474–475
    data arrangement, for SPM, 490
    matrix relationships, 475
    software for, 477
    SPM and SQC methods, 489
Partial least squares regression, 217–219, 221, 229, 234, 409, 470
Pasta NMR, to determine texture of, 382
PCA, see Principle component analysis
PCB, see Partial current boundary
PC loading, 472–473
PCR, see Principal components regression
PCs, variation of variables, 483
PD, see Photodiode
PDA, see Photodiode array
PDS, see Piecewise direct standardization
Peaches, wooliness in, 383, 452–453
Pears, internal breakdown in, 453–458
Pectin degradation, 380
Penicillin fermentation process
    process variables, 500
    product-quality variables, 500
Penicillium digitatum, 90
Peptide tryptophan fluorescence, 337
Phenolic compounds, source and role of, 27–28
Photocathode and photomultiplier, usages of, 155
Photodiode, 68, 432–433, 438, 449, 454–456, 458–459, 463
Photodiode array, 163, 165, 169, 183, 201, 223
Photodiodes and photoresistors, usages of, 155
Photoluminescence, concept of, 253–254, 274
    anisotropy, 261–262
    bimolecular interaction, 256–258
    intramolecular deactivation, 254–255
    stokes shift, 258–261
Photomultiplier tubes, 68, 70, 263, 265–266, 323

Photon migration
   light propagation in diffusive media, 55–60
   mathematics for, 60–67
   time-resolved instrumentation for, 67–73
Phylloquinone K1, 275
Piecewise direct standardization, 243
Pit, 447–448
Planck's law, 145–146
Plant fluorescence, 274–275
Plant food processing, quality determination of
   criteria for, 17–19
   determinants for, 24–29
   fresh and minimally processed food
      first degrees of processing, 20–21
      processing and preservation, 21–24
      raw material, 19–20
   quality changes in, 29–35
Plants, native fluorophores in
   blue-green fluorescence, 275–276
   fluorescence characteristics, 280–282
   fluorescence excitation spectra, 280
   plant fluorescence, 274–275
   red+far-red chlorophyll fluorescence, 276–280
Plant tissues
   FIS in, 310–314
   fluorescence emission spectra of
      blue-green and red+far-red fluorescence, wavelength behavior of, 285–286
      fluorescence ratios, 286–287
      shape of emission spectra, 282–284
      wavelength choice, 284–285
   fluorescence signatures of
      chlorophyll a and ferulic acid concentration, 287–292
      excitation light and tissue structure, 292–294
      photosynthetic activity, 295–296
      tissue structure, 294–295
   multicolor fluorescence imaging application in, 296–297
      developmental stage, 304–306
      fluorescence ratio imaging, 297–304
      preventive measures in, 308–310
      stress or strain effects, 306–308
PLS, see Partial least squares
PLSR, see Partial least squares regression
PMTs, see Photomultiplier tubes
Polycyclic aromatic hydrocarbon, 258–259, 265
Polyphenol oxidases, 30
Polyphenols, source and role of, 27–28
Postharvest processing, quality maintenance of, 11–13
Potassium titanyle arsenate, $KTiOAsO_4$ (KTA), 151
Potassium titanyl phosphate, $KTiOPO_4$ (KTP), 151

Potatoes
   texture, determination, 384
   tuber, spectral signature of, 184–185
PPO, see Polyphenol oxidases
Prediction error sum of squares, 473
PRESS, see Prediction error sum of squares
Principal components regression, 474
Principle component analysis, 132–133, 245
   MSPM and SQC techniques, 470–471, 485
   multivariate statistical technique, 471
   nondestructive spectral data, 481
   nonlinear extensions of, 474
   used to, 492
Prism monochromator, in spectral selection, 152
Proton exchange, 431
Proton NMR and quality evaluation for fruits, 427
Pulsed electric field (PEFs) technique, in food processing, 23, 35
Pulsed lasers, 273
Pulsed mode excitation, 156–157
Pulsed xenon flashlamp FIS devices, 311, 323
Pyrus communis L., 326

Q

QCG, see Quality chain graph
QPMs, see Quality projection matrices
QTH lamp, see Quartz tungsten halogen lamp
Quality
   attributes in fruits, 424–425
   inspection, 392
Quality chain graph, xxxii–xxxiv, xxxvi
Quality projection matrices, xxxiii–xxxiv, xxxvi–xxxvii
Quantum theory, 394
Quartz tungsten halogen lamp, 202–203, 229, 400

R

Radiation transport theory, 45–47
Radiative transport equation, 61–62, 64
Radio frequency identification, 175
Radiofrequency (RF) pulse, magenetic, 429
Rapid acquisition relaxation enhanced, 443–444
RARE, see Rapid acquisition relaxation enhanced
Ratio performance deviation, 216, 222
Rayleigh criterion, 152
R_d, see Diffuse reflectance
Reactive oxygen species, 32
Red+far-red, 326, 329
   and blue-green fluorescence wavelength, in plant tissues, 285–286
   chlorophyll fluorescence, in plants, 276–280, 321
Reflectance measurements, 395

Index

Relaxation maps, 432
Relaxation time
   longitudinal, 429–430
   metabolites, effects of, 431
   signal detection, 432
   transverse, 430
RFID, see Radio frequency identification
RFR, see Red+far-red
RLM, see Run-length matrix
Root mean square error of prediction (RMSEP), 215–217, 219, 222, 224
ROS, see Reactive oxygen species
RPD, see Ratio performance deviation
R_s, see Spectral resolution
RTE, see Radiative transport equation
Rubidium titanyle arsenate, $RbTiOAsO_4$ (RTA), 151
Rubidium titanyle phosphate, $RbTiOPO_4$ (RTP), 151
Run-length matrix, 133

S

Salmonella, 7
SAXS, see Small-angle X-ray scattering spectroscopy
Scanning monochromator systems, 205
Scattering, 393–394
Scattering-dominant material
   angular diffuse reflectance intensity, 406
   depends on, 394
Scopoletin, 276
Score biplot, 496
   clusters, 484
   each day for each cultivar, 485
   fruit samples, 508–509
      relation of, 512
   PCs, 482
SE, see Spin echo
Second harmonic generation, 151
Sensor networks, 367–368; see also Integrated system design, in agro-products
Sensors, in fruits packing line, 108–111
Shewhart charts, 470
SHG, see Second harmonic generation
Shigella, 7
Short-wave NIR, 192–193, 197, 201–202, 204–206, 208, 210, 213–214, 222, 224–225, 228–229, 234
Signal detection, in relaxation, 432
Signal-to-noise ratio (SNR), 70, 154, 156, 163, 166, 198, 208, 222, 274, 323, 385, 400, 402, 440–443, 447, 463–464
   of detector system, 206–207
Single-kernel NIRS technique, 226
Single-photon avalanche diode, 71

Singular vector decomposition, 132
SLC, see Surface lipid content
SMA, see Sub-miniature version A
Small-angle X-ray scattering spectroscopy, 386
Soluble solids content, 4–5, 25, 36, 195–197, 217–219, 221–224, 230, 232–233
SPAD, see Single-photon avalanche diode
Spatial resolution, 441
SPE calculation, 490
Spectral data, comparison of
   normal growth and abnormal new sample, 509
Spectral lamps, emission lines of, 146–147
Spectral resolution, 152–153, 163, 313
Spectrophotometer technology
   detection in, 154–157
   in fruit and vegetable light dispersion monitoring, 160, 163, 165–167
      portable spectrophotometer devices in, 170–174
   light sources in
      blackbody radiation, 145–146
      laser, 147–1151
      optical radiation, 143–145
      spectral lamps and light emitting diodes, 146–147
   and NIRS, 198–199
   spectral selective detection, 152–154
Spectroscopy and chemometrics theory
   calibration techniques, 217–219
   chemometric terms, 215–217
   data preprocessing, 207–208
   multivariate regression, calibration on, 214
   PLSR wavelength selection, 219–222
   spectroscopy of water, 208–212
   sugar in water, 212–214
SPE graph, for abnormal sample during growth, 511
SPE-statistic, 479
Spin echo, 433, 443, 447, 449, 493, 460
Spin–lattice relaxation, 429
SPM, see Statistical process monitoring
SQC, see Statistical quality control
Squared prediction error (SPE) charts, 477, 479, 485–486, 488–489
   sum of variable contributions, growth, 514
   variable contributions, during growth period, 514
SSC, see Soluble solids content; Starch or sugar content
Starch, 381
Starch-based cereal foods
   textural measurement
      MRI spectroscopy, 382
      NIR spectroscopy, 381–382
      NMR spectroscopy, 381
Starch or sugar content, 9, 11, 14

Statistical process monitoring, 470
  batch processes
    apple growth data, MPCA model, 508–515
    fed-batch penicillin fermentation, 506–508
    in food processing, 494
    MPLS models, 501–505
    multivariate charts, 496
    on-line estimation and monitoring, 515–519
    postmortem analysis, multiway PLS-based, 499
    Q-statistic, 498
    quality prediction, 500
    T^2-statistic, 497
    types of, 495
  fault diagnosis in continuous processes
    based on PCA, 485
    detection and diagnosis of, 493–494
    discriminant analysis techniques, 492
    Hotelling's T^2 Charts, 487–488
    multivariate monitoring space, 486
    SPE charts, 488–489
    and SQC, 489–492
    using contribution plots, 492
Statistical quality control, 470, 489–492, 519
Steady-state fluorescence spectrometers, 262–265
Steady-state spatially resolved technique, 397
Stokes shift, in photoluminescence, 258–261
Streak camera (SC) detection technique, 67
Sub-miniature version A, 201
Sulfur-containing compounds, role of, 28
Support vector machines, 103, 480
Surface lipid content, 226
SVD, see Singular vector decomposition
SVMs, see Support vector machines
Sweet cherry, spectral measurements of, 181
SWNIR, see Short-wave NIR

T

TA, see Titratable acidity
TAC, see Time to-amplitude converter
Tasting robot, in foods spectroscopic information, 369–373
TAXT2 texture analyzer, 379
TCSPC, see Time-correlated single-photon counting detection technique
T_d, see Diffuse transmission
Temperature
  freezing point, 458
  thermal equilibrium, 427
Terahertz, 362
Terpenes, source and role of, 27
Texture (food), 378–380
  cell walls of fruit and vegetable, relationship with, 380–381

detection of
  in cereals, 381–382
  in dairy products, 385–386
  in fruits, 382–384
  in nondairy processed foods, 386
  in vegetable, 384–385
Thermal treatment, of foods, 30–33
THG, see Third harmonic generation
Third harmonic generation, 151
Three-dimensional data set projected, 472
THz, see Terahertz
Time-correlated single-photon counting detection technique, 67, 266
Time-resolved emission spectra, 267
Time-resolved laser remittance spectroscopy
  to detect mealiness of apples, 383
Time-resolved reflectance spectroscopy, 73–78
Time-resolved spectrophotometers, 265–268
Time to-amplitude converter, 68
Ti:sapphire laser, 149
Titratable acidity, 4–5
2 T NMR spectrometer, 445
Tobacco leaf, multicolor fluorescence images of, 297–298
Total internal reflection fluorescence (TIRF) spectroscopy, 269
Total transmission, 51
TRES, see Time-resolved emission spectra
TRS, see Time-resolved reflectance spectroscopy, Time-resolved laser remittance spectroscopy
T_t, see Total transmission

U

Ultra-high-frequency (UHF), 174
Ultra-high-pressure processing (UHP), 22
Ultraviolet (UV) radiation, 143
  epidermis, effect on, 293
Universal serial bus (USB), 95

V

Variables calculated, laboratory analysis, 512
VDLUFA forage maize network, 247
Vegetables
  optical properties
    absorption coefficient, 408
    diffuse reflectance profiles, 405–407
    light–tissue interactions, 405
    scattering coefficient, 408
  texture
    cell walls, 380–381
    FTIR, 385
    NIR spectra of, 384

Index

NMR relaxation, 384
potatoes, MRI of, 384
Video signals, in agricultural crops, 92–95;
    see also Agricultural crops,
    machine vision system in
Visible spectroscopy (VIS), 36
    in fruit and vegetable monitoring and mapping
        in apple fruit spectral measurements, 175–180, 185–189
        mapping, GPS technique and data management in, 170–175
        pigment contents and spectral signature and indices, 157–160
        in potato tuber spectral signature measurement, 184–185
        in spectral measurements of carrot root, 181–184
        in sweet cherry spectral measurements, 181
        wavelength range of, 160–170
Visible (VIS) region, 143
VIS spectral signature, of plant canopy, 158–159
Visual quality, of fruits and vegetables, 3–4

W

Watershed algorithm, in grain segmentation, 135
Water-soaking, detection of, 384

Wavelength resolution, 206
Wavelengths, light, 414
WAXS, *see* Wide-Angle X-ray Scattering
Wheat bread, NIR spectra of textures in, 382
White rice, NIR spectra of textures in, 381
Wide-Angle X-ray Scattering, 380

X

X-ray fluorescence (XRF), 359
    and infrared spectroscopy, 359–362 (*see also* Integrated system design, in agro-products)
X-ray techniques, in fruit inspection, 117–118;
    see also Machine vision system

Y

Yttrium aluminum garnet (YAG), 149

Z

Zero boundary condition (ZBC), 64
Zwicki, food texture analyzer, 379